Statistical Thought:
A Perspective and History

Statistical Thought: A Perspective and History

SHOUTIR KISHORE CHATTERJEE

Formerly Professor of Statistics
Calcutta University

OXFORD
UNIVERSITY PRESS

OXFORD
UNIVERSITY PRESS

Great Clarendon Street, Oxford OX2 6DP

Oxford University Press is a department of the University of Oxford.
It furthers the University's objective of excellence in research, scholarship,
and education by publishing worldwide in

Oxford New York

Auckland Bangkok Buenos Aires Cape Town Chennai
Dar es Salaam Delhi Hong Kong Istanbul Karachi Kolkata
Kuala Lumpur Madrid Melbourne Mexico City Mumbai Nairobi
São Paulo Shanghai Taipei Tokyo Toronto

Oxford is a registered trade mark of Oxford University Press
in the UK and in certain other countries

Published in the United States
by Oxford University Press Inc., New York

ISBN 0 19 852531 1

10 9 8 7 6 5 4 3 2

Typeset by Newgen Imaging Systems (P) Ltd, Chennai, India
Printed in Great Britain
on acid-free paper by Biddles Ltd, King's Lynn, Norfolk

"Kālaḥ kalayatām aham..."
"I am Time among all who measure and enumerate..."

—*Gītā, X.30.*

To
The Teacher Supreme
who has been teaching me through many forms

PREFACE

This is *not* a book on history of statistics. Rather it is an attempt to trace the course of development of statistical concepts from the perspective of the present state of the subject. Hence in it chronological details are often relegated to the sidelines and contributions are given importance in the light of their roles on the future course of development.

The book is primarily meant for those who have had sufficient exposure to statistical theory. But since it views the problem of statistical induction in a wider philosophical context and to some extent discusses the impact of developments in statistics on the thought-world of man, a large part of it may be of interest to knowledgeable philosophers and logicians.

The book is divided into two parts. Part I (Perspective) consists of Chapters 1–4, and Part II (History), Chapters 5–10. In Chapters 1 and 2, the statistical approach to induction is viewed as a way of resolving the philosophical problem of induction when certain special conditions hold, and the roles the theory of probability plays in the process are highlighted. In Chapter 3 the distinction between the different interpretations of probability and their implications are discussed. Chapter 4 shows how different approaches to statistical induction have developed as a result of diverse schools embracing different interpretations of probability at different stages of the inductive process. This chapter is much longer than the earlier three and is unavoidably more technical in content—in it the various approaches and sub-approaches, the distinctive ways in which they handle different types of inductive problems, the conflicts of various ideas, and the comparative merits of the various approaches are described in some detail. In Part II of the book, the origin and growth of the different concepts of probability and the concomitant development of the various approaches to statistical induction are described with reference to the present framework of the discipline as delineated in Part I. Although generally the description follows the historical sequence, this rule is violated whenever it is convenient to relate a later development to an earlier one; for this the detailed preview given in Part I comes handy. The violation of the chronological sequence is most flagrant in the last chapter which deals with the developments in the 20th century. Here a classificatory description of selected topics is attempted. Admittedly, the selection and organization of topics in this chapter have been much influenced by the personal views and preferences of the author.

A word or two may be in order about the terminology and conventions that are followed in this book. Generally the term 'statistical induction' is used to cover both inductive inference and behaviour (decision-making). An approach to inference is called 'objective' whenever it interprets the probability model for the observables in frequency terms. A new term 'instantial' (i.e. 'relating to a particular instance') is introduced to describe the type of approach representing likelihood-inference, P-value testing, and fiducial inference. For reasons that are detailed in Chapter 4, the usual parametric Bayesian approach is called 'objective pro-subjective'. In fact the term 'Bayesian' is not used except towards the end—instead, procedures derived through an application of Bayes's Theorem are often called 'posteriorist'. (Not only is it more suggestive—it respects the historical fact that Bayes's early contribution earned universal recognition somewhat later.) As the narrative style is followed in the book, sectional headings are not given (except in the case of large sections). Instead, I have preferred to note the topics discussed in sections, subsections, and sometimes even paragraphs in corresponding parboxes. Sections are numbered as usual; but generally only sections which are large or which discuss somewhat disjoint topics are split into numbered subsections. However, the numbering of examples and displays is always section-wise.

I have had two objectives in mind while embarking on the project of writing this book. The first is pedagogical. I have taught a course on the 'History of Statistical Thought' (based on selections of topics covered in Part II of this book) to the post-graduate statistics students of Calcutta University for more than a decade. The course was introduced with the object of giving the students, who take the usual courses on the various branches of statistics, an integral view of the subject—to ensure that they do not miss the forest because of the trees. I have also given from time to time shorter courses on special topics (selected from both Parts I and II) in refresher programmes for college and university teachers, organized by various Indian universities. I have felt that if the entire material of these courses could be enlarged and organized into a volume, it may be of some benefit to the academic community and may serve as the base for similar and possibly more fruitful courses by others in future. The second objective is somewhat personal. Like most statisticians of my generation, my initial training in the subject was in the frequentist and behavioural tradition. But during the last three decades the very bases of that acquisition have been subjected to questioning by the proponents of the neo-Bayesian schools and this has caused a good deal of personal disconcertment. To settle that conflict at least at the subjective level, I have felt it necessary to undertake an undogmatic and dispassionate appraisal of the issues involved. Furthermore, it has seemed to me that to be fruitful such an effort should be undertaken in the backdrop of the philosophico-historical development of the concepts and viewpoints which have evolved over time to create the present-day scenario of the subject. In fact the writing of this book has been an education to its writer. As may be seen through the pages of the book, I have not identified anyone of the diverse paths as *the* right path in statistics, but have ended up by according legitimacy to all the paths in their right settings.

In writing this book I have freely drawn upon the findings of the primary historical research carried out by various writers. In particular, the two volumes

(1970, 1977) on *Studies in the History of Statistics and Probability* edited respectively by E. S. Pearson and M. G. Kendall, and M. G. Kendall and R. L. Plackett,
the book by Stephen Stigler (1986), the three books by Ian Hacking (1965, 1975,
1990), and the two books by Anders Hald (1990, 1998)have served as the major
source books for this enterprise. As a matter of fact, Hald's encyclopedic second
volume became available midway during the writing of this book and necessitated the reorientation of some of my ideas and led to the partial reorganization
of my plans for the book. I thankfully acknowledge my indebtedness to these
and many other authors from whom I have gathered information and borrowed
ideas, consciously or subconsciously.

While writing this book, I have received help from so many people in so
many ways that it will be hard to mention them all. In particular, Sir David
Cox, Pranab Kumar Sen and Jayanta Kumar Ghosh have encouraged me in my
effort through helpful discussions, enlightening comments, and in other ways. Sir
David read through Part I and portions of Part II of the book and gave various
suggestions. Mainak Mazumdar went through the entire typescript almost at
break-neck speed—his comments have helped to straighten many a kink in my
expressions and removed haziness at various places. Arijit Chaudhuri, Samindra
Sengupta, Nripes Mandal, Kalyan Das, Rahul Mukerjee, and Tathagata Banerjee
read through various portions of the book, offered useful suggestions, and helped
me in various ways. Help by way of occasional discussions, provision of reference
materials and in several other forms was also received from Shyama Prasad
Mukherjee, Late Basudeb Adhikary, Adhir Basu, Bikas Sinha, Gour Mohan
Saha, Uttam Bandyopadhyay, Manisha Pal, Asis Chattopadhyay, Sugata Sen
Roy, Gaurangadeb Chattopadhyay, and Aditya Chatterjee. I am thankful to all
the members of the staff of the Department of Statistics, Calcutta University
and Calcutta Statistical Association for their co-operation during the writing of
this book. Special thanks are due to Priya Ranjan Lahiri, who patiently worked
through the maze of my hand-writing and typed the entire material of the book
in LaTeX form. Lastly, I mention the contribution of the members of the staff of
OUP whose meticulous editing has improved the presentation significantly.

The Council of Scientific and Industrial Research, Govt. of India provided
support under its Emeritus Scientist Scheme for a part (April, 1997–January,
2000) of the period of writing this book.

S. K. Chatterjee
Calcutta Statistical Association
C/o Department of Statistics, Calcutta University
35 Ballygunge Circular Road
Kolkata 700019
INDIA

July 20, 2002

CONTENTS

CONTENTS

xiii

PART I

PERSPECTIVE

1

PHILOSOPHICAL BACKGROUND

Statistical thought

1.1 We regard statistical thought as a single structure in the realm created by the human mind. The structure is not very compact. Rather, it is like a palace which has many disjoint substructures with their separate halls, domes, and pinnacles. Still the thing stands on the same base, has the same purpose, and is identifiable as an integral whole. Our object is to take a look around this structure, examine its foundations and view its different parts, and then to trace its history—to recount how the structure has grown to reach its present dimensions. Our perspective of history is from the standpoint of the present: while reviewing the evolution of various ideas we always keep in mind the structure as it stands now and try to reconstruct how those ideas have been integrated into its different parts.

Naturally it would be proper to start by giving a definitive characterization of 'statistical thought'. As is well known, the word 'statistics' is commonly used in two different senses. Statistics in everyday parlance means numerical records of any sort. (Nowadays, in television or radio commentaries of a cricket match one even hears the person in charge of the score book being referred to as the statistician.) In scientific discussions, on the other hand, statistics often stands for a discipline, a branch of knowledge. It goes without saying that when we speak of statistical thought we are using the word statistics in the latter sense.

In fact, one could have used the phrase 'statistical science' to indicate the discipline. While this is often done, it may nevertheless create a wrong impression. Statistics is not a branch of science in the sense physics or chemistry or botany or geology or economics are. It does not have a specific domain—an aspect of nature or society—as its field of study like those sciences. It is domain-neutral as pure mathematics is, and in our age, its concepts and techniques, just like those of the latter, permeate and sometimes swamp all the natural and social sciences.

Prolongation of inductive reasoning

1.2 Bertrand Russell, once remarked that pure mathematics is 'nothing but a prolongation[1] of deductive logic' (Russell (1961) p. 784). In the same vein the core of statistics can be characterized as 'a prolongation of inductive reasoning'. Deduction and induction both belong to the branch of philosophy known

---

[1]Note that Russell does not say that mathematics is part of deductive logic. Gödel's well-known theorem (1931) (see e.g. Weyl (1949) pp. 219–235) must be partly responsible for this caution in the use of words.

as epistemology or the theory of knowledge. Statistics, if it is a prolongation of inductive reasoning, would naturally be an outgrowth of epistemology.

| Other viewpoints |

Of course some might contend that such a view of statistics would be too narrow or one-sided—there are large chunks of statistical theory which are non-inductive, being purely deductive in nature or concerned with quantification of concepts like 'variability', 'correlation' etc. They would rather describe statistics broadly as the branch of knowledge concerned with drawing conclusions from numerical data, if required, by the application of the theory of probability. But if we accept the latter view we have to include accountancy within statistics, and if application of probability theory be the criterion, much of the kinetic theory of gases and the theory of microparticles and most of actuarial science which requires large-scale direct application of probability theory will come under statistics. In fact, although much of statistical theory is deductive, close examination shows that all the deductive exercises in statistics are related ultimately to solution of problems of induction, 'closeness to material validity' or 'usefulness as a basis of action' of the conclusion being important in all such cases. When we compute the value 0.67 of the coefficient of correlation between the scores in mathematics and languages of even a fixed group of students, say the candidates appearing in the secondary examination conducted by a particular Board in a particular year, somehow at the back of our mind we have the tendency to take it as a measure of the association between mathematical and linguistic abilities of the broad class of students at that level. Whenever there is a general discussion on the association between the two abilities, we would be prone to cite the figure 0.67 as evidence of a moderate positive association. When we compute the GDP at factor cost of India in the year 1994–95 as Rs. 854,103 crores at current prices, we never advance this figure as merely the outcome of a complex exercise in large-scale data collection and arithmetic deduction. Rather we look upon the figure as representing some more or less intangible entity, to be used for comparison with similar figures and for the purpose of planning. And, as regards quantification of concepts, this is really a question of following convenient conventions as in the measurement of entropy in physics, or biodiversity in ecology or rates of inflation and unemployment in economics. Thus whenever statistics comes into the picture, we find there is some form of induction, overt or covert. After all, astronomers from ancient times and physicists and other scientists at least from 16th–17th centuries onwards have been making observations, quantifying concepts, and computing various measures in their own fields. It is extremely doubtful whether, without the element of induction in it, statistics would have gained recognition as a distinct branch of knowledge.

| Value judgements in statistics |

A more subtle objection against regarding statistics as concerned solely with induction and thus relating it to epistemology may be that epistemology does not deal with value judgements. (The branch of philosophy dealing with value judgements has been called 'axiology'—it includes ethics, asthetics etc. See Day (1961), Chapters 2, 6.) Admittedly much of statistical theory is geared to action-taking in practice, and while choosing an action we have to take account of the losses resulting from alternative actions under different situations and this involves value judgement. But then, in all such contexts losses are determined by practical considerations and are mainly judged by subject matter specialists or users. Statisticians generally start by assuming the losses as given and use these ultimately as bases for deciding the extent to which the inductive policy should be conservative or liberal. The rationale behind the setting of the losses in various situations is not generally the responsibility of the statistician and is not part of statistics.

| Specialities of statistics | However, although we may view statistics as a prolongation of inductive reasoning, tackling all sorts of problems of induction is not the concern of statistics, nor do all approaches to problems of induction follow the statistical path. Statistics seeks to find the solution to only special types of problems of induction and only following special modes of reasoning. (These specialities of statistics we would take up in Chapter 2.) Nevertheless to trace the origin and the history of statistical thought we have to start by looking at the problem of induction and the evolution of inductive philosophy in general.

| Means of knowledge | **1.3**   Every moment of our waking life we try to acquire knowledge and base our actions on the outcomes of such efforts; in scientific investigations we pursue this acquisition of knowledge in a more motivated and systematic way. Broadly we have two instruments for acquiring knowledge : (i) directly by perception through the senses whose power is often greatly magnified with the aid of various instruments (when we take recourse to memory or authority the primary source is still perception) (ii) indirectly through inference or reasoning.

Inference again is broadly of two types: *deductive* and *inductive*. Although in any chain of reasoning both types of inference usually coexist, the two types have distinct characteristics. This distinction has been sharply emphasized in Western philosophy at least since the 17th century.[2]

| Deduction | In both deductive and inductive inference we start from certain premises and draw certain conclusions. In the case of deduction the conclusion follows necessarily from the premises in the sense that the negation of the conclusion would be inconsistent with the premises. We can say the premises here are sufficient for the conclusion to hold. In everyday

---

[2]In other schools of thought, e.g. in Indian philosophy, the demarcation between deduction and induction is not very clear-cut.

life we constantly take to deductive reasoning, often without any conscious effort. Thus suppose on the underground, trains are scheduled to run in the afternoon at 10 minute intervals. I have been waiting at a station for the last 20 minutes and no trains have come—I immediately conclude that there has been some kind of disruption in service. Or I know tonight is a full-moon, and the street lights are off; looking out of the window I find it is quite dark; since there is no lunar eclipse tonight and my vision is normal (of course usually I do not consciously recall these facts), I conclude there are clouds in the sky. In our intellectual pursuits we often resort to deductive reasoning more deliberately. This is true of historical studies, legal proceedings, and more transparently, in the rejection of a provisional hypothesis in the face of clearcut counter-evidence in any branch of science. The entire edifice of pure mathematics is built through deductive reasoning. The purpose of such reasoning is merely the explication of what is already contained in the premises, i.e. the knowledge generated by deduction, in the ultimate analysis, is tautological. Bertrand Russell's remark that mathematical knowledge is 'all of the same nature as the "great truth" that there are three feet in a yard' (Russell (1961) p. 786) applies truly to deductive knowledge in general.

| Induction |

Unlike deduction, in the case of *induction* the premises do not imply the conclusion in the necessary sense; they only provide some support to the conclusion. Here the premises are definitely not sufficient, but may or may not be necessary for drawing the conclusion; only, we do not expect the premises and the negation of the conclusion to occur together in practice, at least usually. What in every-day life we call 'common sense' or 'the result of experience' is often a piece of inductive reasoning. Thus on the past two occasions when I had to travel by city bus from my residence to the railway station it took me 35 and 40 minutes respectively. I have to catch a 12 o'clock train. Reasoning inductively I decide it will be all right if I start from my residence at 11 o'clock and take a city bus to the station. Many adages and rules of behaviour handed down to us by society from our childhood are really based on induction. While going out for a walk in the evening, I see the sky is covered with thick clouds and think it prudent to carry an umbrella. This is because on innumerable occasions in the past it has been observed that when there are thick clouds in the sky, it often rains. Both medical diagnosis and treatment of diseases rest heavily on induction based on past observations. Generally in all branches of science positive truths which are discovered empirically are outcomes of induction. A 'reasonable guess' that we often make in practice is also induction. Thus an official is asked to guess the proportion of literate adults in a large tribal village. He takes the voter list of the village and finds that of the first 50 persons in the list 20 are literate. So he thinks it reasonable to give the induced figure of 40% for the required proportion. In any field, what we call 'expert opinion', is almost always inductive reasoning. Thus when an art critic is asked to judge whether a piece of art is the creation of a certain master, or when a physical anthropologist is asked to classify a given skull as male or female, what is wanted is an inductive conclusion based on examination of various features or measurements on the given specimen (which perhaps originated long back in the past) and accumulated past experience about similar specimens, about the truth of the proposition that it belongs to one or other particular category. In a murder case when the prosecuting lawyer argues that the involvement of the accused in the crime is strongly indicated by a preponderance of evidence (say, the discovery of the murder weapon in his possession and his abscondence after the incident), he is reasoning inductively on the basis of observed behaviour of criminals in the past.

| Types of induction | The instances of induction given above cover a wide spectrum. They show that induction may be about some fact relating to an event that will take place in future or it may be about some general law of nature that is presumed to connect one kind of event with another, it may be concerned with drawing a conclusion about a whole on the basis of an examination of a part of it, or its purpose may be to make conjectures about some fact relating to the unknown past or the unobserved part of the present, all on the basis of available evidence.[3]

Also induction may be of two types depending on our approach to it. It may be *open*, the proposition to be induced being suggested by the premises of induction without reference to any preconception (as in the case of guessing the proportion of literate adults); or it may be *hypothetic* starting with a preconceived hypothesis (as in judging whether a particular piece of art is a genuine Rembrandt) which is to be confirmed or rejected in the light of the premises. In the former case induction is used as a tool for discovery, in the latter, as one for testing.[4]

| Characteristics of induction | If we examine the different forms of induction above, we find there are certain characteristics common to all of them:

(i) Among the premises in any problem of induction, there are always some which are empirical, i.e. based on our experience of the perceptual world. This is unlike deduction where all the premises may be postulated abstractly (consistently with each other), as in many branches of pure mathematics. The empirical part of the set of premises in induction, is called *evidence*. Since the evidence relates to the perceptual world, the conclusion also is relevant only in relation to it.

---

[3]Some philosophers exclude one or other of these from induction; but we take a liberal attitude in this regard.

[4]There are philosophers who maintain that non-deductive reasoning geared to a preconceived hypothesis is a distinct form of inference, which is not to be called induction; but we think the distinction is unnecessary.

(ii) In deduction, given the premises, the conclusion necessarily follows from them. Thus if a piece of deductive reasoning is free from fallacy, its conclusion is formally valid. In such a case if the premises are materially valid so is the conclusion, but deduction as such is not concerned with material validity. In induction the premises only lend some support to the conclusion and since the evidence, and hence, the conclusion relate to the contingent (i.e. situated in space and time) world, apart from formal validity, the question of the material validity or validity as the basis of practical action of the conclusion (over and above the same of the premises) naturally arises. Unavoidably an element of uncertainty is always present regarding the truth or correspondence with facts of the conclusion reached through inductive reasoning. As it has been pointed out by Russell (1961) pp. 208–209, this is true of even such a universally accepted proposition as 'Man is mortal'; on the basis of available data we can strictly assert with certainty only that all men born until, say, 150 years ago are mortal. Therefore in the case of every exercise at induction, the question of assessing the degree of uncertainty, or in other words, the extent of support given to the conclusion by the evidence, is relevant.

(iii) In induction, the conclusion transcends the set of premises and therefore, if true, would make a net addition to our knowledge. (Because of this feature, induction, as we have been considering, is sometimes called ampliative induction.) This is unlike deduction where the conclusion is already inherent in the premises. The degree of uncertainty in induction depends partly on the extent of net addition to knowledge ('the inductive leap') implicit in the conclusion.

(iv) In deduction there is a core set of premises which entails the conclusion; it does not matter the least bit if some further premises are added to the core set in a consistent way. To conclude that Ram is mortal, it is enough if we know 'Man is mortal and Ram is a man'; it is superfluous to add further bits of information like 'Ram is tall' or 'Ram is married'. In the case of induction there is no such hard core of the set of premises, particularly of the part of it representing the empirical evidence; we may add or subtract items from the evidence. Of course, thereby the content of the conclusion may change; also the support extended to the conclusion may get stronger or weaker. When the evidence has already been collected it is incumbent on the investigator to base the induction on all the evidence available.

| Systematization of deduction started first |

1.4 Capacity for both deductive and inductive modes of reasoning is ingrained in human intelligence and both have been used by *Homo sapiens* since prehistoric times for the acquisition of knowledge leading to the development of civilization. But broadly speaking for a long time it was only the former which was systematically studied. Ancient Greek philosophy which was greatly influenced by the thinking of Pythagoras, Plato, and Aristotle gave almost an exclusive role to deduction based on concepts and self-evident truths as means of acquiring knowledge and generally maintained that nothing worthy to be called knowledge can be gained out of sense observations. True, Aristotle, although he overemphasized the importance of syllogistic deduction in philosophy, admitted the need of induction ('epagoge' in Greek) for setting up universal first premises. But to him 'induction' meant something else; it was an intuitive process concerned with essence of things, not involving ampliation from empirical evidence. In Indian philosophy there were the ancient logical system of *Nyāya* (logic) as laid down by Gautama and the systems built up by the Buddhist and Jaina schools (all these are deemed to have developed during the period 2nd century B.C.–3rd century A.D.; see Murti (1953) p. 34), and later the thorough-going system of *Navya Nāya* (new logic) as presented by Gangesa (around the 12th century) (see Bhattachayya (1953)). The Indian philosophers, although they did not separate inductive from deductive reasoning, did not overlook the importance of material validity of inferred conclusions altogether and emphasized the need for caution while setting up the first premises to ensure this (see Datta (1953) p. 555). Nevertheless the principal concern of all these systems seems to be the establishment of correct modes for arguing deductively. (A notable exception may be the *Syādvāda* logic of the Jaina system of philosophy which considers a multivalued logic admitting less-than-certain inference (see Mahalanobis (1957)).)

Of course in all civilizations practical sciences like those of medicine and agriculture were developed mainly through inductive generalization starting from observations. But, as far as intellectual investigations are concerned, for a long period of history the emphasis was almost entirely on deduction. This tradition continued right through the medieval period. Medieval philosophy in the West was largely an amalgam of Greek deductive thinking and Christian theology.

| Philosophy of |
| induction begins |

Things began to change only after the Renaissance started in Europe in the 14th century. With the wide-spread dissemination of knowledge made possible by the advent of the printing press, a spirit of enquiry took hold of the people during the period of Enlightenment. They no longer remained satisfied with knowledge derived through ratiocination from old postulates and doctrines; they wanted to discover new facts about nature and acquire new knowledge based on such facts. Copernicus (1473–1543), Kepler (1571–1630), and Galileo (1564–1642) made their epoch-making discoveries in astronomy and dynamics during the 16th and 17th centuries. Newton's (1642–1727) achievements made the triumph of science most comprehensive. Much of this new knowledge about the world was acquired empirically by patient collection of facts combined with bold guessing as to the laws binding the facts together. Methods adopted for setting up premises and verification were often inductive. The same can be said of discoveries in other branches of science, like the properties of magnets or the mechanism of blood circulation in human bodies that were made around this time. Side by side with the advance of science, a large number of instruments like the compound microscope, telescope, thermometer, barometer, and improved versions of clocks that facilitated observation and measurement were also invented during this period. The result of all this was that the indispensability of empirical methods and inductive reasoning for discovering new scientific truths was generally recognized by the end of the 17th century.

As inductive inference came to claim for itself an important role in various kinds of investigations, philosophers started examining the principles underlying the methods of such inference and the question of their justifiability.

| Francis Bacon |

Francis Bacon (1561–1626) is generally regarded as the pioneer in this regard. Through his books *Novum Organum* and *The Advancement of Learning* he sought to develop a method of inductive investigation which would lead to the discovery of true laws of causation behind various natural phenomena. The policy he advocated was a stagewise approach in which more and more fundamental laws are sought to be discovered in the successive stages of the investigation, and in each stage attention is paid to instantial variety as opposed to instantial multiplicity so that spurious hypotheses may get rejected. Bacon was very critical of deduction based on unquestioning acceptance of first premises handed down by authority and wrote confidently about the scope and power of induction. His writings had profound influence on scientific investigations in the 17th and 18th centuries. But it cannot be said he went into the philosophical questions about the basis of induction very deeply.

| John Locke |

The first person who conceived a comprehensive philosophy not so much of induction in particular, but of the empirical approach to knowledge in general, was John Locke (1632–1704), the founder of empiricism and one of the most influential Western philosophers of the modern period. The essence of empiricism is that all our knowledge (with the possible exception of logic and mathematics) is derived from experience; it does not admit innate ideas or principles. Perception, Locke says, is 'the first step and degree towards knowledge'. From such a position, recognition of the importance of induction is but a small distance. In fact, in his book *Essay Concerning Human Understanding* (1690) on epistemology, Locke states at various places certain fundamental principles which apply to inductive reasoning and which, we can recognize, have been part of the general lore of statistical thinking. We mention three of these. (i) Locke emphasizes the need to take in account all available information while drawing an inductive conclusion. 'He that judges without informing himself to the utmost that he is capable, cannot acquit himself of judging amiss'[5] (quoted by Keynes (1921) p. 76). (ii) Locke clearly perceives

---

[5]In recent times this principle (which we have already mentioned in section 1.7) has been emphasized and called the 'principle of total evidence' by Carnap (1947). We can find an echo of this principle e.g. in Fisher (1956) p. 109. Also, clearly this maxim underlies the sufficiency principle of statistics. Of course the crucial thing here is to sort

the element of uncertainty associated with all generalizations based on empirical information. 'Deny not but a man accustomed to rational and regular experiments shall be able to see further into the nature of bodies and guess righter at their yet unknown properties than one who is a stranger to them; but ... this is but judgement and opinion not knowledge and certainty' (quoted by Kneale (1949) p. 113). He even says that 'the degree of assent we give to any proposition should depend upon the grounds of probability in its favour' (see Russell (1961) p. 587).(Although the sense in which 'probability' is meant here is not clear, this is the position taken by statisticians of all persuasions in their approach to inference.) (iii) One unerring mark of love of truth, Locke says, is 'not entertaining any proposition with greater assurance than the proofs it is built upon will warrant' (see Russell (1961) p. 587). (If by 'proofs' we mean the derivation leading to the computation of such things as confidence coefficient, size, etc., then this gives a very general formulation of statistical ethics.) As the above illustrations show, Locke's philosophy has influenced scientific thinking and, in particular, statistical thinking considerably right up to the present day.

| Hume's problem of induction |

Although Locke laid the foundation of an empirical philosophy, as we have noted, he did not examine induction in particular to any depth. The person who did this was David Hume (1711–1776), who in the first volume of his three-volume work *Treatise of Human Nature* (1739) analysed induction and drew attention to the problem that arises when we try to justify it. In Hume's time it was generally thought that induction is concerned with the derivation of natural laws of the causal type (like 'a particular drug gives relief in migraine'). Hume characterized the cause-effect relation between two phenomena as one of constant conjunction and then observed that occurrence of such conjunction in any number of instances does not entitle us to logically conclude that such a conjunction must occur in all instances. '... that those instances of which we have had no experience, resemble those that we have had experience' is not logically necessary since we can conceive a change in the course of nature. To say that it is empirically established is but to beg the question—even if we concede that such generalizations have been found to hold in all problems of induction that we have studied, we cannot conclude that they will hold in all problems that may arise.[6,7] The problem is equally present when we have induction of laws of non-causal type

---

the grain from the chaff and to collect and utilize all the information that is *relevant*, subject to the limitation of the resources available.

[6]Surprisingly, almost the same argument regarding the logical untenability of induction had been voiced by the sceptic school of Indian philosophy, known as *Lokāyata Darśan*. See article by Shastri (1953) p. 173.

[7]As Popper (1963) Chapter I has emphasized, the difficulty is not mitigated if we consider, instead of a universal law, a probabilistic law where the conclusion follows from the premises not always, but usually.

in which we have association of attributes (like 'all animals with divided hooves ruminate'). Even in the case of other types of induction e.g. about some characteristic of an aggregate on the basis of observations made on a part of it, the same problem arises. Also the same is true in the case of hypothetic induction where we look upon the preconceived hypothesis as the conjuction of the set of all its consequences, when we seek to confirm it by verifying a subset of the same; only, when in the case of such induction we reject the hypothesis because one or more of the consequences checked do not tally, does the conclusion logically follow (by the rule known as *modus tollens*).

| Suggested solutions of Hume's problem |

1.5   Hume's problem of induction described above created a major crisis in philosophy.[8] Many philosophers have expressed exasperation by realizing its implication. Russell (1961, pp. 645–647) says that if one accepts Hume, every attempt to arrive at general scientific laws from particular observations would be ill-founded and there can be no such thing as a rational belief or rational line of action. For the last three centuries philosophers of various hues have been grappling with the problem and trying to find various ways of justifying induction. Broadly speaking, they have come out with solutions of three different types to the general problem of induction, which we shall call (i) simplistic (ii) pragmatic and (iii) stoic solutions. Of these the first two maintain that induction being concerned with phenomena in the contingent world cannot be justified solely in terms of abstract rules of logic without making any assumption about the nature of the world as we have it. Generally the assumptions made are about the uniformity (i.e. spatio-temporal invariance) of nature in one form or other. We describe the three types of solution below.

(i) The *simplistic* solution seeks to justify induction deductively. The idea was implicit in Bacon's work and was later made explicit by John Stuart Mill in his *System of Logic* (1843). Mill confines himself to induction as a tool for identifying causes and assumes the causal version of the law of uniformity of nature, which affirms 'Under the same conditions the same cause always produces the same effect'. Mill further assumes that in the situation at hand only a finite set of alternative causes of the phenomenon is possible and develops his well-known canons of induction for identifying the true cause. Following these canons, under certain circumstances, one can discover causal laws of the universal type with deductive certainty. But Mill's methods cannot tackle situations which are staple in statistics—situations where the cause produces the effect very often but not always (e.g. a certain medicine may cure an ailment most of the time). Furthermore, the condition of a known finite set of possible causes

---

[8]Hume tempered his skepticism somewhat by adding a subjective part to his doctrine (see Russell (1961) p. 640) according to which causal necessity is a sort of impression existing in a person's mind—not in the objects themselves. But such a psychological justification of induction has been regarded as cold comfort by many philosophers.

can rarely be met. Also Mill does not allow for the element of uncertainty so characteristic of inductive reasoning. These limitations apart, Mill's suggestion that the law of universal causation may itself be inductively established has given his approach a shaky foundation. It can be said Mill has failed to meet the Humean objection to induction. In recent years L. J. Cohen (1989) has attempted to streamline and extend the Bacon–Mill approach by incorporating a mode of gradation of uncertainty in it. But generally speaking it cannot be said that this 'solution' of Hume's problem has had much lasting influence on the philosophy of induction and in particular on statistical thinking, except that the need for instantial variety as against multiplicity as emphasized by Bacon and Mill has been generally accepted. (This is not to deny that in the canons of induction, one can find the seeds of some of the ideas in Fisher's theory of design of experiments.)

(ii) We call the approach to the general problem of induction by philosophers like Johnson, Keynes, Russell, and Carnap on the one hand and Peirce and Kneale on the other *pragmatic*, because both these groups of philosophers in somewhat different ways seek to justify induction on the basis of the principle that 'proof of the pudding is in the eating'. All of them agree that induction can never be completely freed from the element of risk inherent in it. A common feature is that in whatever form they bring in the assumption of uniformity of nature, it is held as a synthetic (i.e. extralogical) a priori (i.e. non-empirical) postulate. Thus in the words of Johnson (1921) p. xl it is 'neither intuitively self-evident nor experientially verifiable but merely demanded by reason in order to supply an incentive to the endeavour to systemize the world of reality and thus give to practical action an adequate prompting motive'.

Keynes (1921) postulates the 'principle of limited independent variety' according to which the characteristics of the objects in the field of study, however numerous, cohere together uniformly in groups of invariable connection, which are finite in number. Keynes uses this postulate to argue that the truth of any reasonable hypothesis with which the observed instances all conform becomes nearly certain as the number of instances multiply. For projecting the conclusions of an induction, Keynes assumes that 'mere position in time and space' is not a character which affects other characters in any fundamental way.

Russell (1948) analyses these principles further and emphasizes the postulates of causal lines, structural similarity, and analogy which we all habitually assume while dealing with nature and society and which prompt us to make generalizations from observations. Russell traces the origin of these postulates to the phenomenon of adaptation of living beings to their environments. The physical world has certain 'habits', i.e. causal laws; living beings have propensity to acquire habits of inference which help them to adapt themselves to the physical world. The postulates of induction are derived by reflecting on this process and refining it. Russell (1948) p. 507 asserts that 'The forming of inferential habits which lead to true expectations is part of the adaptation to the environment upon which biological survival depends'.

Neither Peirce nor Kneale (1949) stress the postulate of uniformity of nature in developing their solutions to the problem of induction. But Peirce explicitly (see Buchler (1955) Chapter 14) and Kneale implicitly (see Kneale (1949) Part IV) assume that the observed instances represent a random sample from the universe of all possible instances. When the universe is a conceptual collection extending over space and time obtained by repeatedly observing a natural process an indefinite number of times, randomness of the sample really means any part of the universe is interchangeable with any other comparable part and this amounts to assuming uniformity. Within this framework both Peirce and Kneale interpret the mode of induction as a behavioural policy. Peirce holds that the trustworthiness of an inductive inference depends on how often the policy leads to true conclusions. Kneale advocates the policy of choosing the strongest hypothesis (i.e. the hypothesis that leaves the least room for unobserved facts) which is consistent with the observations and asserts that the strongest hypothesis is the one which is most useful in practice for the purpose of prediction.[9]

(iii) According to the *stoic* solution (stoic, because it does not make any allowance for the natural human tendency to induce) of the problem of induction, induction is not only logically untenable as Hume shows, but is altogether unnecessary as an instrument of knowledge. Karl Popper, who has been the principal exponent of this idea, maintains that in every-day life and in scientific investigations, on the basis of observations we do not induce, but rather put up some hypothesis provisionally (See Popper (1963) Chapter 1). We hold on to the putative hypothesis and act according to it so long as it is not *falsified* by further observations. No hypothesis can be logically established on the basis of empirical evidence, but the procedure (*modus tollens*) of rejecting a hypothesis on the basis of counter-evidence is perfectly logical. According to Popper, the progress of science is marked by the successive setting up and falsification of more and more informative hypotheses on the basis of empirical tests. Thus Popper replaces induction by what he calls 'the deductive method of testing'.

Clearly of the three solutions, the stoic solution seems to take up Hume's challenge most squarely. However many philosophers have genuine difficulty with Popper's account of scientific progress; they sincerely believe that some form of induction is an integral part of epistemology.

| Statistical pragmatism | The above is a brief and of course oversimplified account of the origin and evolution of inductive philosophy. Our purpose in giving it has been to help trace the roots of statistics in the broad grounds of general inductive reasoning in later chapters. Since the whole of statistics is concerned with the analysis of and drawing inference from empirical

---

[9]In Indian philosophy the objection of the sceptic school to inductive inference had been met by pointing out that rejection of induction leads to contradiction and impossibility of practical life. This standpoint is close to the pragmatic solution. See Bhattacharyya (1953) pp. 141–142.

evidence and gearing our action to such inference, it has to start with an abnega-
tion of the Humean objection to induction from such evidence. In its special
setting it seeks to get round the objection by invoking probability calculus in various ways depending on the point of view adopted. As we proceed to examine the
foundations and the structure of statistical thought in the next three chapters,
it will be seen that the statistical approach in all its forms tries to justify itself
explicitly or implicitly in terms of practical success and is thus most akin to the
pragmatic solution to Hume's problem described above. In fact, had the history
of scientific thought followed logical steps, one could have said that the statistical approach is really an outgrowth of the pragmatic solution specialized to a
particular area. As it happened, however, statistical ideas generally arose out
of the need to solve real-life problems and the pace of their development often
outstripped the relatively measured pace at which philosophy progressed.

2

STATISTICAL INDUCTION—WHEN AND HOW?

Two approaches in statistics

2.1   In the first chapter we put forward the view that in statistics we are always ultimately concerned with some form of induction. But at the same time we quali-fied that thesis by noting that only special types of problems of induction come within the ambit of statistics, and further, in statistics such problems are sought to be solved in certain special ways. We now proceed to elaborate on these points.

It is best to state at the outset that statisticians are a divided community—although all of them take more or less a pragmatic stand regarding induction, they do not speak in one voice about the scope and basis of the methods of their discipline. Broadly, there are two schools of thinking giving rise to two different approaches in statistics:

(A) *Objective*   (B) *Subjective.*

The difference between the two schools is caused by two distinct views about the nature of probability which is brought in to quantify uncertainty in various stages of induction by both. The details of this distinction will be brought out later as we proceed in this chapter and also when we consider the different interpret-ations of probability and their bearing on statistical induction in the next two chapters. For the present it will be sufficient to say that in the objective approach generally, some at least of the probabilities considered are purported to represent the pattern of long-term behaviour of the outcomes of repeatable experiments in the external world. In the subjective approach on the other hand the prob-abilities involved are not representations of anything in the external world—*all* probabilities reflect the subjective expectation of the statistician about the pos-sible realizations of unknown entities which have to be handled in the process of induction. The main body of statistical theory as developed by R. A. Fisher, J. Neyman and E. S. Pearson, A. Wald, and even the developments due to a large class of Bayesians, can all be said to pertain to the objective school. The staunchest advocates of the subjective school have been Bruno de Finetti and his associates and followers. In the following we shall discuss the problem of statistical induction in a unified way as far as possible, and then point out where the approaches of the two schools are irreconcilably divergent.

Constituents of statistical induction

2.2   We noted in the preceding chapter that in all problems of induction (as indeed in any problem of inference in general) we start from a set of initial premises to reach certain conclusions. The initial premises of induction usually comprise (apart from the basic rules of logic) an empirical part which we agreed to call *evidence* and some *assumptions* which reflect our reading of conditions obtaining in the situation at hand.[1] The assumptions we introduce in any particular problem are those which seem plausible to us; they are brought in from outside to facilitate the inductive reasoning. As regards the aspects of the physical reality about which we intend to infer through induction, we say these represent our *inductive concern*. There are distinctive features of the type of evidence we handle, the type of assumptions we make, and also, to a large extent, the inductive concern we have in mind, in the case of *statistical induction*. These we discuss below.

| Meaning of inductive inference |

At this stage we make one point clear about the sense in which we would be using the term 'inference' in the context of statistical induction. Often the purpose of induction is to choose one or other of a set of possible actions in a practical situation. But even there the choice of the action suggests some kind of inference implicitly made about aspects of the physical reality of interest. Thus when we decide to reject a lot of items produced in a factory on the basis of inspection of a sample, we imply that the proportion of sub-standard items in the lot is judged to be too high. In the following, whenever we speak of inference or conclusion in the context of statistical induction, we would mean both explicit inference in which the conclusion takes the form of a statement and implicit inference in which the conclusion is in the form of some decision about practical action.

| First feature— evidential domain |

Taking the case of evidence first, in all kinds of problems of induction the evidence represents a collection of *contingent* facts—its content is never fully known before it is collected. A primary feature of the evidence in statistical induction is that the set of all possible alternative forms (two or more) that it may take is clearly defined beforehand. We call this set the *domain of evidence*. We give two examples of practical situations in which, the evidence being anecdotal in nature, the domain of evidence is not well-defined and which, therefore, are not suitable for statistical induction.

(i) Suppose a stone pillar with some inscriptions on it has been discovered in a remote part of central India and we want to judge whether it represents an edict of Emperor Ashoka (period of reign 268–226 B.C.). The empirical evidence would be facts such as the type of the script, language, and content of the inscription, whether other edicts of Ashoka have been found in neighbouring

---

[1] Since the choice of the assumptions depends on our judgement, all induction is in this sense to some extent subjective. However, the level at which subjectivity enters and also its extent vary from approach to approach.

areas, whether historical sources confirm that the part of the country was within Ashoka's empire, etc.

(ii) An office worker is concerned that his boss may have become displeased with him and he wants to infer about this on the basis of whatever data he can collect. Evidence here would consist of such vague things as the boss's demean-our, the look in his eyes, the tone of his response to a request, his reported remarks to some colleague and so on and so forth.

Lack of definiteness regarding the possible forms of evidence in such cases may be due to absence of any definite idea about the features that are to be observed, vagueness of the features observed, limitations of our senses or measuring devices, and similar reasons.[2]

Evidence as observations

Generally in statistical induction the empirical evidence is represented as a sequence of observations on a number of observable characters relating to some common *physical reality*. As examples consider the following:

(i) If we toss a coin 100 times and for each toss observe the character 'the face shown up', we will have a sequence of heads and tails of length 100 as our evidence. Here the given coin represents the physical reality to which all the observations relate. If instead, the sequence is generated by throwing once each of 100 coins taken from a batch coming out of a mint, the batch would represent our underlying physical reality.

(ii) We take one by one 50 households from among the 800 resident households in a township; for each household taken, we observe its size. Here the group of resident households of the township is the relevant physical reality.

(iii) We take a sample of patients suffering from hypertension out of all such patients registering themselves for treatment in a clinic during a year and, for each, observe the systolic and diastolic blood pressures (S_b, D_b) before treatment and also the same (S_a, D_a) one hour after the administration of a particular drug. Here each observation is a quad-ruplet (S_b, D_b, S_a, D_a) and the physical reality is represented by the combination of the population of all such patients and the particular drug considered.

It is clear that each character in any such situation would have a set of possible 'values' (in an extended sense) as its *domain*. The domain of the evidence here would be the Cartesian product of the domains of all the characters observed (or a

---

[2]Sometimes, however, through careful analysis of each component of anecdotal evid-ence, a set of clear-cut alternative possibilities can be visualized and in that case the problem may be amenable to statistical treatment. In particular this is true of certain kinds of judicial problems. For a detailed application of such reasoning in the context of a particular legal case, see Kadane and Schum (1996).

subset of this product). For the domain of the evidence to be clearly defined, it is necessary that the domain of each observable character should be well-defined.

| Types of characters |

As regards type, a character may be one-dimensional or multidimensional (being an ordered tuplet of a number of elementary one-dimensional characters as in example (iii) above). A one-dimensional character may be a qualitative character or attribute with a set of alternative categories for its domain; or it may be a quantitative character or variate with a set of numbers for its domain. In practice all variates assume only discrete values, either because of their very essence (e.g. when their values arise as a result of counting as in example (ii) above) or because they must be expressed as multiples of smallest unit measurable. In the latter case, when the smallest measurable unit can be conceptually made arbitrarily small, we often regard the variate as continuous with some interval on the real line for its domain (e.g. each of the variables S_b, D_b, S_a, D_a in example (iii) can be regarded as continuous when measured as lengths on a mercury column instrument). Sometimes we may have a character which is semi-quantitative in the sense that although its intensity cannot be measured on any standard scale, different individuals in a group can be linearly or partially ordered by a person according to it (e.g. different performers according to the quality of their rendition of a musical piece, different coloured specimens according to their brightness etc.). In such a case the observation would be on the group as a whole and the domain of the character would be the set of all such orderings possible. (In the case of linear order, for instance, an observation would be in the form of a permutation of all the individuals and the domain would be the set of all such permutations.) With the advance of instrumentation, nowadays we often come across observations in more complex forms represented by graphs or photo-images. (This is common in medical diagnosis, remote sensing etc.) Usually in such cases one derives various numerical measures to represent aspects of the graph or image and treats the derived figures as observations on characters. (This, of course, involves shedding part of the information, but that is more or less inherent in all forms of observation. Even in the coin-tossing experiment we note only the face coming up and ignore, e.g. the angle of orientation of the axis of the face.) However, with the development of computers one can visualize the possibility of regarding the entire graph or image as the observation of a complex character. But whatever form the observation on a character may take, its domain will have to be clearly defined in all cases.

| Second feature—order unimportant |

The second feature of the evidence in statistical induction arises because quite often from experience and common sense we can identify within the set of characters subsets in which the members are analogous in the sense that they are at par with each other in relation to the underlying physical reality. Characters in such a subset all have the same domain and they differ only in respect of the spatio-temporal order in which they are considered. If spatio-temporal order has no bearing on our ultimate inductive concern in such a context, we ignore the order of presentation of the observations within each such subset. Thus referring to section 2.4 we may ignore the serial numbers of the tosses in example (i), of the households observed in example (ii) (if the sampling of households is in some sense unselective with regard to household size), and of the patients observed in example (iii) (if the order of arrival of the patients at the clinic has no bearing on our study, i.e. we are not interested in any kind of trend or seasonality). In de Finetti's words, within each subset '. . . the analogy leads us to make the conclusions depend only, or, at least mainly on *how many* "analogous" events occur out of some given number and not on *which* of them occur' (see de Finetti (1975) p. 206). When all the characters observed are analogous to each other (in the objective approach in such a case we often regard the observations to have been obtained by observing the same character repeatedly, but this point of view is frowned upon in the subjective approach, see Lad (1996) p. 19) this means we replace the original evidence by its summarized form as given by a frequency table.[3]

| Operations generating observations |

To generate the sequence of observations in a problem of statistical induction we have to define in unambiguous terms not only the corresponding characters but also the operation of observing each character. Of course each such operation will involve the physical reality to which the characters relate. Thus in example (i) of section 2.4 for each toss we have to specify the actual experimental conditions of tossing the coin and observing the outcome. Generally, when an observation is obtained by performing an experiment, all the practical details of the experiment (whether of the 'controlled' type or of the 'observational' type) will have to be clearly laid down. When to observe a character we have to select an individual from a group or population, we have to specify the method of selection as well. Thus in example (ii), the method of selection of the successive households out of all households resident in the township (whether purposive or random, with replacement or without replacement etc.) has to be specified. The same is true of example (iii), although here the population comprising all hypertensive patients to be registered in the clinic during the year is not available ready at hand at the time of each selection, and as such rules which do not require initial serial

---

[3]When the different realizations of evidence obtained by permuting analogous observations are logically at par with each other, the desirability of using only procedures invariant under such permutations demands that we replace the set of observations by their distinct values and the corresponding frequencies, which together represent the *maximal invariant* (see Ferguson (1967) pp. 243–247). Replacement of the evidence by the frequency table is standard also in what in Chapter 1 (section 3) we called covert induction: usual measures like mean, variance, etc. are all permutation invariant, and hence, computable solely from the frequency table.

numbering of all individuals in the population (like taking every 10th patient coming) may have to be adopted.

Third feature— extensibility, repeatability	The third feature of evidence in statistical induction can be described in terms of two conditions about the sequence of operations leading to the observations.

The first of these may or may not hold explicitly in the objective approach depending on whether or not the problem is one of prediction, but is an essential part of the subjective approach; the second is imperative in the objective approach but is not invoked in the subjective approach.

(a) *Extensibility.* This requires that the sequence of operations generating the evidence can *in principle* be extended to include one or more additional operations. These additional operations, if they were performed, would give rise to some further observations whose inclusion would result in an extension of the evidence. Thus we can conceive that in example (i) the coin could be tossed once again, in example (ii) another household could be selected (from amongst all the 800 households or only the remaining 750 households according as selection with or without replacement was being followed), and in example (iii) another patient registering at the clinic could be taken and treated. In the objective approach such additional operations are of relevance when induction involves the prediction of future observables. This form of extensibility may be called forward extensibility. In the subjective context we may have also backward or collateral extensibility involving unobserved characters belonging to the past or the unknown part of the present. For reasons that will be apparent later, some form of the extensibility assumption is invariably made in the subjective approach.

It should be clearly understood that the additional operations conceived would always relate to the same physical reality as the original evidence. Also the characters associated with the additional observables would have clear-cut domains and naturally therefore the conceptually extended evidence would also have a well-defined domain.

In this connection we mention that for certain kinds of theoretical studies we have to assume that the scope for extensibility is unlimited, i.e. any number of extra observations can be added to the evidence. This means that an infinite sequence of potential observations can be conceived. In this case we say the evidence is *infinitely extensible.* Obviously then, the number of observations to be included in the evidence can also be indefinitely increased, always leaving some additional unrealized observables.

(b) *Repeatability.* This condition, which applies *only* in the objective approach and *not* in the subjective one, means that *conceptually* one can make independent repetitions (i.e. repetitions whose outcomes are causally unrelated) of the entire sequence of operations leading to the evidence (the extended evidence, if extensibility is being assumed) any number of times, positing oneself to the stage where no evidence is available before every such repetition. The repetitions conceived here are supposed to be independent in the sense that the realization we would get in one would not causally affect the actual realization in another. The independent copies of the evidence varying over the evidential domain can be looked upon as constituting an infinite 'conceptual' or 'hypothetical' population. Thus we presume in example (i) of section 2.4 that we can repeatedly perform the set of 100 tosses and note the outcomes and in example (ii) that we can repeatedly draw a sample of 50 households from the entire population of 800 households according to the prescribed procedure, observe their sizes and replace the sample in the population before the next draw. In example (iii) the natural process according to which successive hypertensive patients get registered at the clinic and some among them are selected, treated, and observed as laid down in the procedure cannot be re-enacted in practice; but we imagine that this can be done again and again to generate independent copies of the evidence. (In certain cases, e.g. in the coin-tossing problem of example (i), forward extensibility may be construed as implied by repeatability; but the essence of the former is that in it some performable new operations are visualized. The difference between the two conditions becomes clear if we consider the case of without-replacement sampling in example (ii).)

The above two conditions in appropriate contexts give us scope for induction; as any unrealized additional observables conceived or the conceptual population of independent copies of the evidence relate to the same physical reality as the realized evidence, the latter throws some light on the former. More details will come up as we proceed.

| Fourth feature— uncertainty assessable | The fourth feature of the evidence in statistical induction is that until the operations generating the observations (including additional observations if extensibility |

is assumed) are performed, the evidence (extended evidence) that would be realized remains uncertain and this uncertainty can be assessed in terms of probability. The words 'uncertainty' and 'probability' in this context, however, have different meanings in the objective and subjective approaches.

| Meanings of uncertainty | In the objective case repeatability is assumed. Here uncertainty about the evidence means we visualize that |

if the set of operations generating the evidence were repeated, the evidence coming out would vary unpredictably from repetition to repetition. Sometimes this happens naturally, because we are always in the dark about any causal connection between the conditions prevailing before the operations and the evidence realized, and these initial conditions are bound to vary imperceptibly from repetition to repetition. Referring to section 2.4, this happens, e.g. in the coin tossing experiment of example (i) and also in example (iii) if the patients arrive and are selected in an haphazard manner. Often, however, to ensure unpredictability, we deliberately befog any possible causal connection by bringing in an element of blindfolding or randomization in the operations whenever selections are to be made. Thus in example (ii) we may sample the serial numbers of the households to be observed by drawing tickets from a lot or reading from a random sampling numbers table. In controlled experiments where a set of units are subjected to different treatments before observation, allocation of the units to the treatments is similarly randomized. Thus uncertainty in the evidence may arise due to one or more of the causes, natural variation, errors of measurement, sampling variation (incorporating any randomness deliberately introduced), unforeseen contingencies etc. (In the above as well as in the following, in the context of the objective approach, what we say about the evidence applies also to the extended evidence when extensibility holds.)

In the subjective approach there is no question of repetition of observations. Here uncertainty only means absence of knowledge about the evidence and extended evidence, before the generating operations are performed. Because of this, the scope for induction is somewhat wider in the subjective than in the objective approach. We consider two examples to illustrate.

(i) Suppose the operations consist in determining the digits that occur in the ten positions starting from the 101st, in the decimal expansion of π. A person has no idea about these digits before the operations leading to their determination (actual computation or consultation of some available source) are undertaken and so they are subjectively uncertain; but repeated determination will always lead to the same values (provided they are correct), and hence, there is no uncertainty in the objective sense here.

(ii) Suppose a traveller who is innocent of any knowledge of geography, suddenly finds himself in a foreign country about whose climate or people he has no idea whatsoever. He is subjectively uncertain about the numbers of official holidays declared in the country for, say, the first six months of the next year. But once these numbers are read from the calendar, they will remain fixed no matter how many further times he may look at the calendar.

In the subjective approach uncertainty is handled in the same manner, whatever may be the cause from which it arises.

| Meaning of probability | As regards probability which expresses the uncertainty about the observables, it is given radically different interpretations in the objective and subjective approaches. We will consider details of these in Chapter 3. For the present we give only brief indications of the difference. In the former, roughly speaking, we assume in effect that the unpredictable variation of the evidence is such that the relative frequency with which it would belong to any meaningful set (the meaningful sets, technically called measurable sets, are those which are of practical interest and are theoretically easy to handle) in the evidential domain would tend to stabilize

around some idealized value, if the number of repetitions were made indefinitely large. The basis of this assumption, which we call *frequential regularity*[4] is our experience with diverse types of particular repetitive experiments. For any set of interest the probability that the uncertain evidence will belong to it is identified with the corresponding idealized long-term relative frequency. Probabilities, so defined, of all meaningful sets in the evidential domain determine a probability distribution over the domain and this gives an objective representation of the evidential uncertainty.

In the subjective approach probability exists only in one's mind and may vary from person to person. For a particular person the probability of any set of interest represents the person's degree of belief in the materialization of the event that the evidence (extended evidence) generated through the operations when they are performed would belong to that set. In practice this degree of belief can be quantified introspectively, e.g. by ascertaining the maximum price one is prepared to pay outright for a unit gain contingent on the actual real- ization of the event. Ideally one should attach numerical degrees of belief to different sets of interest in a consistent or *coherent* manner. (In the objective case the relative frequencies of different sets and hence their idealized values are automatically coherent and therefore one need not keep a watch over this point.) Coherent probabilities for different meaningful sets in the domain define a prob- ability distribution over it. Note that since uncertainty here means absence of knowledge, such a probability distribution may cover evidence extended back- ward or collaterally to involve unobserved characters belonging to the past or the present. To facilitate statistical induction in the subjective approach generally it is assumed that the probability distribution over the extended evidential domain is *exchangeable* (fully or partially) in the sense that when analogous observa- tions are permuted among themselves the distribution remains unchanged. This assumption, which is clearly in conformity with the second feature of evidence discussed in section 2.5 above, allows us to incorporate our experience and com- mon sense knowledge about the operations in the build-up of the probability distribution and helps us to infer about the unobserved characters belonging to the past, present, or future, on the basis of knowledge of those observed. In situ- ations where the evidence is infinitely extensible so that any number of further observations can be added, we even assume *infinite exchangeability*, i.e. exchange- ability, over the extended evidence whatever the number of further observables added.

It should be noted that the centuries-old postulate of 'Uniformity of nature' makes its appearance in different garbs through both the assumption of fre- quential regularity under indefinite repetition in the objective case and that of infinite exchangeability of observables in the subjective case. In the former

---

[4]This is commonly called *statistical regularity*. But since this regularity is the basis of the objective approach to statistics, to call it 'statistical' would be to put the cart before the horse.

it is presumed that repetition could be continued over time indefinitely under identical conditions and in effect that the idealized relative frequency of a set would be same for *all* sequences of repetitions of the operations; in the latter, it is presumed that to whatever length the sequence of observables be extended over time, symmetry or permutation invariance of the joint probability distribution in respect of analogous observables continues to hold.

| Assumptions imposed | **2.3** We next take up the features of the assumptions extraneously brought in as part of the premises in stat-istical induction. We have noted above that in the statistical mode of reasoning we always suppose that the uncertainty about the (extended) evidence can be expressed in terms of the probability distribution of the observables. The extraneous assumptions are made with regard to the particulars of this prob-ability distribution. The particular form of distribution assumed represents the probability model for the problem. The basis of bringing in any such assumption, i.e. of choosing a particular probability model in any situation is past experience, intuition born of familiarity with the field, background knowledge of scientific laws which underlie the data-generating process, and sometimes plain convention and the ease of analytical handling.

| Probability model— objective | In the objective approach, while building the model a broad principle which is often invoked is that causal independence of observables implies their probabilistic independence. This determines the extent of factorizabillity of the joint probabil-ity distribution. Furthermore, the functional forms of factors involving analogous observables are supposed to be same. But in this approach the assumptions made never completely specify the probability distribution—the 'value' of some *para-meter* remains unspecified. The parameter is supposed to represent the unknown state of nature, i.e. the unknown aspects of the physical reality with (a part of) which we are concerned in the situation at hand. In parametric models the para-meter is finite-dimensional, i.e. real or vector-valued; in nonparametric models it is infinite-dimensional, i.e. it cannot be represented by a finite-dimensional para-meter. Thus in section 2.4 in example (i), the standard objective model assumes that the sequence of tosses represents independent and identical Bernoulli trials (trials with two complementary outcomes) with an unknown probability p of get-ting heads in a trial. Here $p \in (0, 1)$ represents the idealized value of the relative frequency of heads in an indefinitely long sequence of tosses. In example (iii) it is customary to regard the observables on the quadruplet (S_b, D_b, S_a, D_a) to be independently and identically distributed, i.e. the set of observables represents a random sample from some 4-variate parent population. In the standard para-metric model this parent population is taken as 4-variate normal with some or all of the means, variances, and covariances unknown. In the nonparametric case the distribution in the parent population itself is taken as an unknown 'function-valued' parameter.

Objectivists, in general, agree that the choice of a realistic probability model in a particular situation is justifiable ultimately in terms of the long-term behaviour of the relative frequencies on repetition. A particular probability model is chosen in a situation on the presumption that in similar situations the same model with certain values of the parameter has been found to give ideal representation. As regards the unknown parameter, some objectivists regard its value in any situation as just an unknown but fixed entity in its domain (*parameter space*); there are others who impose a subjective probability distribution over the parameter space representing the a priori degree of credence given to different statements about the value of the parameter. These different attitudes result in a parting of ways even among those following the objective approach.

| Probability model—subjective |

Out-and-out subjectivists do not admit the objective existence of quantities which cannot be measured and observed even in principle. (However, such a quantity or rather its probability distribution may exist subjectively in the mind of a person.) Thus in the subjective approach there is no scope for bringing in an objectively existent but unknown parameter in the final probability model—the joint probability distribution of all the observables (including additional observables) is supposed to be fully specified by the model. (Even when fictitious parameters are initially introduced to facilitate model building, ultimately they are summed or integrated out.) This joint probability distribution gives the statistician's prior belief that the observables would realize values in various sets. One remarkable point of difference with the objective approach is that, since the observables relate to the same physical reality, subjectively they are not regarded as probabilistically independent even when there is apparent causal independence between them, provided they throw light on each other. Thus in the case of tossing a coin, even though in no conceivable objective sense can the outcomes of different tosses causally affect each other, since they involve the same coin knowledge of a few outcomes gives us some subjective idea about others and therefore they are regarded as dependent. (In this connection Keynes's ((1921) p. 343) illustration with the extreme case where the coin is known to have either both faces heads or both faces tails, is illuminating.) However, as noted earlier, symmetry or exchangeability with regard to analogous observables is generally incorporated in the model. Even then, a subjectivist statistician has a very tall task—he has to construct the model in its entirety mentally through intensive introspection. In practice, subjectivists often confine themselves to conventional frameworks.

| Inductive concern |

2.4 We now come to an examination of the nature of the inductive concern (i.e. the aspect of the physical reality about which we want to infer) in statistical induction. The purpose behind the investigation is represented in terms of the inductive concern; it determines the type of the inferential statement we intend to make or the action we deem appropriate. Thus logically it precedes the collection of the evidence and the introduction of the model. But we cannot describe the inductive concern in a problem of statistical induction in any precise manner without first bringing in the observables and the model characterizing the uncertainty about them. That is why we take it up now.

Objective case

The nature of the inductive concern is radically different in the objective and subjective approaches. In the objective approach, as noted above, the model always involves an unknown 'parameter' representing unknown aspects of the physical reality of interest. (Here we are using the term 'parameter' in a very broad sense. It may well be that in a situation we are undecided about the form of the model and that itself has to be induced from the data. The unknown form of the model is here regarded as a parameter of interest.) In statistical induction we want to infer in respect of this parameter (or some function of it) and the problem of induction is of different types depending upon the type of the inference we have in mind. Thus in point estimation we want to determine a reasonable working value for the parameter (i.e. choose a point in the parameter space) in the light of the evidence, whereas in the case of interval estimation we want to give a range for its value (i.e. choose a set in the parameter space). These are what we have called in Chapter 1 problems of open induction. In the case of hypothetic induction we have a preconceived hypothesis about the value of the unknown parameter (specifying that it lies in a given sub-set in the parameter space); on the basis of the available evidence we want to declare the hypothesis as either tenable or untenable. Such hypothesis testing, in its turn, can be regarded as the simplest form of general action-oriented problems (decision problems) in which we have to choose one out of a set of possible actions and different actions are appropriate to different degrees for different values of the unknown parameter. In the objective approach if forward extensibility is assumed there would be one or more potential observations whose values are 'future contingents' and we may want to predict sets in which these would lie. But such prediction is always done by utilizing (at least implicitly) the information about the unknown parameter contained in the evidence and the level of accuracy of prediction depends on the value of the parameter. Thus we can say that in the objective approach to statistical induction we are always concerned with questions about the unknown parameter involved in the model or in other words, the unknown parameter represents our inductive concern. (Note that in the objective approach, backward extensibility is not assumed; unobserved characters belonging to the past, if they affect the observed characters, may be treated as unknown parameters.)

Subjective case

In the subjective approach on the other hand, the probability model representing the uncertainty about the values of the observables before they are realized is supposed to be completely known—it involves no unknown parameter. Inference in statistics consists in forming opinion about one or more of the unrealized additional observables in the light of the observables already made (since extensibility is assumed such additional observables can always be visualized). This opinion is expressed in terms of the subjective (conditional) probability distribution of the additional observables given the values of the realized observations; when exchangeability holds, determination of this distribution becomes easier. Thus in the coin-tossing example if the outcomes of 100 tosses of a coin have been observed, in the objective approach we suppose there is an unknown probability p of head associated with the coin and want to infer about this p on the basis of the evidence. On the other hand, in the subjective approach the statistician gives merely his subjective assessment about the possible outcome if the coin were tossed one more time, ten more times, etc. Thus here one is not concerned with knowing about or attuning one's action to any presumed fixed aspect of the physical reality being studied, or even with the 'prediction' of the values of some unobserved additional characters.[5] Rather one has to formulate one's own subjective opinion about the values of the additional characters. The inductive concern here is represented by the set of additional observables of interest. Varieties of inference problems here would correspond to the different ways in which this subjective opinion may be given expression to or put to use. Thus one may make a statement about the subjective expectation or prevision of an unobserved quantity or about the probability content of a sub-set, or one may compare the subjective expectations of gains accruing from different actions to choose the most advantageous action. By the very nature of the approach, the inference made would depend on the person who makes it.

| The two viewpoints |

According to subjectivists it is pointless to speak about a population parameter which according to them is a 'metaphysical' quantity for which there is no operationally definable measurement process that can be carried out in finite time (see de Finetti (1937) pp. 141, 149).[6] On the other hand, objectivists hold that, even if the personal element present in the subjective approach be ignored, confining oneself only to inference about one or more additional observables and not attempting any kind of generalization on the basis of evidence is taking too narrow a view of induction. Generalizations are not only intellectually more satisfying, but inferences about particular instances become more plausible when they are derived from general conclusions. Roughly, we can say, objectivists give more emphasis to generalization in induction whereas subjectivists are concerned exclusively with

---

[5] See de Finetti (1974) p. 70, (1975) p. 202.

[6] Apparently, subjectivists would admit of parameters only in cases like drawing from an urn, finite population sampling etc., where these are in principle observable; but even in such cases they interpret these in terms of additional characters; see de Finetti (1937) pp. 141–142, Lad (1996) pp. 211–226.

'particular induction',[7] or as it has been called, *eduction* (see Johnson (1924) Chapter IV).

More fundamentally, if we look upon induction as a tool for knowing about the world, the difference between the objective and subjective approaches in statistical induction is rooted in a difference in point of view as regards epistemology. Objectivists admit two kinds of knowledge about the world—knowledge of facts and knowledge of general connections between facts (see Russell (1948) pp. 421–433). The latter kind of knowledge is represented by natural laws of both the universal and the probabilistic type (such probabilistic natural laws have to be considered when, as Russell (1948), p. 482 says, there is a hopeless entanglement of too many causal lines). A statement about the value of a parameter in a probability model is not a presumptuous attempt at quantifying a metaphysical entity; rather in general, it is a concise way of giving expression to an approximate probabilistic law of nature. Thus when we say that the probability p of head for a coin is $\frac{1}{2}$, we mean, 'if the coin is thrown any large number N of times, *usually* we would find the number of heads near around $\frac{1}{2}$N'. Similarly, denoting the mean yields per unit area of a crop under two treatments A and B by μ_A and μ_B, when we say that $\mu_A - \mu_B \geq 10$ we mean, 'if large numbers of similar plots are sown with the crop under treatments A and B, the mean yield for A would *usually* exceed that of B by about 10 or more units'. Both these statements represent probabilistic natural laws expressing hypothetical connections between facts. Of course following Hume (cf. Chapter 1), one can contend that logically such laws state nothing more than conjuctions of events which supposedly are found to occur in succession in most cases. Nevertheless, pragmatically, discovery and recognition as 'knowledge' of such laws have made advancement of science possible. Agronomists, engineers, and doctors all rely on such laws. That human civilization has progressed so far through adaptation to, and successful handling of nature on their basis is proof enough of their value (Russell (1948) p. 507).

Proponents of the subjective approach on the other hand adopt a positivist-cum-semi-idealist position as regards our knowledge of the world. Regarding empirical facts they take the positivist standpoint—facts are admitted only if they are operationally verifiable. However, regarding laws of nature they maintain that these exist only in our mind. This, they say, is true even of deterministic laws of nature which are based on masses of evidence collected over long periods that make them acceptable to everybody with subjective certainty. They say it is pointless to search for any external reality underlying such universal acceptance (de Finetti (1937) p. 152). Similarly, as regards probablistic laws of nature, they do not deny their practical utility, but hold that these express only the opinion or belief of the person concerned and they allow some room for personal predilections in their formulation. (Objectivists disagree with all this. According to them beliefs generate expectation as to what is going to happen in future. If under

---

[7]The controversy as to whether generalization or particular induction is more basic is an old one in philosophy. See Kneale (1949) p. 45.

the same circumstances the same percepts always generate the same expectation which usually are realized, we are justified in assuming the existence of a real connection; see Russell (1948) p. 425.) We can say that the subjectivists accept Hume *in toto*—both as regards the logical untenability of induction and as regards the importance of psychological induction in practice (see section 1.9 and, in particular, footnote 8 in Chapter 1).[8] (Hume (1739) Book I part iii; see the discussions in Kneale (1949) pp. 223–224, Russell (1961) pp. 640–645, de Finetti (1975) p. 201, and also Lad (1996) p. 23.) They gear their entire programme towards the systematization of the latter. (It may be noted that those subjectivistic schools of philosophy who do not regard individual minds as working in isolation solely on the basis of sense data and the rules of logic but admit of other universal principles (like the unity of all minds or the existence of a priori synthetic knowledge) can explain interpersonal agreement about laws of nature more directly; their position would be close to the objective one.)

| Uncertainty re conclusion |

2.5   In Chapter 1 we noted that an element of uncertainty is ingrained in every conclusion reached through induction. This uncertainty is with regard to the validity of the conclusion—validity either in the sense of correspondence with reality or usefulness as a basis for action in practice. A speciality of the statistical way of induction is that here an attempt is made to quantify this uncertainty in terms of probability.

The objective approach to statistical induction is very clear-cut in this regard. Since in this case our inductive concern is represented by some unknown parameter involved in the asssumed probability model, our conclusion here takes the form of a statement or implies a presumption about the true value of this parameter. The parameter here represents some aspect of the unknown physical reality being studied—an aspect epitomizing an approximate probabilistic law. Therefore how 'closely' the conclusion about the parameter corresponds to reality, closeness being interpreted in terms of practical consequence, is a pertinent question. The uncertainty, or its positive complement, the trustworthiness of the conclusion, is assessed in terms of probability and reported by making some form of assertion involving probability along with the conclusion. Those objectivists who regard the parameter as an unknown fixed quantity invoke the repeatability condition for this purpose; those objectivists, who express the belief in different values of the parameter in subjective probability terms, do not use repeatability at this stage (they generally use it for model building in any case), but rather exploit Bayes's theorem for this purpose. We will consider further details in Chapter 4.

In the subjective approach to induction, there are no parameters in the model—one is interested only in the unknown values of unobserved additional

---

[8]Interestingly, Popper accepts only the first but dismisses the second (see Popper (1963) Chapter 1).

characters outside those included in the realized evidence. The inductive concern of the investigator is attended to by the statistician who unravels his conditional subjective probability distribution for these unobserved characters and the inductive conclusion is a statement based on this distribution. Thus the conclusion is in the form of a subjective opinion and probability is undetachably involved in it. (For example, the conclusion may be of the form: 'on the basis of experiments made on 20 migraine patients, I am personally 90% convinced that the incoming patient will be relieved of migraine by taking this medicine'.) If the investigator has arrived at such a personal opinion on the basis of the available evidence by arguing in a consistent or coherent way, we cannot raise any question about the material validity of such an opinion, since no spatio-temporally invariant norms (parameter or causal law) against which we could judge such an opinion can be postulated in this approach. At best we can counter the opinion by another opinion but never establish one over the other. Verification here is possible only when the additional characters have been actually observed, but that would mean altering the ground conditions represented by the evidence under which the earlier opinion had been formed.

| Large body of evidence | If infinite extensibility of the evidence is assumed we can say something more about validity both in the objective and subjective cases. Under this assumption, the number of observations in the evidence can be increased indefinitely. In the objective case, it can be shown that for any reasonable induction procedure the uncertainty about the material validity of the conclusion as measured directly or indirectly by probability, asymptotically tends to zero. In the subjective case, if infinite exchangeability of the sequence of observations holds and some plausible conventions about the form of the inductive conclusion are agreed upon by two persons, their subjective opinions about an un-observed character, even though initially widely different, will tend to converge as the number of realized observations increases.[9] This means, when a sufficiently large body of evidence is available, a consensus will emerge among different persons and such a consensus will provide a valid basis for agreed practical action, which can be regarded as a substitute for material validity. Thus when we have sufficient evidence, whether we follow the subjective or the objective approach, validity of our conclusion is established in some sense.

A different form of uncertainty arises in statistical induction when we have doubts about the appropriateness of the model assumed. As a result, here there is a sort of secondary uncertainty—uncertainly about the statement of uncertainty that is attached to the conclusion. Such doubts arise when we are not sure about the model and other assumptions in the objective case and about the opinion we hold in the subjective case. Such secondary uncertainty is assessed by studying the stability of the conclusion (including the uncertainty statement)

---

[9]See e.g. Skyrms (1996) and also Cohen (1989) p. 67.

when the premises are varied. Such studies are often called studies of robustness or sensitivity.

Design problem   **2.6** Before concluding this chapter we address ourselves briefly to an aspect of statistical induction which we have so long kept in the background: the mode of collection of the evidence. The importance of this has been recognized by scientists since ancient times and emphasized by philosophers at least since the time of Francis Bacon. In modern times R. A. Fisher among others has been instrumental in developing a substantial theory relating to modes of collection of suitable evidence for statistical induction.

An investigator generally embarks on an inductive study with its purpose more or less vaguely conceived in terms of the inductive concern. The cost (in terms of money, time etc.) that he can afford for the study is limited by the resources and the situation at hand. The collection of evidence to be undertaken should ideally be optimally geared to both these considerations. Depending upon the field of investigation, evidence in statistical induction may be collected through an 'observational study' or 'controlled experiment'. In the former case we just observe the characters of interest on a process or phenomenon that occurs naturally, without influencing it in any way; this is common in fields like astronomy, geology, or the social sciences. In the latter, as in fields like agriculture, biology, medicine, and industry, we artificially manipulate various conditions to influence the process or phenomenon studied in the desired way. Naturally we have greater scope for controlling the uncertainty in the inductive conclusion in the case of controlled experiments. We can say finite population sampling in statistics lies midway between an observational study and a controlled experiment;[10] in this case we cannot influence the characters observed but we can select the units on which the characters are to be observed and by that means we can control the uncertainty in the conclusion to some extent. (If we regard the population considered as the physical reality and the observations on the selected units as those on characters associated with that physical reality, by selecting the units, in a way, we influence those characters.) Various considerations arise with regard to collection of evidence, especially in the case of controlled experiments and finite population sampling.

(i) *Identification of factors.* In most cases we have some prior knowledge on whose basis we can identify certain factors affecting the variability of the characters to be observed. These factors may be qualitative or quantitative. Also we have some scope for manipulating these factors or recording their 'values' at the time of observation. Thus in testing the efficacy of some drug in the treatment of a disease, the response of the patient may depend on qualitative factors like gender, stage of development of the disease, and a quantitative factor like age.

---

[10]See Mahalanobis (1950).

(ii) *Manipulation of factors.* While collecting the evidence (especially in the case of controlled experiments and sampling from a finite population), we can often manipulate some of the factors influencing the characters before embarking on collection of evidence; sometimes prior manipulation of the factors is not possible but these can be observed as part of the evidence. All this, if cleverly done, reduces the uncertainty in the inductive conclusion. This is because, as we have noted, in statistical induction the uncertainty is determined by the probability model assumed and when the evidence collected is engineered in this way, we can plausibly assume models which are more informative as regards the inductive concern and therefore lead to less uncertain conclusions. Thus in an agricultural study on the comparative yields of paddy under two treatments to be carried out on a set of field plots, if we know the fertility pattern in the field, we can compare the treatments within blocks formed of pairs of similar plots. Again we know that here the yield of a plot is affected by the factor 'number of plants in the plot', which cannot be manipulated beforehand but can be observed while observing the yields. The blocking and the measurement of the concomitant character allow us to assume a model in which the relevant variability in the yields is considerably smaller. Since ultimately we are going to answer the question 'how would the yields of the two treatments compare if they were applied to the *same* plot?' to ward off lurking variability within a block we randomize the treatments within each block. This justifies the assumption of symmetry (exchangeability in the subjective approach) in the model, thus reducing any secondary uncertainty that might arise due to the model being inappropriate. Similarly in the case of sampling from a finite population, prior stratification of the population, random selection within each stratum, and recording of auxiliary characters go towards reduction of inferential uncertainty.

In the context of induction of universal causal laws, philosophers like Bacon and Mill (see Cohen (1989)) had long ago emphasized the importance of instantial *variety* as a basis of induction. Techniques like blocking, stratification etc., can also be looked upon as extension and detailed working out of the same principle in the case of statistical induction. Thus the blocks in an agricultural experiment may be construed as representing different soil types, weather conditions etc. and if comparison of treatments is made separately for each block the conclusion becomes valid across a wide range of conditions.[11] Another requirement of induction stressed by philosophers is that there should be strong *analogy* between the observed instances and any new instance to which the conclusion is going to be projected.[12] As noted above, randomization makes the comparison of treatments within blocks similar to the potential comparison that would arise if they could be applied to the same plot.

---

[11]Fisher (1937) p. 110 mentions this 'wider inductive basis' in the context of factorial experiments.

[12]See e.g. Copi and Cohen (1996) p. 460.

(iii) *Size of the evidence.* Apart from deciding on the way in which different factors influencing the characters should be taken into account, another point that has to be settled regarding the collection of the evidence is the number of observations (of each type) that should be made. It is a principle of induction in general that under the same circumstances, as the number of instances is increased, the uncertainty of the conclusion diminishes.[13] On the other hand, as mentioned earlier, in practice there is always a limit to the resources and time available. Two things mainly decide the volume of the evidence we should collect in any problem of induction. First, the extent of variability in the observations can be judged from parallel studies undertaken in similar problems. (In the context of general scientific induction Kneale (1949) p. 112, gives the example: to determine the properties of a newly discovered chemical substance such as deuterium, the heavy isotope of hydrogen, a few experiments will suffice, for we know that properties of such chemical substances are very uniform; on the other hand, to know the mating habits of a newly discovered species of birds, it will be prudent to collect a large body of evidence to ensure a fair degree of certainty about the conclusion.) Second, the extent of uncertainty that may be tolerated in a problem of induction is determined by the type of use to which the conclusion is going to be put in practice and the losses that would result from different types of wrong conclusions. When the damage caused by a wrong conclusion would be immense (for instance, when a big dam is going to be constructed or a new drug for treatment of cancer is being proposed) one has to be extremely cautious and an adequately large volume of evidence must be amassed before an inductive conclusion is ventured. In less serious cases hasty conclusions may be permissible. Another consideration may be sometimes of importance. Even for the same form of model the informativeness as regards the inductive concern may depend on the particular sequence of observations realized. (For instance, when we are testing a drug for migraine and all the first 10 patients on which it is tested report positive relief, we are surer about the efficacy of the drug than when of 20 patients 12 report relief, 8 no relief, and these two types of responses come up in a mixed order.) When the observations can be taken sequentially, such differential informativeness in different observation sequences can be cashed in to collect a smaller or larger body of evidence depending on the sequence one encounters in practice. This generally results in substantial saving in the cost.

All the above represent what are known as design problems of statistics which cover considerations like choice of the type of experiment, the type of model, actual allocation of design points, etc. As would be evident from the discussion, choice of the design, i.e. a particular mode of collection of an adequate body of evidence, would involve various subjective judgments. Nevertheless the theory for this that has been developed with regard to controlled experiments is mostly based on the objective approach. This is because, as so far developed, the theory of designs requires the introduction of design-independent parameters

---

[13]See Russell (1912) p. 104.

(like treatment effects) in terms of which a common formulation of the problem of induction for different designs (like completely randomized, randomized block, Latin squares etc.) can be given and also counterfactual consideration of designs, other than that through which the evidence has been actually collected, can be undertaken. (Reviews of the objective theory of design selection are given by Pukelsheim (1993) and Chaloner and Verdinelli (1995) from the non-Bayesian and Bayesian points of view respectively.) The theory of design of controlled experiments constitutes a major portion of the theory of statistical induction. As of now very little work has been done towards developing such a theory at the interpersonal level following the subjective approach.[14]

---

[14] As regards the theory of sample surveys, the situation is less one-sided. Since in the case of finite population sampling, parameters are admitted even under the subjective set-up (see footnote 6), a purely subjective formulation of the theory is conceivable here (see e.g. Lad (1996) Chapter 3 for some developments along these lines).

INTERPRETATION OF PROBABILITY:
VARIOUS NUANCES

| Interpretation of |
| probability |

3.1 In the preceding chapter we saw that the concept of probability is organically associated with statistical induction. It enters the process of induction in at least two stages. Firstly, we express the uncertainty about the empirical evidence before it is collected in the form of a probability model. Secondly, the inductive conclusions that we reach are derived, justified, and expressed in terms of probability. For this, formally we make use of the rules of the probability calculus. But to use probability for statistical induction, apart from formal operations on probability, we must consider also the semantics of probability, i.e. the empirical interpretation of probability statements. As we briefly noted earlier, probability, however, has no universally accepted interpretation. Different interpretations of probability with different practical implications are invoked in various stages in a variety of ways by different schools of statistics, giving rise to a multiplicity of approaches and sub-approaches to statistical induction. In this chapter we consider various interpretations of probability and certain related aspects of the probability calculus; in the next chapter we would outline in general terms their use in different modes of statistical induction.

| Objective and |
| subjective |

Probability semantics is an intricate and extensive topic of philosophy. Here we shall take up only those aspects which have bearing on the problem of statistical induction, and that too, to the extent required.

We can say that broadly there are two types of interpretation of probability: (i) *objective* and (ii) *subjective*. Also, roughly speaking, we can say that these two have their roots in two opposing schools of philosophy that have tussled with each other since ancient times, namely, the realistic and the idealistic schools. According to the former, the external world has existence independently of ourselves and there are definite objects in it corresponding to our various sense perceptions. According to the latter, the only data we have are our own sense perceptions and the presumption of existence of external objects behind these is logically untenable; the external world exists only in the minds of beings. Objective interpretation of probability regards probability of an event in a set-up as something pertaining to the external world, irrespective of what we know about the set-up, and hence, can be said to correspond to a realistic standpoint. Subjective interpretation, on the other hand takes the stand that probability exists only in the cognitive faculty of human beings and therefore can be said to have affinity with the idealistic position. (These statements are, however, made in a loose sense. As we shall see, all forms of objective interpretation of probability involve human judgment one way or other. Also, no proponents of the subjective viewpoint go so far as to deny the existence of the external world; their stand can be said to be semi-idealistic, or simply, anti-realistic; see Festa (1993) p. 53.) We shall see in later chapters that both types of interpretation have been advanced since very early in the history of development of the concept of probability.

3.2 Objective interpretation of probability

| Long-term relative frequency |

3.2.1 In the case of objective interpretation, probability can be defined only in the context of a *random experiment* or *trial*. By a random experiment means an operation whose outcome is unpredictable beforehand and which is capable of indefinite repetition under identical conditions, at least conceptually (cf. section 2.6). Thus drawing a ball blindly from an urn containing balls of different colours, tossing a die, observing whether in a community a new-born child arbitrarily chosen would survive through its first year, or measuring the length of life of an electric bulb arbitrarily taken from a production line are all random experiments. The totality of possible outcomes for such a random experiment constitutes the domain of outcomes or the *outcome space* and probabilities are assigned to various meaningful (measurable) sets in this outcome space; these sets are said to represent the events for the experiment. It has been empirically found that for many random experiments the relative frequency of occurrence of an event tends to become stable as the experiment is repeated a large number of times. On the basis of this empirical finding, in the objective interpretation, probabilities of events are related to the long-term relative frequencies of their occurrence. There are varieties of objective interpretation corresponding to different ways of relating probabilities to the relative frequencies.

| Von-Mises–Reichenbach |

3.2.2 The most straight-forward way of relating the probability $P(A)$ of an event A to the relative frequency of A would be to suppose that there is an infinite sequence of causally independent, identical repetitions of the experiment. If we represent the occurrence of A by 1 and non-occurrence by 0, from the outcomes of the sequence, we get an infinite binary sequence. Denoting by f_n the number of 1s upto the n-th term of this sequence, $P(A)$ is defined to be some form of limit of the relative frequency f_n/n. Although the idea was floated and naively pursued earlier, in modern times it was sought to be put on a rigorous footing by von Mises (1928) and Reichenbach (1935). Von Mises defined probability only with reference to what he called a *collective*. In the context of a random experiment a collective can be thought of as representing an infinite sequence of elements each of which belongs to the outcome space, satisfying certain conditions. Given any event A we can get a binary sequence from such a collective by replacing each element in it by a 1 or 0 according as it belongs to A or not. The collective must be such that, whatever A, (i) as $n \to \infty$, the ratio f_n/n for the binary sequence tends to some fixed number p, (ii) property (i) holds with the same p for any sub-sequence of the binary sequence derived through place selection, i.e. by a rule in which whether or not an entry in the sequence would be taken is determined solely by its position and the values of the preceding entries in the sequence ((ii) is called the property of *irregularity*). Given a collective, $P(A)$ is equated to the p of the binary sequence corresponding to A. Presumably, associated with a random experiment there would be a collective with appropriate ps for different As. Conceptual difficulty arises in the case of von Mises' interpretation of probability because it is not clear how the notion of a mathematical limit which is appropriate for mathematically defined sequences would apply to the sequence f_n/n which is only empirically defined with reference to a collective. Specifically, since any value of f_n/n for a finite n is compatible with any value of the limit p, how is one to verify in practice whether for a particular event A, $P(A)$ has a given value by examining a finite number of initial elements of the collective?[1] To get round this difficulty, Reichenbach considered a less demanding notion of empirical limit in which the limiting value is some number which is very nearly approached by f_n/n whenever n is large in a practical sense (i.e. 'not far short of the maximum that our means of observation enable us to reach'—Russell (1948) p. 364). Also Reichenbach postulated an axiom to the effect that if f_n/n is found for all n up to a certain maximum and if, say, throughout the last half of these values of n, f_n/n differs from the fraction p by less than a small number ϵ, then it may be posited that the same will hold for all subsequent n, and hence for the limit. This axiom would allow the determination of $P(A)$ to a close approximation on the basis of empirical data. But it cannot be said that this ploy completely removes the vagueness in the interpretation of probability as the limit of relative frequencies.

Properties (i) and (ii) of a collective in von Mises' formulation in effect mean that for any event A the binary sequence generated by a collective can be regarded as the records of the outcomes of a sequence of independent and

---

[1]No parallel should be drawn, as is sometimes done, between the limiting process here and that in the definition of a physical entity like the density of a fluid at a particular point (at a particular time), although in the latter case also it is not possible to reduce the volume element of the fluid around the given point arbitrarily. Firstly, in the case of fluid density, the observations are not usually given as a sequence—our experience of the physical world lets us presume the existence of the limit for the volume element approaching zero notionally. Secondly, for the same volume element we expect two different determinations of the density to tally up to observational errors. We make a determination for a small volume element and take that as an approximation to the limit. In defining probability, the existence of the limit is to be judged by studying the initial relative frequencies of the sequence. Also, we are never sure that repetition will not give us conflicting values of relative frequencies.

identical Bernoulli trials. In recent years Dawid (1984) has sought to generalize this to data sets conforming to probability models representing trials which may be dependent and non-identical; but the problem of empirical verification of assertions regarding the limit in a logically tenable way remains.

| Fisher and Russell |

3.2.3 Apart from the difficulty of defining an infinite empirical sequence by extension, in the von Mises–Reichenbach type interpretation of probability, for any event A, the limiting value of the relative frequency would be relative to a particular arrangement of the elements of the collective—if these are rearranged the limit may change. (In a binary sequence in which infinitely many 0s and 1s occur, we can always rearrange the terms so that f_n/n tends to any value in $[0, 1]$.) Intuitively, however, it seems that the probability of an event should depend only on the collection of elements and not on their arrangement. When the outcomes of successive repetitions of a random experiment are recorded over time, the temporal arrangement of the outcomes is something which is purely contingent; it would be restrictive to make the probability of an event defined for a random experiment to depend on that contingent fact. When a random draw is made from an urn containing a finite number of balls of different colours, the probability of drawing a white ball can be naturally equated to the proportion of white balls in the urn. Fisher extended that idea to conceive the outcome of a random experiment as a random draw from the imaginary infinite population that would be generated if the experiment were repeated an infinite number of times and interpreted the probability of any event in the experiment as its relative frequency in the infinite population. (Is there a faint reflection of the Hegelian conception of the Absolute (see Russell (1961), p. 702) in Fisher's postulation of an imaginary population?) This does away with the necessity of regarding the outcomes in a particular sequence, but only apparently. Since the relative frequency of A cannot be directly defined for an infinite population, one has to think in terms of limit of the relative frequencies for a sequence of finite populations whose sizes increase indefinitely (Fisher (1956) p. 110) and for the construction of such a sequence of populations one has to fall back on a particular order in which the individual outcomes are to be included. Russell (1948) pp. 371–372 suggested that we can extend the urn model (which, according to him, gives the finite frequency theory of probability) to interpret the probability of an event for any random experiment by bringing in an inductive principle to the effect that *most* sequences of certain kinds which exhibit a property in a large number of finite initial segments will have the property throughout. The asserted probability of the event, according to Russell, will then be a 'probable probability' value. But such an approach, besides involving a notion of a sort of 'second order probability', will turn our programme of basing induction on probabilistic reasoning upside down.

| Formalistic
development |

3.2.4 As the above discussion shows, the problem of giving a constructive definition of probability in terms of frequencies is beset with seemingly unresolvable conceptual difficulties. To avoid these, a formalistic development of probability as an undefined entity satisfying certain axioms was advocated. Kolmogorov (1933) chalked out a development of probability defined on the class of events as a normalized measure subject to the usual axioms and subsequently more or less similar formulations were given by others. But so long as probability is looked upon as an undefined entity obeying certain axioms, we can only get conditional propositions with regard to it; to use the results of probability theory in the case of a real-world random experiment, we have to establish a correspondence between the undefined concept and entities in the real world. Kolmogorov suggested that this correspondence is provided by the interpretation clause (see also Cramér (1946)): If the probability of an event A in a random experiment is $P(A) = p$ then (a) when the experiment is performed under identical conditions a large number n of times and A occurs in f_n of these, one can be 'practically certain' that f_n/n will differ very slightly from p and (b) when p is very small one can be 'practically certain' that A will not occur in a single performance of the experiment. Neyman (1952) replaced (a) by something a little more sophisticated: when n is large if the performances are grouped into sets of equal size, the numbers of occurrences of A (successes) in such sets 'usually behave as if they tended to reproduce' the binomial law with the overall success-rate as the parameter. However, qualifications like 'it is practically certain' or 'they usually behave' bring in an element of circularity in the interpretation clause and make the correspondence more tenuous than in the case of say, geometrical entities like points, lines etc. and figures drawn on the blackboard. Further, some authors seem to feel that the interpretation clause is too crucial to be left outside the axiomatic formulation of any theory of probability meant to serve the purpose of statistics (see Hacking (1965) p. 8, whose formulation we shall briefly consider later).

| Problem of instantiation |

3.2.5 There is another conceptual difficulty with important practical implications, about the interpretation of probability as an idealized relative frequency— that of instantiation to a particular case. If a gambler is interested only in the next throw of a loaded die will it satisfy him if we say, 'the probability of getting a 6 in the next throw is $\frac{1}{4}$ in the sense that the next throw is a member of a long sequence of throws in which the relative frequency of 6 approaches $\frac{1}{4}$'? Will somebody about to undergo a by-pass surgery on the heart, feel reassured if he is told 'your operation is a member of a long sequence of operations in which the relative frequency of success tends to $\frac{19}{20}$'?

| A priori principle |

Russell (1948) pp. 341–344, 381 implies the principle that when all the important features of a particular case have been taken into account while defining the sequence of identical trials, the relative frequency for the latter can be taken to measure the probability in the particular case (see also Cohen (1989) pp. 99–101). Thus if we want to find the probability that a particular male aged 30, say, Ram, will survive to age 60, we should consider Ram's present state of health, occupation, family history etc. and try to find the relative frequency of survival to age 60 in a population, as far as possible, of similar individuals. If we extend a term introduced by Keynes (1921) in another context (see below), the features of the particular case that are taken into account, represent the 'weight of evidence' here. Fisher (1956) pp. 32–33, 57 gives a similar a priori principle (but viewed from the opposite angle) when he lays down the requirement: the subject of a probability statement must not only belong to a set (reference class) of which a known fraction fulfils a certain condition, 'but every subset to which it belongs and which is characterized by a different fraction, must be unrecognizable'.

| Popper's propensity | Karl Popper (1959b), however suggests a somewhat dif-
ferent conceptual solution to the problem. He advocates
that the probability of any event for an experimental arrangement or chance set-up be interpreted as a dispositional property of the set-up, which he calls *propensity* and which, he says, gets reflected in the long-term value of the relative frequency of the event when the set-up is worked upon repeatedly.[2] Thus 'the probability of getting a 6 by throwing a loaded die under particular conditions is $\frac{1}{4}$' means when the die is thrown under the given conditions the set-up has the propensity $\frac{1}{4}$ of generating a 6. (It is not a property of the die alone but of the set-up as a whole, since if the gravitational field were changed, say the throw were performed on the surface of the moon, the propensity may change.) To test a probability statement we have to test a sequence of repetitions of the experiment. Popper draws an analogy between propensity and an electrical field which we can test only by placing a test body in the field and measuring the effect of the field on the body (see also Hacking (1965) p. 10).[3] But as far as practical determination is concerned, it is not always clear which sequence of performances should be regarded as repetitions of the same chance set-up, or

---

[2]In other words since the relative frequency in *any* large number of repetitions of the experiment tends to become stable around the same fraction, we can presume an underlying causal tendency in the chance set-up associated with that fraction. This is akin to the structural postulate for validating scientific inference put forth by Russell (1948) pp. 464, 492. According to the latter, when a group of complex events in more or less the same neighbourhood, all have a common structure and appear to be grouped about a central event, it is usually the case that they have a common causal ancestor.

[3]Some writers have argued that propensity–probability cannot work backward, and hence in general is not relevant in situations where Bayes theorem applies (see e.g. Eerola (1994) p. 9). But if propensity is looked upon as a dispositional property of a chance set-up, this objection does not seem to stand. When items are produced by different machines with different proportion defectives and an item taken blindly from the aggregate produce is found to be defective, we can look upon 'selection from the aggregate until a defective item comes up' as representing the chance set-up and very well consider the propensity of it to have come from a particular machine.

from another viewpoint, the problem of deciding on the appropriate reference class remains.

| Subjectivity unavoidable | **3.2.6** From the above, it is clear that none of the standard objective interpretations of probability is completely free from subjectivity. They involve human judgement in one form or other, be it in pronouncing on the irregularity of a collective, the identity of the conditions in the repetitions of an experiment, or the appropriateness of a reference class in a particular instance. Even in the simple case of an urn model randomness of a draw is a matter of judgement. This is understandable, since when probability is used in practice one has to satisfy oneself as to whether the conditions assumed in developing the theoretical concepts are approximated in the practical field, and this cannot but be done subjectively.

3.3 Subjective interpretation of probability

| Natural domain and types | **3.3.1** Coming to the subjective-type interpretation of probability, here probability is regarded as a measure of uncertainty where 'uncertainty' is taken as meaning nothing but 'absence of full knowledge'. Therefore such a probability is relevant in the case of not only the outcome of a yet-to-be-performed random experiment, but any statement about whose truth or falsity we are not absolutely certain. Thus we may speak of the subjective probability of a newly discovered piece of art being the creation of a particular master or the probability of the result of a particular boxing bout having been fixed up in advance (cf. section 2.7). So the natural domain of a subjectively interpreted probability function is a class of propositions, which is closed under appropriate logical operations. True, sometimes it is convenient to follow the language of objective probability and speak of the subjective probability of an 'event'; but there we should understand the word 'event' in the wider sense of a proposition which may not be related to any random experiment.

Even within the subjective conception of probability there are various nuances. Broadly two types of subjective probability have been considered: (a) *impersonal* and (b) *personal*. Their distinction is brought out below.

| Impersonal subjective | **3.3.2** In the case of impersonal subjective probability whose principal exponents have been Keynes (1921) and Jeffreys (1931, 1939)(see also Koopman (1940a,b)), the probability of an uncertain proposition A can be defined only in relation to another proposition H which represents the body of knowledge given and the appropriate notation for it is the dyadic notation $P(A \mid H)$. According to Keynes the probabilities of different propositions are not always comparable whereas Jeffreys postulates that the probabilities of any two propositions relative to the same body of knowledge can

be simply ordered subject to the usual conditions of trichotomy and transitivity. In any case, several of the rules of operation on probabilities follow from suitably postulated axioms. Under certain further assumptions like existence of finite sets of propositions which are symmetrically related to the given body of knowledge, some of the probabilities can be expressed numerically on the scale $[0,1]$ and the standard probability calculus developed. Keynes and Jeffreys both look upon the theory of probability as an extension of classical logic which has only two possible truth values 0(false) and 1(true) for a proposition to a generalized logic with a continuum $[0,1]$ of truth values. Accordingly $P(A \mid H)$ gives a measure of the strength of the logical relation in which A stands to H or the degree of logical support which H provides in favour of A. For this reason this variety of subject-ive probability has been called 'logical' or 'necessary' probability.[4] However, as Jeffreys (1957) p. 23 remarks, 'logic would not be of much use without a mind to appreciate it' and both Keynes and Jeffreys agree that in practice $P(A \mid H)$ may be interpreted as the 'degree of belief' which *any rational person* who is in pos-session of knowledge H, should entertain about the proposition A (see Keynes (1921) p. 4, Jeffreys (1957) p. 22, (1961) p. 34). As the probability $P(A \mid H)$ is specific not to any particular person, but to the body of knowledge given by H, we say it represents subjective probability of the *impersonal* kind (this is sometimes called 'credibility'). Generally the probability $P(A \mid H)$ would apply to A in all instances where H holds, e.g. for all throws of a die, for all males of a certain age following a certain occupation, and so on. In actual practice when we want to find such a probability for a proposition A in a particular instance, a problem similar to the instantiation problem noted in the case of objective probability arises: one has to include as much information about the instance as possible in H. It was in the context of this kind of probability that Keynes (1921) coined the phrase 'weight of evidence' to denote the total information (both favourable and unfavourable) about A contained in H.

In fact frequency probability instantiated to a particular case, as considered in subsection 3.2.5, can be looked upon as a form of impersonal belief probability. However since it is applicable only when the particular case is one of a large number of similar cases, it is narrower in scope than the general concept.

| Personal subjective |

3.3.3 If subjective probability is considered as rep-resenting an idealistic conception of probability, the *personal* version of subjective probability would be akin to the extreme form of idealism known as solipsism. The latter denies the existence of even other minds (see e.g. Jeffreys (1957) p. 174). In the case of personalistic subjective

---

[4]Carnap (1950) gave a formulation in which a priori measures are assigned to the dif-ferent logically possible worlds in the given situation and the probability of a proposition is defined on the basis of the measures carried by those worlds in which the proposition is true. Such probabilities have been called probabilities of logical relation (see Cohen (1989) pp. 74–79).

probability, whose development has been pioneered by Ramsey (1926) and de Finetti (1937), probability is specific to a particular person at a particular time. Any one at any time is free to give one's own intuitive assessment of probability representing his personal degree of belief to any uncertain proposition. There is no limitation or scope for hedging here; it is supposed that in principle a person can assign probability to any uncertain proposition when called upon to do so. In assigning a probability one would naturally draw upon one's current stock of knowledge (conscious or subconscious), and hence, it is no longer necessary as it was in the impersonal case to mention explicitly the body of knowledge to which the probability relates. Also there is no instantiation problem here as a person directly assigns probabilities to particular propositions like 'the next turn of the roulette wheel would show this specified number' or 'Ram, the 30 year old lorry driver, who lives at such-and-such address would survive to age 60' and does not have to arrive at these via the probabilities of general propositions. However, one thing is incumbent on a person assigning probabilities: he must assign probabilities to different propositions in a logically consistent or *coherent* way. This is crucial because among the propositions 'probabilized' there would be many which are logically dependent on others and this naturally puts restrictions on the probabilities that can be assigned.

| Operational |
| determination |

Operationally, here the probability $P(A)$ of an uncertain proposition A to a person is defined to be a number $0 \leq p \leq 1$, such that currently the person considers ps to be a fair price or one-time premium for an insurance policy against A, under which the insurer gives the policy holder an amount s if A turns out to be true but nothing otherwise, subject to the proviso that s is not too high.[5] The same thing can be expressed in betting language by saying that the person is prepared to bet either on or against A according to a contract under which when A is true the bettor-on wins an amount $(1 - p)s$ from the bettor-against whereas when A is false the latter wins an amount ps from the former, provided again s is not too large. Thus the person considers $p/(1 - p)$ which is the ratio of the possible loss to the possible gain of the bettor-on to be the fair betting odds on A. Similarly $(1 - p)/p$ is the fair betting odds against A. The quantity s, which was the amount insured in the context of insurance, becomes the total stake in the betting context.

| Coherence |

When probabilities are assigned to a number of propositions (some of which are logically dependent on others) by a person in this way, the assignment will be coherent if any two bets on the same proposition derived in two different ways give the same odds. This is often expressed in a round-about way by saying that there is no package of bets

---

[5]This proviso is required because of the diminishing marginal utility of money: utility of an amount s remains proportional to s only for low values of s.

corresponding to several propositions, in which the person can be duped into choosing sides in such a way that, whatever the truth values of the propositions, he is sure to lose a positive amount in the aggregate. Thus if a person assigns probabilities p and q to A and not-A respectively, so that the odds are $p/(1-p)$ on A and $q/(1-q)$ on not-A, we must have $q = 1 - p$ for betting against not-A is same as betting on A. Here unless $q - 1 - p$, the person is sure to lose the total amount $(p + q - 1)s$ if he is made to bet *on* both A and not-A when $p > 1 - q$, and the total amount $(1 - p - q)s$, if he is made to bet *against* both A and not-A when $p < 1 - q$. (In either case, in race track jargon, it is said that a Dutch book can be made against the person.) We will see that operationally coherence boils down to minding that the probabilities assigned to different propositions obey the usual axioms of probability (except, possibly, that of countable additivity).[6]

More sophisticated operational methods for determining probabilities have been discussed by a number of authors. Many of them are set in a decision situation where there is a collection of exhaustive and mutually exclusive states and there is uncertainty as to the state which is actually true. Obviously we can identify each state with an uncertain proposition. Also available is a set of actions one of which is to be chosen by a person. Depending on which state is true, each action results in a definite consequence. It is supposed that the person can consistently order the actions according to his personal preference, taking into account his degrees of belief about the truth of the different states and the desirability of the different consequences. Savage (1954) shows that if the set of states, the set of actions, and the chosen ordering of the actions satisfy certain plausible conditions, this implies the simultaneous existence of subjective probabilities associated with the different states and utilities attached to the different consequences, such that in effect the person prefers an action to another if and only if the expected utility of the former exceeds that of the latter. A variant of this development is given by Anscombe and Aumann (1963). Degroot (1970) also starts from a collection of states, but instead of considering the ordering of actions, supposes that the person can qualitatively order different propositions represented by suitable subsets of states according to his degree of belief. Certain further assumptions regarding the structure of this order enable him to quantify the subjective probabilities of the propositions. It is to be noted that in all these developments probabilities are assigned coherently to an entire class of relevant propositions and not piecemeal to individual propositions as in

---

[6]An alternative approach to coherence would be to start with uncertain or random quantities for each of which there are two or more possible values, each value being assumed with some probability according to one's personal judgement; the problem then reduces to the determination of one's *previsions* or subjective mean values for the different quantities in a coherent manner. An uncertain proposition A can then be equated to a random quantity with its value same as the truth value of A and the coherence condition for prevision translated to one for probability. This is the approach followed by de Finetti (1974).

the betting approach described above. Various complications come in because of this, but coherence is automatically taken care of.

| Personalistic viewpoint—criticism, rejoinder |

3.3.4   It is clear from the above that the theory of personalistic subjective probability permits a sort of *laissez faire* regime in which every person is allowed to live and act in his own world of belief so long as he is coherent in his set of beliefs. But a person's beliefs are more or less influenced by his biases and interests. Someone may have dreamt that he is going to win a lottery and have become personally so sure about it as to stake a considerable fortune in purchasing 1000 tickets in a lottery in which one in a million wins. More seriously, a crank may go out in the cold in light clothing or promenade on the guard rails of a bridge in the belief that there is negligible probability of his coming to grief through such behaviour.[7] How can society discourage such wayward beliefs if self-consistency is all that is required for a belief to be admissible? In answer, personalists bring in the principle of *exchangeability* (briefly introduced in section 2.2 and to be further considered in a formal way in the following), according to which analogous trials can be taken to be exchangeable so that the probability of the occurrence of certain numbers of different types of events in these is same, whichever the actual trials in which the events take place. When exchangeability holds for several analogous trials all but one of which have been already observed (either by the person himself or by others whose reports he accepts), it is imperative that *a person must condition on the available observations while assessing the probability in the remaining trial*. It is found that when such a policy is followed, whatever the initial beliefs of different persons, their personal probabilities converge as the number of available observations increases. In this way a consensus is bound to emerge as experience accumulates on a certain kind of events. As mentioned in section 2.4, de Finetti (1937), p. 155, goes so far as to assert that even established natural laws reached through induction based on innumerable past observations represent nothing but propositions to which everybody ascribes very high personal probabilities. According to him, it is meaningless to see the existence of a deterministic cause–effect relation in external nature behind such a law. The phenomenon of frequential regularity of an event observed for many random experiments, which is the starting point of objectivist theories of probability, is likewise explained by de Finetti as something to which everybody attaches a high subjective probability. In fact de Finetti (1937) points out that if a sequence of random experiments is exchangeable to a person and if f_n/n is the relative frequency of events of a certain type in n such experiments, then for any $\epsilon > 0$ the person can take the subjective probability of the proposition $|(f_n/n) - (f_{n'}/n')| < \epsilon$ arbitrarily close to 1 by making n, n' sufficiently large.

---

[7]Fréchet eloquently drives home this point; see Lad (1996) p. 358. Lad, however, quotes Fréchet only to discount the latter's view-point.

In de Finetti's opinion it is pointless to search for any objective probability behind this phenomenon.[8]

| Plurality of terms and attitudes |

3.4   In the above we have considered two broad conceptions of probability which have been called objective and subjective probability. The terminology here, however, is far from standardized and various other names have been used by different authors to denote more or less the same concepts. Thus, what we have called objective probability has been termed frequency probability, chance, aleatory probability, physical probability, casual probability, possibilité, and probability$_2$. Similarly, ignoring differences in shades of meaning, subjective probability has been called belief probability, credibility, epistemic probability, probabilité and probability$_1$ (some of these terms have been used mostly in connection with impersonal subjective probability). But apart from such terminological plurality, different groups of statisticians and philosophers evince different degrees of dogmatism as regards their attitudes towards the two concepts. There are some (many classical statisticians among them) who think that only the objective interpretation of probability is sensible; there are others like de Finetti, who take the diametrically opposite position and admit only the subjective version. There are still others—philosophers like Russell, Carnap, and Hacking among them—who think that probability has a dual interpretation; the subjective sense is appropriate in certain contexts and the objective in others (see Russell (1948) pp. 339–344, Carnap (1950) pp. 23–36 and Hacking (1975) pp. 11–17). In fact, as we will see, there are even thinkers who are of the opinion that there exist forms of uncertainty which cannot be quantified on the $[0, 1]$-scale in terms of either of the above two concepts of probability. But before going into that we would first briefly outline the probability calculus appropriate for objective and subjective probability.

---

[8]An interesting different view of subjective probability has been suggested by Cox and Hinkley(1974)(see Problem 10.6 on p. 408). According to it a person while assigning his own subjective probability to an uncertain proposition may think of a large number of uncertain propositions such that (i) they belong to quite unrelated fields and therefore can be regarded by the person as representing independent trials, (ii) the person assigns roughly the same personal probability to each of them, (iii) the individual truth-values of the propositions and hence the relative frequency of true propositions can be ascertained only at the end after each one has been assigned a personal probability. Under these conditions by the Weak Law of Large Numbers (section 3.9) the person must have a very high personal probability that the relative frequency of true propositions would lie within a close neighbourhood of p (the average probability of the individual propositions). But this relative frequency is objectively verifiable. Thus the person himself and others can judge *objectively* whether the person's assignment of personal probabilities was *right* or *wrong*. (The author is indebted to Sir David Cox for clarifying the view-point further in a personal communication.)

3.5   Axiomatic development

Two conventions—
events

3.5.1   As regards the axiomatic development of the probability calculus, remarkably, there is broad agreement between the objective and subjective schools described in the preceding sections. This is the reason why we find so much in common among the different approaches to statistical induction so far as the mathematics is concerned.

Firstly, as already mentioned in sections 3.3 and 3.4, there are two conventions regarding the domain of definition of a probability function. In one, we start from a universe or space Ω of possible alternatives, one and only one of which is true, and define the probability of every set belonging to some non-empty class $\mathcal{A}$ of sets $A \subset \Omega$. Ω here may be the outcome space of a random experiment or the set of all alternative states that may obtain in an uncertain situation. In the other, $\mathcal{A}$ is just a class of propositions which may not consist of any constituent elements. (The first convention can be translated in terms of the second by associating a proposition with each set $A \subset \Omega$ in an obvious manner.) In either case we call the members A of $\mathcal{A}$ as events. It is convenient to assume that the class $\mathcal{A}$ of events is closed under the set-theoretic (logical) operations of complementation (negation) and union (disjunction) and thus forms a field (Boolean algebra). It is often, but not always, further assumed that closure under countable union (countable disjunction) holds and in that case $\mathcal{A}$ is a σ-field (Boolean σ-algebra).

Absolute probability
axioms

3.5.2   Broadly, two types of axiomatic formulation have been considered. In the first, whose full-scale development following the set-theoretic convention about $\mathcal{A}$ was given first by Kolmogorov (1933), one starts from a monadic function $P(A), A \in \mathcal{A}$, representing *absolute probability*. Taking $\mathcal{A}$ to be a field, $P(\cdot)$ is supposed to satisfy the three axioms:

(i)$_\text{a}$ $0 \le P(A), \quad A \in \mathcal{A}$,
(ii)$_\text{a}$ $P(\Omega) = 1$
(iii)$_\text{a}$ if $A_1, A_2 \in \mathcal{A}$ are mutually exclusive, $P(A_1 \cup A_2) = P(A_1) + P(A_2)$.

An event A is called *sure* (*improbable*) when $P(A) = 1(0)$. Clearly the axioms imply that for A to be a sure (improbable) event it is sufficient (but not necessary) that $A = \Omega(\phi)$. For mathematical convenience, most probabilists, following Kolmogorov, take $\mathcal{A}$ to be a σ-field and include a further axiom known as the axiom of continuity. The finite additivity axiom (iii)$_\text{a}$ and the continuity axiom together are equivalent to the axiom of countable additivity:

(iii)$'_\text{a}$ if $A_n \in \mathcal{A}, n = 1, 2, \ldots$ is a sequence such that for $m \ne n, A_m, A_n$ are mutually exclusive, then $P(\cup_1^\infty A_n) = \sum_1^\infty P(A_n)$.

Under this, a probability function becomes simply a normalized measure. However, it is possible to stop with the finite additivity axiom and live without the property of countable additivity for a probability function; de Finetti (1974), vol. 1, p. 119, in particular, prefers such a scheme. But one must be careful while operating with probabilities which are only finitely additive: for instance, it would be permissible then for the union of a countable collection of improbable events to have positive probability, and even to be a sure event and for the limit of the probabilities of a decreasing sequence of events not to be equal to the probability of the intersection.

After the absolute probability function $P(\cdot)$ is defined subject to the above axioms, the conditional probability of A given C for any $A, C \in \mathcal{A}$ with $P(C) > 0$, is *by definition* taken as

$$P(A \mid C) = P(A \cap C)/P(C). \tag{3.5.1}$$

The multiplicative law

$$P(A \cap C) = P(C) \cdot P(A \mid C) \tag{3.5.2}$$

then follows trivially. The standard results of probability theory including Bayes's Theorem can then all be deduced from the axioms.

If we follow the convention under which $\mathcal{A}$ is just a class of propositions, the form of the axioms and the derived results remain the same; only Ω has to be replaced by any logically necessary proposition or tautology, and the signs $\cup, \cap$ have to be understood as representing disjunction and conjunction of propositions (more often denoted by $\vee$ and $\wedge$). Such a development is contained e.g. in the formal system proposed by Popper (1938) (see Popper (1959a) p. 319).

In the set-theoretic formulation, given a probability function $P(\cdot)$ for a space Ω and a class of events $\mathcal{A}$, a real-valued random variable is defined as a measurable function mapping Ω into the real line, i.e. as a function for which any interval of possible values on the real line represents an event. The joint distribution of any number of such random variables can be determined from, and the probabilities of, meaningful statements involving them evaluated in terms of the basic probability function $P(\cdot)$. In the case of formulation in terms of propositions on the other hand, each statement about every random variable is to be regarded as a proposition in its own right and all such propositions have to be probabilified in a consistent manner. This virtually means, in any context, *all* primary random variables that one will have to handle have to be visualized and their joint distribution brought in at the outset.

Conditional probability axioms

3.5.3 In the second type of axiomatic formulation, one starts with a dyadic function $P(A \mid C)$ which is called the *conditional* or *relative probability* of the event A given the conditioning event C. In the set-theoretic context such a development was suggested by Rényi (see Rényi (1970) and also Lindley (1965) Part-1). Taking $\mathcal{A}$ to be a field here, $P(A \mid C)$ is defined for all $A \in \mathcal{A}$, and for $C \in \mathcal{C}$, where $\mathcal{C} \subset \mathcal{A}$ is a subclass of non-empty sets satisfying some conditions.[9] $P(\cdot \,|\, \cdot)$ is supposed to be such that the following counterparts of the axioms for absolute probability hold (we assume that, for all conditional probabilities involved in the following, $C \in \mathcal{C}$):

(i)$_c$ $0 \leq P(A \,|\, C)$, $A \in \mathcal{A}$
(ii)$_c$ $P(A \,|\, C) = 1$ if $C \subset A$, $A \in \mathcal{A}$
(iii)$_c$ if $A_1, A_2 \in \mathcal{A}$ are mutually exclusive given C,

$$P(A_1 \cup A_2 \,|\, C) = P(A_1 \,|\, C) + P(A_2 \,|\, C).$$

As earlier, taking $\mathcal{A}$ to be a σ-field, we can introduce a continuity axiom to replace (iii)$_c$ by the countable additivity condition:

(iii)$'_c$ if $A_n \in \mathcal{A}, n = 1, 2, \ldots$ is a sequence such that for any $m \neq n, A_m, A_n$ are mutually exclusive given C, then

$$P(\cup_1^\infty A_n \,|\, C) = \sum_1^\infty P(A_n \,|\, C).$$

What is new here is an additional axiom embodying the multiplicative theorem in the form

(iv)$_c$ if $A, B \in \mathcal{A}$ and $C \in \mathcal{C}$ are such that $B \cap C \in \mathcal{C}$,

$$P(A \cap B \,|\, C) = P(B \,|\, C) \cdot P(A \,|\, B \cap C).$$

Other developments are as in the case of absolute probabilities.
When $\Omega \in \mathcal{C}$, writing

$$P(A \,|\, \Omega) = P(A), \quad A \in \mathcal{A}, \tag{3.5.3}$$

$P(A)$ is found to behave as an absolute probability such that (3.5.1) holds whenever $P(C) > 0$. But it may happen that $\Omega \notin \mathcal{C}$. For example, taking $\Omega = \{a_1, a_2, \ldots\}$ = a countable infinite set, we may attach a positive weight w_i to $a_i, i = 1, 2, \ldots$ with $\Sigma w_i = \infty$ and define $\mathcal{A}$ to be the field (σ-field) comprising all finite (countable) subsets of Ω; then for any *non-empty finite* set $C = \{a_{i_1}, a_{i_2}, \ldots, a_{i_n}\}$ we can set up

$$P(A \,|\, C) = \sum_{i:a_i \in A \cap C} w_i \bigg/ \sum_{i:a_i \in C} w_i, \quad A \in \mathcal{A}$$

Then, clearly $P(A \,|\, C)$ satisfies (i)$_c$ $-$ (iv)$_c$, but $P(A \,|\, \Omega)$ is not defined. Thus the formulation in terms of conditional probabilities is more general than that in terms of absolute probabilities.

---

[9]Rényi (1970) requires $\mathcal{C}$ to be a class of non-empty sets, closed under finite union, such that Ω can be represented as a countable union of members of $\mathcal{C}$.

As we saw while discussing the impersonal subjective interpretation of probability in section 3.3, in the context of propositions it is natural to work with the probability $P(A \mid C)$ of an event (proposition) A relative to a given body of knowledge C. Rudiments of an axiomatic formulation based on such a dyadic probability function are, in fact, contained in the work of Keynes (1921) (see also Jeffreys (1931, 1939) and Russell (1948) p. 345, who refers to C. D. Broad as his source). A full-scale formal development of conditional probability axiomatics which contains the formulation in terms of propositions is given by Popper (1959a) pp. 326–348. The axioms are broadly the same as above with (ii)$_c$ to be read as '$P(A \mid C) = 1$ if C implies A' and with $\cup$ and $\cap$ understood as before in the sense of disjunction and conjunction. However, a noteworthy point of difference is that here $P(A \mid C)$ is defined for all pairs of propositions A, C, provided C is not self-contradictory. The rest of the development is as earlier.

| Motivation behind axioms |

3.5.4 Naturally, to trace the motivation behind the different axiom-systems, we have to refer to the practical interpretation of probability. We recall that in the case of objective interpretation, probabilities are idealizations of relative frequencies in a sequence of repetitions of a random experiment. The axioms (i)$_a$–(iii)$_a$ are really formalizations of properties possessed by relative frequencies and are, therefore, the natural ones to assume in the context of random experiments with sets of outcomes as events. The motivation for the definition (3.5.1) of the conditional probability $P(A \mid C)$ comes from the consideration of relative frequencies for the 'conditional experiment', which consists in performing the original random experiment repeatedly until C occurs and ignoring all but the last performance. The same consideration applies for the system of axioms (i)$_c$–(iv)$_c$ based on conditional probabilities if absolute probabilities can be defined as in (3.5.3); when this is not possible, the conditional probability system is a sort of idealization.

In the case of the impersonal subjective interpretation of probability, Keynes (1921) and Jeffreys (1931, 1939) develop the axioms (i)$_c$–(iv)$_c$ starting from certain more basic axioms involving a qualitative simple ordering of probabilities which have intuitive appeal (see also Koopman (1940a,b)).

When probability is interpreted in personalistic subjective terms, as already noted in subsection 3.3.3, the key principle is that of coherence. To explain this for each bet let us define a 'directed bet' by choosing a side, i.e. by deciding whether we will bet 'on' or 'against' the concerned event. Coherence means if a combination (i.e. simultaneous acceptance) of several directed bets can be interpreted as a bet on some recognizable event, then the odds for the combination must tally with those on the latter. Various probability axioms and relations follow from this.

To illustrate, take the additivity axiom (iii)$_a$ for absolute probability. Let the odds on the mutually exclusive events A_1, A_2, and their union $A_1 \cup A_2$ be $p_1/(1 - p_1)$, $p_2/(1 - p_2)$ and $p/(1 - p)$ respectively. Consider the combination of the two directed bets corresponding to betting *on* each of A_1 and A_2. Taking 1 as the total stake for each bet, as A_1, A_2 are mutually exclusive the gains are as follows:

Table 3.1. *Gains under different events*

	Event		
	A_1	A_2	not-$(A_1 \cup A_2)$
Bet on A_1	$1 - p_1$	$-p_1$	$-p_1$
Directed bet (1)	$1 - p_1$	$-p_1$	$-p_1$
Bet on A_2	$-p_2$	$1 - p_2$	$-p_2$
Directed bet (2)	$-p_2$	$1 - p_2$	$-p_2$
Bet on $A_1 \cup A_2$	$1 - p$	$1 - p$	$-p$
Combination of (1) and (2)	$1 - p_1 - p_2$	$1 - p_1 - p_2$	$-p_1 - p_2$

Clearly the combination can be looked upon as a bet on $A_1 \cup A_2$. In order that this may give the same odds as the original bet on $A_1 \cup A_2$, we must have $p_1 + p_2 = p$ which establishes axiom (iii)$_a$.

Here the conditional probability $P(A \mid C)$ comes from the odds for a fair 'conditional bet' $A \mid C$, which is activated only if C holds, i.e. which by definition is a contract under which the bettor-on gets [gives] some small amount proportional to $1 - P(A \mid C)$ $[P(A \mid C)]$ when A [not-A] as well as C holds and no transaction takes place if C fails to hold. To see how relation (3.5.2) comes, let the odds for the bets on C, $C \cap A$ and $A \mid C$ be $p/(1 - p), p'/(1 - p')$ and $p''/(1 - p'')$ respectively. Let the directed bets (1) and (2) consist respectively in betting *on* $C \cap A$ and *against* $A \mid C$. Consider the combination of the these directed bets. Taking as before 1 as the total stake for the bets $C, C \cap A$ and $A \mid C$, the gains are as given in the following table.

Table 3.2. *Gains under different events*

	Event		
	$C \cap A$	$C \cap (\text{not-}A)$	not-C
Bet on C	$1 - p$	$1 - p$	$-p$
Bet on $C \cap A$	$1 - p'$	$-p'$	$-p'$
Directed bet (1)	$1 - p'$	$-p'$	$-p'$
Bet on $A \mid C$	$1 - p''$	$-p''$	0
Directed bet (2)	$-(1 - p'')$	p''	0
Combination of (1) and (2)	$p'' - p'$	$p'' - p'$	$-p'$

Clearly the combination can be looked upon as a bet on C with total stake p''. In order that it may give the same odds as the original bet on C, we must have

$p'/(p'' - p') = p/(1 - p)$, i.e. $p' = p.p''$ which is same as (3.5.2) (see De Finetti (1937) pp. 103–104, 108–109).

Defining conditional probabilities $P(A \mid C)$ in terms of conditional bets as above, the system of axioms (i)$_c$–(iv)$_c$ can similarly be obtained.

Note that the definition of the conditional bet, and hence the assessment of $P(A \mid C)$ are with reference to the situation where 'whether C holds or not' is not known and not necessarily to the situation obtained after the truth of C has been confirmed. We would see later that this gives rise to some philosophical questions regarding the implication of the personalistic conditional probability $P(A \mid C)$ as used in certain modes of statistical induction.

Probabilistic independence

3.6 One of the most important problems associated with the systematization and application of the theory of probability is, in the language of Kolmogorov (1933) p. 9, 'to make precise the premises which would make it possible to regard any given real events as independent'. In the objective interpretation of probability, for a sequence of repetitions of a random experiment under identical conditions if the relative frequency of an event A among those instances in which the event C occurs tends to be equal to the same among all instances, intuitively, it seems all right to say that A is independent of C. If $P(C) > 0$, this, when idealized, gives

$$P(A \cap C)/P(C) = P(A \mid C) = P(A).$$

Hence we call the two events A, C to be independent (the relation is symmetric) if

$$P(A \cap C) = P(A).P(C). \tag{3.6.1}$$

The same definition is also natural for the personalistic subjective interpretation if one regards A as independent of C when one's odds on A in the conditional bet on A given C as defined earlier is same as his odds without reference to C. Extending (3.7.1) we formally define a set of events $A_1, A_2, \ldots, A_n$ to be independent if for any $2 \leq r \leq n$, and $1 \leq i_1 < i_2 < \cdots < i_r \leq n$,

$$P(\cap_{j=1}^{r} A_{i_j}) = \prod_{j=1}^{r} P(A_{i_j}). \tag{3.6.2}$$

After defining the independence of events we define the independence of trials and random variables. If there are n trials with Ω_i and $\mathcal{A}_i$ representing the outcome space (some logically necessary proposition) and the class of events for the ith trial $i = 1, \ldots, n$, then, if whatever $A_i \varepsilon \mathcal{A}_i$, for the compound trial, the events

$$A_1 \times \Omega_2 \times \cdots \times \Omega_n, \ \Omega_1 \times A_2 \times \cdots \times \Omega_n, \ldots, \Omega_1 \times \Omega_2 \times \cdots \times A_n$$

are independent, we say that the set of trials is independent. Also if the observation of each of the random variables $X_1, \ldots, X_n$ is regarded as a trial and this set of trials is independent, we call the set of random variables independent. An infinite sequence of trials or random variables is said to be independent if for it any finite selection of members forms an independent set.

| When independence? | The concept of probabilistic or stochastic independence is of great importance in statistical induction under the objective set-up. Sometimes it is the subject of our conclusion (or as we called it in Chapter 2, a part of our inductive concern) and has to be tested in the light of the evidence. More often, however, as we noted in section 2.3, it is assumed as part of the premises, simplifying the construction of the probability model for the observations. But in this connection, a question that naturally arises is: under what qualitative situations can we assume that the random variables corresponding to the observations are independent? Long-accumulated experience with various types of random experiments shows that when causal independence seemingly holds, the condition in terms of relative frequencies leading to the formulation of (3.6.1) is approximately realized. Hence when probability is objectively interpreted, causal independence is taken as implying probabilistic independence, i.e. if the outcome of one trial does not seem to causally affect the outcome of another, the two trials are taken as independent. But when probability is subjectively interpreted, even if the outcomes of trials seem to be causally independent, i.e. do not seem to affect each other, it may happen that the trials relate to the same physical reality so that their outcomes have some common causal ancestor. In that case knowledge of the outcome of one trial may throw light on the prospect in another and thus change our assessment of the probability of an event in the latter. For example, if the same coin is tossed a number of times the different tosses, though causally independent, are not probabilistically independent in the subjective sense: the outcomes of some of the tosses may indicate the bias of the coin and give a person some idea about what to expect in the other tosses (cf. section 2.3). Likewise, if a certain treatment is being tried on a number of patients of a certain type sequentially, the nature of the response in the earlier trials tells us about what to expect in the later ones and thus the trials are not probabilistically independent in the subjective sense. Under the subjective dispensation, two trials are to be regarded as independent only if the knowledge of the outcome of one does not make a person change the evaluation of probabilities for the other. Clearly, independence of trials in this sense can be assumed only in rare situations: even if one is tossing different coins in different trials it may happen that all the coins have come out from the same mint!

| Identical distribution | **3.7** In the objective set-up, apart from independence the principle that helps to simplify the probability model for the observations on the characters is that of identical trials or random variables: whenever the conditions under which two trials are performed or

two random variables are observed are analogous, we take that their probability laws are same. Thus when n real- or vector-valued random variables $X_1, X_2, \ldots, X_n$ are not only independent but follow identical probability laws, their joint cumulative distribution function (cdf)

$$P(X_1 \leq x_1, X_2 \leq x_2, \ldots, X_n \leq x_n) - F(x_1, x_2, \ldots, x_n)$$

has the form

$$F(x_1, x_2, \ldots, x_n) = \prod_{i=1}^{n} F_1(x_i), \qquad (3.7.1)$$

where $x_1, \ldots, x_n$ each varies over the common domain of $X_1, \ldots, X_n$ and $F_1(\cdot)$ stands for the common marginal cdf. When $F(x_1, x_2, \ldots, x_n)$ possesses a probability mass function or density function (pmf or pdf) $f(x_1, x_2, \ldots, x_n)$, (3.7.1) can be expressed as

$$f(x_1, x_2, \ldots, x_n) = \prod_{i=1}^{n} f_1(x_i), \qquad (3.7.2)$$

$f_1(\cdot)$ being the marginal mass or density function corresponding to $F_1(\cdot)$.

Exchangeability

3.8 In the subjective set-up, as noted above, typically independence of trials or random variables cannot be assumed. However, analogy between trials or characters as judged by a person is very much a part of his information and he would like to utilize that while constructing the probability model. How can analogy divorced from independence be incorporated in the model? As already mentioned briefly in sections 2.2 and 2.4, this is done by bringing in the principle of *exchangeability* of analogous trials or random variables. Speaking in terms of random variables $X_1, X_2, \ldots, X_n$, these are called exchangeable if, for all permutations $(i_1, i_2, \ldots, i_n)$ of $(1, 2, \ldots, n)$ for the joint cdf

$$F(x_1, x_2, \ldots, x_n) = F(x_{i_1}, x_{i_2}, \ldots, x_{i_n}), \qquad (3.8.1)$$

and when the joint pmf or pdf exists, for the latter,

$$f(x_1, x_2, \ldots, x_n) = f(x_{i_1}, x_{i_2}, \ldots, x_{i_n}), \qquad (3.8.2)$$

holds with $x_1, \ldots, x_n$ each varying over the common domain of $X_1, \ldots, X_n$.

From the philosophical point of view exchangeability means that the spatio-temporal arrangement of the observations has no significance. If some of a number of variables have been already observed while certain others remain unobserved, under exchangeability, since the unobserved variables resemble those observed, a person can judge the probabilistic behaviour of the former from the values of the latter. (This, however, does not conflict with Hume's thesis (section 1.4), since from the personalistic viewpoint the resemblance is purely subjective—it is not presumed to have existence in external nature.)

But exchangeability has a technical implication also. Let us call a sequence of random variables $X_1, X_2, \ldots$ *infinitely exchangeable* when every finite set of variables chosen from the sequence is exchangeable. There is a famous Representation Theorem due to de Finnetti (1937), according to which, *if and only if the sequence $X_1, X_2, \ldots$ is infinitely exchangeable, for any n, we can write for all $x_1, \ldots, x_n$,*

$$F(x_1, x_2, \ldots, x_n) = \mathbb{E}_{F_1}\{F_1(x_1)F_1(x_2)\ldots F_1(x_n)\}, \qquad (3.8.3)$$

where $\mathbb{E}_{F_1}$ denotes mean value with respect to some probability distribution over the space of all marginal cdfs $F_1(\cdot)$. If this distribution assigns positive mass only to a countable set of choices of F_1, $\mathbb{E}_{F_1}$ would be representable as a weighted sum; generally $\mathbb{E}_{F_1}$ would be a generalized Lebesgue–Stieltjes integral with respect to some measure over a function space (see de Finetti (1937), p. 132). (3.8.3) means under infinite exchangeability $F_1(\cdot)$ itself is random and given $F_1(\cdot)$ conditionally, $X_1, X_2, \ldots, X_n$ behave as if they were independently and identically distributed (i.i.d.) random variables with a common cdf $F_1(\cdot)$. When the pmf or pdf exists, (3.8.3) becomes

$$f(x_1, x_2, \ldots, x_n) = \mathbb{E}_{F_1}\{f_1(x_1)f_1(x_2)\ldots f_1(x_n)\}, \qquad (3.8.4)$$

$f_1(\cdot)$ corresponding to the 'random' $F_1(\cdot)$. For example, when $X_1, X_2 \ldots$ are binary random variables each assuming one of the values 0, 1 for any $\delta_i = 0, 1, i = 1, 2, \ldots, n$,

$$P(X_1 = \delta_1, X_2 = \delta_2, \ldots, X_n = \delta_n) = \mathbb{E}_p\{p^{\Sigma_1^n \delta_i}(1 - p)^{n - \Sigma_i^n \delta_i}\}, \qquad (3.8.5)$$

$\mathbb{E}_p$ being mean value with respect to p which has some distribution over $[0, 1]$. Therefore given p, the sequence $X_1, X_2 \ldots$ behaves like an independent Bernoulli sequence with p for the probability of success. Thus an infinitely exchangeable sequence of random variables can be looked upon as a conditional i.i.d. sequence whose distribution involves a *notional* 'parameter' (p in (3.6.5) and generally $F_1(\cdot)$ in (3.8.3) and (3.8.4)) which itself is random having a non-degenerate distribution. We would see later that because of this fact inductive inference under the personalistic subjective set-up sometimes has the same formal structure as objective inference about some notional parameter.

| Approximate probability |

3.9 The issue of approximate evaluation of probabilities often assumes importance in course of the practical application of probability in statistical induction. Typically, we have a real-valued random variable T_n whose distribution involves certain known parameters denoted by n and possibly certain other parameters denoted by θ which may be partly or fully unknown. We require to evaluate or at least to have an idea of the order of probabilities $P(t' < T_n < t'')$ for certain choices of $-\infty \le t' < t'' \le \infty$. (In applications, T_n usually arises as a function of a number of sample observables; n stands for one or more sample size.)

But direct evaluation of these probabilities is impossible, either because they are analytically intractable, or because they depend on some of the unknown parameters θ. In such cases, when the known value of n is propitious (typically when the sample size(s) are large), it is often possible to ascertain the order (close-to-zero or close-to-one) and even to find the values of the probabilities approximately for suitable choice of t', t'', even when the probabilities depend on some unknown parameters. This is achieved by invoking certain limit theorems of probability theory. (These limit theorems also have great mathematical interest and epistemic value; see Gnekenko and Kolmogorov (1954) p. 1.)

Considering for convenience the case when n is positive-integer-valued (e.g. when T_n is based on a sample of size n), we consider a strictly monotonic function $h_n(\cdot)$, generally depending on n, such that the distribution of $h_n(T_n) = Z_n$ converges to the known distribution of some random variable Z, as n becomes large. (For example, we may have $h_n(t) = (t - a_n)/b_n$, where a_n and $b_n (>0)$ are known numbers depending on n.) Then a statement of the form $t' < T_n < t''$ can be translated into an equivalent statement of the form $z' < Z_n < z''$ and for large n the probability of the latter can be approximated by the known probability of the same statement about Z.

The concept of convergence that is ultimately most useful for this purpose is that of *convergence in law or distribution*. A sequence of real-valued random variables $Z_n, n = 1, 2, \ldots$ is said to converge in law to the random variable Z if as $n \to \infty$

$$F_n(z) = P(Z_n \le z) \to P(Z \le z) = F(z) \quad \text{(say)} \tag{3.9.1}$$

at all points z where the cdf $F(z)$ is continuous. (We do not insist that (3.9.1) hold for all z as that would make the concept unnecessarily restrictive.) This is denoted by $Z_n \overset{L}{\to} Z$ or $Z_n \overset{L}{\to} F(\cdot)$. (Note that the concept is defined in terms of the cdfs of $Z_n, n = 1, 2, \ldots$ and Z, regardless of whether their joint distribution is meaningful or not.)[10] The definition is equivalent to saying that as $n \to \infty$

$$P(z' < Z_n < z'') \to P(z' < Z < z'') \tag{3.9.2}$$

for all F-continuity intervals (z', z'') i.e. for all such intervals for which z', z'' are continuity points of $F(z)$.

Two forms of the limiting distribution are of special interest from the point of view of application: (i) when $F(z)$ represents a degenerate distribution with total mass concentrated at a single point z_0 and (ii) when $F(z)$ is continuous at all $-\infty < z < \infty$.

In case (i) (3.9.1) is equivalent to saying that for any $\epsilon > 0$

$$P(z_0 - \epsilon < Z_n < z_0 + \epsilon) \to 1 \quad \text{as} \quad n \to \infty. \tag{3.9.3}$$

---

[10]Convergence in law can be defined in terms of suitable metrics on the space of all cdfs. One such metric is that proposed by Paul Levy; see Gnedenko and Kolmogorov (1954) p. 33.

In this case, for large n, we take it for granted that Z_n lies in the neighbourhood of z_0 with probability close to 1. (Note, however, that since z_0 is a point of discontinuity of $F(z)$, we may not have $P(Z_n \leq z_0) \to F(z_0)$.) When (3.9.3) holds we also say that Z_n *converges in probability* to z_0 and write $Z_n \overset{P}{\to} z_0$. (The concept of 'convergence in probability' can be widened to cover the convergence of the sequence Z_n to a random variable Z which is not necessarily degenerate when the distribution of $Z_n - Z$ is defined for every n by requiring that $Z_n - Z \overset{P}{\to} 0$. There is also a stronger concept of convergence called 'almost sure convergence' of Z_n to Z applicable when $Z_1, Z_2 \ldots$ and Z have a joint distribution, see Loeve (1960) p. 151.) The simplest general theorem giving a sufficient condition for convergence in probability is the Weak *Law of Large Numbers* (LLN) for i.i.d. random variables. According to it, if $X_1, \ldots, X_n, \ldots$ is a sequence if i.i.d. random variables with the mean μ of the common distribution finite, then the mean of the first n random variables $\bar{X}_n = (1/n)\sum_1^n X_l \overset{P}{\to} \mu$. (More general forms of the LLN involving X_ls which may not be i.i.d. as also many other results on convergence in probability are available in the literature; for some of these see Feller (1950), Chapter 10 and Cramér (1946), Chapter 20.)

In case (ii) we have a happy situation in that (3.9.2) holds uniformly for all $-\infty \leq z' < z'' \leq \infty$. There for large n we can approximate $P(z' < Z_n < z'')$ by $P(z' < Z < z'')$ for every interval (z', z''). The simplest and most widely known general result of this type is the *Central Limit Theorem* (CLT) for i.i.d. random variables. According to it for a sequence of i.i.d. random variables $X_1, X_2, \ldots, X_n, \ldots$ with the mean μ and variance σ^2 finite for the common distribution, we have

$$Z_n = \sqrt{n}\frac{(\bar{X}_n - \mu)}{\sigma} \overset{\mathcal{L}}{\to} \Phi(\cdot) \tag{3.9.4}$$

where $\Phi(z) = \int_{-\infty}^z e^{-u^2/2}du/\sqrt{(2\pi)}$ is the cdf of the standard normal distribution $\mathcal{N}(0,1)$. Generally a random sequence u_n is said to be asymptotically $\mathcal{N}(\mu_n, \sigma_n^2)$ when $(u_n - \mu_n)/\sigma_n \overset{\mathcal{L}}{\to} \mathcal{N}(0,1)$. Thus by (3.9.4) $\bar{X}_n$ is asymptotically distributed as $\mathcal{N}(\mu, \sigma^2/n)$. Then for large n we can take $\bar{X}_n$ as approximately distributed like this. (More general versions of CLT and other results on convergence in law are available in the literature; see e.g. Cramér (1946) Chapters 17, 28.)

Of course while approximating $P(z' < Z_n < z'')$ by $P(z' < Z < z'')$ for large n, it would be reassuring if the closeness of approximation can be gauged in terms of an upper bound on the difference $| P(z' < Z_n < z'') - P(z' < Z < z'') |$. Although in some special cases such upper bounds (decreasing in n) are available (e.g. in the case of the CLT the bounds given by the Berry–Esseen Theorem; see Loeve (1960) p. 288), in practice we have often to approximate $P(z' < Z_n < z'')$ by $P(z' < Z < z'')$ for large n whenever (3.10.2) holds, even when no such bounds are known. Nowadays with the advent of computers, there is a growing practice of using computer-based simulation methods to get *some* idea about the order of approximation in such situations.

Non-standard probabilities

3.10   Can all forms of uncertainty be quantified in terms of probability obeying the standard probability calculus? If not, can we always compare the uncertainties of two doubtful statements and order these according to the degree of uncertainty? If we stubbornly confine ourselves only to uncertain events defined for random experiments, or at the other extreme, leave the assessment of all types of uncertainty to the subjective judgment of a particular person at a particular time, clearly the answer to the first question is positive. Therefore the questions are relevant when we consider at the impersonal level the uncertainty of doubtful propositions which do not describe the outcomes of random experiments. As we would see in the next chapter, the questions assume special importance with regard to propositions which arise as the conclusions of statistical induction. Diverse positions have been taken by different statisticians, probabilists and philosophers with regard to the above questions.

Fisher, although avowedly an objectivist in his views about the nature of probability, admits that there are forms of uncertainty which cannot be directly expressed in terms of classical probability but can be judged indirectly by other means. In particular, he suggests indicators like likelihood and level of significance as means of gauging the uncertainty of conclusions about unknown population parameters drawn from samples. Such indicators of uncertainty do not, in general, obey the rules of standard probability calculus (Fisher (1935) p. 40, (1944) pp. 9–10, (1956) pp. 43–45). We will consider Fisher's viewpoint in more detail in the next chapter.

Keynes (1921), Chapter III, in effect expresses the view that the uncertainty of a proposition relative to a given body of evidence generally has a multi-dimensional representation, and hence can be linearly ordered and quantified in terms of probability only under restrictive assumptions. For example, the uncertainty of an inductive generalization based on a number of observed instances may have the three components: number of supporting instances, degree of heterogeneity (i.e. variety) of those instances, and the scope of the generalization—uncertainty decreases as the first two increase and the third decreases. In such cases the degree of uncertainty of two generalizations based on different bodies of data can be compared only when the directions of divergence in the different components do not conflict with each other. Koopman (1940b) restricts himself to the comparison of uncertainties of what he calls 'experimental propositions', i.e. propositions whose truth or falsehood can be ascertained by a single crucial experiment. This means propositions (induced through confirmation and not contradiction) that express contemplated natural laws or historical or juridical conclusions, which we can never hope to settle for certain, are excluded from consideration. (Recall that de Finetti admits of only particular and not general induction so that non-experimental propositions expressing natural laws lie outside his scheme of induction.) Even among experimental propositions, Koopman postulates a partial and not a complete order relation in terms of their degree of uncertainty and thus leaves open the possibility that some pairs of uncertain statements remain incomparable. Only under certain special conditions does Koopman suppose that the uncertainty of a proposition given a body of evidence is appraisable as probability. Although Jeffreys (1939) starts by assuming that the uncertainties of different propositions given the same data are always comparable, he has to bring in additional assumptions (and thereby restrict the scope for making probability statements) to allow for variation in the data and quantification of uncertainty.

Some philosophers have also expressed the view that if the uncertainty about propositions (especially non-experimental propositions) is expressed as probability, it would be unrealistic to insist that such a probability be always subject to standard probability axioms (see Cohen (1989) pp. 17–21, 65–66). Thus when the evidence is jejune it may be quite reasonable to say that both A and not-A (representing e.g. whether an accused is guilty or not guilty) have low probabilities, or in other words, the issue is very much undecided. Also, sometimes we may want to interpret the conditional probability $P(A \mid C)$ as the probability of the relation 'C implies A' (e.g. probability of the accused being guilty given the evidence is same as the probability that the evidence implies the guilt of the accused). But for a probability obeying standard axioms this is not generally permissible—the relation 'C implies A' can be contraposed to get the equivalent version 'not-A implies not-C' but generally $P(\text{not-}C \mid \text{not-}A)$ cannot even be computed from $P(A \mid C)$ alone using the axioms. Again there are philosophers who think that probabilities associated with the results of induction are different in nature from both the objective and subjective varieties discussed earlier. According to them inductive probabilities cannot in general be quantified on the $[0, 1]$ scale, but, at best, propositions can be partially ordered on their basis (Kneale (1949) pp. 214, 244–245, Day (1961) pp. 51–52, 131). In their view induction is to be pursued as a policy and a mode of induction which is rational and reliable, makes the conclusion attain a high inductive probability. In other words, 'the probability of the conclusions of induction depends on the justification of induction and not vice-versa' (Kneale (1949) p. 225). In particular, the negation of an inductive conclusion may not be itself an inductive conclusion and therefore may not have any probability associated with it. We would see that this view point, while being in direct opposition to that adopted in the Bayesian approach (whether of the objective or subjective brand) to induction is not far removed from the stand taken by some of the objective schools of statistical induction. Certain other forms of non-standard probability have been proposed in recent years. Notable among these are imprecise interval-valued probabilities (see Dempster (1968), Shafer (1982) and Walley (1991)). But induction based on these, as so far developed, takes precise Bayesian induction as the starting point and is actually an extension and weakening of the latter. We will touch upon such induction marginally when we consider the robustness of Bayesian procedures in the next chapter.

4

BEARING OF INTERPRETATIONS OF PROBABILITY
ON STATISTICAL INDUCTION

Approaches to statistical induction

4.1 In Chapter 2 we distinguished in general terms between the objective and subjective approaches to the problem of statistical induction. We noted there that the distinction between the two approaches arises from differing views about the nature of probability. Now that we have discussed at some length the various alternative interpretations of probability in Chapter 3, we can look at the distinctive features of the approaches more closely.

As already observed, probability enters as part of the premises of statistical induction through models representing the uncertainty about the evidence before it is collected. In the objective approach this probability is generally given an objective interpretation in terms of long-term relative frequency one way or other. Here the probability distribution of the observables comprising the evidence involves some unknown parameter and therefore is never fully known. The unknown parameter epitomizes our inductive concern—the aspect of the physical reality which is of interest to us; we intend to infer about it through induction. In the subjective approach, on the other hand, the probability distribution of the observables is of the personalistic subjective kind. The agent carrying out the induction (the experimenter or the statistician) possesses complete knowledge about the distribution of the observables—realized and unrealized—and no unknown parameter comes into the picture. The inductive concern is represented by the unrealized observables and the agent analyses the situation to unravel his own views about those in the light of the realized observations.

Variety of objective approaches

However, probability is invoked in statistical induction also to justify the mode of induction and to give directly or indirectly an idea about the uncertainty associated with the inductive conclusion. In the purely subjective approach there is more or less a straight rule for this based on Bayes's Theorem. In the objective approach, on the other hand, at this stage the road divides out into a number of alleys corresponding to different modes of induction. We would see that such parting of ways occurs because although generally speaking all objectivists start with an objective probabilistic model, some of them implicitly or explicitly allow non-objectivistic interpretations of probability at one or other point in course of developing the argument. There are broadly three modes of induction giving rise to three distinct sub-approaches within the objective approach. We call these (i) objective–behavioural (ii) objective–instantial and (iii) objective–pro-subjective. The reason for such nomenclature will be clear as we proceed. For the present we mention only that of the three, the objective–behavioural approach is most uncompromising in adhering to the objective interpretation of probability. In this respect its stance exactly counterpoises that taken in the purely subjective approach which uncompromisingly sticks to the personalistic subjective interpretation.

4.2 Objective–behavioural approach

| Inductive behaviour | **4.2.1** The objective–behavioural approach to statistical induction is based on the theory developed by J. Neyman, E. S. Pearson, and later A. Wald. The hallmark of this approach is the emphasis on inductive behaviour or action and not so much on explicit inductive inference. As Neyman (1950, 1952) puts it, in practice what ultimately matters is not whether what we conclude through induction is true or not, but how we regulate our actions in the light of what we observe by referring to the permanencies which seem to exist in nature. Thus when we see dark clouds in the sky, we take cover; deliberation on whether or not it would rain is of little practical interest. Induction, according to this school, consists in choosing an action out of a set of possible actions on the basis of observations according to some rule. Choice of the action which implicitly stands for inductive inference here is a plunge, 'an act of will', and there is no room for doubt or uncertainty here. This is true even if the action consists in making an assertion about some unknown parameter θ which represents our inductive concern—we proceed as if the truth of what we assert is 'known for certain'.

| Choice of inductive rule | Of course the rule which determines the action on the basis of the observations is crucial here. It has to be chosen so that the result of applying it in practice is satisfactory. In the objective approach we assume conceptual repeatability (cf. section 2.2) of the set of operations generating the observations (the sampling experiment) and the observables vary unpredictably from repetition to repetition. Therefore the action determined by feeding the values of the observables into the rule and the consequence of taking it are also variable. Under the objective probability regime, frequential regularity holds and the long-term pattern of variation is represented by the probability distribution of the consequence on repetition of the sampling experiment. The performance of a rule is to be assessed by studying this probability distribution in different situations. (Do we find in this sidelining of truth and emphasis on expediency a resemblance with the pragmatic philosophy of William James and the instrumentalist philosophy of John Dewey which came up in the United States in the latter half of the 19th and the first half of the 20th century? See Russell (1961) Chapters 29, 30.) We then take some measure based on the probability distribution of the consequence as an index of performance of the

inductive rule and choose the latter so that this performance index is in some sense optimum.

| Assurance relates to rule |

In this approach the performance index of the chosen rule is stated as a measure of assurance along with the inductive conclusion. But it is to be clearly understood that the measure of assurance attaches to the rule and not to the conclusion which, as noted above, takes the form of an action. It is somewhat like the rating given to a business house by a credit-rating agency—the actual showing of the house in a period is unpredictable and may or may not be up to the rating given, but still the investor has nothing but the rating to go by when making the investment.

| Point estimation |

4.2.2 We next consider specific types of induction. We consider a sampling experiment $\mathcal{E}$ whose outcome is represented by the set of observables X with domain (sample space) $\mathcal{X}$. The distribution of X is characterized by the parameter θ about which we only know that it belongs to the parameter space Θ. To discuss the problem of point estimation, for the sake of simplicity, we consider the case when the parameter of interest is a real-valued function $\xi = \xi(\theta)$ of θ. In the case of point estimation we want a working value for the unknown ξ. The rule is represented by an *estimator* T which is a statistic or function of the set of observables X. The random consequence of using a particular estimator T is given by the error $T - \xi$ in the *estimate* (for the sake of brevity we use the same notation for the estimator and the estimate which is the value of the estimator). We require the probability distribution under conceptual repetition of the sampling experiment (the sampling distribution) of this error to be centred at and as concentrated around 0 as possible. This means, according to a standard interpretation that the *bias* $E(T|\theta) - \xi$ be 0 (the estimator be *unbiased*) and the *sampling variance* $\mathrm{Var}(T|\theta)$ be as small as possible, whatever the value of $\theta \in \Theta$. Various principles and techniques for choosing a T for which these conditions are exactly or approximately realized have been proposed and developed. When the bias can be disregarded, the sampling variance or its square root, the standard error (estimated, if unknown), is here advanced as a measure of assurance. Naturally this relates to the estimator and not to the estimate obtained in any particular instance.

In this context a requirement called *consistency* applicable to a method of estimation that can be used for samples of increasing sizes is usually imposed. If the method applied to a sample of size n produces the estimate T_n for ξ, then consistency is said to hold if as $n \to \infty$ T_n converges in probability to the true value of ξ (section 3.10). (A sufficient condition for this is that the bias $E(T_n|\theta) - \xi \to 0$ and $\mathrm{Var}(T_n|\theta) \to 0$.) In such a case for large n we can presume that T_n lies close to ξ with a high probability. However, although the position of one using an inconsistent general method of estimation would be logically untenable, clearly the requirement has not much practical bearing in the case of estimation from a small sample of fixed size. Another notion called *Fisher-consistency*, which is free from this shortcoming but is applicable only when the observables from a random sample from some population (or a collection of such samples), was espoused by R. A. Fisher (see Rao (1965) pp. 281–282). A method of estimation is Fisher-consistent if when applied on the entire population (i.e. on an infinitely large sample) it produces the true value of the parameter. Thus, whereas the earlier notion relates to the sequence of distributions of the estimate, Fisher-consistency stresses the functional form of the estimator.

Interval estimation | In the case of interval estimation of ξ, the inductive rule gives two statistics T_1, T_2 ($T_1 \leq T_2$) (the values of T_1 and T_2 represent the lower and upper confidence limits) and we intend the confidence interval (T_1, T_2) between the limits to straddle across the true value of ξ, while being itself as short as possible. To assess the performance of a rule we have to study the probability of ξ being covered by the interval as also the 'shortness' of the latter in some sense, for different values of $\theta \in \Theta$. The basic theory for choosing an optimum rule here was developed by Neyman (1937). The standard procedure here is to confine oneself to rules for which the lower bound to the probability of coverage (for variation in θ over Θ) is fixed at a stipulated value $1 - \alpha$ (the confidence level) and then to choose within such a class a rule for which the interval is optimally 'short'.

In many problems involving observations on continuous variates, the solution is obtained by starting with a suitable real-valued function (pivotal function) $U = U(X, \xi)$ of the observables and the parameter of interest such that the distribution of U is continuous and same irrespective of $\theta \in \Theta$. Two numbers u_1 and u_2 can then be suitably determined so that

$$P(u_1 < U < u_2 | \theta) = 1 - \alpha \quad \text{for all } \theta \in \Theta. \tag{4.2.1}$$

The inequalities

$$u_1 < U(X, \xi) < u_2 \tag{4.2.2}$$

are then inverted to get a confidence interval with confidence level $1 - \alpha$ in the form $T_1 < \xi < T_2$. (When the coverage probability is $1 - \alpha$ for all $\theta, 1 - \alpha$ is also called the confidence coefficient.) $U(X, \xi), u_1, u_2$ here are to be chosen so that such inversion is possible and the confidence interval is optimally 'short'. In standard situations (e.g. when the distribution of U is continuous and bell-shaped) the last requirement ensures that the confidence interval for different levels are nested, i.e. one corresponding to a lower level is included in that corresponding to a higher.

Example 4.2.1. Suppose on the basis of a random sample from the normal population $\mathcal{N}(\mu, \sigma^2)$, in which the mean μ and variance σ^2 are both unknown, we want a confidence interval for μ. Denoting the mean and the standard unbiased estimate of σ^2 based on the sample by $\bar{X}$ and s^2, it is known that $\sqrt{n}(\bar{X} - \mu)/s$ has a known distribution ('Student's' t-distribution with $n-1$ degrees of freedom) whatever (μ, σ^2). If the interval $(t_1 = -t_{(\alpha)}, t_2 = t_{(\alpha)})$ includes the middle $(1-\alpha)$ probability in this symmetric distribution, the inequalities

$$-t_{(\alpha)} < \sqrt{n}(\bar{X} - \mu)/s < t_{(\alpha)} \qquad (4.2.3)$$

can be inverted to get a confidence interval with confidence level $1 - \alpha$ in the form

$$\bar{X} - t_{(\alpha)}s/\sqrt{n} < \mu < \bar{X} + t_{(\alpha)}s/\sqrt{n}. \qquad (4.2.4)$$

It can be shown that this interval is optimal in various senses.

In the general case when X represents a large sample of size n, it is often possible to find an estimator $T = T(X)$ consistent for ξ and a statistic $s^2 = s^2(X) > 0$ such that, whatever θ, as $n \to \infty (T - \xi)/s \xrightarrow{\mathcal{L}} Z$ (section 3.10) where Z is distributed as $\mathcal{N}(0, 1)$. There in large samples $(T - \xi)/s$ can be taken as a pivotal function and $(T - z_{(\alpha)}s, T + z_{(\alpha)}s)$, where $P(-z_{(\alpha)} < Z < z_{(\alpha)}) = 1 - \alpha$ can be interpreted as a confidence interval for ξ at an approximate confidence level $1 - \alpha$.

After the confidence limits are obtained, along with the confidence interval the confidence level is stated as a measure of assurance. The confidence level, of course, refers to the rule and not to its application in any particular instance. As regards the confidence statement $T_1 \le \theta \le T_2$, in practice one has to proceed as if it is true. This point has been stressed again and again by Neyman (e.g. in Neyman (1952) pp. 209–211, 235).

| Hypothesis testing | Problems of estimation, both of the point and interval type, are what in Chapter 1 (section 1.3) we called open

induction. Some philosophers regard hypothetic induction more important than open induction for the progress of science in any area (except possibly in the initial stages), since one can give free play to one's imagination in framing the hypothesis (see Day (1961) p. 273). Popper even goes to the extreme of denying the tenability of all inference except of the hypothetic kind. In hypothetic induction the problem is one of testing a pre-conceived null hypothesis H_0 (null because prior to the investigation our attitude to it is neither positive nor negative) about the parameter θ which assumes value in the parameter space Θ. H_0 has the form $\theta \in \Theta_0$ where Θ_0 is a *proper* subset of Θ. Inference consists patently in choosing one of the two actions: acceptance and rejection of H_0. As we adopt the behavioural attitude, we proceed as if $\theta \in \Theta_1 = \Theta - \Theta_0$ or $\theta \in \Theta_0$ according as H_0 is rejected or accepted. The rule for induction would be represented by a dichotomy of the sample space $\mathcal{X}$ into a rejection or *critical region* $\mathcal{X}_1$ and the complementary acceptance region $\mathcal{X}_0 = \mathcal{X} - \mathcal{X}_1$ and we reject or accept H_0 according as $X \in \mathcal{X}_1$ or $X \in \mathcal{X}_0$. The performance of a rule is assessed by studying $P(X \in \mathcal{X}_1|\theta), \theta \in \Theta_0$, i.e. the probability with which it commits the error of rejecting H_0 when it is true (the *first kind of error*) and $1 - P(X \in \mathcal{X}_1|\theta), \theta \in \Theta_1$ i.e. the probability with which it commits the error of accepting H_0 when it is not true (*the second kind of error*). The standard practice is to fix a small fraction α (*level of significance*) as upper bound to the former and to minimize the latter (i.e. to maximize the power $P(X \in \mathcal{X}_1|\theta)$) globally over Θ_1, as far as possible, subject to other requirements. Various techniques based on the famous *Neyman–Pearson Lemma* (see e.g. Lehmann (1986) pp. 72–78) and its proliferations (which are really adaptations of methods of calculus of variations) have been proposed for realizing such conditions. In the context of testing it is customary to state the level of the test rule along with the conclusion. This, along with the power function over Θ_1, represents the measure of assurance.

In large samples, the problem of test construction is often solved approximately by setting up a test criterion which converges in law (section 3.9) to a known distribution for all θ belonging to the boundary of Θ. In such a context the requirement that for a fixed level the power of the test rule against every alternative of interest should tend to one as sample size increases is generally imposed on the test rule. Such a test rule is said to be *consistent*, (i.e. sensitive in large samples) against the relevant alternatives.

In this approach to testing, after Θ is partitioned into two complementary subsets, one of which would be Θ_0 and the other Θ_1, some care is necessary in framing Θ_0, i.e. in choosing the subset to be designated as Θ_0. The general rule is: frame Θ_0 so that accepting Θ_0 when θ lies outside but close to Θ_0 is not so serious an error as rejecting Θ_0 when it is true. For example, when for a normal population $N(\mu, 1)$ we have to choose between $\mu \leq 0$ and $\mu > 0$, and declaring $\mu > 0$ would entail undertaking some expensive measures while declaring $\mu \leq 0$ would entail keeping the option open for the present, it would be proper to choose $\mu \leq 0$ as H_0. Such a policy is vindicated in terms of power if power is a continuous function of θ because of the restriction put on the probability of first kind of error.

Assurance across heterogeneous situations

4.2.3 Operational meaning of measures of assurance like confidence level, level of significance etc. in the objective–behavioural approach may be provided in terms of the objective relative frequency interpretation of probability. In this connection two points are noteworthy. We discuss these with reference to the problem of interval estimation although the same considerations would arise in the context of testing as well. 'The level of confidence of an interval estimate is $1 - \alpha$' means directly that if the sampling experiment were repeated conceptually *under the same conditions as that actually performed* and the same rule of interval estimation followed, the long-term relative frequency of cases in which the interval would contain the true value would not fall below $1 - \alpha$. But indirectly it also means that if in various problems of estimation (involving possibly different parameters) the investigators maintain the same confidence level $1 - \alpha$, then in the long run at least $100(1 - \alpha)\%$ of all confidence statements would turn out to be correct. Thus the force of the measure of assurance of a rule comes from applying similar rules in many different problems and not merely from repeated conceptual application in the same problem (see Neyman (1937), and also Neyman and Pearson (1933) and Hacking (1965) for discussion of the same point in the context of testing). The attitude of the statistician here can be likened to that of a doctor who treats various patients with their diverse ailments and achieves a high overall cure-rate.

Relevance of conditioning

The other point concerns the interpretation of the phrase 'under the same conditions as that actually performed'. It works somewhat in the opposite direction compared to that above—instead of widening the scope of the assurance given by the confidence statement it puts the focus on narrower subclasses. The point is best discussed with reference to an example.

Example 4.2.2. Suppose we want to estimate the mean daily consumption (μ) of rice per individual for all the residents of a city on the basis of a random sample of n residents. It is known that the residents of the city naturally group themselves into two communities, the first comprising all whose staple diet is rice and the second comprising the rest. The rice-eating community is broadly represented by a particular linguistic group and its proportion ρ in the entire population is known from the population census. The n individuals in the sample are post-stratified and out of the n, S are found to belong to the rice-eating community. Note that before the sampling is performed S is a random variable whose distribution depends on n and ρ and is known. If $\bar{Y}_1$ and $\bar{Y}_2$ denote respectively the mean rice intake of the individuals in the sample who belong to the two communities (rice-eating and other) a natural estimator for μ would be $\bar{\bar{Y}} = \rho\bar{Y}_1 + (1 - \rho)\bar{Y}_2$. $\bar{\bar{Y}}$ is unbiased for μ and for large n, assuming normality, an approximate 95% confidence interval for μ would be given by

$$(\bar{\bar{Y}} - 1.96 \text{ (S. E. of } \bar{\bar{Y}}), \ \bar{\bar{Y}} + 1.96 \text{ (S. E. of } \bar{\bar{Y}})). \qquad (4.2.5)$$

But what should one take here for the S. E. of $\bar{\bar{Y}}$? We can take unconditional S. E. given by $\sqrt{\text{Var}(\bar{\bar{Y}})}$ or the conditional S. E. given by $\sqrt{\text{Var}(\bar{\bar{Y}}|S)}$. In the first case the interval (4.2.5) will include μ roughly in 95% of all repetitions of the sampling experiment. In the second case, the same will hold for 95% of all repetitions in which the value of S remains same as that realized and hence a fortiori for all repetitions when S is allowed to vary ($\bar{\bar{Y}}$ is unbiased in both situations). Note that here the distribution of S in no way depends on the value of the unknown μ. We have $\text{Var}(\bar{\bar{Y}}) = E \text{Var}(\bar{\bar{Y}}|S)$ where the expectation is taken over S. If we use the conditional S. E., the span of the confidence interval would depend on how many members of the first community have entered the random sample, whereas if we use the unconditional S. E. the span would be constant irrespective of the composition of the sample. (Alternatively, if we use an interval of fixed span, in the former case the confidence coefficient would depend on S, whereas in the latter it would become rigidly fixed independently of the sample.)

Generally, when we can identify a statistic S whose distribution does not depend on the unknown parameter θ (*ancillary statistic*), we may require that the confidence level be maintained not merely unconditionally but conditionally for each possible value of S. Then the interval would have to take account of the realized value of S and would be sometimes short and sometimes wide depending on it (or in the case of an interval of fixed span, the confidence coefficient would be sometimes high and sometimes low, depending on S). This makes the rule more relevant to the situation at hand. The issue is very similar to the instantiation problem of direct probability discussed towards the end of section 3.3. When we want the probability that Ram, who is a 30-year old male mine worker by profession, would survive to age 60, we relate it to the relative frequency of survival to age 60 among all 30-year old male mine workers and that makes the probability more relevant. Only, there the features of Ram are known and the uncertainty arises because the event considered is a future contingent. In the present case S is a random variable whose distribution is free from θ, but whose value is determined only after the sampling experiment and the uncertainty arises because the parameter of interest has an unknown value. The non-dependence of the distribution S on θ is important because otherwise S may carry some relevant information and fixing S conditionally would be wasteful.

In recent years the role of conditional inference procedures with their data-dependent measures of assurance (like confidence coefficients) under the behavioural approach has attracted some attention (see Cox (1958), Kiefer (1977), Berger (1985b) and Lehmann (1986) Chapter 10). Through these we can to some extent gear our rule of inference to the special features of the sample actually realized. (Other modes of inference with data-dependent measures of assurance have also been proposed to realize the same objective within the behavioural framework; see e.g. Chatterjee and Chattopadhyay (1992, 1993).)

| Decision Theory |

4.2.4 The objective–behavioural approach to induction reaches its culmination in the *Decision Theory* developed by Wald in the 1940s. Given the observables $X \in \mathcal{X}$ whose distribution depends on the unknown parameter $\theta \in \Theta$, the problem of induction here (now called a decision problem) is posed as that of choosing one action a out of a set of possible actions $\mathcal{A}$. The preferability of the consequences of choosing various actions under different states of nature θ is represented by a *loss function* $L(\theta, a)$ whose domain is $\Theta \times \mathcal{A}$. The bulk of the theory is developed assuming $L(\theta, a)$ is real-valued and there the consequence (θ, a_1) is preferable to (θ, a_2) (the action a_1 is preferable to the action a_2 under θ) if and only if $L(\theta, a_1) < L(\theta, a_2)$. (If there is a linear order among the consequences (θ, a) and certain additional assumptions involving random choice of such consequences hold, it has been shown (see e.g. Berger (1985a) Chapter 2) that $L(\theta, a)$ can always be taken as real-valued.) More generally if $L(\theta, a)$ is vector-valued we have a partial order among the consequences (θ, a) corresponding to the appropriate partial order among the vectors $L(\theta, a)$.

The straight-forward representation of a rule of induction (*decision rule*) here would be a mapping $d(\cdot) : \mathcal{X} \rightarrow \mathcal{A}$ for which, given the observation x, the prescribed action is $d(x)$. The performance index of the rule d according to the objective interpretation of probability would be the expected loss $E_\theta L(\theta, d(X)) = R_d(\theta)$ under θ, i.e. the long-term average loss resulting from application of the rule d in conceptual repetitions of the sampling experiment. The function $R_d(\theta), \theta \in \Theta$, is called the *risk function* of d. Considering the case when $L(\theta, a)$, and hence $R_d(\theta)$ is real-valued, for any θ the rule d_1 performs better than d_2 if and only if $R_{d_1}(\theta) < R_{d_2}(\theta)$. For θ varying over Θ we get from this a partial order among rules d, a rule d_1 being better than d_2 if $R_{d_1}(\theta) \leq R_{d_2}(\theta)$ for all $\theta \in \Theta$, with strict inequality for at least one θ. A rule which is in this sense undominated, i.e cannot be improved upon is called *admissible* (Pareto-optimal, if we use the language of welfare economics).

In any decision problem if we refuse to step beyond the ambit of the knowledge that $\theta \in \Theta$, the utmost that we can do as regards the choice of a rule is to identify the class of admissible rules. Although this class can sometimes be narrowed further by imposing requirements like invariance ('symmetries present in the problem should be reflected in the rule'), generally to choose a particular rule we must introduce some more or less subjective principle or convention. Least subjective perhaps is the minimax principle which enjoins the choice of a rule minimizing $\sup_{\theta \in \Theta} R_d(\theta)$. But a minimax rule is often not very satisfactory because there exist rules which, though having slightly larger supremal risk, have lower risk for most part of Θ. Another option is to conventionally limit consideration to rules for which the risks at specified points are subject to given bounds. Other choices would require the subjective introduction of a weight function over Θ, but these are best discussed in the context of the objective–pro-subjective approach. When $L(\theta, a)$ is vector valued, $R_d(\theta)$ also would be a vector. Corresponding to the partial order of the loss vectors there would be a natural partial order among the risk vectors. Hence we can extend the idea of admissible rules. Of course minimaxity will not be appropriate but other conventions for choosing a particular rule can be defined as before.

In all cases after a decision rule is chosen, the corresponding risk function gives us a measure of assurance associated with the rule. It may be readily seen that classical problems of inference like those of point estimation, interval estimation, and testing are actually particular incarnations of the general decision problem corresponding to special forms of $\mathcal{A}$ and $L(\theta, a)$. (Specifically, for the problem of interval estimation, a two-dimensional loss with one component taking care of the inclusion of the true value and the other, the shortness of the interval, may be called for.) From the general viewpoint, to fix the confidence level of an interval estimate or the level of significance of a test means nothing but the introduction of subjective conventions for narrowing the field of choice of the rule.

| Post-experimental randomization |

4.2.5 If we adopt the objective–behavioural approach to induction with its interpretation of probability as long-term relative frequency we cannot but allow post-experimental randomization into our procedure. In the case of testing, when the boundary between the acceptance region $\mathcal{X}_0$ and the critical region $\mathcal{X}_1$ in the sample space has positive probability under H_0, such randomization may be brought in to make the first kind of error exactly equal to a specified level α. From the practical point of view more important is the random re-ordering of sets of observations that goes with certain test rules in which the inference is not invariant under observational rearrangement.[1] In the framework of decision theory, incorporation of randomization in the decision rule leads to what are called randomized decision rules. For such a rule, unlike in the case of decision rules considered above (non-randomized decision rules), observation of X does not in general give a determinate action in $\mathcal{a}$, but rather a probability distribution over $\mathcal{a}$. The actual action to be taken is found as the outcome of a subsequent random trial determined by the observation. Definition and interpretation of risk and its use for choosing the rule follow earlier lines. Introduction of randomized rules enlarges the class of available rules and endows it with convexity properties which helps theoretical treatment. In practical terms minimax rules sometimes turn out to have randomized form.

| Criticisims of the behavioural approach |

4.2.6 The objective–behavioural approach is well-entrenched in statistical theory and is generally regarded as representing the traditional viewpoint. Because the justification of the mode of induction in it rests heavily on the idea of frequencies in conceptual repetition of the sampling experiment, it is also known as the frequentist approach. It has been criticized by the proponents of other approaches on various counts.

Specifically one aspect of the objective–behavioural approach, which all other schools find unpalatable, is that it often fails to take account of the special features of the actual data at hand. (Conditioning and other data-dependent modes of inference can retrieve the situation sometimes, but an element of arbitrariness often gets into the procedure through such devices.) Fisher, whose views we shall discuss in connection with the objective–instantial approach, regards the behavioural viewpoint as the product of artificial import of ideas of acceptance sampling of industry and commerce into statistical induction and quite foreign to the requirements and spirit of scientific research. Fisher ridicules the use of post-experimental randomization by wondering 'how the research worker is to

---

[1]For example, Scheffé's test for the Behrens–Fisher problem; see Scheffé (1943) and also Anderson (1984) section 5.5.

be made to forget' that 'the judgment of significance has been decided not by the evidence of the sample, but by the throw of the coin'. Further, he regards the idea of loss as totally inappropriate in the context of quest for knowledge about nature through scientific research (Fisher (1956) pp. 5, 76–77, 97–103). In all this, Fisher's view point can be considered as particularization to statistical induction of the criticism of the pragmatist instrumentalist point of view with its emphasis on expediency, made by those schools of philosophy which regard the search for truth as the ultimate ideal (see Russell (1961) pp. 480, 779–782).

But apart from Fisher, the votaries of even those approaches which consider it useful to regard induction as a process of choosing an action (see Savage (1961) p. 177) and consequently the concept of loss as meaningful, find fault with the objective–behavioural way of looking at the problem. The butt of their criticism is that within it the in-built element of uncertainty is artificially torn apart from the inductive conclusion and tagged to the rule of induction, substituting conceptual long-term relative frequency for credibility. Thus de Finetti (1974) p. 216 notes that according to Neyman's attitude towards induction it would be illogical for someone travelling by air to take air travel insurance. (If you decide that the risk is negligible, why insure? On the other hand if you decide the risk is substantial, why travel at all?) It is no use trying to stifle the sense of uncertainty one continues to feel about the decision even after it has been chosen according to an optimum rule with controlled risk. According to de Finetti the dilemma arises from supplanting the very natural question, 'what decision is appropriate in the given state of uncertainty?' by the theoretical one, 'what decision is appropriate given the accepted hypothesis?'

4.3   Objective–instantial approach

Instance at hand emphasized

4.3.1   In the objective–behavioural approach to statistical induction as described above, the concept of repeatability of the sampling experiment is invoked not only for model postulation but also to justify the mode of induction. In the objective–instantial approach, which we are going to describe now however, the justification of the mode of induction is geared to the *particular instance* at hand—no appeal to what would happen if the sampling experiment were repeated is directly made at the stage of drawing conclusions. We will see that this is also true of the objective–pro-subjective approach to be considered in the next section; but unlike the latter, in the objective–instantial approach nothing is postulated about the unknown parameter θ beyond that it assumes its value in a specified domain.

For many (but not all) standard problems of inference the mathematical deductions from the postulated probability model, on which modes of induction rest, are often the same in the objective–behavioural and objective–instantial approach. These deductions require the derivation of the sampling distributions of various statistics, or of pivotal functions which involve both the observations and the parameter. Because of this, the two approaches together are commonly referred to as the *sampling–theory approach*. (However, as we will see, pure likelihood inference under the instantial approach does not require any sampling theory.)

Contrast with behavioural approach | But there are fundamental differences between the objective–behavioural and instantial approaches in respect of the attitude to induction and interpretation of the conclusion. In this connection we note the following:

(i) Unlike in the objective–behavioural approach, in the objective–instantial approach the emphasis is not on deciding to take any action in practice, but rather on gaining knowledge about the true state of things. It is pointless to argue, as the diehard behaviouralists do, that to make an assertion about the true state of things is also a form of action-taking, for as Hacking (1965) p. 31 notes, 'deciding something is the case' is quite different from 'deciding to assert it'. Thus the objective–instantial approach can be said to be more scientific than technological in spirit. In this way it can be looked upon as being in the tradition of pioneer philosophers like Francis Bacon and John Locke rather than of the latter-day pragmatists.

(ii) Of course, if some conclusion is accepted as true with a high degree of assurance, we can use it as the basis of action in practice. But the same inductive knowledge can be the basis of action in diverse fields in various ways—some of them not even conceived by the person making the induction. Thus consider the case of a new drug whose effects are being studied by a research scientist attached to the laboratory of a pharmaceutical company. The conclusion of the study may have different bearings on the action to be taken by (a) the scientist whose line of further investigation would depend on it, (b) the company whose business decisions could be determined by it, and (c) the Government whose policies as to health care, drug control, etc. would take shape on that basis. This means there cannot be a uniquely defined loss function in this approach. In this regard we have already noted the viewpoint of Fisher in subsection 4.2.6.

(iii) We saw that in the objective–behavioural approach the measure of assurance stated relates to the rule of induction. As regards the conclusion we entertain no qualms or doubts and proceed as if it is true and final. But in the present approach we do not stubbornly deny the uncertainty that we inevitably feel about the inductive conclusion and try to give an assessment of the degree of support that can be attached to it. This is not a before-the-trial assessment as in the behavioural case, but an after-the-trial one, and hence can take account of any special features of the data at hand.

(iv) However, the after-the-trial measure of support that we attach to the inductive conclusion in the objective–instantial approach cannot be directly interpreted as a probability in the objective sense. When it can be expressed and manipulated as a probability, the probability is of the nature of credibility or (impersonal) subjective probability. It may not always be expressible as a probability on the [0, 1] scale subject to the standard axioms (see section 3.5). (As we shall see this is in sharp contrast to the objective–pro-subjective approach.) In particular, if a certain degree of support is given to an inductive conclusion, it may not follow that the negation of that conclusion would also be inductively valid with support derived through complementation. As we noted towards the end of section 3.10, this is in conformity with what has been emphasized by many philosophers. However, prospective conclusions here can, in general, be partially ordered according to their degrees of support. In such a case the degree of support can be likened to a form of qualitative probability (cf. section 3.10; see also Hacking (1965) Chapter III).

4.3.2 The basic set-up of inductive inference in the objective–instantial approach is same as earlier: we have a random experiment $\mathcal{E}$ whose prospective outcome is some sort of a random observable X assuming values in the sample space $\mathcal{X}$; the distribution of X is determined by an unknown parameter θ about which we want to infer. *Nothing is known about θ beyond that θ lies in a given parameter space Θ.* (We would see that for certain developments under this approach, even full knowledge about Θ is not presumed.) All inference along with the concomitant statement of assurance here is made assuming the realized value x of X as given.

Forward and backward instantiation

In section 3.2 we noted that advocates of objective probability generally agree to the following form of instantiation which we now call 'forward' instantiation of probability. If under each of a set of mutually incompatible antecedent propositions (hypotheses about θ) the probabilities of various contingent consequent propositions involving X (events) are known, then given one of the antecedent propositions holds, the credibility that any consequent *would be true* in a particular instance is measured by its probability. We cited the views of Russell and Fisher among others in support of such instantiation. (This forward instantiation principle has been called 'the frequency principle' by Hacking (1965), p. 135.) In the instantial approach the realized value x of X is given and therefore we can think of consequents (implied by such a value) that may be *deemed to be actually true*. Given the truth of any such consequent and knowing its probabilities under various antecedents we have to judge the support to be given to the latter. This requires a kind of backward instantiation of probability. Compared to forward instantiation, in direct backward instantiation we can order the uncertainties about various antecedents but generally cannot quantify them.

In everyday life we often resort to this kind of backward instantiation. Thus a rise in the diastolic pressure of a person very often causes a feeling of heaviness in the nape of neck, although such a feeling may be the result of many other causes recognizable or not. When a patient complains of such a feeling, the doctor suspects a rise in the patient's diastolic pressure. On the negative side, if a candidate performs poorly at the interview, the selection board would be reluctant to declare him fit for the job, as a prospectively successful candidate would with a high probability perform well at the interview.

Such backward instantiation is a feature of all forms of instantial arguments to induction. In contrast, in the behavioural approach one posits oneself prior to the experiment and relates the statement of assurance to the inductive rule to avoid backward instantiation. We would see later that both the objective–pro-subjective and subjective approaches sanction only forward probabilistic arguments. (Strictly speaking, with the subjective interpretation of probability no need for instantiation arises.)

| Modes of instantial inference |

However, the backward instantiation problem may be tackled in different ways (and can even be seen as one of forward instantiation by a change of perspective) depending on what consequent and possible antecedents we consider, and how given a consequent we analyse the support to different antecedents. This has given rise, even within the instantial approach, to different subsidiary lines of development or modes of inference representing what we call (1) likelihood inference, (2) extended *modus tollens*, and (3) fiducial inference. All these owe a lot for their explication to R. A. Fisher who has been the principal spokesman for the objective–instantial viewpoint. Before considering the different modes of inference, however, it would be convenient to discuss certain basic concepts.

| Likelihood and likelihood ratio |

4.3.3   We start with an example.

Example 4.3.1. Suppose the unknown proportion θ of white balls in an urn containing only white and black balls can have only three possible values $\theta_1 = \frac{3}{4}, \theta_2 = \frac{1}{2}$ and $\theta_3 = \frac{1}{4}$. Two balls are drawn randomly from the urn with replacement; the first turns out to be white, the second black. The probabilities of this outcome resulting from the antecedents $\theta_1, \theta_2, \theta_3$ are respectively $\frac{3}{16}, \frac{1}{4}$ and $\frac{3}{16}$. On their basis we can affirm that the degrees of support lent by the observed outcome of the experiment to the propositions that true θ is equal to θ_1, θ_2, or θ_3 are in the ratios $3 : 4 : 3$. Suppose in another experiment a coin for which the unknown probability τ of head can be either $\tau_1 = \frac{3}{4}$ or $\tau_2 = \frac{1}{2}$ is tossed once. The outcome of the toss is a head. Given this we can say the degrees of support to the truth of $\tau = \tau_2$ and $\tau = \tau_1$ are in the ratio $2 : 3$. Not only that; it seems legitimate to say that our relative support to $\theta = \theta_2$ versus $\theta = \theta_1$ on the result of the first experiment is twice the same to $\tau = \tau_2$ versus $\tau = \tau_1$ on the result of the second experiment.

In the general case suppose the outcome of the random experiment $\mathcal{E}$ is the value of a (real- or vector-valued) random variable X which is either discrete with pmf or continuous with pdf $f(\cdot|\theta)$ (see section 3.7) defined over the sample space $\mathcal{X}$ for each $\theta \in \Theta$. Then given that X has the realized value x (when we regard X as continuous, this is to be interpreted as 'X lies in an infinitesimal neighbourhood of x'), the *likelihood* of θ is defined as

$$L_{\mathcal{E}}(\theta|x) = f(x|\theta) \text{ up to a multiplicative factor free from } \theta, \qquad (4.3.1)$$

so that log-likelihood of θ is

$$\wedge_{\mathcal{E}}(\theta|x) = \log f(x|\theta) \text{ up to an additive term free from } \theta. \qquad (4.3.2)$$

(In the following, unless essential, we keep the subscript $\mathcal{E}$ suppressed in the notation.)

The qualifications in the definition of L and $\wedge$ mean that there is no question about their absolute values having any significance. What is meaningful is the likelihood-ratio (LR) function

$$\rho(\theta_2, \theta_1|x) = L(\theta_2|x)/L(\theta_1|x) = f(x|\theta_2)/f(x|\theta_1), \qquad (4.3.3)$$

or the log-likelihood-ratio (LLR) function

$$\begin{aligned} \delta(\theta_2, \theta_1|x) &= \log \rho(\theta_2, \theta_1|x) \\ &= \wedge(\theta_2|x) - \wedge(\theta_1|x) \\ &= \log f(x|\theta_2) - \log f(x|\theta_1), \end{aligned} \qquad (4.3.4)$$

which has domain $\Theta \times \Theta$ (excluding those meaningless points (θ_1, θ_2) at which both $f(x|\theta_1)$ and $f(x|\theta_2)$ are 0) and range $[0, \infty]$ or $[-\infty, \infty]$ under usual conventions. Given the consequent that X has the realized value x, intuitively $\rho(\theta_2, \theta_1|x)$ represents the degree of support this lends to the antecedent proposition that 'true θ is θ_2' relative to the proposition 'true θ is θ_1'. Similarly $\delta(\theta_2, \theta_1|x)$ gives on the logarithmic scale the excess (positive or negative) of support lent to the former over the latter. As Example 4.3.1 above suggests, ρ and δ can be considered as absolute measures of *comparative* degrees of support lent to any two possibilities for the true value of the parameter, whose values can be compared across different experiments possibly with different parameters.[2] In this regard we can liken δ to the difference in two temperature readings made with a given thermometer whose zero has been arbitrarily marked.

We emphasize the limitations of the likelihood ratio. It is a function defined on $\Theta \times \Theta$ and legitimately measures only comparative degrees of support for ordered pairs (θ_1, θ_2); it tells us nothing absolute about singular propositions like 'true θ is equal to a particular point in Θ' and a fortiori nothing about

---

[2]It is common to confine the use of likelihood to situations where, even though the experiments may be different, the same parameter θ is involved. But for comparing LRs this seems to be an unnecessary restriction.

propositions like 'true θ lies in a given subset of points in Θ' or 'true θ is not equal to a specified point in Θ' (cf. Barnard (1967)).

Apart from these general limitations, we must have the underlying model suitably well-behaved in order that the likelihood ratio may be meaningful as a measure of relative support; freakish models may lead to a bumpy LR function and wayward inference. To illustrate consider a random sample of observation $x_1, \ldots, x_n$ from the mixture population with density

$$\frac{1}{2\sqrt{(2\pi)}}e^{-(x-\mu)^2/2} + \frac{1}{2\sigma\sqrt{(2\pi)}}e^{-(x-\mu)^2/(2\sigma^2)}, \quad -\infty < x < \infty,$$

with the parameter (μ, σ) varying over $\infty < \mu < \infty$, $0 < \sigma < \infty$. It may be readily seen that if we fix (μ_1, σ_1) arbitrarily and set $\mu_2 = x_i$ for any $i = 1, 2, \ldots, n$, then taking σ_2 close to 0 we can make the LR $\rho((\mu_2, \sigma_2), (\mu_1, \sigma_1))$ indefinitely large.

So far we have been considering the LR function given the value x of the observable X taken in full. If $T = T(X)$ is a statistic with pmf or pdf $g(\cdot|\theta)$ we may define the LR and LLR functions given $T = t$ as in (4.3.1)–(4.3.4). In particular,

$$\delta(\theta_2, \theta_1|t) = \log g(t|\theta_2) - \log g(t|\theta_1). \tag{4.3.5}$$

Generally $\delta(\theta_2, \theta_1|t)$ for $t = T(x)$ would differ from $\delta(\theta_2, \theta_1|x)$ and naturally, since t and $g(\cdot|\theta)$ are obtained from x and $f(\cdot|\theta)$ by sacrificing possibly a part of the information, $\delta(\theta_2, \theta_1|x)$ would give a fuller picture than $\delta(\theta_2, \theta_1|t)$. However, if T is such that $f(x|\theta)$ can be factored as

$$f(x|\theta) = g(T(x)|\theta) \cdot h(x), \tag{4.3.6}$$

where $h(x)$ is same for all $\theta \in \Theta$, from (4.3.4)–(4.3.6) it follows that $\delta(\theta_2, \theta_1|T(x)) = \delta(\theta_2, \theta_1|x)$ for all θ_1, θ_2 and thus the same picture would emerge from knowledge of T alone as from X taken in full.

Sufficient statistic and partition

When a factorization such as (4.3.6) holds T is called a *sufficient statistic* for Θ. Clearly, if T is a sufficient statistic, the conditional distribution of X given $T(X) = t$, is same for all $\theta \in \Theta$ and conversely.

Example 4.3.2. Consider an experiment $\mathcal{E}_n$ which consists in throwing n times a coin with unknown probability $\theta \in (0, 1)$ of a head. If $X_i(=0, 1)$ denotes the number of heads shown up at the ith throw, $i = 1, \ldots, n$, the random observable here is $X = (X_1, \ldots, X_n)$. The pmf of X is

$$\theta^{\sum x_i}(1 - \theta)^{n - \sum x_i} = \theta^{T(x)}(1 - \theta)^{n - T(x)}, \tag{4.3.7}$$

where $T(x) = \sum x_i$ is a sufficient statistic. Here given $T(X) = t$, conditionally, all the $\binom{n}{t}$ realizations of X (corresponding to the different choices of the throws in which the heads would occur) are equiprobable irrespective of θ.

Any statistic $T = T(x)$ induces a partition of the space $\mathcal{X}$. The slices (disjoint subsets) of the partition correspond to different values (in the discrete case) or different infinitesimal ranges of values (in the continuous case) of T. When T is sufficient for Θ, the conditional distribution of X within any such slice is same for $\theta \in \Theta$. We may also call any partition of $\mathcal{X}$ for which this property holds as sufficient for Θ. A sufficient partition (or a statistic which induces it) is called minimal sufficient if there is no coarser partition (whose slices are unions of the slices of the former) which is also sufficient.

| Sufficiency principle |

All schools of statistical inference explicitly or implicitly accept the following principle known as the *sufficiency principle*: *in any problem of inference if a minimal sufficient statistic $T(x)$ can be identified then the observation x can be replaced by the summary observation $T = T(x)$ and the information in x over and above that in T can be disregarded.* The reasoning behind this is that since the conditional distribution of X given the value of T is wholly known, knowing the latter we can generate an observation which is logically equivalent to x. We can express the same thing in terms of a minimal sufficient partition as: knowing the particular slice of such a partition to which X belongs, we need not bother about the actual location of X within the slice. (We have already noted that when T is (minimal) sufficient the LR function is same whether based on X or T; sufficiency principle is thus automatically realized in the definition of the LR function.[3])

| Minimal sufficiency of LR partition |

For any inference situation of the type we consider here, the LR function ρ (or equivalently the LLR function δ) induces a partition of the sample-space $\mathcal{X}$: two points x, x' belong to the same slice of the partition if and only if $\rho(\theta_2, \theta_1|x) = \rho(\theta_2, \theta_1|x')$ for all θ_1, θ_2. Interestingly, this partition turns out to be minimal sufficient. In the discrete case this may be seen by fixing θ_1 and writing $\rho(\theta, \theta_1|x) = r(x, \theta)$ so that $f(x|\theta) = r(x, \theta) \, f(x|\theta_1)$. Then under any θ for any slice S of the partition $r(x, \theta)$ is same, say $r_S(\theta)$ for all $x \in S$, and therefore, the probability content of S is $r_S(\theta) \sum_{x \in S} f(x|\theta_1)$. Hence the conditional distribution of X given $X \in S$ is free from θ and this is true for every S, which means the partition is sufficient. The minimality follows by noting that by the nature of the partition we cannot have the functions $r_S(\theta)$ identical for two distinct slices S. Essentially the same proof with appropriate modifications will go through in the continuous case.

---

[3]Of course the sufficiency principle can be invoked to replace the original data x by the minimal sufficient statistic $T(x)$ only when there are no questions about the validity of the model. When the model itself is in doubt, the conditional distribution of X given the value of $T(x)$ provides valuable information on it and it is inadvisable to go for $T(x)$ disregarding x.

It should be noted, however, that for any x the LR function $\rho(\theta_2, \theta_1|x)$ over $\Theta \times \Theta$ gives something more than the label of the particular slice of the minimal sufficient partition, to which x belongs. This is seen by noting that if we are given the set of realizations of $\{(\rho(\theta_2, \theta_1|x), \theta_1, \theta_2 \in \Theta\}$ for x varying over $\mathcal{X}$ then to represent the information in the original set-up, it would be enough if we specify $f(x|\theta_1), x \in \mathcal{X}$ for one arbitrarily fixed θ_1 (or more precisely, the probabilities of the different slices of the partition under one such θ_1). On the other hand, obviously, mere specification of $\rho(\theta_2, \theta_1|x), \theta_1, \theta_2 \in \Theta$ corresponding to the particular x realized, does not give all of the information in the original set-up.

| Likelihood principle |

4.3.4 We are now in a position to state the following principle known as the *likelihood principle* which, as we will see, forms the basis of all likelihood inference: *once the experiment $\mathcal{E}$ has been performed and a particular observation $x^{(0)}$ on X has been realized, inference about θ should be based solely on the realized LR function $\rho(\theta_2, \theta_1|x^{(0)}), \theta_1, \theta_2 \in \Theta$ which carries all relevant information about θ provided by $\mathcal{E}$.* The principle is implicitly stated by Fisher (see e.g. Fisher (1956) pp. 66–67; but as we would see below, Fisher did not stick to the principle faithfully in all parts of his theory of inference) and has been later elaborated by others (e.g. Barnard (1949, 1967), Alan Birnbaum (1962) and Basu (1975); see also Berger (1985a) pp. 27–33). The principle has an implication which is in direct conflict with the thinking in the objective–behavioural approach: if two different experiments with different set-ups but involving the same parameter θ happen to give the same LR function at their realized observations, then the same inference about θ would be appropriate in the two cases. (If the experiment $\mathcal{E}$ is kept fixed, obviously the likelihood principle is same as the sufficiency principle; in this form it is sometimes called the 'weak' likelihood principle. When for inferring about the same θ different possible $\mathcal{E}$s are considered, naturally the principle goes beyond sufficiency and in this case it is called the 'strong' likelihood principle. In the following, whenever we speak of the likelihood principle we mean its strong version.)

Example 4.3.3. For the same coin as in Example 4.3.2, define the alternative experiment $\mathcal{F}_r$ which consists in throwing the coin until the rth head appears. The random observation $Y = (Y_1, Y_2, \ldots, Y_N)$ is a vector in which $Y_i = 0, 1$ according as the ith throw turns up *a head or a tail* and $N = N(Y) = r$, $r + 1, \ldots, Y_N = 0, \sum_1^N Y_i + r = N$. At any realization y the pmf is

$$\theta^r(1 - \theta)^{\sum_1^N y_i}, \quad \text{where } N = N(y). \tag{4.3.8}$$

Now consider any realized observation $x^{(0)}$ for $\mathcal{E}_n$ with $\sum_1^n x_i = r$ and $y^{(0)}$ for $\mathcal{F}_r$ with $N(y^{(0)}) = n$ so that $\sum_1^N y_i = n - r$. From (4.3.7)–(4.3.8) it readily follows that

$$\rho_{\mathcal{E}_n}(\theta_2, \theta_1|x^{(0)}) = \rho_{\mathcal{F}_r}(\theta_2, \theta_1|y^{(0)}). \tag{4.3.9}$$

According to the likelihood principle the same inference about θ would be appropriate whether we have observed $x^{(0)}$ for $\mathcal{E}_n$ or $y^{(0)}$ for $\mathcal{F}_r$. In other words, if a sequence of throws has given r heads and $n - r$ tails with the last throw a head, it does not matter whether initially we started with a scheme in which just n throws were to be made in any case or one in which throws were to be continued until the rth head was observed.

In any inference situation the datum consists of information about (i) the sample space $\mathcal{X}$, (ii) the model represented by $f(x|\theta), \theta \in \Theta$, and (iii) the realized observation $x^{(0)}$. As noted above, given $\mathcal{X}$ and Θ, even if we know all possible realizations $\rho(\theta_2, \theta_1|x), \theta_1, \theta_2 \in \Theta, x \in \mathcal{X}$ of the LR function, we must know $f(x|\theta_1)$, $x \in \mathcal{X}$, for at least one θ_1 to retrieve the model in full. According to the *principle of total evidence* (see Carnap (1947), Wilkinson (1977)) where it is called *relevance principle*, and also Cox's comments on Wilkinson's paper), which is considered basic to all induction, *any inductive inference must utilize all relevant information about θ available*. (We informally discussed this principle in section 1.3, and also in section 1.4 with reference to Locke; it has been repeatedly emphasized by Fisher, e.g. in Fisher (1956) pp. 55, 109.) Even if our inference is geared to the particular instance at hand, there is no reason to exclude the input from our knowledge of the experimental scheme as represented in the model, unless we regard that information as irrelevant. Thus in the ultimate analysis, adoption of the likelihood principle is a matter of subjective judgement or preference.

In this connection we mention that sometimes it is argued (see Birnbaum (1962), Basu (1975)) that the likelihood principle follows from certain fundamental principles about which there is not much controversy. The argument can be described in terms of Example 4.3.3 above. Consider a 'mixture experiment' defined as follows: generate an observation on the random variable Z which assumes the values 1 and 2 each with probability $\frac{1}{2}$; choose the experiment $\mathcal{E}_n$ or $\mathcal{F}_r$ according as $Z = 1$ or 2 and perform the chosen experiment. If T and U denote the number of heads and tails for the performed experiment, (T, U) is minimal sufficient for the mixture experiment. Thus for the mixture experiment by the sufficiency principle we can discard all information about the observation except that contained in (T, U), and whenever (T, U) is same, we should make the same inference. Note that (T, U) is same whether $\mathcal{E}_n$ has been chosen and $\sum_1^n X_i = r$ or $\mathcal{F}_r$ has been chosen and $N(Y) = n, \sum_1^n Y_i = n - r$. After (T, U) has been recorded, if we can identify whether $\mathcal{E}_n$ or $\mathcal{F}_r$ was performed, i.e. we can have knowledge of the value of Z, we invoke the 'conditionality principle', which implies that the mixture experiment should give the same inference as the conditional experiment for which the ancillary Z is held fixed at its realized value. Therefore it is concluded that when $(T, U) = (r, n - r)$ we should make the same inference irrespective of whether this has been reached through $\mathcal{E}_n$ or $\mathcal{F}_r$. The weakness in the argument lies in the fact that, given (T, U), we can identify the experiment performed (for $\mathcal{E}_n, T + U = n$, and for $\mathcal{F}_r, T = r$) *except when $(T, U) = (r, n - r)$*. Thus, even if we brush away the general comments made above, this argument suffers from logical inconsistency (see Durbin (1970), Chatterjee (1989)).

After this somewhat long discussion of background concepts, we now take up the different modes of instantial inference.

| Likelihood-based point estimation |

4.3.5   We first discuss likelihood inference. Consider the problem of point estimation. If we want a point estimate of θ itself, when nothing beyond $\theta \in \Theta$ is known, the logical course would be to search for a value $\hat{\theta} \in \Theta$ which is supported by the data more than or as much as every other possible value. When such a $\hat{\theta}$ exists we have in terms of LLR function $\delta(\theta, \hat{\theta}|x) \leq 0$, or equivalently,

$$\wedge(\theta|x) \leq \wedge(\hat{\theta}|x) \quad \text{for all } \theta \in \Theta. \tag{4.3.10}$$

A $\hat{\theta}$ for which this holds is called a *maximum likelihood estimate* (MLE). When we want to estimate a function $\xi = \xi(\theta)$, the MLE is taken as $\hat{\xi} = \xi(\hat{\theta})$.[4]

When θ is real- or vector-valued, $\hat{\theta}$ is an interior point of Θ, and $\wedge(\theta|x)$ has the usual smoothness properties, we can translate (4.3.10) in terms of the derivatives of $\wedge$. For simplicity, we suppose θ is real-valued, Θ is an open interval on the real line and $\wedge(\theta|x)$ has a unique maximum in Θ. Then $\hat{\theta}$ is the solution of the equation (the maximum likelihood equation)

$$\frac{d \wedge (\theta|x)}{d\theta} = 0. \tag{4.3.11}$$

When dependence on x is to be emphasized, we write $\hat{\theta} = \hat{\theta}(x)$, which considered as a function over $\mathcal{X}$ represents an estimator of θ. However, all the present development is in relation to a fixed x and we are not concerned with things like the bias and the sampling variance of the estimator $\hat{\theta}(\cdot)$, which were so important in the objective–behavioural approach.

But how are we to attach a measure of assurance to $\hat{\theta}$? Naturally, thinking in terms of relative likelihood we would feel more assured about the estimate $\hat{\theta}$, the more sharply $\wedge(\theta|x)$ drops off from its highest value as θ moves away from $\hat{\theta}$.

---

[4]In general, however, the best supported value $\hat{\theta}$ may not be always a good working value for the unknown θ in practice. For example suppose θ can have 6 possible values: 10 and the five values 1.2, 1.1, 1.0, 0.9, 0.8. Of these, suppose given the realized observation, $\hat{\theta} = 10$ has the highest likelihood, and the LR of the five other values relative to $\hat{\theta}$ are all less than but close to 1. Also, suppose in practice there would be great loss if the presumed working value differs from the true value of θ by more than 50%. In this situation the experimenter may prefer to choose 1.0 instead of 10 as the working value. But it should be noted that in such a situation we have more knowledge than that represented by Θ alone.

By Taylor expansion in the neigbourhood of $\hat{\theta}$ we have by (4.3.11)

$$\delta(\theta, \hat{\theta}|x) = \wedge(\theta|x) - \wedge(\hat{\theta}|x)$$

$$= \tfrac{1}{2}(\theta - \hat{\theta})^2 \wedge'' (\hat{\theta}|x) + \text{higher order terms}, \qquad (4.3.12)$$

where $\wedge''(\theta|x)$ denotes the second derivative with respect to θ. Hence $- \wedge'' (\hat{\theta}|x)$ which is positive ($\hat{\theta}$ being the maximal point) and represents the magnitude of the curvature at the maximum, can be taken as a measure of the assurance we can associate with $\hat{\theta}$. But it should be noted that although ρ and δ are pure numbers (and have absolute significance), the unit of $- \wedge'' (\hat{\theta}|x)$ is (unit of $\theta)^{-2}$.[5]

In contrast to the behavioural approach, here the measure of assurance relates to the actual estimate $\hat{\theta}$ and not to any estimator and can be known only after x has been observed. In practice we may want to have some idea about the prospective assurance before the experiment e.g. for choosing the experimental design. For this, Fisher suggested replacing $\hat{\theta}$ by θ and x by X in $- \wedge'' (\hat{\theta}|x)$ and taking the mean value under θ

$$I(\theta) = -E_\theta \wedge'' (\theta|X), \qquad (4.3.13)$$

which he called the 'amount of information' (AI) in X under θ. Under broad conditions it can be shown that when X represents a large sample from some population and θ is true, the ML estimator $\hat{\theta}(X)$ is approximately normally distributed with mean θ and variance $\{I(\theta)\}^{-1}$; the statement is true even if we replace $\{I(\theta)\}^{-1}$ by the estimated variance $\{- \wedge'' (\hat{\theta}|X)\}^{-1}$ or $\{I(\hat{\theta})\}^{-1}$. This makes $- \wedge'' (\hat{\theta}|X)$ or $I(\hat{\theta})$ a natural measure of assurance to be attached to the MLE in large samples. All this, however, means we are no longer confining ourselves to the likelihood principle or the particular instance at hand.

$I(\theta)$, defined by (4.3.13), is the AI in the entire sample X which follows the distribution $f(\cdot|\theta)$ and may be denoted by $I_f(\theta)$. When, instead of the entire sample, we consider a statistic $T = T(X)$, using the notation in (4.3.2) the AI based on it can be denoted by $I_g(\theta)$. Under broad regularity conditions $I_g(\theta)/I_f(\theta) \leq 1$, the equality holding if and only if T is a sufficient statistic. The ratio I_g/I_f can be regarded as a measure of *efficiency* of T. From the above it is clear that the AI of the ML estimator $\hat{\theta}(X)$ as computed on the basis of its approximate large sample normal distribution, would equal $I_f(\theta)$ and thus $\hat{\theta}(X)$ would be fully efficient in large samples. In small samples, however, unless

---

[5]Interestingly, from the philosophical point of view, Kneale (1949) pp. 226–246 has advocated the maximum likelihood method as the only rational method of primary induction. Kneale's argument is that if we adopted any hypothesis other than $\hat{\theta}$ we would be 'allowing a larger field of possibility' for occurrences that we have not found. In the same vein Kneale says that $1 - L(\theta|x)/L(\hat{\theta}|x)$ as a function of θ represents the 'extravagance' involved in departing from the most rational hypothesis and gives us an idea about the acceptability of the inductive conclusion.

$\hat{\theta} = \hat{\theta}(X)$ itself is sufficient, the AI computed from the exact distribution of $\hat{\theta}$ may fall short of $I_f(\theta)$. In such a case it sometimes happens that there is an ancillary statistic S (a statistic whose distribution does not depend on θ) such that $\hat{\theta}$ and S together are sufficient. Fisher suggested that there we should compute the AI based on the conditional distribution of $\hat{\theta}$ given S and state this conditional AI which depends on S as a measure of assurance. (The mean value of the conditional AI then equals the AI in $\hat{\theta}$ and S together which is equal to the total AI in X.) This is exactly like the conditioning principle which we discussed in terms of the variance of a point estimator in the objective–behavioural approach in subsection 4.2.3. It has been shown that under broad conditions $- \wedge'' (\hat{\theta}|X)$ gives at least approximately a measure of the conditional AI here and in large samples $\hat{\theta}(X)$ is approximately normal with mean θ and estimated variance $\{- \wedge'' (\hat{\theta}|X)\}^{-1}$ in the conditional set-up as well. Further, even when an exact ancillary statistic S as above does not exist, in large samples we can take $- \wedge'' (\hat{\theta}|X)$ as an approximate ancillary and proceed similarly; see Efron and Hinkley (1978) and Hinkley (1980).

When θ is vector-valued (4.3.11) would be replaced by a set of equations, one corresponding to each component of θ. Instead of the negative scalar $\wedge''(\hat{\theta}|x)$ we would have a negative definite matrix of second derivatives. The AI function would be represented by a positive definite matrix. The observations made above would all apply *mutatis mutandis*.

Likelihood based interval estimation and testing

If we want to find not a single working value for θ but a set of values which would contain the true value with a high degree of assurance, we choose a threshold number $\rho_0(0 < \rho_0 < 1)$ for the LR and take

$$\{\theta : \rho(\theta, \hat{\theta}|x) \geq \rho_0\}$$

or equivalently

$$\{\theta : \delta(\theta, \hat{\theta}|x) \geq \log \rho_0\}. \qquad (4.3.14)$$

The interpretation of the set is simply that it contains just those points θ for which the LR (representing the degree of support) relative to the MLE does not fall short of ρ_0; besides this, we do not attach any other measure of assurance to the set as a whole (cf. Fisher (1956) pp. 71–72, Basu (1975) Part II; see also Edwards (1972) who calls (4.3.14) a 'support set'). When θ is real-valued and the likelihood function $L(\theta|x)$ gives a bell-shaped curve, (4.3.14) reduces to an interval straddling $\hat{\theta}$. When, further, the required smoothness conditions are satisfied we can invoke (4.3.12) and the large-sample approximations discussed earlier to get the limits of the interval in the simplified form $\hat{\theta} \pm \sqrt{(-2\log \rho_0)}/\sqrt{I(\hat{\theta})}$. From the large sample normality of $\hat{\theta}(X)$ it follows that these limits have the same form as the large sample confidence limits for θ based on the MLE, the confidence level being determined by ρ_0. (Specifically, $\rho_0 = \frac{7}{50}$ corresponds roughly to the 95% confidence level.) But, of course, such an interpretation would be quite foreign to the spirit of the instantial approach.

When θ is a vector representing several real parameters, it may happen that we are interested only in a proper subset of the parameters (the complementary subset gives the nuisance parameters). Then to get a set estimate of the parameters of interest one option would be to take the projection of the set (4.3.19) on the appropriate subspace (see Kalbfleish and Sprott (1970) who also discuss other options).

Similarly, for testing a null hypothesis $H_0\colon \theta \in \Theta_0$ against the alternative $\theta \in \Theta_1 = \Theta - \Theta_0$, we choose a critical number ρ_0 as above and reject H_0 if and only if for every $\theta \in \Theta_0$ there is a $\theta' \in \Theta_1$, such that the degree of support of θ relative to θ' as measured by the LR is less than ρ_0. Since $\rho_0 < 1$, it is readily seen that when the relevant maxima exist this is equivalent to the rejection rule

$$\max_{\theta \in \Theta_0} L(\theta|x) / \max_{\theta \in \Theta} L(\theta|x) < \rho_0, \qquad (4.3.15)$$

$$\text{or } \max_{\theta \in \Theta_0} \rho(\theta, \hat{\theta}|x) < \rho_0. \qquad (4.3.16)$$

(When Θ_0 is a singleton $\{\theta_0\}$, H_0 is the simple hypothesis $\theta = \theta_0$. Then, in the numerator of (4.3.15) and in (4.3.16), θ is to be replaced by θ_0 and '$\max_{\theta \in \Theta_0}$' is to be dropped.) Clearly (4.3.16) enjoins rejection if and only if there is no point common to Θ_0 and the set estimate (4.3.14). The rule (4.3.15) has the same form as the rejection rule of the standard likelihood-ratio test in the behavioural approach (see e.g Lehmann (1986) p. 16); but of course the interpretation of the assurance is radically different in the two cases.

How should we choose the ρ_0 in (4.3.14) and (4.3.16)? It is entirely a matter of convention. Although as noted above in large samples certain values of ρ_0 would correspond roughly to conventional values of the confidence level in interval estimation or level of significance in testing, there can be no justification for sticking to those values here. (Royall (2000), in the context of testing a simple hypothesis against a simple alternative, notes that use of conventional sampling theory values may be misleading here and suggests the use of much larger values of ρ_0.) Ultimately choice of ρ_0 is a matter of subjective judgement which has to be deliberately made by the experimenter taking all aspects of the situation into consideration. As Hacking (1965) p. 111 observes in the context of testing, 'There is no more a general answer to the question than there is to the question whether to reject a scientific theory which has powerful but not absolutely compelling evidence against it.'

As we remarked in subsection 4.3.3, the LR is a reasonable index of relative support only for well-behaved models. When the underlying model is not well-behaved, inference based on the LR may be counter-intuitive. In particular, examples of non-standard models are available, for which the MLE is a poor solution to the problem of point estimation and has a wayward sampling behaviour (see Ferguson (1982) and Lecam (1986) pp. 623–625).

The key concept in likelihood inference is the LR $\rho(\theta_2, \theta_1 | x)$ and the principle that θ_2 is considered as better supported than θ_1 by x if $\rho(\theta_2, \theta_1 | x) > 1$. This basic idea has been seized upon by Hacking (1965) p. 71 (who subsumes it within his 'law of likelihood') as the definition of comparative support. Hacking takes this as the starting point both of a logic of inference and of a purported postulational 'definition' of probability (which we can put as 'probability is an undefined term which is such that when we use it to define comparative support as above subject to standard axioms, the best supported hypothesis relating a model with an observation is the one which in practice we generally accept').

| Extended *modus tollens* | **4.3.6**   The mode of instantial inference we call *extended modus tollens* applies exclusively to the problem of testing of hypotheses. In this case we have to judge whether in the light of the realized value x of the random observation X we should recommend the rejection of a null hypothesis H_0: $\theta \in \Theta_0$, and if so, how strongly. It is not essential that the entire parameter space Θ of which Θ_0 is a subset be clearly specified at the outset; it is enough if we keep in mind some alternative possibilities for θ outside of Θ_0. The actual alternatives that may hold may be explored only after H_0 is rejected.

In elementary logic the form of argument called *modus tollens* is well-known (see e.g. Copi and Cohen (1994) pp. 305–306). It is a kind of mixed syllogism in which we have as the first premise a conditional proposition which states that if a certain antecedent is true the negation of a consequent cannot occur. Then there is a second premise which states that such a negation actually occurs. Hence we get the conclusion that the antecedent is not true. This mode of argument forms the very basis of the hypothetico-deductive method which in recent times has been espoused by Karl Popper as *the* method of science (section 1.5). In the extended version of the argument as used in instantial inference, the antecedent is H_0 and the consequent is a suitably chosen event such that under H_0 its negation is not totally ruled out but can occur with a more or less low probability. The occurrence of this negation then makes us face the disjunction, 'either H_0 is not true or an event with a low probability has occurred'. The interpretation clause associated with objective probability (see part (b) of the clauses in Kolmogorov's formulation in subsection 3.2.4) lays down that we can be practically certain that an event with a very low probability would not occur in a single performance of the experiment. Therefore the disjunction suggests that we reject H_0. This idea, which has a long history, was extensively used by R. A. Fisher, who sought to give with it a sort of gradation of the force of rejection when rejection is prescribed.

To carry out such a programme of hypothesis testing we start with a (real-valued) test criterion $T(\cdot)$ defined over $\mathcal{X}$ which satisfies the following:

(a) $T(x)$ represents in a plausible sense a measure of discrepancy between H_0 and the observation x; for simplicity we assume that greater the value of $T(x)$, more the discrepancy.[6]

(b) The distribution of $T = T(X)$ is same for all $\theta \in \Theta_0$ and this common distribution is known so that given the realized value $x^{(0)}$ of X, $P^{(0)} = P_{H_0}(T \geq t^{(0)}) =$ the probability of getting a value of T equal to or more extreme than the observed value $t^{(0)} = T(x^{(0)})$, can be determined.[7]

Condition (a) implies that if we recommend rejection of H_0 because the measure T of divergence realizes the value $t^{(0)}$, actually we are committing ourselves to a rule which prescribes rejection at least for all values of $T \geq t^{(0)}$. The event $T < t^{(0)}$ here may be taken as the consequent appropriate for the application of extended *modus tollens*. The probability $P^{(0)}$ of its negation is called the observed level of significance, tail probability or P-value. Rejection of H_0 is suggested if $P^{(0)}$ is too low, generally less than some critical value like 0.05 or 0.01. Conventionally, $t^{(0)}$ is called 'significant' or 'highly significant' according as $P^{(0)} < 0.05$ or < 0.01. If $P^{(0)}$ is not small, we say, 'there is no reason to reject H_0', or accept H_0 'provisionally' (Fisher (1956) pp. 46, 99). 'Acceptance' here means merely that H_0 is considered to be included in the class of hypotheses that seem tenable. (Thus, for example, when the hypothesis of normality of population is not rejected it may mean that the distribution in the population is possibly normal, logistic, or double exponential etc., if the T we have used is sensitive only against asymmetry.) As 'acceptance of H_0' is always provisional, considerations similar to those in the behavioural case (subsection 4.2.5) apply in framing H_0.

Fisher (1956), pp. 43–44, suggests that values of the observed level of significance can be interpreted as a sort of 'rational and well-defined measure of reluctance to the acceptance of hypotheses', the smaller the value the greater the reluctance. When the $P^{(0)}$ for a particular null hypothesis H_0 is low, other null hypotheses H_0' which are different from H_0 (H_0, H_0' may be mutually incompatible or overlapping, or H_0' may be broader than H_0) may be tested on the basis of different criteria on the same observation $x^{(0)}$ to see if they give higher P-values (see Fisher (1944) pp. 82–85). Thus, according to this line of thinking the P-value is a tool for qualitatively ordering various hypotheses according to their plausibility in the light of the data.

---

[6]In other cases 'the probability of getting a value equal to or more extreme than that observed' in (b) has to be suitably modified.

[7]This condition may be relaxed by requiring that an upper bound to the tail probability $P(T \geq t|\theta)$, over $\theta \in \Theta_0$, which monotonically decreases to 0 as t increases, be known for each t.

| Questions that arise | From the above it is clear that in the present mode of instantial inference hypothesis-testing is conceived |

as something more than a mere two-action problem. Often the term 'test of significance' is used to emphasize the distinction. But a number of questions naturally arise in this connection.

Firstly, how should we choose the measure of divergence $T(\cdot)$ to start with? Fisher, who endorsed the above approach to testing, extensively used it to judge various hypotheses for which natural test criteria suggest themselves from a study of sampling distributions of standard estimates. These criteria mostly coincide with those involved in the optimum behavioural test-rules derived through a consideration of power against suitable alternative hypotheses (subsection 4.2.2). Fisher, while spelling out his attitude to alternative hypotheses, seems to have spoken in two voices. On the one hand, opposed as he was to the behavioural approach, he expressed himself strongly for testing 'a single hypothesis by a unique body of observations' and disposing of such a hypothesis without any prior 'reference to, or consideration of, any alternative hypothesis which might be actually or conceivably brought forward' (Fisher (1956) pp. 21, 42). On the other, he had to admit that in choosing the grounds of rejection one will 'rightly consider all points on, which, in the light of current knowledge, the hypothesis may be imperfectly accurate and will select tests, so far as possible, sensitive to those possible faults, rather than to others' (Fisher (1956) p. 47; see also pp. 48, 90). In any case, it is intuitively clear that even if one does not subscribe to the behavioural viewpoint, one must take into account some relevant alternative hypotheses in choosing the criterion $T(\cdot)$ (see Cox (1958)). No general theory under the instantial paradigm is available, although under broad conditions, in large samples at least, the likelihood–ratio criterion (see (4.3.15)) generally fills the bill.

Other questions arise in connection with the interpretation of P-values. Generally a P-value is computed with reference to a specific sampling experiment having a given sample size and a model. (When a natural ancillary statistic can be recognized for the model, Fisher would have the P-value based on the conditional distribution of the criterion given the realized value of the ancillary and that would be specific to this value.) Can we logically compare the P-values of different test criteria based on different experiments for judging the relative departures from corresponding null hypotheses?

It can be shown than under H_0 the random variable P has a (possibly discrete) uniform distribution over $[0, 1]$ in the sense that *if α is a possible value of P* (the possible values may form a discrete set) $\text{Prob}_{H_0}(P \leq \alpha) = \alpha$. (cf. Kempthorne and Folks (1971) pp. 222–224, 316–317.) Hence, if two P-values are both insignificant, their relative largeness tells us nothing meaningful about the relative supports given to the respective null hypotheses. (If both null hypotheses are true the P-values represent independent observations on random variables which have (roughly) the same distribution.[8]) When one P-value is insignificant and the other significant, there is no problem since we reject only the latter null hypothesis. When both are significant we have to study the distribution of P under alternative hypotheses that can be supposed to hold in the two cases. When one P-value is decisively less than the other, we can claim with high assurance that the distribution of the criterion in the first case is more shifted towards 0 than in the second. But these distributions depend on factors like sample size, value of ancillary statistic, etc., which are unrelated to our primary inductive concern represented by θ; 'departures from the null hypothesis' should be studied in terms of θ, and θ alone. For example, if we are testing $H_0 : \mu = 0$ against $\mu > 0$ for a normal population $\mathcal{N}(\mu, \sigma^2)$ with both parameters unknown, on the basis of a sample of size n the distribution of P for the standard t-test when a $\mu > 0$ is true would depend on the degrees of freedom and the noncentrality parameter $\sqrt{(n)}\mu/\sigma$ both of which depend on n. Possibly for this reason some authors have spoken against comparing significant P-values from different experiments and for different hypotheses (e.g. Barnard (1967) pp. 28–29). However, if the distribution of P under alternatives functionally depend on the parameter in the same way, there need not be any qualms about such comparison. Thus if we have independent samples of *same* size from two normal populations $\mathcal{N}(\mu_1, \sigma_1)$ and $\mathcal{N}(\mu_2, \sigma_2)$ with all parameters unknown, then we may compare significant P-values for testing $\mu_1 = 0$ and $\mu_2 = 0$ respectively, provided we agree that 'departure from the null hypothesis' in the two cases is measured by $\mu_i/\sigma_i, i = 1, 2$.

But what about the comparison of P-values of tests of different hypotheses on the same data? If two P-variables are not independently distributed under relevant alternatives, then the value of one being smaller than the other does not plausibly entitle us to conclude that the distribution of the former is more shifted towards 0 than the latter. Yet as mentioned earlier, Fisher made such comparison of multiple P-values based on the same data time and again (see Fisher (1944) Chapters 4, 7).[9] The issue would be resolved if some absolute

---

[8]From another point of view if we say that a large insignificant P-value makes us more assured about the truth of H_0, we would be committing a fallacy akin to that of 'affirming the consequent' in elementary logic (see e.g. Copi and Cohen (1994) p. 305).

[9]Barnard (1967) p. 32 suggests a somewhat unusual frequency interpretation of the probability distribution of a P-value to get round this difficulty. According to him *at any particular time*, one's view of the world is represented by a large number of hypotheses for each of which there is a P-value based on some appropriate data-set. If all these hypotheses were true, the different P-values would be roughly independent observations on the same random variable which is uniformly distributed over $[0, 1]$. (Thus conceptual independent repetitions are considered not of the same trial over time but over different trials at the same time so that there remains scope for judging different hypotheses on the same data-set at different times.) If there are too many

meaning could be attached to a significant P-value. (As noted above, no such meaning can be attached to a large value of P which is insignificant.) Examining the arguments based on significant P-values in various concrete examples in the writings of Fisher (in particular the discussion in Fisher (1956) pp. 38–40), it is seen that in every situation one can visualize some alternative hypothesis H_1 such that $P_{H_1}(T \geq t^{(0)})$ is very nearly equal to 1. So, if we denote the event $(T \geq t^{(0)})$ (the consequent) by $E^{(0)}$ we can interpret $P^{(0)} = P_{H_0}(T \geq t^{(0)})$ (roughly) as the LR $\rho(H_0, H_1 | E^0)$ of H_0 relative to H_1 given that the event $E^{(0)}$ has been realized. Our reluctance about the acceptance of H_0 gets stronger as this LR gets smaller. So if we agree that the numerical value of a LR has absolute meaning (cf. subsection 4.3.3 and in particular, footnote 2), the same can be attributed to $P^{(0)}$ under these conditions. (Of course, the realization of $x^{(0)}$ only implies that of the event $E^{(0)}$ and therefore to base the LR on the occurrence of $E^{(0)}$ may be wasteful of information. But if $E^{(0)}$ has been based on a suitable criterion $T(\cdot)$, the wastage would not be excessive.) However, even with such an interpretation it does not seem we can say anything further than the relative order of the P-values compared.[10]

| Fiducial inference | **4.3.7** The two forms of instantial inference described above both seek to tackle the problem of backward instantiation frontally. Because of this both allow only qualitative comparison of the degrees of uncertainty in different inferential conclusions, but in neither of them do we have any scope for direct quantification of that uncertainty on a numerical scale. Fiducial inference attempts such a quantification by looking upon the problem of backward instantiation as one of forward instantiation through a change of perspective; but as we would see, situations where we can consistently apply the type of argument it involves seem to be rather limited.[11]

| Motivation and description | We start with a slightly adapted version of an example given by Hacking (1965) pp. 136–137.

---

values that do not conform to this distribution at any time, we have to suspect departure in the case of some of the hypotheses. But such a 'view of the world' attitude seems far removed from the position taken in the instantial approach.

[10]However, from the objective–behavioural point of view, Bahadur (1971) has studied the almost sure limit of ratios of the transform $-\log P$ in large samples and developed a method of asymptotic comparison of efficiencies of test procedures on that basis.

[11]Fisher borrowed the term 'fiducial' from the theory of surveying, where it is used with reference to something which is taken as 'the fixed basis of comparison'. He called the particular form of argument he proposed as fiducial, apparently because in it the observation is taken as the fixed basis for comparing the uncertainties of different inferences regarding the parameter. A probability derived by fiducial argument is elliptically called 'fiducial probability' (see Fisher (1956) p. 51).

Example 4.3.4. Suppose a sealed box contains a ball of unknown colour which may be either white or black. An urn contains 100 balls of the same two colours. It is known that 95 of these are of the same colour as the boxed ball and 5 are of the other colour. One ball, blindly drawn from the urn, turns out to be white. Given these data, it seems plausible to state that the probability of the proposition h: 'the boxed ball is white' being true, is 0.95. In what sense is such a statement justified and what 'probability' is meant here?

To answer the above questions, following Hacking, we proceed as follows. Let E = the event that any ball drawn blindly from the urn would be of the same colour as the ball in the box.

Clearly the event E is unobservable so long as the box remains sealed. However, whatever be the colour of the ball in the box, before drawing the ball from the urn we have $P(E) = 0.95$. Suppose a ball has been drawn but its colour has not yet been noted. Let i be the proposition, 'the particular ball drawn from the urn is of the same colour as the ball in the box'. By the forward instantiation principle, using $p(\cdot)$ temporarily to denote probability in the sense of credibility (given the above general data),

$$p(i) = 0.95. \tag{4.3.17}$$

Let c denote the colour of the ball drawn from the urn in the instance at hand. Intuitively, it seems, the credibility of the proposition i is, in no ways, affected by our knowledge of the value of c, i.e.

$$p(i|c = \text{white }) = p(i|c = \text{black }) = p(i). \tag{4.3.18}$$

In other words, the knowledge of c is *irrelevant* for the evaluation of the credibility of i. (But this does not mean, that if C denotes the random character representing the colour of any ball to be blindly drawn from the urn and something is known about the colour of the boxed ball, C and the event E are stochastically independent. In fact, when the boxed ball is white (black), $P(E|C = \text{white}) = 1(0)$ and $P(E|C = \text{black}) = 0$ (1). Similarly when the colour of the boxed ball is randomly determined and it is white or black with probabilities $\frac{3}{4}$ and $\frac{1}{4}$. $P(E|C = \text{white}) = \frac{3}{4}, P(E|C = \text{black}) = \frac{1}{4}$. Irrelevance holds only with regard to the instantiated credibility statements involving $p(\cdot)$ in (4.3.18) in the total absence of knowledge regarding the colour of the boxed ball.) In our particular case $c = \text{white}$. From (4.3.17)–(4.3.18) we get

$$p(i|c = \text{white }) = 0.95.$$

But, given $c = \text{white}$, the two propositions h and i are equivalent. Hence obviously it is legitimate to write

$$p(h|c = \text{white }) = 0.95,$$

which is what our hunch told us.

Fiducial argument, so far as it can be used in the general case, may be developed by extending the line of reasoning above. Suppose the distribution of the random observable X is determined by the unknown parameter $\theta \in \Theta$, which is real or vector-valued. (If Θ is discrete we can represent it by a set of integers; if Θ is a continuum, we take it to be an open set of a Euclidean space.) We know nothing about θ beyond that any point in Θ can be its value (Fisher (1956), pp. 51, 55–56, emphasizes this as a primary requirement for the applicability of the fiducial argument; for if θ has an a priori distribution (see section 4.4), according to him, there is no scope for utilizing that information in the fiducial argument and disregarding that information we would be violating the 'principle of total evidence' (see subsection 4.3.4).) Suppose further that we can identify a real or vector-valued function (unobservable) $U(x, \theta)$ defined on $\mathcal{X} \times \Theta$ with an open set $\mathcal{U}$ for range such that the following hold:

(i) whatever $\theta \in \Theta$ be true, $U(X, \theta)$ has the same distribution; this implies that for every θ, the set $\{U(x, \theta) : x \in \mathcal{X}\} = \mathcal{U}$ and $U(x, \theta)$ gives a mapping (possibly, many-one and different for different θ) from $\mathcal{X}$ to $\mathcal{U}$,

(ii) whatever x, $\{U(x, \theta) : \theta \in \Theta\} = \mathcal{U}$, and for each x, $U(x, \theta)$ represents a one-one correspondence (possibly different for different x) between $\mathcal{U}$ and Θ.

A function satisfying these two conditions is called a *pivotal function* (cf. subsection 4.2.2) or simply a *pivot*. (Thus in Example 4.3.4 above, if the random variable X is 0 or 1 according as the ball drawn from the urn is white or black, so that the distribution of X is determined by a parameter θ which assumes two values, say θ_0, θ_1 according to whether the boxed ball is white or black, we can take $U(x, \theta)$ such that $U(x, \theta_x) = 0$, $U(x, \theta_{1-x}) = 1$, $x = 0, 1$. Then the events E and '$U(X, \theta) = 0$' are same and whatever θ, $U(X, \theta)$ assumes the values 0 and 1 with probabilities 0.95 and 0.05 respectively).

Condition (ii) means that if X, and hence, $U(X, \theta)$ has a discrete distribution so that $\mathcal{U}$ is countable, Θ also must be countable. Although artificial problems (like the one in Example 4.3.4) of this type can be constructed, in most standard problems with discrete X, Θ happens to be a continuum. Apparently because of this, Fisher (1956), p. 50, explicitly lays down that for the fiducial argument to apply, X must be a continuous variate. In the following we confine ourselves exclusively to this case and take it for granted that $U(X, \theta)$ possesses a density function.

Condition (ii) implies that for any $x \in \mathcal{X}$ and $u \in \mathcal{U}$ the equation $U(x, \theta) = u$ in θ has a unique solution, say, $\theta = \theta(u|x)$. We assume further that the pivot $U(x, \theta)$ has the same dimension as θ and is smooth enough so that, for any x, $U(x, \theta)$ lies in an infinitesimal neighbourhood of u if and only if θ lies in a similar neighbourhood of $\theta(u|x)$ and the volume element of the latter can be expressed in terms of the same of the former and the appropriate Jacobian.

Let us temporarily write θ^* for the true value of θ. By condition (i), before the experiment, for any $u \in \mathcal{U}$ the value of

$$P(u \leq U(X, \theta^*) \leq u + du),$$

is known irrespective of θ^* (inequalities in the case of vectors hold coordinate-wise). Now suppose the experiment has been performed but the realized value x of X has not yet been noted. By the forward instantiation principle, using as before $p(\cdot)$ to denote credibility of a particular proposition, we have

$$p(u \leq U(x, \theta^*) \leq u + du) = P(u \leq U(X, \theta^*) \leq u + du) = \varphi(u) \prod du, \quad (4.3.19)$$

where we denote the known pdf of $U = U(X, \theta^*)$ by $\varphi(u)$. (We write $\prod du$ to indicate the product differential so as to cover the vector case as well.)

Let $x^{(0)}$, the actual value of x in the particular instance, be now noted. If the knowledge $x = x^{(0)}$ be regarded as *irrelevant* for judging the credibility of any proposition about the value of $U(x, \theta^*)$, we get

$$p(u \leq U(x, \theta^*) < u + du | x = x^{(0)}) = p(u \leq U(x, \theta^*) \leq u + du)$$
$$= \varphi(u) \prod du. \quad (4.3.20)$$

Now obviously

$$p(u \leq U(x, \theta^*) \leq u + du | x = x^{(0)}) = p(u \leq U(x^{(0)}, \theta^*) \leq u + du | x = x^{(0)}). \quad (4.3.21)$$

(This also follows from the standard axioms of intuitive probability; see axiom R in Koopman(1940b).)

Thus we have

$$p(u \leq U(x^{(0)}, \theta^*) \leq u + du | x = x^{(0)}) = \varphi(u) \prod du. \quad (4.3.22)$$

But by our assumptions, the proposition $u \leq U(x^{(0)}, \theta^*) \leq u + du$ is equivalent to the proposition that θ^* lies in an appropriate infinitesimal neighbourhood in Θ located at $\theta = \theta(u|(x^{(0)})$ with volume element $|J(U(x^{(0)}, \theta)/\theta)|\prod d\theta$, $|J(\cdot)|$ denoting the appropriate Jacobian. As u varies over $\mathcal{U}$, by condition (ii), $\theta = \theta(u|x^{(0)})$ spans the entire space Θ and the u corresponding to any θ is $U(x^{(0)}, \theta)$. Hence, by (4.3.22), we get that the fiducial (credibility) distribution of θ^* given $x^{(0)}$ has the form

$$\varphi(U(x^{(0)}, \theta))|J(U(x^{(0)}, \theta)/\theta)| \prod d\theta. \quad (4.3.23)$$

| Choice of pivot |

Choice of pivot $U(x, \theta)$ is of course crucial for the above argument. In many problems more than one option is available with regard to this. One requirement may be that $U(x, \theta)$ be so chosen that irrelevance of $x^{(0)}$ leading to the statement (4.3.20) is justifiable. However, Fisher did not look at the problem from this angle (the importance of irrelevance in the argument was emphasized by Hacking (1965); we shall refer to Hacking's formulation of the meaning of irrelevance and then discuss the question from an alternative point of view later), but instead required certain conditions that must hold and laid down certain procedures that must be followed for determining $U(x, \theta)$ (see Fisher (1956) pp. 49–56, 70, 80, 83–84, 119–120, 167–173). His primary requirement was that the maximum likelihood (ML) estimator $\hat{\theta}$ (which is supposed to exist as an explicit statistic $T = T(X)$, naturally with dimension and range same as those of θ) be a minimal sufficient statistic either (i) by itself or (ii) in combination with a complementary statistic S which is itself ancillary.

In the simpler case (i), if θ is real, and the cdf $F(t|\theta)$ of T is monotonic in θ (approaching the values 0 and 1 as θ approaches its limits, for every fixed t), he prescribed that $F(T(x)|\theta)$ be taken as the pivotal function (often equivalent simpler transforms are available). When θ has dimension two or more he advocated a step-by-step approach as illustrated in the following.

Example 4.3.5. Given a sample $X_1, \ldots, X_n$ from the population $\mathcal{N}(\mu, \sigma^2)$ with (μ, σ^2) unknown, the ML estimator, as represented by $\bar{X} = \sum X_i/n, V/n$ where $V = \sum (X_i - \bar{X})^2$, is minimal sufficient. Since marginally V/σ^2 is distributed as a χ^2_{n-1}, we can take it as the pivotal component for generating the marginal fiducial distribution of σ^2. When V and σ^2 are fixed, $\bar{X}$ is distributed as $\mathcal{N}(\mu, \sigma^2/n)$ and hence, $\sqrt{n}(\bar{X} - \mu)/\sigma$ which is distributed as a standard normal variable Z, can be taken as the pivotal component for generating the fiducial distribution of μ. Thus, writing $\bar{x}^{(0)}$ and $v^{(0)}$ for observed values, the joint fiducial distribution of μ, σ^2 is such that

$$\text{marginally } \sigma^2 \overset{d}{=} v^{(0)}/\chi^2_{n-1},$$

$$\text{given } \sigma^2, \text{conditionally } \mu \overset{d}{=} \bar{x}^{(0)} + \sigma Z/\sqrt{n}, \tag{4.3.24}$$

where χ^2_{n-1} and Z are independently distributed and $\overset{d}{=}$ stands for 'is distributed as'.

In case (ii), Fisher's prescription is that the ancillary complement S to the ML estimator T be fixed and the same procedure as above based on the conditional distribution of T given S, be followed. This is in conformity with the principle of forward instantiation laid down by Fisher (section 3.3) in situations where smaller reference sets in the sample space are recognizable (the samples for which S takes the value observed in the particular instance constitute a reference set here).

But apart from the need of logically justifying the rules of procedure prescribed by Fisher, these do not solve the problem of choice of the pivot and determination of the fiducial distribution in many situations.[12] It seems some

---

[12]Thus Dempster (1963) has shown that in Example 4.3.5 if we re-parametrize to $\delta = \mu/\sigma$ and σ and follow Fisher's step-by-step procedure determining first the marginal

restrictions on the type of the problem and the choice of the pivotal function must be imposed to make fiducial inference valid and unambiguous.

| Interpretation | A question that naturally arises in the context of fiducial distribution of θ is whether such a distribution can be manipulated like any ordinary probability distribution to derive the distribution of functions of θ, or in other words, whether fiducial probability obeys the standard axioms of probability (section 3.5). A related question is about the interpretation of fiducial probability.[13] In the above we have derived a fiducial probability distribution by making a one-one transformation (specific for the $x^{(0)}$ realized) on the (ordinary) probability distribution of $U(X, \theta)$; from this it follows that all standard formal operations on the former would be permissible. Also we have made it clear that fiducial probability is to be interpreted as credibility or a sort of impersonal subjective probability. As it is specific to a particular $x^{(0)}$, no objective frequency interpretation can be vouchsafed for it. This seems to have been Fisher's intended position. (Thus Fisher (1956) on p. 100 speaks of fiducial probability as giving 'the weight of the evidence actually supplied by the observations' and calls 'the frequency of events in an endless series of repeated trials which will never take place' as meaningless; similarly on pp. 120–121 he says that fiducial probability provides information which is 'logically equivalent

---

distribution of δ and then conditional distribution of σ given δ, we get a different joint fiducial distribution of μ, σ. The step-wise procedure has also been considered by Wilkinson (1977) who brings in what he calls a 'cross-coherence' condition which, when satisfied, would validate it. This precludes certain stepwise constructions, but the condition is not always easy to verify. What is disconcerting is that the step-wise procedure suggested by Fisher may sometimes end up with pivotal components whose joint distribution involves the parameter, thus violating the crucial requirement (i) on a pivot. Thus it may be checked that the joint distribution of the components of the alternative pivot based on a sample from $\mathcal{N}(\mu, \sigma^2)$ considered by Dempster (1963) involves the parameter δ. Similarly the joint distribution of the components of the pivot based on a sample from the bivariate normal population with unknown parameters, derived by Fisher (1956) pp. 169–171 involves the population correlation coefficient ρ. In such situations it seems Fisher is prepared to give requirement (i) on the pivot a go-by, but that, in our opinion, would obscure the argument.

[13]Wilkinson (1977) holds out the thesis that fiducial probability is non-coherent in the sense that it does not obey Kolmogorov axioms. But Wilkinson's development is rigidly wedded to the confidence or frequency interpretation—he would admit a fiducial probability statement only if it has a direct confidence interpretation and in general this precludes many types of manipulation on the fiducial distribution. True, Wilkinson equates 'inferential probability' with frequency probability and allows instantiation as in (4.3.19) whenever applicable. But he ignores the point that after the actual observation $x^{(0)}$ has been realized, the crucial step from (4.3.19) to (4.3.20) cannot be logically sustained without bringing in some postulates like that of irrelevance.

to the information which we might alternatively have possessed' if the population 'had been chosen at random from an aggregate specified by the fiducial
probability distribution'. A similar observation is made in Fisher (1959).)[14] The
impersonal subjective nature of fiducial probability has been emphasized and
made explicit notably by Hacking (1965).

We now consider two examples to illustrate how manipulation on the fiducial
probability leads to the solution of specific problems of statistical induction.

Example 4.3.6. Consider the same set-up as in Example 4.3.5. Suppose we are
interested in inferring about μ alone. From the joint fiducial distribution of μ
and σ as given by (4.3.24) writing $V/(n-1) = s^2$ and $s^{(0)}$ for the realized value
of s it immediately follows that marginally

$$\sqrt{n}(\mu - \bar{x}^{(0)})/s^{(0)} \overset{d}{=} Z/\{\chi^2_{n-1}/(n-1)\}^{\frac{1}{2}} \overset{d}{=} t_{n-1}, \qquad (4.3.25)$$

where t_{n-1} denotes 'Student's' t with $n-1$ d.f. (The fiducial distribution of μ as
given by (4.3.25) could also be obtained by starting with the real-valued pivotal
function $\sqrt{n}(\bar{X} - \mu)/s$, but then it will be troublesome to justify this choice of
the pivotal function.) If we are interested in an interval estimate for μ, from
(4.3.25), using notations same as in Example 4.2.1, we get

$$P(\bar{x}^{(0)} - t_{(\alpha)}s^{(0)}/\sqrt{n} < \mu < \bar{x}^{(0)} + t_{(\alpha)}s^{(0)}/\sqrt{n}) = 1 - \alpha. \qquad (4.3.26)$$

(We no longer stick to our earlier practice of denoting fiducial probability by
$p(\cdot)$ and making the conditioning on $\bar{x}^{(0)}$ and $s^{(0)}$ explicit.) The fiducial interval
(4.3.26) has the same form as the behavioural confidence interval for μ obtained
in (4.2.4) but the interpretations are radically different. The formal similarity
of the two intervals in the initial stages of development led people to think that
the two forms of reasoning are equivalent and created a good deal of confusion,
misunderstanding, and acrimony.

Example 4.3.7. (Behrens–Fisher problem.) Suppose we have independent
samples of sizes n_1, n_2 from two normal populations $\mathcal{N}(\mu_1, \sigma_1^2)$ and $\mathcal{N}(\mu_2, \sigma_2^2)$
with all parameters totally unknown. From each sample we can get the two-
dimensional fiducial distribution of corresponding (μ, σ^2) as in (4.3.24), and
hence, a representation of the marginal fiducial distribution of μ as given
by (4.3.25). Writing t_{n_1-1}, t'_{n_2-1} for two independent 'Student's' ts with d.f.
$n_1 - 1, n_2 - 1$ respectively, we derive that fiducially

$$\{(\mu_1 - \mu_2) - (\bar{x}_1 - \bar{x}_2)\}/e \overset{d}{=} \frac{e_1}{e}t_{n_1-1} + \frac{e_2}{e}t'_{n_2-1} = \bar{t}(n_1 - 1, n_2 - 1; e_1/e_2)$$

$$= \bar{t} \text{ (say)}, \qquad (4.3.27)$$

where we have suppressed the superscript for distinguishing realized values and
where $e_i^2 = s_i^2/n_i, i = 1, 2$, $e^2 = e_1^2 + e_2^2$. The distribution of the random variable $\bar{t}$

---

[14]But, unfortunately, at other places (e.g. Fisher (1956) pp. 51, 56, 59) Fisher's
language is somewhat ambiguous causing a confusion between the evaluative meaning
of probability and its descriptive interpretation.

has been tabulated for different values of $n_1 - 1, n_2 - 1$ and e_1/e_2 (Fisher and Yates (1963) pp. 60–62). Given $\bar{x}_1 - \bar{x}_2, e_1$ and e_2, using the tables, we can determine an interval estimate for $\mu_1 - \mu_2$ with a specified fiducial probability content.

If in a particular situation $\bar{x}_1 - \bar{x}_2 > 0$, (4.3.27) implies that the distribution of $\mu_1 - \mu_2$ is shifted to the right of 0. There we may want to know whether the shift is significant enough to make the null hypothesis H_0: $\mu_1 - \mu_2 = 0$ (or $\mu_1 - \mu_2 \leq 0$) untenable in the light of the data. For that, remembering that the distributions of $\bar{t}$ and $-\bar{t}$ are same, we compute the fiducial probability

$$P(\mu_1 - \mu_2 \leq 0) = P\left(\frac{\bar{x}_1 - \bar{x}_2}{e} + \bar{t} \leq 0\right) = P\left(\bar{t} \geq \frac{\bar{x}_1 - \bar{x}_2}{e}\right), \qquad (4.3.28)$$

and reject the null hypothesis if this is less than some specified level α. Denoting the point in the $\bar{t}$ distribution with upper tail α by $\bar{t}_{<\alpha>}$, this means we reject H_0 at specified level α if and only if $(\bar{x}_1 - \bar{x}_2)/e > \bar{t}_{<\alpha>}$. Thus the test is formally similar to a right-tailed test based on a test criterion in the behavioural approach.[15]

In the context of such uses of fiducial probability, objections have been raised time and again that on repetition of sampling with $\mu_1 - \mu_2$ fixed, the relative frequency of the interval estimate based on (4.3.27) covering the value of $\mu_1 - \mu_2$ would not generally approach the stated fiducial probability. Similarly when H_0: $\mu_1 - \mu_2 = 0$ holds, the relative frequency of rejection of H_0 in repeated samples will not approach the stated level α. No such anomaly can arise, however, if we keep in mind that fiducial probability represents credibility based on the particular instance at hand and no relative frequency interpretation is licensed in its case. A fiducial probabilist can justifiably claim that if sampling were repeated, with each new sample he would revise his fiducial probability statement, at each stage considering the aggregate of all samples to date as the instance at hand.[16]

| Question of irrelevance | **4.3.8**   We now examine the assumption of irrelevance which was so crucial in passing from (4.3.19) to (4.3.20) |

---

[15] One slight point of distinction is that although in fiducial theory we take whichever of $\bar{x}_1 - \bar{x}_2$ and $\bar{x}_2 - \bar{x}_1$ is positive in the numerator of the test criterion and compare the latter with the right hand α-tail point, we get the level α for the test. A similar procedure in the behavioural case would have given a level 2α. This is because in the fiducial case we are concerned only with the instance at hand and visualize no repetition.

[16] The same defence would serve against criticisms such as that of Stein (1959) who pointed out that if on the basis of observations on independent random variables distributed as $\mathcal{N}(\mu_i, 1), i = 1, \ldots, n$ one manipulated the joint fiducial distribution of the μ_i's to get a lower fiducial limit to $\sum \mu_i^2$ with a high probability, on repetition of sampling the relative frequency of instances where the lower limit would be exceeded would become negligible for large n.

in the fiducial argument. Hacking (1965), who pointed out the logical necessity of making the assumption, heuristically laid down a condition on the likelihood–ratio function, which would validate the assumption. A close examination of Hacking's condition, however, shows that it is always satisfied when the pivot, besides satisfying conditions (i) and (ii), is such that we can write $U(X,\theta) = v_\theta(T)$, where T is a sufficient statistic of dimension same as that of θ (and hence, of $U(X,\theta)$) and for each θ the function $v_\theta(t)$ is invertible and has the usual smoothness properties ensuring that the density of U and T can be expressed in terms of each other.[17] This means, apart from the heuristic nature of the condition, it does not lead to a unique pivot in all problems.[18]

We shall look at the question of irrelevance from the point of view of invariance and give a justification for it only for problems in which there exists a kind of comprehensive natural symmetry. Our development will closely follow that of Fraser (1961) who showed that in many problems of inference invariance considerations lead us to a uniquely determined pivot. (However, Fraser did not consider irrelevance and instead of interpreting fiducial probability as a sort of impersonal subjective probability as we have done, he sought to view it as a kind of artificially conceived frequency distribution of possible values for the parameter 'relevant to' the particular realized observation.)

Consider once again Example 4.3.4. It is clear that if we rename the colours, calling white as 'black' and black as 'white', the set up and the definition of the event E and the proposition i remain same. Only $p(i|c = \text{white})$ now changes to $p(i|c = \text{black})$ and vice versa. Hence the equality of the first two members in (4.3.18) is immediate. Now note that '$c = \text{white}$' and '$c = \text{black}$' are complementary propositions. In this particular case by the standard axioms of intuitive probability (see the 'Axiom of Alternative Presumption' and Theorem 11 in Koopman (1940a)), from this we get that the common value of the first two

---

[17]Hacking required that for any two points $\theta_1, \theta_2 \in \Theta$, the likelihood ratio given $X = x^{(0)}$ is the ratio of the corresponding probabilities of $U(X,\theta_i)$ lying in appropriate neighbourhoods of $U(x^{(0)}, \theta_i), i = 1, 2$, where the appropriate neighbourhood of $U(x^{(0)}, \theta_i)$ is derived from the 'experimental' neighbourhood of $x^{(0)}$ taking account of the value of θ_i, and hence, has its volume element in general dependent on θ_i.

[18]For example in the case of a sample of size n from the bivariate normal population $\mathcal{N}_2(0, 0, \sigma_1^2, \sigma_2^2, \rho)$ with variables x, y, the minimal sufficient statistic is represented by the sample correlation coefficient r and the standard unbiased estimates s_1^2, s_2^2 of σ_1^2, σ_2^2. The functions $\sqrt{(n)}s_1/\sigma_1, \sqrt{(n)}s_2\sqrt{(1-r^2)}/\left(\sigma_2\sqrt{(1-\rho^2)}\right), \sqrt{(n)}\left(rs_2/\sigma_2 - \rho s_1/\sigma_1\right)/\sqrt{(1-\rho^2)}$ have their joint distribution free from the parameters and can be taken as the components of a pivot (see Fisher (1956) p. 172; Fisher, however, finds fault with this pivot because it is not derived by a stepwise procedure). This pivot satisfies Hacking's condition, but so does the alternative pivot obtained by interchanging the roles of x, y in it. Indeed, in the corresponding multivariate situation, Mauldon (1955) derived an infinity of alternative pivots all meeting the Hacking condition.

members in (4.3.18) is equal to $p(i)$. If as earlier, we think in terms of the random variable X which is 0 or 1 according as the ball drawn is white or black and a parameter θ which is θ_0 or θ_1 according as the boxed ball is white or black, renaming of white as black and viceversa amounts to effecting the transformation $x \rightarrow 1 - x, x = 0, 1$ in the space of X and a corresponding transformation $\theta_\ell \rightarrow \theta_{1-\ell}, \ell = 0, 1$ in the space of θ. We now generalize these ideas.

As before let the distribution of X assuming values in $\mathcal{X}$ be determined by the value of the (real- or vector-valued) parameter $\theta \in \Theta$. We assume that θ is an identifying parameter, i.e. distinct θs represent distinct distributions of X. Suppose the set-up is invariant in the sense that there exists a class G of one-one (measurable) transformations g of $\mathcal{X}$ onto itself which forms a *group* (i.e. is closed under multiplication (composition) and inversion).[19] With respect to G, the sample space $\mathcal{X}$ can be partitioned into disjoint sets of transitivity or *orbits* such that the orbit S_x containing the point x consists of exactly those points which are obtainable from x by applying the transformations in G. We assume G is such that the following hold:

(a) For every g, corresponding to each θ, there is a unique $\theta' = \bar{g}\theta \in \Theta$ such that when X follows θ, gX follows $\bar{g}\theta$. Then as θ is an identifying parameter $\theta \rightarrow \bar{g}\theta$ is a one-one transformation of the parameter space Θ onto itself; further, the class $\bar{G}$ of $\bar{g}$'s corresponding to all the g's in G, forms a group. We can assume, if necessary by redefining G, that the mapping $G \rightarrow \bar{G}$ which preserves the group operations, is one-one.[20]

(b) G is uniquely transitive on each orbit S_x in the sense that given any $x_0, x_1 \in S_x$ there exists a unique $g_1 \in G$ such that $g_1 x_0 = x_1$. (This means that fixing any x_0 in S_x and matching every $x_1 \in S_x$ with the g_1 which gives $g_1 x_0 = x_1$, we get a one-one correspondence between the points in S_x and the transformations in G.)

(c) $\bar{G}$ is uniquely transitive on Θ, i.e. given any $\theta_0, \theta_1 \in \Theta$, we can find a unique $\bar{g}_1 \in \bar{G}$ such that $\bar{g}_1\theta_0 = \theta_1$. (So fixing θ_0 and matching θ_1 with $\bar{g}_1$, we get a one-one correspondence between Θ and $\bar{G}$, and hence, between Θ and G.)

First consider *the case when the entire sample space $\mathcal{X}$ constitutes a single orbit* so that (b) means G is uniquely transitive on $\mathcal{X}$. We can often reduce the set-up to realize this when a minimal sufficient statistic T with dimension same as θ exists. Every transformation g will then have its counterpart for T. Here if we invoke the sufficiency principle (subsection 4.3.3) and take T as our basic observable, in many problems condition (b) will hold with G uniquely transitive

---

[19]When X is continuous we assume that the transformations g are smooth enough to permit the derivation of the density function of gX from that of X in the usual way.

[20]If the mapping is in many-one, the subclass G_0 of elements $g \in G$ whose map is the identity of $\bar{G}$ forms a normal subgroup and we can then replace G by the quotient group G/G_0 to get a one-one mapping.

on the space of T. We continue to describe the problem in terms of X and $\mathcal{X}$ with the understanding that these may stand for T and the space of T.

Here the problem in terms of X is formally same as that in terms of gX with the role of θ being taken over by $\bar{g}\theta$ in the latter case (the transformation $X \to gX$ is like renaming the colours in Example 4.3.4). The standing of any possible value of θ vis-à-vis a realized value $x^{(0)}$ is same as that of $\bar{g}\theta$ vis-à-vis the realized value $gx^{(0)}$. Hence for logical consistency we require that our instantial credibility distribution of θ given the realized observation $x^{(0)}$ be such that, whatever g, when $x^{(0)}$ is replaced by $gx^{(0)}$ and $\tilde{\theta}$ is used to denote the corresponding θ, the resulting distribution of $\tilde{\theta}$ is same as what would be obtained by calling $\bar{g}\theta$ as $\tilde{\theta}$ in the original distribution of θ. In other words, replacement of $x^{(0)}$ by $gx^{(0)}$ and substitution of $\bar{g}\theta$ for θ in the original distribution will leave the latter unchanged. From (4.3.21) we get that this will hold if the pivot satisfies besides (i) and (ii) (see subsection 4.3.13) the additional condition (iii) $U(x,\theta) = U(gx, \bar{g}\theta)$ for all $x \in \mathcal{X}, \theta \in \Theta$, whatever $g \in G$.

We call a pivot satisfying (i)–(iii) an *invariant pivot*. It can be shown that when the set-up is invariant with respective to a group G in the above sense, there exists an essentially (i.e. upto one-one transformations) unique invariant pivot $U(x,\theta)$.[21]

Let $U(x, \theta)$ be an invariant pivot. Since the set-up with observable X and parameter θ is formally same as that with observable gX and parameter $\bar{g}\theta$, and when $x^{(0)}$ is the realized observation in the former, that in the latter is $gx^{(0)}$, we have, using notations as in (4.3.19)–(4.3.20), by virtue of condition (iii),

$$p(u \leq U(x, \theta^*) \leq u + du | \text{obsn.} = x^{(0)})$$

$$= p(u \leq U(gx, \bar{g}\theta^*) \leq u + du | \text{obsn.} = gx^{(0)})$$

$$= p(u \leq U(x, \theta^*) \leq u + du | \text{obsn.} = gx^{(0)}). \qquad (4.3.29)$$

(The first equality is required for logical consistency; the second follows from condition (iii). We write 'obsn.' in place of x to avoid confusion.)

---

[21] Write (G) for the set of all transformations $(g)(x, \theta) = (gx, \bar{g}\theta)$ of $\mathcal{X} \times \Theta$ onto itself. Although G on $\mathcal{X}$ and $\bar{G}$ on Θ are transitive, (G) on $\mathcal{X} \times \Theta$ is generally not. The orbits of (G) may be labelled by the points $\{(gx_0, \theta_0) : g \in G\}$ or alternatively $\{(x_0, \bar{g}\theta_0) : \bar{g} \in \bar{G}\}$ where $x_0 \in \mathcal{X}$ and $\theta_0 \in \Theta$ are arbitrarily fixed. From the conditions (i)–(iii) it follows that $U(x, \theta)$ is an invariant pivot if and only if it is a maximal invariant (see Lehmann (1986) Chapter 6) under (G). Such a maximal invariant, which is essentially unique, may be constructed as follows: for a fixed θ_0, for every θ, let $k_\theta \in G$ be such that when X follows θ, $k_\theta X$ follows θ_0 (i.e. $\bar{k}_\theta \theta = \theta_0$); then set up $U(x, \theta) = k_\theta x$ (cf. Fraser (1961)).

This is true for all $g \in G$. By our assumption, as g moves over $G, gx^{(0)}$ assumes every point in $\mathcal{X}$ as value. Thus (4.3.29) implies

$$p(u \leq U(x, \theta^*) \leq u + du | x = x')$$
$$= p(u \leq U(x, \theta^*) \leq u + du | x = x^{(0)}) \quad \text{for all } x' \in \mathcal{X}. \tag{4.3.30}$$

We now introduce the following postulate about credibility:

The set-up is such that if, for any proposition h about x and θ^*, the credibility of h given $x = x'$ has a constant value for all $x' \in \mathcal{X}$ then the credibility of h given $x \in \mathcal{X}$ (which is always true) has the same value.[22]

Invoking this postulate, from (4.3.30) we get the irrelevance condition (4.3.20).

To illustrate these ideas consider the set-up of Example 4.3.5. In terms of the minimal sufficient statistic $(\bar{x}, v)$ the set-up is invariant under all transformations

$$g_{a,b}(\bar{x}, v) = (a + b\bar{x}, b^2 v), \quad -\infty < a < \infty, \ b > 0,$$

with corresponding transformations $\bar{g}_{a,b}(\mu, \sigma^2) = (a + b\mu, b^2 \sigma^2)$ on the parameter space. The group of all transformations $g_{a,b}$ is uniquely transitive on the space of $(\bar{x}, v)$; the same is true for the transformations $\bar{g}_{a,b}$ on the parameter space. Also

$$\left(\sqrt{n} \frac{(\bar{x} - \mu)}{\sigma}, \frac{v}{\sigma^2} \right) \tag{4.3.31}$$

represents an invariant pivot here. Hence irrelevance holds and the fiducial distribution of μ, σ^2 (which was earlier obtained by proceeding stepwise) can be legitimately obtained by inverting the pivot (4.3.31).

Next consider *the case when the sample space $\mathcal{X}$ breaks up into more than one orbit* under G. For any such orbit S_x, by definition, the events $X \in S_x$ and $gX \in S_x$ are identical. Hence $P(X \in S_x | \theta) = P(X \in S_x | \bar{g}\theta)$ which, by the unique transitivity of $\bar{G}$ on Θ, implies that the orbits of $\mathcal{X}$ represent an ancillary partition. Then, given an observation $x^{(0)}$, we confine ourselves to the conditional set-up given $X \in S_{x^{(0)}}$ and proceed as earlier, substituting $S_{x^{(0)}}$ for $\mathcal{X}$. If the pivot $U(x, \theta)$ satisfies the invariance condition (iii), the fiducial distribution of θ can be deduced from the conditional distribution of $U(X, \theta^*)$ and irrelevance validated arguing as before.

Thus when the set-up is invariant with respect to a group of transformations G as above, irrelevance follows in a plausible way. An additional advantage here is that given G the invariant pivot is uniquely defined upto equivalence. However,

---

[22]When $\mathcal{X}$ is finite, this follows from the standard axioms of intuitive probability (cf. the 'axiom of alternative presumption' and Theorem 13 in Koopman (1940a); however, Koopman (1941) is wary about extending the axiom to the general case of an infinite collection of conditioning propositions. In our case $\mathcal{X}$ is infinite, but the context is limited to the conditioning propositions $x = x', x' \in \mathcal{X}$. Since we want credibility to follow the same rules as probability, the postulate is a natural one to introduce.

there are problems in which there exists more than one group G such that the set-up is separately invariant under each.[23] In such a case one must identify as part of the model the group which is most natural from the practical point of view and work with the corresponding invariant pivot.[24]

In conclusion we note that fiducial inference does not obey the likelihood principle across experiments. This is because even though the minimal sufficient statistics specific to two different experiments may have the same form and give the same value at particular realizations, the sampling distributions of those, and hence of pivots based on those would be generally different (see Anscombe (1957), Basu (1975)). (However, as we have already remarked, whether or not we regard the likelihood principle as something sacrosanct is a matter of preference.)

The different forms of objective–instantial approach as delineated above rely on arguments which reduce problems of induction to those of backward or forward instantiation of probability and do not seem to have a philosophy central to them like the objective–behavioural approach (and, as we would see, this approach is certainly not as monolithic as the other two approaches that would follow). Nevertheless, the various modes of instantial reasoning seem to catch the essence of how we often infer inductively in various situations in everyday practice.

4.4   Objective–pro-subjective approach

| Subjective probability in input |

4.4.1   We now come to the third approach to statistical induction under the broad objective category, namely, the objective–pro-subjective approach. Just as in the case of the other two forms of objective approach, in this also the probability model for the observables may have a frequency basis. (Actually, the methods developed would be applicable whenever the probability model with its parameters has agent-independent interpretation. This means that the probabilities based on the model may have straight-forward objective interpretation; or these may be physical probabilities (propensities) which can be determined only as the limit of subjective probabilites (Good (1983) pp. 15, 70); or these may even be impersonal subjective probabilities which can be measured approximately through frequencies (Jeffreys (1961), Box and Tiao (1973)). (Incidentally, Savage (1954) p. 60 subsumes necessary, i.e. impersonal subjective probability under the 'objective' category.) But in this form of objective approach subjective probability (impersonal or personal) generally comes into the picture not only as

---

[23]Thus for a sample from the bivariate normal population with means zero (see Footnote 18), the problem is invariant under transformations of (x, y) separately by all nonsingular lower as well as upper triangular matrices.

[24]Starting from this idea Fraser (1968) developed his theory of structural probability and related inference in which the model is specified by the underlying probability distribution together with a group of transformations.

the output of the process of induction for assessing the uncertainty in the conclusion (this, as we have seen, is also true of certain forms of objective–instantial approach), but as a deliberate input in the process. Thus, although the approach may be model-wise objective, it has a pronounced leaning towards the subjective interpretation of probability. The basic tenet of this form of objective approach (in contrast with the other two forms considered earlier) is that all forms of uncertainty, whether about outcomes of yet-to-be-performed repeatable random experiments or about unknown values of parameters which we can never conceivably ascertain, can be, in theory, quantified in terms of probabilities which obey standard axioms.

| Posterior distribution |

4.4.2   Let as earlier, the outcome of the sampling experiment $\mathcal{E}$ be represented by a real- or vector-valued random variable X. The distribution of X over the sample space $\mathcal{X}$ (which is a countable set when X is discrete and a continuum when X is continuous) is represented by the pmf or pdf $f(x|\theta), x \in \mathcal{X}$. The function $f(x|\theta)$ would be fully determined if the value of the unknown parameter θ could be ascertained. In the following, for the sake of simplicity, we assume θ is real- or vector-valued. The domain of θ is the parameter space Θ which is a countable set or a continuum according as θ is discrete or continuous.

The key result for all modes of inference and decision in the objective–prosubjective approach is Bayes's Theorem which is relevant only when θ can be regarded as the realization of a random variable θ at least in the subjective sense. (In the subjective case we do not distinguish between θ and its unknown realization.) Further, the probability distribution of θ over Θ must be known (prior to the observation of X). Let this distribution, which is called the *prior* distribution of θ, be represented by the pmf or pdf $\pi(\theta), \theta \in \Theta$. Here $\pi(\theta)$, when specified, constitutes an additional premise in the process of induction, which is not available and is generally considered meaningless in the behavioural and instantial approaches. Knowing $\pi(\theta)$ and $f(x|\theta)$, for any realized observation $x^{(0)}$ we can find by Bayes's Theorem the conditional distribution of θ given $X = x^{(0)}$ (when X is continuous, as usual, '$X = x^{(0)}$' is to be interpreted as 'X lies in an infinitesimal neighbourhood of $x^{(0)}$'). This conditional distribution which is called the *posterior* (i.e. subsequent to the realization of $x^{(0)}$) distribution of θ has its pmf or pdf (depending on whether θ, and hence, θ is discrete or continuous) given by

$$\pi\left(\theta|x^{(0)}\right) = \frac{\pi(\theta)f(x^{(0)}|\theta)}{\bar{f}_\pi(x^{(0)})}, \quad \theta \in \Theta, \tag{4.4.1}$$

where

$$\bar{f}_\pi(x) = \sum_{\theta \in \Theta} f(x|\theta)\pi(\theta) \quad \text{or} \quad \int_\Theta f(x|\theta)\pi(\theta)\,d\theta \tag{4.4.2}$$

according as θ is discrete or continuous ($d\theta$ standing for the appropriate product differential in the latter case, if θ is a vector).[25] In the objective–pro-subjective approach all inductive conclusions are derived from the posterior distribution $\pi(\theta|x^{(0)})$ of θ, which can be looked upon as a sort of master conclusion here.[26]

Sufficiency and likelihood principles

Several features of this approach to induction follow immediately from the expression for the posterior distribution. Firstly, when a sufficient statistic exists so that $f(x|\theta)$ factorizes as in (4.3.6), the factor not involving θ cancels out in (4.4.1) and the posterior distribution based on the full data reduces to that based on the sufficient statistic. Thus the sufficiency principle (subsection 4.3.3) is realized in this approach automatically and does not require to be justified or imposed as in certain other approaches. Secondly, the data $x^{(0)}$ enters the posterior probability only via $f(x^{(0)}|\theta)$ (which represents the likelihood, see (4.3.1)). Hence, fixing a $\theta_0 \in \Theta$ such that $f(x^{(0)}|\theta_0) > 0$ in (4.4.1), we can substitute $f(x^{(0)}|\theta) = \rho(\theta, \theta_0|x^{(0)}) \cdot f(x^{(0)}|\theta_0)$ to get the posterior distribution in terms of the LR function (see (4.3.3)). This means any mode of induction based on the posterior distribution would obey the likelihood principle (subsection 4.3.4) (even across experiments if the prior $\pi(\theta)$ depends only on our inductive concern and does not change with the experiment). A fall-out of this is that in this approach one need not bother about any rule for stopping (which determines the number of observations sequentially by prescribing after each observation whether sampling should stop or continue). Similarly if the experiment involves a censoring mechanism which determines whether an observation taken should be admitted in the sample or discarded but all the sample observations happen to be uncensored, one need not bother about censoring provided it is non-informative (see Berger and Wolpert (1988)). All this is in sharp contrast with the behavioural approach where sampling distributions of criteria are in general drastically affected by such modifications of the sampling scheme.

Choice of prior

4.4.3 To derive the posterior distribution (4.4.1), of course one would require the prior distribution $\pi(\theta)$. How would one get this?

---

[25]Since any point x where $\bar{f}_\pi(x) = 0$ can be excluded from $\mathcal{X}$ at the outset, without any loss of generality, we can assume that the denominator in (4.4.1) is positive.

[26]In view of this, as the posterior distribution is obtained through Bayes's Theorem, this approach could have been called 'objective Bayesian'. However, in the literature methods based on Bayes's Theorem have been called objective Bayesian when the prior is 'objectively' determined, either as a frequency probability distribution from an initial experiment (Wilkinson (1977) p. 121), or more commonly, on the basis of some well-defined rationality principle relating to the model (Berger (1985a) p. 110, Press (1989) p. 47, Howson (1995) pp. 12–13). In our case the term 'objective' refers to the model (which is supposed to have a frequency basis) and not to the prior.

Frequentist

Before addressing the problem, let us first of all dispose of a case which, strictly speaking, is not relevant to the context of the objective–pro-subjective approach. It occurs when the experiment is a two-stage one: the sampling experiment giving the observation X with distribution $f(x|\theta)$ is preceded by an initial random experiment which fixes the value θ (unobservable in the particular instance considered) of the random parameter θ. The initial experiment is (at least conceptually) repeatable and the prior distribution $\pi(\theta)$, which has a frequency interpretation, can be guessed from the nature of the experiment or from records of direct determination of the values of θ made in the past. (For example, in the sampling inspection set-up, X may be the number of defectives in a random sample from a lot with proportion defective θ; the lot itself comes from a production line for which the long-run frequency distribution of proportion defective in lots follows a given pattern. Similarly for a particular year X may represent the yields for a random sample of experimental plots from a population in which the mean yield is θ; θ itself depends on various climatic and environmental conditions, and hence varies randomly over the years.) In such a situation the derivation of the posterior distribution (4.4.1) or any characteristic based on it is simply an exercise in deductive inference.[27] The prior distribution being objective, the posterior probabilities also would have objective interpretation. Of course to apply these to the instance at hand, one has to take to forward instantiation (subsection 4.3.2). Statisticians of all hues more or less agree that in such a situation, Bayes's Theorem is the right tool for solving the problem of inference (see Fisher (1956) p. 55, Fisher (1959) and Neyman (1952) pp. 162–165). (However, even here, Lecam (1977) p. 144 expresses the opinion that unless the entire problem presents itself over and over again, it may be preferable to work assuming θ is conditionally fixed.)

In this context it may be noted that for such a two-stage random experiment the posterior probabilities can also be interpreted as long-term relative frequencies and the contrast between inferential statements reached through the behavioural approach and the present approach can be brought out sharply. Thus suppose the two-stage experiment is repeated a large number N of times giving rise to pairs (θ, x), where $\theta \in \{\theta_{(1)}, \theta_{(2)}, \ldots, \theta_{(r)}\}$ and $x \in \{x_{(1)}, x_{(2)}, \ldots, x_{(s)}\}$. The (θ, x)-pairs can be classified according to both θ and x to generate a fictitious two-way frequency table as in Table 4.1.

Suppose a statement of the form $\theta < t(x)$ is made for every observed (θ, x)-pair, i.e. for every cell of the above table according to both the behavioural and the present approach and the same threshold of assurance γ—confidence level in the first and posterior probability in the second—is attached to the statement in the two approaches. (Here $t(\cdot)$ is a suitable function which may be different in the two cases.) This means that in the first case the long-term proportion of pairs (θ, x) for which the corresponding statement is correct is at

---

[27]However, Fisher (1956) p. 109 says that the solution of the problem still has an element of induction if the prior distribution $\pi(\theta)$ has to be determined empirically.

Table 4.1. *Frequency table based on N observations on* (θ, x)

θ	x	$x_{(1)}$	$x_{(2)}$	$\cdots$	$x_{(s)}$	
$\theta_{(1)}$		f_{11}	f_{12}	$\cdots$	f_{1s}	$f_{1\cdot}$
$\theta_{(2)}$		f_{21}	f_{22}	$\cdots$	f_{2s}	$f_{2\cdot}$
$\vdots$		$\cdot\cdot$	$\cdot\cdot$	$\cdot\cdot$	$\cdot\cdot$	$\cdot$
$\theta_{(r)}$		f_{r1}	f_{r2}	$\cdots$	f_{rs}	$f_{r\cdot}$
		$f_{\cdot 1}$	$f_{\cdot 2}$	$\cdots$	$f_{\cdot s}$	N

least γ *for every row* of the table, whereas in the second case the same holds *for every column*. (Of course thereby both types of statement would hold at least in the proportion γ in the entire aggregate of (θ, x)-pairs.)

Subjective—impersonal and personal	In the usual situations where we want to infer on the basis of empirical evidence, θ cannot be interpreted as above as a random variable in the frequency sense.

θ can at best be considered as a random variable in the subjective sense (i.e. as an unknown quantity for which uncertainty arises solely due to absence of knowledge; cf. sections 2.2 and 3.3). To derive the posterior distribution (4.4.1), we then have to supply the prior probability distribution $\pi(\theta)$ on the basis of our subjective knowledge. Recall the distinction between two forms of subjective probability, impersonal and personal, that we drew in section 3.3. The prior distribution $\pi(\theta)$ here may similarly represent either impersonal subjective probability based on all knowledge that is publicly available, or personal subjective probability specific to a particular person at a particular time. Since combination of subjective probability of any kind with objective probability would generate subjective probability of the same kind, accordingly the posterior distribution $\pi(\theta|x)$ (hereafter we write x for $x^{(0)}$) also would have either an impersonal or personal subjective interpretation.

Impersonal prior	**4.4.4** In the impersonal case the prior is endogenously determined by the set-up of the problem and any

information about which there is a general consensus, on the basis of certain principles of rationality laid down a priori. Such an impersonal prior is appropriate in situations where some broad public policy would be based on the inductive conclusion and it is therefore desirable that the conclusion should be as 'objective' or agent-independent as possible. Different rationality principles have been proposed and these have given rise to varieties of impersonal priors.

Non-informative
prior—group-theoretic

First, there are what are called *non-informative* or *informationless* or *vague* priors appropriate for situations where no information beyond that contained in the set-up is available. If certain symmetries are discernible in the model, sometimes those can be exploited to generate the prior. Thus, as in subsection 4.3.8, suppose the set-up is invariant in the sense that there is a group G of one-one transformations of $\mathcal{X}$ onto itself and an induced group $\bar{G}$ of corresponding one-one transformations $\bar{g}$ of Θ onto itself such that if X follows θ, gX follows $\bar{g}\theta$. This means, considering the case of continuous X, under the usual smoothness assumptions, that for all $g \in G$

$$f(x|\theta) = f(gx|\bar{g}\theta)\left|J\left(\frac{gx}{x}\right)\right|, \qquad (4.4.3)$$

where J denotes the Jacobian. It is natural to require here that the posterior distribution be *invariant* in the sense that the distribution of θ given x, be recoverable from that of $\bar{g}\theta$ given gx by applying the inverse transformation $(\bar{g})^{-1}$ on θ. (For example if x and θ are temperatures in the Centigrade scale and gx and $\bar{g}\theta$ are same in the Fahrenheit scale, we would want the distribution of the parameters θ in °C given the reading x in °C to be the same as what would be obtained by first deriving the distribution of the parameter in °F given the reading gx in °F and then expressing the parameter in the Centigrade scale.) This means $\pi(\theta|x)$ must satisfy for all $\bar{g} \in \bar{G}$

$$\pi(\theta|x) = \pi(\bar{g}\theta|gx)\left|J\left(\frac{\bar{g}\theta}{\theta}\right)\right|. \qquad (4.4.4)$$

In the case of continuous θ, from (4.4.1)–(4.4.3) it readily follows that (4.4.4) is true if $\pi(\theta)$ satisfies for all $\bar{g} \in \bar{G}$

$$\pi(\theta) = \pi(\bar{g}\theta)\left|J\left(\frac{\bar{g}\theta}{\theta}\right)\right|, \qquad (4.4.5)$$

which means a priori $\bar{g}\theta$ has the same distribution as θ. In the case of discrete X(and/or θ) the Jacobian will not appear in (4.4.3) (and/or (4.4.4)–(4.4.5)). When $\bar{G}$ is transitive on Θ, sometimes the form of $\pi(\theta)$ follows from the invariance condition (4.4.5).

In some cases it is convenient to apply the condition in stages. Thus suppose θ is a vector which splits into two complementary subvectors, say, θ_1 and θ_2 with 'marginal' domain Θ_1 of θ_1 and 'conditional' domain $\Theta_{2(\theta_1)}$ of θ_2 given θ_1. Suppose further there exists a subgroup $\bar{G}_1 \subset \bar{G}$ such that when we transform θ by $\bar{g} \in \bar{G}_1$ the θ_2 component remains unchanged, and for every θ_1 there exists a subgroup $\bar{G}_{2(\theta_1)} \subset \bar{G}$ such that given θ_1 when we transform θ by $\bar{g} \in \bar{G}_{2(\theta_1)}$ the θ_1 component remains same. We can then regard $\bar{G}_1$ as operating on Θ_1 and, given $\theta_1, \bar{G}_{2(\theta_1)}$ as operating on $\Theta_{2(\theta_1)}$. Writing $\pi(\theta) = \pi(\theta_1)\cdot\pi_2(\theta_2|\theta_1)$ as the product

of the marginal and conditional densities, we may consider the invariance of
$\pi(\theta_1)$ under $\bar{G}_1$, and given θ_1, the invariance of $\pi_2(\theta_2|\theta_1)$ under $\bar{G}_{2(\theta_1)}$. When
transitivity holds, from this, we can sometimes determine the factors $\pi(\theta_1)$ and
$\pi_2(\theta_2|\theta_1)$ separately. The ultimate solution for $\pi(\theta)$ obtained by such a stagewise procedure may differ from that obtained through a single-stage application
of (4.4.5).

We illustrate all this through an example (for further details and applications
see Berger (1985a) sections 3.3, 6.6).

Example 4.4.1. Consider the set-up of Example 4.3.5 with the observables
$X_1, \ldots, X_n$ representing a random sample from $\mathcal{N}(\mu, \sigma^2)$. Take $\theta = (\mu, \sigma)$ with
$\Theta = \{\theta : -\infty < \mu < \infty, 0 < \sigma < \infty\}$. Clearly the set-up is invariant under any
transformation $g_{a,b}, -\infty < a < \infty, 0 < b < \infty$, where

$$g_{a,b} \, x_i = a + bx_i, i = 1, \ldots, n,$$
$$\bar{g}_{a,b}(\mu, \sigma) = (a + b\mu, b\sigma). \qquad (4.4.6)$$

Sets of all $g_{a,b}$'s and $\bar{g}_{a,b}$'s give G and $\bar{G}$ respectively.

The prior (4.4.5) now takes the form

$$\pi(\mu, \sigma) = \pi(a + b\mu, b\sigma) \cdot b^2. \qquad (4.4.7)$$

For any (μ, σ) we may take $b = 1/\sigma, a = -\mu/\sigma$ (this implies transitivity of $\bar{G}$
over Θ) to get the form

$$\pi(\mu, \sigma) = \pi(0, 1) \cdot \frac{1}{\sigma^2} = \frac{\text{const.}}{\sigma^2}. \qquad (4.4.8)$$

Whatever positive number we may choose for the constant, the posterior (4.4.1)
remains same. However, although the denominator in (4.4.1) is well-defined, the
total integral of (4.4.8) over Θ is ∞, so that here the prior is not a probability
distribution; we call it an *improper* prior. (We would consider the question of
interpretation of such priors later.)

Alternatively, we may resort to a two-stage determination of the prior. Split-
ting θ as μ and σ, we factorize $\pi(\mu, \sigma) = \pi_1(\mu) \cdot \pi_2(\sigma|\mu)$. The elements of the
subgroup $\bar{G}_1 = \{\bar{g}_{a1} : -\infty < a < \infty\}$ of $\bar{G}$ operate transitively on μ over the
domain $-\infty < \mu < \infty$ without affecting σ. Hence applying (4.4.5) on $\pi_1(\mu)$ we
get $\pi_1(\mu) = \pi_1(a + \mu)$ which implies $\pi_1(\mu) = \pi_1(0)$. Also given μ, σ varies over
$(0, \infty)$ and the elements of the subgroup $\bar{G}_{2(\mu)} = \{\bar{g}_{(1-b)\mu,b} : 0 < b < \infty\}$
of $\bar{G}$ operate transitively over $0 < \sigma < \infty$ keeping μ same. This gives
$\pi_2(\sigma|\mu) = \pi_2(b\sigma|\mu) \cdot b$, so that $\pi_2(\sigma|\mu) = \pi_2(1|\mu) \cdot (1/\sigma)$. If we take $\pi_2(1|\mu)$
as same for all μ, we get

$$\pi(\mu, \sigma) = \frac{\text{const.}}{\sigma}. \qquad (4.4.9)$$

(4.4.9) also is improper but it differs from (4.4.8).

Using (4.4.9) and substituting

$$f(x_1,\ldots,x_n|\mu,\sigma) = (2\pi\sigma^2)^{-n/2}e^{-n(\bar{x}-\mu)^2/(2\sigma^2)-v/(2\sigma^2)}, \quad v = \sum(x_i - \bar{x})^2,$$

in (4.4.1), we see that the posterior joint distribution of μ,σ is such that marginally σ is distributed as $\sqrt{v}/\chi_{n-1}$, and given σ, and hence independently of $\sigma,\sqrt{n}\ (\mu-\bar{x})/\sigma$ is distributed as $\mathcal{N}(0,1)$. From this it follows that $\sqrt{n}(\mu-\bar{x})/\{v/(n-1)\}^{\frac{1}{2}}$ is distributed as 'student's' t_{n-1}. These results are exactly same as those obtained in Examples 4.3.5 and 4.3.6 through fiducial reasoning.[28] If, instead of (4.4.9), we use the prior (4.4.8) we get a similar result but the d.f. of χ^2 and t now becomes n.

Fiducial distribution as posterior

The identity of the fiducial distribution of (μ,σ) and the posterior distribution of (μ,σ) under the noninformative prior (4.4.9) in the above example is not something fortuitous. Consider a general set-up with continuous X and θ in which there is a sufficient statistic of dimension same as θ; we regard the sufficient statistic as our basic observable and denote it by the same symbol X and its domain by $\mathcal{X}$. Suppose G on $\mathcal{X}$ and $\bar{G}$ on Θ are transitive and $\mathcal{X},\Theta,G$ and $\bar{G}$ are isomorphic to each other. We may then regard x and θ as elements of G and perform the group operations (multiplication and inversion) on these. If the set-up is invariant, taking $\bar{g} = \theta^{-1}$ in (4.4.3) and denoting the identity in $\bar{G}$ (i.e. Θ) by $\bar{e}$, we get

$$f(x|\theta) = f(\theta^{-1} \cdot x|\bar{e}) \cdot \left| J\left(\frac{\theta^{-1} \cdot x}{x}\right) \right|. \qquad (4.4.10)$$

The probability element of X under θ is obtained by supplying the differential dx in (4.4.10). As noted in section 4.3, in this case there exists an (essentially) unique invariant pivot which can be taken in the form $U(x,\theta) = \theta^{-1} \cdot x$ (see footnote 21 in which we can set $\theta_0 = \bar{e}$). The probability element in the distribution of $U(X,\theta)$ is obtained by making the transformation $x \to u = \theta^{-1} \cdot x$ in the probability element of X under θ and by (4.4.10) this is just $f(u|\bar{e})\,du$. The fiducial distribution is the distribution of θ derived from this by the back transformation $u \to \theta = x \cdot u^{-1}$ and has probability element

$$f(\theta^{-1}x|\bar{e}) \cdot \left| J\left(\frac{\theta^{-1} \cdot x}{\theta}\right) \right| d\theta. \qquad (4.4.11)$$

---

[28]In fact Jeffreys (1940) sought to provide a Bayesian justification of the fiducial solution to the Behrens–Fisher problem of Example 4.3.7 in this way (this was disapproved by Fisher), by assuming that independently each pair of parameters (μ,σ) is subject to a prior of the form (4.4.9).

On the other hand by (4.4.10), the posterior distribution of θ under the prior $\pi(\theta)$ has probability element

$$\frac{f(\theta^{-1}\cdot x|\bar{e})|J(\theta^{-1}\cdot x/x)|\pi(\theta)\,d\theta}{\int_{\Theta} f(\theta^{-1}\cdot x|\bar{e})|J(\theta^{-1}\cdot x/x)|\pi(\theta)\,d\theta}. \tag{4.4.12}$$

If in this we take

$$\pi(\theta) = c(x)\cdot\left|J\left(\frac{\theta^{-1}\cdot x}{\theta}\right)\right|\cdot\left|J\left(\frac{\theta^{-1}\cdot x}{x}\right)\right|^{-1}, \tag{4.4.13}$$

the factor $c(x)$ which depends only on x cancels out between numerator and denominator and (4.4.12) reduces to (4.4.11) and the posterior distribution of θ becomes same as the fiducial distribution. It is easy to check that in Example 4.4.1, (4.4.13) reduces to (4.4.9).[29]

| Non-informative prior *á la* Jeffreys |

The above group-theoretic approach to the determination of a non-informative prior is, however, of limited scope: for many problems a group G which represents the symmetries of the set-up and can be used to define uniquely a meaningful prior as above is not available. To remedy this, Jeffreys suggested an alternative approach for deriving non-informative priors based on the 'amount of information' (AI) function for the set-up. When θ is real-valued and appropriate regularity conditions hold (cf. (4.3.13)), the AI function exists and has the form $I(\theta) = -E_\theta \partial^2 \log f(X|\theta)/\partial\theta^2$. Jeffrey's suggestion was to take

$$\pi(\theta) = \text{const.}I^{\frac{1}{2}}(\theta). \tag{4.4.14}$$

We may intuitively motivate this as follows: The relative amounts of information at different θ-points provided by the model $f(x|\theta)$ are given by the relative values of $I(\theta)$. Therefore as soon as we choose the model $f(x|\theta)$ we tacitly presume that a point θ is subjectively more or less probable according as $I(\theta)$ is high or low. (If our subjective assessment of prior probabilities is totally at variance with values of $I(\theta)$, the model would be regarded as hardly appropriate. But of course one could counter this by saying that in practice the model is usually chosen by considering only the nature of the data-generating process.) In any case the prior (4.4.14) has the advantage that if we re-parametrize by replacing θ in the model by a one-one transform η, where $\theta = h(\eta)$ is a smooth function (this is simply a re-parametrization, not necessarily a transformation of the set up as earlier), the AI function changes to $I(h(\eta))|h'(\eta)|^2$ so that whether we determine the prior

---

[29]It may be shown that generally if $\pi(\theta)$ is the right Haar density over Θ, (4.4.13) holds. The proof of this is contained in the proof of Result 3 in Berger (1985*a*) pp. 410–412.

from this new AI function or effect the transformation $\theta \to \eta$ in $\pi(\theta)$ given by (4.4.14), we get the same result (this accounts for the index $\frac{1}{2}$ in (4.4.14)).

Example 4.4.2. Consider a discrete random variable X subject to the binomial model

$$f(x|\theta) = \binom{n}{x} \theta^x (1-\theta)^x, \quad x \in \{0, 1, \ldots, n\}, \qquad \theta \in (0, 1). \tag{4.4.15}$$

Here the set-up is invariant under the group G consisting of the identity and the transformation $g \cdot x = n - x$ (of order 2) with $\bar{g}\theta = 1 - \theta$; but this does not define a prior uniquely via (4.4.5). Here $I(\theta) = n\{\theta(1-\theta)\}^{-1}$. Hence (4.4.14) gives the non-informative prior $\pi(\theta) = \text{const.}\{\theta(1-\theta)\}^{-1/2}$.

When θ is multidimensional, say, $\theta = (\theta_1, \ldots, \theta_k)'$ and usual regularity conditions hold, instead of a scalar AI function we have a positive definite information matrix $(I_{ij}(\theta))$, where $I_{ij}(\theta) = -E_\theta \partial^2 \log f(X|\theta)/\partial\theta_i\partial\theta_j, i, j = 1, \ldots, k$. Jeffreys suggested for this case the generalized form

$$\pi(\theta) = \text{const.}|I_{ij}(\theta)|^{1/2}. \tag{4.4.16}$$

This is also equivariant under one-one re-parametrization.[30]

Example 4.4.3. Consider the set-up of Example 4.4.1. $X_1, \ldots, X_n$ is a random sample from $\mathcal{N}(\mu, \sigma^2)$ with $\theta = (\mu, \sigma), \Theta = \{\theta : -\infty < \mu < \infty, 0 < \sigma < \infty\}$. It is readily seen that here

$$I(\theta) = n \begin{pmatrix} 1/\sigma^2 & 0 \\ 0 & 2/\sigma^2 \end{pmatrix}$$

Hence (4.4.14) gives the improper prior $\pi(\mu, \sigma) = \text{const.}/\sigma^2$, which coincides with the form (4.4.8) obtained earlier by the group-theoretic approach.

Non-informative priors derived as above depend on the experiment (through $\mathcal{X}$ with respect to which G is defined in the group-theoretic formulation and more explicitly, through the model $f(x|\theta)$ in Jeffreys' formulation). This may look somewhat anomalous since θ which represents our inductive concern is more primary than the experiment projected. (For example, our prior assessment about the value of θ of the probability of a head for a coin, should not depend on whether we plan a binomial or a negative binomial experiment with it.) Box and Tiao (1973) p. 44 make the interesting observation that a non-informative prior does not represent total ignorance but 'an amount of prior information which is small relative to what the particular projected experiment can be expected to provide' and as such non-informative priors appropriate for different experiments may be different. Whether we accept this explanation or not, it is clear that the use of such a non-informative prior would make the

---

[30] A motivation for (4.4.14) and (4.4.16) can be given by interpreting these as flat priors in terms of a sort of Riemannian metric; see Kass and Wasserman (1996), section 3.6.

posterior distribution (4.4.1) depend on the experiment and thus the likelihood
principle (subsection 4.3.4) may not hold across experiments.

| Interpretation of improper priors |

In the above we saw that non-informative priors, particularly in situations where Θ is unbounded, often turn out to be improper, i.e. the total measure of Θ becomes $+\infty$. Although the posterior distribution of θ based on such an improper prior may be a proper probability distribution, how are we to live down the dubious antecedent of such a posterior? At least three different explanations for the use of the improper priors have been advanced. *Firstly*, there is the limiting explanation. Considering the case of a continuous θ, this means that the improper prior density $\pi(\theta)$ can sometimes be interpreted as a sort of pseudo-limit based on a sequence of proper prior densities $\pi_n(\theta), n = 1, 2, \ldots$ in the sense that for a suitable sequence of scalars $c_n > 0, \pi(\theta) = \lim c_n \pi_n(\theta)$ for every θ. Of course the posterior (4.4.1) would be same whether it is based on $\pi_n(\theta)$ or $c_n \pi_n(\theta)$, so that the posterior computed from $\pi(\theta)$ can be interpreted as the limit of the posteriors based on the proper priors $\pi_n(\theta)$. (For example, when X is subject to the location model $f(x - \theta)$ with $-\infty < \theta < \infty$, invariance considerations lead us to the improper prior $\pi(\theta) = 1$. Taking the sequence of rectangular priors $\pi_n(\theta) = (2a_n)^{-1}, -a_n < \theta < a_n$ where $a_n > 0$ and $a_n \to \infty$, we can choose $c_n = 2a_n$.) However, whether or not we regard this as vindicating an improper prior, we would see in the context of testing that this ploy does not always work. *Secondly*, we have the pragmatic interpretation suggested by Savage (1962) and extensively worked upon by Box and Tiao (1973). According to it, $\pi(\theta)$ is really a proper density which has the form assumed over some finite range in which the likelihood $f(x|\theta)$ is non-negligible and suitably tapers off to zero as θ moves beyond that range. This of course presumes that we have a priori knowledge about the range beyond which the likelihood is negligible. A variant of this position is that taken by Berger (1985a) p. 90, according to which an improper prior is regarded as an *ad hoc* approximation to some proper prior. *Thirdly*, we can invoke the Renyi axiom system for probability (subsection 3.5.3) with Θ and other intractable subsets of Θ excluded from the class of conditioning events. Thus the improper density $\pi(\theta) = 1, -\infty < \theta < \infty$ can be interpreted as implying a uniform conditional distribution of θ given $\theta \in [a, b]$ for every finite interval $-\infty < a < b < \infty$. Of course this will mean the posterior distribution also can be represented only in terms of conditional probabilities (given that θ lies in one or other tractable subset) and we can no longer speak of the absolute posterior distribution given $\theta \in \Theta$.[31]

---

[31] It has been shown that sometimes posteriors based on improper priors can also be generated by finitely additive proper priors (Heath and Sudderth (1978)). On the other hand it has been found (Guglielmi and Melilli (1998)) that an attempt to construct the latter through truncation and limit often yields meaningless finitely additive priors in which the entire mass lies at the boundary of Θ.

Apart from the difficulty of interpretation, improper priors sometimes give rise to certain technical problems. These arise generally in the case of multi-dimensional θ because $\pi(\theta)f(x|\theta)$ cannot be regarded as representing the joint probability distribution of θ and X in the usual sense and as such standard manipulation of all types cannot be performed on it. Specifically, one has to be careful while deriving the marginal posterior distribution of a subvector of θ (see e.g. Robert (1994) p. 131). Another kind of difficulty with non-informative priors in general in the case of multi-dimensional θ is that a prior which looks fairly non-informative when we are interested in the full parameter vector θ, may imply the assumption of a contorted form of prior belief when we are interested only in certain functions of θ. This may lead, though unintentionally, to rather lop-sided inference about such functions. Thus on the basis of a sample from $N_k(\theta, I_k), \theta = (\theta_1, \ldots, \theta_k)'$, we may be interested in inferring about only the function $\eta = \sqrt{\sum \theta_i^2}$. For inferring about θ, the non-informative Jeffreys prior here is $\pi(\theta) = $ const. But this implies in terms of η, assuming a prior density of the form const.η^{k-1} which unduly favours large values of η. To get round such difficulties sophisticated extensions of the Jeffreys non-informative prior which take account of the differential importance of various parameters with reference to a particular inference situation, have been proposed and these have been called *reference priors* (Bernardo (1979), Robert (1994) pp. 117–120).

Partially informative prior

So far we have been considering determination of impersonal subjective priors to suit situations where we have no information about θ beyond the fact that $\theta \in \Theta$. However, sometimes it so happens that there is general consensus about certain features of the prior distribution. In particular there may be agreement about the prior mean values of certain functions (e.g. θ, θ^2 etc. when θ is real-valued). There we would want a prior which meets such constraints but incorporates as little as possible of further information. Rationality principles drawing upon concepts of Information Theory for choosing *partially informative* impersonal priors to suit such situations, have been considered by Jaynes (1983) and others. The first step here is to choose an initial prior $\pi_0(\theta)$ (by consideration of a transformation group or otherwise) which can be regarded as information-less. Then subject to the given constraints, one chooses the intended prior $\pi(\theta)$ so that its divergence or dissimilarity vis-à-vis $\pi_0(\theta)$ as represented by the 'Kullback–Leibler information measure' of π with respect to π_0 (see Rao (1965) pp. 47, 131)

$$-E_\pi \left\{ \log_e \frac{\pi_0(\theta)}{\pi(\theta)} \right\} \tag{4.4.17}$$

is minimized. In the particular case when $\Theta = \{\theta_1, \theta_2, \ldots, \theta_L\}$ is finite, the natural choice for π_0 is the uniform pmf $\pi_0(\theta_\ell) = L^{-1}, \ell = 1, \ldots, L$ and minimization of (4.4.17) is same as maximization of the entropy

$$-\sum_{\ell=1}^{L} \pi(\theta_\ell) \log_e \pi(\theta_\ell),$$

which may be interpreted as a measure of the 'degree of ignorance' about θ as represented by $\pi(\theta)$. Because of this and also because the negative of (4.4.17) is sometimes called entropy of $\pi(\theta)$ in a loose sense,[32] generally priors derived by minimizing (4.4.17) are called *maximum entropy* priors. It has been found that if a proper distribution $\pi(\theta)$ can be obtained in this way as a solution, then it would have the form

$$\pi(\theta) = \text{const.} e^{\sum_1^m \lambda_i \tau_i(\theta)} h(\theta), \tag{4.4.18}$$

where m is the number of linearly independent functions of θ whose mean values are determined by the constraints (leaving aside the constraint equating the total measure of Θ to 1), $h(\theta) \geq 0, \tau_1(\theta), \ldots, \tau_m(\theta)$ are functions of θ, and $\lambda_1, \ldots, \lambda_m$ are to be adjusted to meet the constraints. In the literature a pmf or pdf having the form (4.4.18) is said to belong to the *exponential family* with an m-dimensional sufficient 'statistic' $(\tau_1(\theta), \ldots, \tau_m(\theta))$.

Example 4.4.4. When the model involves a location parameter $\theta, -\infty < \theta < \infty$ and the problem is invariant under all translations, we can take $\pi_0(\theta) = 1$ as the non-informative prior density over the real line. There, if available knowledge suggests the constraints $\int_{-\infty}^{\infty} \theta \pi(\theta) = \mu, \int_{-\infty}^{\infty} (\theta - \mu)^2 \pi(\theta) = \sigma^2$, minimization of (4.4.17) gives the density of $N(\mu, \sigma^2)$ as the solution for $\pi(\theta)$ (see Rao (1965) pp. 131–132, Berger (1985) p. 93, Robert (1994) p. 94).

Vindication of impersonal Bayesian analysis

The determination of an impersonal subjective prior (whether non-informative or partially informative) rests on certain accepted rationality principles. In any situation there should exist a unique such prior as the correct choice in the context. The normative approach followed leaves little scope for afterthought or disagreement here. Nevertheless, possibly because impersonal subjective priors often turn out to be improper, one can discern a tendency to regard Bayesian analysis based on such priors as an 'objective' technique which yields conclusions agreeing with our 'common sense' and which is vindicated inductively by studying the consequences of repeated application in various situations (cf. Box and

---

[32]It is well known that if π_0 and π both represent proper distributions over Θ, (4.4.17) is always non-negative. However, if π_0 is improper and π proper, a situation obtaining often when Θ is unbounded, (4.4.17) may assume negative values and may be even unbounded below. Since (4.4.17) is a directional measure of 'dissimilarity' and not a 'distance', this may not be entirely counter-intuitive; nevertheless one feels somewhat uncomfortable in recommending the minimization of such a measure. However, this poses no special problem in respect of the formal determination of $\pi(\theta)$.

Tiao (1973) p. 2). In particular, sometimes the procedures are judged in the light of behavioural reasonableness criteria like admissibility. From the philosophical point of view, however, this seems to be a bit anomalous and as showing up one's lack of faith in the stand originally adopted. If one accepts that θ is random with its distribution correctly given by the prior chosen, given the outcome of the experiment performed, any posterior probability statement stands on its own right. Then why bother about what rule the conclusion is based on and what would happen if the experiment were repeated and the same rule applied with θ held fixed? (However, the position of a behaviouralist who uses the prior merely as a weight function to generate a suitable rule and studies its frequentist behaviour is less vulnerable; see Welch and Peers (1963).)

Personal prior

4.4.5   We next come to the form of the objective–pro-subjective approach in which the prior input $\pi(\theta)$ represents *personal* subjective probabilities: it reflects the honest personal belief of the agent (experimenter or statistician) at the time considered. Since $\pi(\theta)$ originates in personal belief, naturally, no question of existence of a unique correct prior arises. It is conceded, or rather, proclaimed that subjectivity or personal opinion is inevitable in the process of induction and therefore such opinion should be deliberately and exogenously introduced as an input in the process. We will see that such a stance is very close to that taken in the purely subjective approach.

How should one determine $\pi(\theta)$ in practice? The only restrictions are: $\pi(\theta)$ should be such that $\bar{f}_\pi(x)$ appearing in the denominator of (4.4.1) and defined by (4.4.2) must be positive for all x which one regards physically realizable and $\pi(\theta)$ must represent a (finitely additive) proper probability distribution (this latter requirement is expressed by saying that $\pi(\theta)$ should give coherent probabilities; subsection 3.3.3). Apart from these, there is unlimited scope for pluralism here. The personal beliefs of the agent involved regarding the unknown parameter θ would have to be operationally extracted by the statistician by repeated (self-) questioning on various propositions as discussed in section 3.3. Care has to be taken that the answers to various such questions are mutually coherent. After enough features of the personal probability distribution of θ as believed by the agent, are quantified in this way, some convenient $\pi(\theta)$ which fits those to a close approximation is chosen. There is a lot of freedom as regards the choice of the features, the questions, and the method of fitting here. Of course some self-discipline is desirable to prevent complete anarchy and is also helpful operationally.

Natural conjugate priors

In the most popular approach one confines oneself to a reference family of priors $\pi(\theta|\lambda)$ having a common form and differing only in respect of λ which represents certain *hyper-parameters* (i.e. 'parameters' characterizing the prior distribution of the parameter θ). The numerical values of the hyperparameters are adjusted so that the prior realizes the ascertained features.[33] For certain models a convenient reference family is generated by regarding the kernel (the part obtained by deleting any θ-free factor) of $f(x|\theta)$ as a function of θ over Θ. We can then replace functions of x appearing at various places in the kernel by hyperparameters which possibly vary over wider domains and normalize suitably, if necessary, after bringing in an additional factor depending only on θ. Often it so happens that when a prior belonging to such a family is combined with the likelihood $f(x|\theta)$ in accordance with (4.4.1), the posterior derived also is another member of the same family and thus has the same form as the prior. When this occurs the family is said to represent the *natural conjugate priors* for the problem.

Example 4.4.5. Let X follow the binomial distribution with $f(x|\theta) =$ const. $\theta^x(1-\theta)^{n-x}, x = 0, 1, \ldots, n$. Considering the kernel $\theta^x(1-\theta)^{n-x}$ as a function of θ over $0 < \theta < 1$ and replacing x and $n - x$ by $\alpha - 1$ and $\beta - 1$ where $\alpha > 0$ and $\beta > 0$ are hyperparameters, we get the family of beta priors

$$\pi(\theta) = \text{const. } \theta^{\alpha-1}(1-\theta)^{\beta-1}, \quad 0 < \theta < 1, \tag{4.4.19}$$

where $0 < \alpha < \infty, 0 < \beta < \infty$. It is easily checked that the posterior obtained by combining any prior belonging to this family with $f(x|\theta)$, also belongs to the same family. Thus (4.4.19) constitutes the family of natural conjugate priors here.

Example 4.4.6. Let $X_1, \ldots, X_n$ be a sample from $\mathcal{N}(\mu, \sigma^2)$. Writing $\bar{x} = \Sigma x_i / n$, $v = \Sigma(x_i - \bar{x})^2$, we have

$$f(x_1, \ldots, x_n|\mu, \sigma^2) = \text{const. } e^{-n(\bar{x}-\mu)^2/(2\sigma^2)-v/(2\sigma^2)}(\sigma^2)^{-n/2}.$$

Now looking at the kernel in this as a function of μ and σ^2 and replacing $\bar{x}$ by λ_1, the n appearing in the exponent by λ_2, v by λ_3 and the $\frac{1}{2}n$ appearing in the power of σ^2 by $\lambda_4 + 1$, where $\lambda_1, \lambda_2, \lambda_3, \lambda_4$ are hyperparameters and bringing in an additional factor $1/\sigma$ we get the prior

$$\pi(\mu, \sigma^2) = \text{const.} \frac{1}{\sigma} e^{-\lambda_2(\mu-\lambda_1)^2/(2\sigma^2)-\lambda_3/(2\sigma^2)}(\sigma^2)^{-\lambda_4-1}, \tag{4.4.20}$$

---

[33] A variant of this which does not require ascertainment of any features of the prior is called the *empirical Bayes approach*. In it one somehow estimates λ by $\hat{\lambda}(x)$ on the basis of the observation x and the model for the unconditional distribution of X given λ. Then we draw conclusions from the pseudo-prior $\pi(\theta|\hat{\lambda}(x))$ by formally treating it as a true prior. But such an approach is behavioural in spirit. Procedures based on it have to be justified on the basis of their frequentist properties. As the prior depends on x the posterior probabilities here cannot be interpreted as conditional probabilities given x.

$-\infty < \mu < \infty, 0 < \sigma^2 < \infty$, where for the integral to converge, we require $-\infty < \lambda_1 < \infty, \lambda_2 > 0, \lambda_3 > 0, \lambda_4 > 0$. Clearly this means the prior distribution of μ, σ^2 is such that given σ^2, conditionally μ is $\mathcal{N}(\lambda_1, \sigma^2/\lambda_2)$ and marginally λ_3/σ^2 follows a gamma distribution with shape parameter λ_4. It may be checked that the posterior distribution of μ, σ^2 also has the same form so that (4.4.20) represents a natural conjugate family.

Generally when the distribution of X belongs to the exponential family, we can write the kernel of $f(x|\theta)$ in the 'elaborate' exponential form

$$e^{\sum_{i=1}^{m} T_i(x)\tau_i(\theta)}, \qquad (4.4.21)$$

where $\tau_i(\theta), i = 1, \ldots, m$, are *linearly* independent. (But $T_i(x), i = 1, \ldots, m$, may not be linearly independent and some of these may even be identical. Also, for a fixed-sample size model some of these may be functions of the sample size, and hence, constant. Thus the dimension of the sufficient statistic for the above exponential family may be less than m, the number of functions T_i.) Replacing $T_i(x)$ by the hyperparameter $\lambda_i, i = 1, \ldots, m$, and bringing in an additional factor $h(\theta) > 0$, we get the natural conjugate family of priors

$$\text{const. } e^{\sum_{i=1}^{m} \lambda_i \tau_i(\theta)} h(\theta), \quad \theta \in \Theta, \qquad (4.4.22)$$

where $h(\theta)$ should be chosen and $\lambda_1, \ldots, \lambda_m$ given values so that the total integral (sum) is finite and may be equated to 1 by choosing the 'const.'. This has the same form as the impersonal subjective prior (4.4.18) derived by the principle of maximum entropy. (But note the difference in attitude (4.4.18) was introduced normatively and had a unique status in the context considered, whereas (4.4.22) is suggested tentatively as approximating the personal beliefs of a particular agent at a particular time.) In practice, the hyperparameters $\lambda_1, \ldots, \lambda_m$ have to be adjusted so that (4.4.22) realize the features of the prior as ascertained in the particular case. Usually the mean values of certain functions of θ are identified as the features of the prior. Standard choices of these functions require adjustment of the λ_i's so that certain lower order moments (means, variances etc.) or certain quantiles (median, quartiles etc.) have specified values.[34]

---

[34]More generally, for a given model, any family of priors for which the posteriors also belong to the same family is called a conjugate family. When a natural conjugate family exists, a wider such family may be constructed by taking all finite mixtures of its members.

The question of robustness

In the case of a personal subjective prior there does not exist any unique correct prior, and hence, strictly speaking, there is no possibility of error as regards the choice of the prior. Nevertheless, there may remain some doubt about the form of $\pi(\theta)$ worked upon, not only because another person or even the same person at another time may regard a different form to be more appropriate, but also because almost always there is some uncertainty about the closeness of fit of the form chosen to the actual belief probability distribution, particularly at the tails. For this reason it is always advisable to study the robustness of the posterior (or conclusions based on it) against departures from the form of the prior chosen (cf. section 2.5). (It should be noted, however, that this kind of robustness study is somewhat different from that undertaken in the context of the objective–behavioural approach where one may be in doubt about the appropriateness of the assumed model in representing objective reality. Here the doubt is about the appropriateness of the subjective input exogenously brought into the induction process.) Broadly there are two main approaches for studying and ensuring robustness here. In one, an insensitive prior is chosen so that the posterior is not too much affected by slight departures from it. Usually a heavy-tailed prior is found to be more insensitive in this sense. For example, in the case of a location parameter θ a Cauchy or t-form prior is found to yield a posterior which is more robust than a normal prior. An extension of this idea leads us to what are called *hierarchical priors*. In this case one starts with a family of priors $\pi(\theta|\lambda)$ characterized by the hyperparameter λ (often the natural conjugate family associated with the model $f(x|\theta)$) and then compounds by assuming a non-informative distribution for λ over its domain. Because of the non-informative element introduced, the resulting prior is generally heavy-tailed compared to the individual $\pi(\theta|\lambda)$s. In the second approach to robustness, some class Π of possible priors $\pi(\theta)$ is considered all through; the class of posteriors and derived conclusions resulting from the priors in Π are studied. The variation in these tells us about the extent to which these can be relied upon (Berger (1985a) section 4.7). Operationally, it is same as if we are dealing with imprecise probabilities which we briefly mentioned in section 3.11.

When the observable X represents a random sample from some fixed population, under broad conditions it is found that as the sample size increases, the influence of the prior on the posterior diminishes. Thus in the case of large samples it matters but little what prior we start with—the posterior and conclusions derived from it are practically same for all choices (provided, of course, the priors considered encompass, i.e. assign positive probabilities to all θ-values that may be true).

A unified approach

4.4.6 We next come to the methods appropriate for solving different types of problems of induction in the objective–pro-subjective approach. As noted earlier, the posterior distribution

(4.4.1) here represents a sort of master conclusion—answers to all types of problems of induction are derived from it. Thus unlike in the behavioural case where different types of problems require piecemeal methods, here we have a unified approach towards problems of all types. As in the instantial case, all inferences here are made specific to the realized value x and one could say that there is little point in studying how the procedures behave as the experiment is repeated; more data would accumulate through such repetition and the procedures themselves would get modified by the process.

| Point estimation | To get a point estimate or working value of θ in the case of a real θ, we can take any feature of $\pi(\theta|x)$ measuring |

its location, like mode, mean, or median. To take the mode means maximizing $\pi(\theta)f(x|\theta)$ with respect to $\theta \in \Theta$; the resulting estimate is often called the generalized maximum likelihood estimate (it reduces to MLE if $\pi(\theta) =$ a constant). We are led to the mean or the median when we regard the problem of point estimation as a decision problem (to be considered below) with squared error or the absolute error representing the loss. After a point estimate is obtained, the degree of assurance that we can attach to it depends on the degree of concentration of the posterior distribution around it. As usual this may be inversely measured by the root mean squared error, the mean value being computed with respect to the posterior distribution. One operational advantage in this approach is that the posterior distribution being fully known, the posterior root mean squared error is a computable quantity and does not require any secondary estimation as is the case with the behavioural approach where the S.E. often involves unknown parameters. If instead of θ we want to estimate a real-valued function $\xi(\theta)$ we may first derive the posterior distribution of $\xi(\theta)$ from $\pi(\theta|x)$ and then work on its basis. All this generalizes straightaway to the case of a vector θ.

| Large sample approximation | We note here a kind of approximate result that is available in large samples under broad conditions. Considering for simplicity again the case of a real θ, suppose |

we have a random sample of size n with observation x from the population corresponding to θ. Let $\hat{\theta}_\pi = \hat{\theta}_\pi(x)$ be the posterior mode given x under the prior π. When appropriate smoothness conditions held, $\hat{\theta}_\pi$ would be the solution of the equation

$$\frac{d}{d\theta} \log \pi(\theta|x) = 0. \tag{4.4.23}$$

It has been shown that for large n under broad conditions the *posterior distribution* of $(\theta - \hat{\theta}_\pi)$ can be approximated by the normal distribution with mean 0 and variance

$$\left\{ -\frac{d^2}{d\theta^2} \log \pi(\hat{\theta}_\pi|x) \right\}^{-1}, \tag{4.4.24}$$

for almost all observations x; further, for large n, it does not matter what prior π (within a broad class) we take here (since corresponding different solutions of (4.4.23) become equivalent upto order $n^{-1/2}$ and (4.4.24) tends to be same upto order n). This result is a version of what is known in the literature as the *Bernstein–von Mises Theorem* (see Lecam and Yang (1990) Chapter 7 and Ghosal *et al* (1995) where a more general formulation is given).

The similarity of (4.4.23) with the ML equation (4.3.11) and of (4.4.24) with the estimated asymptotic variance of the MLE (subsection 4.3.5) is obvious; in fact, the two equations and variance expressions are formally same in the case of uniform prior. However, it should be remembered that the asymptotic normality here refers to the posterior distribution of θ given the observations x; in the case of MLE, the same refers to its sampling distribution, x being replaced by the corresponding random variable X.

Interval and set
estimation

The determination of an interval or set estimate for θ on the basis of the posterior distribution is also a straightforward exercise; we have only to demarcate an interval or set in Θ whose posterior probability content is not less than the specified level $1 - \alpha$. Such a set is called a *credible set* for θ (irrespective of whether the prior is impersonal or personal, although the term 'credibility' is generally used to mean impersonal subjective probability) and $1 - \alpha$ is called the *credibility level* of the set. Thus if C_x is a credible set at level $1 - \alpha$, we would have the direct probability interpretation

$$P_\pi \left(\theta \in C_x | x \right) \geq 1 - \alpha, \qquad (4.4.25)$$

P_π indicating that the (posterior) probability is computed under the prior π. (Contrast this with the roundabout interpretation of the confidence level of a confidence set in the behavioural approach as discussed in section 4.2. Here θ is random while x is fixed at the realized value, whereas for a confidence set θ is a fixed unknown entity, the randomness coming from X. Note also the similarity of the set estimate in the present approach and that obtained through fiducial reasoning in the instantial approach of section 4.3.) In practice, subject to (4.4.25) we would naturally like to minimize the size of the set C_x. This is achieved by taking C_x in the form

$$C_x = \{\theta : \theta \in \Theta \cdot \ \pi(\theta|x) \geq c\}, \qquad (4.4.26)$$

where c is chosen as large as possible subject to (4.4.25). Such a credible set is called HPD (Highest Posterior Density) credible set at level $1 - \alpha$. If θ is real-valued with an interval for Θ and $\pi(\theta|x)$ is bell-shaped, this reduces to an interval. From what we said earlier it follows that for many standard problems the HPD credible interval for a real parameter under an appropriate non-informative prior, is *formally* same as the corresponding confidence or fiducial interval.

Hypothesis testing

Testing a null hypothesis $H_0: \theta \in \Theta_0$ against $H_1: \theta \in \Theta_1 = \Theta - \Theta_0$ in the objective–pro-subjective approach is quite straight-forward. Given x, we find $\pi(\theta|x)$ for an appropriate prior $\pi(\theta)$ and hence compute $P_\pi(\theta \in \Theta_i|x), i = 0, 1$. H_0 is rejected or accepted for the given x according as the ratio $P_\pi(\theta \in \Theta_0|x)/P_\pi(\theta \in \Theta_1|x)$ is too low or too high compared to some threshold. The threshold may be taken equal to 1 or chosen suitably high or low depending on the relative gravities of inappropriately declaring that H_0 (or H_1) is true. (The actual choice of the threshold may be justified in terms of loss values through decision theory.) As a spin-off, here $P_\pi(\theta \in \Theta_i|x)$ has direct interpretation as the probability of H_i being true, $i = 0, 1$. As we saw in testing of significance under the instantial approach, the tail probability or P-value cannot be so interpreted, although when P is significant the degree of its smallness is an indicator of our reluctance to accept H_0. Another advantage is that unlike in the behavioural and instantial approaches (subsections 4.2.2 and 4.3.6), here we do not have to bother about which hypothesis $\theta \in \Theta_i, i = 0, 1$, is treated as the null hypothesis, for their posterior probabilities complement each other.

In this approach the problem of testing H_0 is somewhat delicate when θ is continuous and Θ_0 is a singleton $\{\theta_0\}$ so that H_0 is the point null hypothesis $\theta = \theta_0$ and H_1 is $\theta \neq \theta_0$. There π has to be so chosen that there is a positive mass say, h_0 representing $P_\pi(\theta = \theta_0)$. Denoting $1 - h_0$ by h_1 and the conditional density of θ given $\theta \neq \theta_0$ by $g_1(\theta)$ we then have

$$P_\pi(\theta = \theta_0|x) = \frac{h_0 f(x|\theta_0)}{h_0 f(x|\theta_0) + h_1 \int_{\theta \neq \theta_0} f(x|\theta)g_1(\theta)\, d\theta}. \qquad (4.4.27)$$

All this of course means, the choice of the prior is dictated by the testing problem at hand, which is logically somewhat anomalous. On the other hand it may be said that framing a point null hypothesis $\theta = \theta_0$ itself indicates that we have some prior information about θ which ascribes a special status to θ_0 and that should be taken into account while choosing π. In any case, since h_0 and $g_1(\theta)$ both have to be chosen, this brings in another element of arbitrariness with regard to the choice of the prior.

When no special information is available it has been suggested that we take $h_0 = h_1 = \frac{1}{2}$.[35] As regards $g_1(\theta)$, taking this equal to an improper non-informative prior creates difficulties. (For example $c_1 g_1(\theta)$, gives different results for different values of $c_1 > 0$. Also, the limit of results obtained for a sequence of proper priors tending to an improper $g_1(\theta)$ may not tally with that obtained for $g_1(\theta)$ directly.) Hence, generally, the use of only proper $g_1(\theta)$s is advocated.

---

[35]The posterior versus prior odds ratio of H_0 against H_1, i.e. $\{P_\pi(\theta = \theta_0|x)/P_\pi(\theta \neq \theta_0|x)\}/\{h_0/h_1\}$, or as it is called, the *Bayes factor* here reduces to $f(x|\theta_0)/\int_{\theta \neq \theta_0} f(x|\theta)g_1(\theta)d\theta$. To reject or accept H_0 according to whether this is too low or too high, is of course equivalent to taking $h_0 = h_1 = \frac{1}{2}$.

In the case of various problems involving real θ, for testing $H_0 : \theta = \theta_0$ versus $H_1 : \theta \neq \theta_0$, numerical studies have been made with $h_0 = \frac{1}{2}$ and different types of proper $g_1(\theta)$ (Cox and Hinkley (1974) pp. 395–398, Berger (1985a) pp. 148–156). These show that for realizations x giving significant values of the tail probability of standard test statistics, values of $P_\pi(\theta = \theta_0 | x)$ are generally much higher, particularly when the sample size is very large. These studies tend to create an impression as if standard tests for point null hypothesis in the frequentist formulation are unduly tilted towards rejection. However, we should remember that $h_0 = \frac{1}{2}$ is hardly a reasonable choice since 'null hypotheses are usually known in advance to be false and the point of significance tests is usually to find out whether they are nevertheless approximately true' (a remark made by Good (1983) p. 49 in a different context). Also, as we remarked in the context of our discussion of tests of significance as considered by Fisher (subsection 4.3.6), while testing H_0, often we do not have in our view the entire Θ but rather those alternatives under which the probability of rejection is close to 1. This seems to suggest that for comparison with tail probabilities we should have h_0 much less than $\frac{1}{2}$ and $g_1(\theta)$ should be skewed towards faraway alternatives. Another way to reduce the conflict between the two modes of inference would be to set more stringent norms for interpreting the realized tail probability P for larger sample sizes, i.e. to affirm for the same value of P a lower degree of significance when the sample size is larger (see Cox and Hinkley (1974) p. 397).

Decision problem

If we have a general decision problem (cf. section 4.2) with action space $\mathcal{a}$ and a real-valued loss function $L(\theta, a)$ defined on $\Theta \times \mathcal{a}$, we can solve it by extending the above ideas. Thus, after choosing a prior $\pi(\theta)$ and determining the posterior $\pi(\theta|x)$ as given by (4.4.1) on its basis, we can evaluate the *posterior expected loss* given x

$$\rho_\pi(a|x) = E_\pi(L(\theta, a)|x) = \sum_{\theta \in \Theta} L(\theta, a)\, \pi(\theta|x) \quad \text{(for discrete } \theta\text{)}$$

$$\text{or} \int_\Theta L(\theta, a)\, \pi(\theta|x) d\theta \quad \text{(for continuous } \theta\text{)},$$

for every action $a \in \mathcal{a}$. We now invoke the *rational expectation principle* (this is called the principle of rationality of type I by Good (1983) p. 153) according to which: *given an observation x an action a_1 is preferred to another action a_2 if and only if the posterior expected loss given x of a_1 is less than that of a_2.* (If the posterior probability distribution of θ had a frequentist interpretation this principle could be justified by interpreting the expectation as long-term average on repeated observation. Here, as the posterior distribution, is a subjective probability distribution, the principle has to be extraneously brought in. Usually it is assumed that loss (= $-$utility) and subjective probabilities are so defined that the principle holds. The principle induces a total ordering of the actions in $\mathcal{a}$ for each x.[36]) If given x an action a_x which minimizes $\rho_\pi(a|x)$ over a exists, that would represent the most preferred action. (Unlike in the behavioural case, here we do not have to determine a decision rule over entire $\mathcal{X}$; however, when a_xs corresponding to different xs exist, piecing these together, we get such a rule.)

Discussion   **4.4.7** Before concluding our discussion of the objective–pro-subjective approach we mention some of the plus and minus points that can be voiced for and against it.

On the positive side, the most attractive feature of this approach is that as in the instantial approach here also we work under the conditional set-up given the particular instance at hand, but unlike in the latter, here we have a unified philosophy for tackling all types of problems of induction. Uncertainty of induction here relates directly to the conclusion reached and is expressed straightaway in terms of probability. Also in many situations significant prior information is available and in this approach we have a ready means of utilizing that information. In the other approaches either there is no scope for such utilization, or in course of attempting that, we encounter heavy weather. In particular, this is true of information which restricts θ to only a part of its natural domain. Also, as noted earlier, when the sampling experiment involves (knowingly or unknowingly) an optional stopping rule or some form of censoring which does not throw any light on the inductive concern, that is readily taken care of in the present approach whereas that may engender serious complications in the other approaches. Another advantage is that in this approach 'data snooping' in a direct manner is permissible—if a parametric function is suggested by inspecting the data, one can infer about it as usual. For example, while comparing several treatments, one can test the significance of and estimate the difference between the effects of the two treatments having the highest and lowest mean yields in the sample observed. Use of the usual procedures of inference in such cases would be sacrilegious in the sampling theory approach as the sampling distribution of the corresponding statistics would be affected. (Roundabout multiple comparison procedures (see e.g. Scheffé (1959)) have been developed in the sampling theory approach to tackle such problems but these always involve some sacrifice in terms of the strength of the conclusion.) Generally speaking, the great advantage of pro-subjective approach is its flexibility. Even when the evidence is a collection of different types of data coming in various ways from diverse sources, in this approach induction is usually feasible without too much headache; in the other approaches, one would often be hopelessly bogged down in such situations.

---

[36] As we mentioned earlier (see section 3.3), in certain developments of personal subjective probability, one starts from such a total ordering of actions as a primitive notion and shows that this implies the existence of subjective probabilities and utilities (or losses) in conformity with the rational expectation principle. In the application of decision theory it is always the other way round.

On the negative side, firstly, the ambiguity and element of arbitrariness as regards the choice of the prior always stare us in the face in this approach. True, some subjectivity is unavoidable in induction but it is one thing to have a plank of uncertain strength at the plinth level of a house and another to have it at the rooftop. In the behavioural approach if we feel that the level of significance chosen for a test is too large, we can always lower it without much trouble. In the pro-subjective case if we want to revise the prior originally chosen later, the whole structure tumbles down; buttressing with robustness studies may be laborious and not always satisfactory. Also, compared to the other objective approaches, here, if a personal subjective prior is being chosen, there is much more scope for dishonest manipulation to reach a preconceived conclusion.

Another difficulty may be discussed with reference to the decision problem. To follow the pro-subjective approach, as we have seen, amounts to effecting a total ordering of all actions $a \in a$ in terms of their posterior expected losses. Our only concern, however, is to choose a best action in a—to import subjective information in the form of a precisely defined prior for achieving that seems to be more than what is necessary. Some advocates of this approach contend that the prior defined need not be precise; it is enough if we can say that it belongs to some a priori class Π of possible priors as in certain robustness studies. Then posterior probabilities, and hence expected losses would be interval-valued and actions would only be partially ordered (Good (1983) pp. 152–153). This is similar to the second approach for studying the robustness of Bayesian procedures mentioned earlier (subsection 4.4.5). But such a strategy may make the choice of a best action also ambiguous. Using the notation (4.4.26), what we want is an action a_x such that $\rho_\pi(a_x|x) \leq \rho_\pi(a|x)$, whatever $a \in a$, for all π belonging to a class of priors. It is doubtful whether in general such a class can be meaningfully defined independently of x.

The derivation of the posterior distribution (4.4.1) rests heavily on the applicability of the multiplication formula for joint probability when probability and conditional probability are subjectively determined e.g. on the basis of fair betting odds (section 3.3). An examination of the proof of this as given in subsection 3.5.4 (see also de Finetti (1937) p. 109 and Lindley (1965) Part I, pp. 34–36) shows that what is proved is if a person specifies all three of $P(A \cap C), P(C)$, and $P(A|C)$ in this way, then these must satisfy $P(A \cap C) = P(C) \cdot P(A|C)$. But it is conceivable that the person may refuse to specify any 'conditional bet' for A given C right now. (After all, it is quite rational to say 'no comments' to certain questions!) But even if we leave aside this point, the definition of $P(A|C)$ in terms of a conditional bet about a future event A made at the present time when whether C would or would not occur is not yet settled, leaves open the possibility that the person would revise the odds on A later after learning that C has actually taken place. The Dutch-book arguments refer to synchronic bets made at a particular time. Unless a person makes an affirmation about any contingent event at the present time and also a commitment to stick to such affirmation in future, Bayes's theorem cannot be used to update the belief probability distribution of θ on the basis of observations on X. (This point originally made by Hacking in 1967 has been the source of a lot of controversy; see the discussion in Howson (1995) and also Lad (1996) pp. 164–165.) In practice it is not rare to have a situation like this: the prior is formulated on the basis of the initial body of evidence, but after the data are collected they are found to be strongly inconsistent with that prior; the initial evidence is then reviewed and it is decided that its earlier interpretation was faulty; the prior is then replaced by a revised one under which the data would be more probable. However, such a course violates established Bayesian credo. It seems these questions all relate to the basic tenet of the objective–pro-subjective approach, namely, that all forms of uncertainty over all times can be quantified in terms of probability obeying the usual calculus. One has to accept this tenet unquestioningly if one is to follow the objective–pro-subjective path to induction.

4.5   Subjective approach

| Distinguishing features |

4.5.1   In Chapters 2 and 3 we discussed in general terms the special features of the purely subjective approach to statistical induction, which distinguish it from different forms of objective approach. It would be convenient to recapitulate those before taking up further details of the subjective approach. As earlier, we denote the sampling experiment (i.e. the set of operations generating the observations) by $\mathcal{E}$. The outcome of $\mathcal{E}$ is represented by the observation of a set of real or vector-valued random variables which we now denote in full by $X_1, \ldots, X_n$.

(i) As we emphasized in Chapter 2, in the subjective approach we do not assume that $\mathcal{E}$ is even conceptually repeatable. However, we invariably assume that there is scope for extending the evidence, i.e. there exist certain further observables $X_{n+1}, X_{n+2} \ldots$ beyond those covered by the sampling experiment performed, where of course all observables, realized or unrealized, relate to the same physical entity.

(ii) In the subjective approach uncertainty or randomness (as quantified in terms of probability) with regard to the observables occurs exclusively due to the absence of knowledge about their values before observation. Only subjective probability of the personal kind is admitted here. The probability distribution of any set of observables thus represents only the opinion or degree of belief about their values as held by the agent (experimenter or statistician) at the time considered and does not have any existence outside of the mind of the agent. Proponents of the subjective approach, deny the existence of objective probabilities and would interpret even a probability derived from consideration of frequencies in personal subjective terms (cf. de Finetti (1937) pp. 153–154, Savage (1954) pp. 51, 56). This uncompromising attitude as regards the nature of probability is a characteristic which marks the purely subjective approach

apart from the objective–pro-subjective approach, discussed in the preceding section. Clearly this attitude can be regarded as antipodal to the
position taken in the objective–behavioural approach which admits only
the frequency interpretation of probability.

(iii) The probability model representing the distribution of all the random
variables considered (prior to observation) is supposed to be completely
known to the agent. No *unknown* parameters are involved in the model
(out-and-out subjectivists do not admit parameters since no measurement operations can be specified for these). There is therefore no question
of inferring about such parameters. This is a distinctive feature of the subjective approach as opposed to all forms of objective approach. The fully
specified model would of course be agent-specific—different persons may
have different models for the same experimental situation. (Sometimes
however, it is helpful to construct the model for the observables in stages,
first given certain parameters, and then, summing or integrating over
those parameters with respect to some known prior personal probability
distribution. As we briefly mentioned in section 3.8 and would elaborate further in the present chapter, when the random variables observed
form part of an infinitely exchangeable sequence of random variables,
the model in the subjective case can always be constructed in terms of
such notional parameters. Naturally, these notional parameters are also
agent-specific.[37])

(iv) Induction, in the subjective approach, is ultimately concerned with one or
more of the unrealized observables, say, $X_{n+1}, \ldots, X_{n+m}, m \geq 1$. No general causal laws are conceived for the phenomena of the external world.
What passes by the name of a general law is merely a consensus to which
all knowledgeable people attribute a very high personal probability. Thus
the subjective approach is concerned solely with *particular induction* or
eduction; no question of general induction arises here (cf. sections 2.4
and 3.3). For example, given the yields of an agricultural treatment in a
number of plots, here a person does not attempt any conclusion about
the mean of any conceptual population of yields, but merely gives his
assessment of the prospective yield of a similar plot (or the average yield
of, say, a hundred or thousand similar plots) under the treatment. (Even
when the model is constructed in stages through notional parameters, it
would be usually meaningless to communicate the inductive conclusion

---

[37]One situation in which such two-stage construction of the model would particularly
facilitate thinking is when one has to make a personal choice of the design for an
observational programme. Thus when several treatments are to be compared on the
basis of either a completely randomized or randomized block experiment, one can think
in terms of notional effects of treatments (which for a particular agent would be same
and subject to the same prior for both experiments) and construct comparable models
for the two prospective experiments.

in terms of such parameters as these would vary from person to person. Since in the subjective approach, the observables, once realized, do not have any meaningful probability distribution, the conclusion here has to be ultimately expressed in terms of the unrealized observables. In the following development we generally follow Lad (1996) to discuss the problem of induction straightaway in terms of the unrealized observables.[38])

It may be remarked that although subjectivists in general deny only the existence of probability outside of the agent's mind, the subjectivist position would suit even an out-and-out empirical idealist who does not admit the existence of independent material causes behind sense perceptions and for whom the external world does not exist outside one's mind. Instead of interpreting that all the observables relate to the same external real entity, he would say that they seemingly do so.

Conditional distribution of interest

4.5.2   In the subjective approach, given the values $x_i^{(0)}$ realized by the observed random variables $X_i, i = 1, \ldots, n$, induction about the prospective value of some function $Y = Y(X_{n+1}, \ldots, X_{n+m})$, (like $X_{n+1}, \sum_{n+1}^{n+m} X_i$, or $\max_{n+1 \leq i \leq n+m} X_i$) of the unrealized random variables $X_{n+1}, \ldots, X_{n+m}, m \geq 1$ is achieved in a comprehensive way as soon as the agent's conditional personal probability distribution of Y is ascertained. (Although our language and notation may suggest that $X_{n+1}, \ldots, X_{n+m}$ belong to the future, the discussion applies equally to situations where these belong to the unobserved past or present.) Denoting the pmf or pdf of the conditional personal probability distribution of Y, given $X_i = x_i^{(0)}$, $i = 1, \ldots, n$, by

$$f(y|x_1^{(0)}, \ldots, x_n^{(0)}), \quad y \in \mathcal{Y}, \tag{4.5.1}$$

where $\mathcal{Y}$ is the domain of Y, the agent's task thus is to derive (4.5.1) in a coherent way. To derive this the agent can follow the operational procedure for determining the personal probability of a proposition in terms of fair betting odds as described in Chapter 3 (subsection 3.3.3). When $\mathcal{Y}$ consists of a finite small number of values, the propositions assigning each such individual value to Y may be judged in a coherent way. Otherwise, propositions corresponding to suitable classes (intervals) of values may be judged and the histogram so obtained may be graduated to get (4.5.1).

The programme chalked out above looks very much like an exercise in applied psychology. When an expert's opinion is sought on, say, the likely demand of a product in the market in the coming month or the likely price of a stock on the following day, it seems the expert's mind works implicitly somewhat along the

---

[38]In this regard the position of Savage (1954) and Lindley (1965) seems to be somewhat less extreme; for instance, Savage (1954), Chapter 7, considers 'partition problems' in which the loss function involves unobservable parameters.

above lines. However, generally such a programme of statistical induction would be unattractive for several reasons. Firstly, the experimenter's or statistician's words are never treated like oracles; he has not only to arrive at the inductive conclusion but also to convince others about its reasonableness. Secondly, while it is impossible to banish subjectivity altogether from the inductive process, to avoid the charge of unfair manipulation it is desirable that all subjective inputs in the process be laid bare, as far as possible, before making the observations. In other words, the procedure to be followed for drawing conclusion should be spelt out before the values $x_1^{(0)}, \ldots, x_n^{(0)}$ are observed—a principle which is clearly violated in the above programme. And thirdly, before the experiment giving the values of $X_1, \ldots, X_n$, the experimenter must have had some belief about the joint distribution of $X_1, \ldots, X_n, X_{n+1}, \ldots, X_{n+m}$, and hence (or may be, directly), about the joint distribution of $X_1, \ldots, X_n, Y$. Direct determination of (4.5.1) leaves no scope for drawing upon that belief. For these reasons usually the following alternative programme of induction is considered. We would see later that this has the added advantage that it allows us to exploit our judgement about the symmetries among different observables and thus helps us to attain inter-personal concurrence to some extent as regards the conclusion.

Let $f(x_1, \ldots, x_n, y)$ represent the agent's joint probability distribution (pmf or pdf) of $X_1, \ldots, X_n, Y$ before realization of $x_1^{(0)}, \ldots, x_n^{(0)}$. This may be ascertained directly or may be obtained through marginalization after ascertaining the joint distribution of $X_1, \ldots, X_n, X_{n+1}, \ldots, X_{n+m}$ as represented over the appropriate domain by

$$f(x_1, \ldots, x_n, x_{n+1}, \ldots, x_{n+m}).\tag{4.5.2}$$

(We use $f(\cdot)$ generically to denote the pmf or pdf of any distribution; different distributions are distinguished, unless it is otherwise essential, solely by the arguments of $f(\cdot)$.) To ascertain any joint distribution we have to proceed as earlier by finding coherently the fair betting odds for different propositions about the values of the random variables involved.

Given $f(x_1, \ldots, x_n, y)$ over the entire domain the marginal joint distribution of $X_1, \ldots, X_n$ is represented by

$$f(x_1, \ldots, x_n) = \int_y f(x_1, \ldots, x_n, y)\, dy,\tag{4.5.3}$$

where the integral will be replaced by a sum in the discrete case. (Clearly, without loss of generality, we can take the domain $\mathcal{Y}$ of Y as same, whatever $x_1, \ldots, x_n$.) Then, coherence of the joint distribution $X_1, \ldots, X_n, Y$ implies that given any $(x_1, \ldots, x_n)$ for which $f(x_1, \ldots, x_n) > 0$, the conditional distribution of Y is represented by

$$f(y|x_1, \ldots, x_n) = \frac{f(x_1, \ldots, x_n, y)}{f(x_1, \ldots, x_n)}, \quad y \in \mathcal{Y}.\tag{4.5.4}$$

The expression (4.5.4) gives the conditional distribution for all relevant points $(x_1, \ldots, x_n)$; in particular (4.5.1) can be obtained by plugging in the observation $(x_1^{(0)}, \ldots, x_n^{(0)})$ in it. Note that (4.5.4) really follows from the definition of conditional probability in terms of conditional bets and the multiplication formula for compound probability (subsection 3.5.4). The random variable Y cannot be generally considered as an antecedent 'cause' of $(X_1, \ldots, X_n)$ here. Nevertheless, since different realizations of Y represent mutually exclusive and exhaustive events, it is customary to say that (4.5.4) follows by Bayes's Theorem, and by the same token, to call the subjective approach to induction as subjective Bayesian. We would follow this convention and call (4.5.1) as the *posterior* distribution of Y.

As noted above, the derivation of (4.5.1) via (4.5.4) gives one an opportunity to incorporate one's pre-experimental beliefs about the joint distribution of all the variables. Of course ascertainment of $f(x_1, \ldots, x_n, y)$ over its entire domain may be considerably more difficult than that of $f(y|x_1^{(0)}, \ldots, x_n^{(0)})$ over the domain of y. On the other hand the exercise of determining of $f(x_1, \ldots, x_n, y)$ coherently over the entire domain would help one to know one's mind better and put a check on wayward assessments one might make looking at $f(y|x_1^{(0)}, \ldots, x_n^{(0)})$ alone. However, the difficulty about applying the conditional probability assessed prior to the occurrence of the conditioning event to the post-event situation, which we mentioned at the end of subsection 4.4.7 in the context of the pro-subjective approach, again pops up its head here. One has to commit that one would abide by one's earlier affirmation in the post-event scenario (in this connection see also de Finetti (1975) p. 203).

| Types of induction | **4.5.3** After ascertaining the posterior distribution |

4.5.3 After ascertaining the posterior distribution (4.5.1) of Y (for simplicity, we now take Y to be real-valued), one can manipulate it in the traditional way to derive various measures and answer different inductive questions about Y (cf. Geisser (1980)). Thus if one is interested in point prediction of the prospective value of Y given the observations $x_1, \ldots, x_n$ (now onwards we write x_i for $x_i^{(0)}, i = 1, \ldots, n$), use may be made of the subjective mean value (*prevision*)

$$E(Y|x_1, \ldots, x_n), \tag{4.5.5}$$

or of alternative measures like the median or the mode of Y based on (4.5.1).[39] If one wants to predict by an interval, given any credibility level $0 < 1 - \alpha < 1$,

---

[39]However, de Finetti (1974) pp. 98, 207 does not sanction the use of the posterior distribution of Y, and in particular, of the prevision (4.5.5) for predicting Y. According to him, in prediction we try to hit the unknown true value of Y whereas the prevision of Y is merely the price we are prepared to pay now for an uncertain prospective gain Y (provided, it is not too large); the predicted value must lie in the domain of Y, but the prevision may not; also, unlike what is desirable for the predicted value, the prevision of a non-linear transform of Y may not equal the same transform of its

we can find (y_ℓ, y_u) such that

$$P(y_\ell \leq Y \leq y_u | x_1, \ldots, x_n) \geq 1 - \alpha. \tag{4.5.6}$$

When (4.5.1) is bell shaped, the shortest such interval can be found (as in sub-section 4.4.6) by taking only those points y at which the ordinate does not fall below some suitably chosen threshold. By the very nature of the approach there is no absolute standard for comparing the inductive abilities of two persons. We can make empirical comparison when the actual value of Y can be observed; at the most, we can evaluate the prediction of each person in terms of the posterior distribution of the other (see Lad (1996) Chapter 8).

In the subjective approach no unknown parameters are postulated and there-fore there is no question of testing hypotheses in the traditional sense. However, multiple-action decision problems are very much relevant. If a denotes the set of available actions the loss resulting from the choice of an action $a \in a$, may be determined by the unknown true value y of the unrealized quantity Y. Thus the appropriate notation for the loss function here would be $L(y, a)$ defined on $\mathcal{Y} \times a$. (Contrast this with the loss function $L(\theta, a)$ for the decision problem in the behavioural (subsection 4.2.4) and pro-subjective (subsection 4.4.6) approaches.) For example, a may be the set of possible values of the quantity of a product for which a store has to place orders in advance to meet the demand Y in the coming month; $x_1, \ldots, x_n$ may be the quantities of the product actually sold in the preceding n months. The rational course here would be to compute for each $a \in a$ the expected (or prevised) loss

$$\rho(a | x_1, \ldots, x_n) = E\{L(Y, a) | x_1, \ldots, x_n\}, \tag{4.5.7}$$

given $x_1, \ldots, x_n$ on the basis of (4.5.1) and invoke the rational expectation prin-ciple (subsection 4.4.6) to choose an order value a for which this is minimized (provided such a value exists).

| Exchangeability |

4.5.4 So far our development has been quite general and we have not assumed anything special about the structure of the model representing the joint distribution of the observables $X_1, \ldots, X_n, X_{n+1}, \ldots, X_{n+m}$. As we saw in sections 3.6–3.7 in the objective set-up great simplification in the form of the model occurs through incorporation of the properties of stochastic independence and identity of distribution of ana-logous random variables: whenever two or more random variables are causally unrelated they are assumed to be stochastically independent and whenever they are considered distinguishable only in terms of their spatio-temporal locations,

---

prevision. Nevertheless, in the subjective statistical context, prediction or forecasting is the primary problem of practical interest and it does not seem that there is any way to proceed without using the prevision or some other measure based on the posterior distribution of Y for that purpose.

they are taken to be identically distributed. This reduces the joint pmf or pdf of the random variables to the product of a number of factors having identical functional forms.

In the subjective case the concept of stochastic independence becomes irrelevant both from the practical and the theoretical point of view. As we discussed in section 3.6, subjectively viewed, stochastic independence of random variables relating to the same physical reality would be a rare occurrence in practice—knowledge of the values assumed by some of the variables would generally influence our beliefs about others, even when they seem to be causally unrelated to the former. And theoretically, when stochastic independence holds the problem of induction becomes vacuous. Thus when Y is stochastically independent of $(X_1, \ldots, X_n)$, from (4.5.4) we get that the conditional distribution of Y given the observation $(x_1, \ldots, x_n)$ is same as the marginal distribution of Y, so that there is no scope for learning from experience, or as de Finetti (1975) p. 208 expresses it, 'the problem of inference does not exist'.

However the analogy or symmetry present among the observables is very much a part of the subjective judgement of the experimenter and one would naturally like to draw upon that at the time of induction. This is done by assuming that the joint distribution of all observables (those to be observed and those to be induced about) before the experiment is *exchangeable*. We already introduced the concept of exchangeability in a preliminary way in section 2.2 and in more detail in section 3.8. As defined in section 3.8, the joint distribution of $X_1, \ldots, X_n, X_{n+1}, \ldots, X_{n+m}$, given by (4.5.2), is exchangeable if each X_i has the same domain (before the experiment), say, $\mathcal{X}_1$ and

$$f(x_1, \ldots, x_n, x_{n+1}, \ldots, x_{n+m}) = f(x_{i_1}, \ldots, x_{i_n}, x_{i_{n+1}}, \ldots, x_{i_{n+m}}), \qquad (4.5.8)$$

for $x_i \in \mathcal{X}_1$, $i = 1, \ldots, n + m$ and any permutation $(i_1, \ldots, i_n, i_{n+1}, \ldots, i_{n+m})$ of the subscripts $(1, \ldots, n + m)$. The condition (4.5.8) would have considerable implication in restricting and simplifying the form of the conditional distribution (4.5.4).

Example 4.5.1. Suppose we have a finite population of known size N in which an *unknown* (and hence, subjectively random) number A of units are of a certain type. We can think of a lot of size N in which A units are defectives. A can have the values $a = 0, 1, \ldots, N$. The units are to be drawn randomly *without* replacement; the random variable $X_i = 1$ or 0 according as the ith draw produces a defective or not, $\sum_1^N X_i = A$. Clearly one can regard the random variables $X_1, X_2, \ldots, X_N$ as exchangeable. Suppose only the first n $(1 \leq n < N)$ draws are actually made. Let as usual x_i $(= 1$ or $0), i = 1, \ldots, N$ denote the values of $X_1, \ldots, X_N$; of these only $x_1, \ldots, x_n$ are actually realized. Write

$$P\left(\sum_1^N X_i = A = a\right) = \omega(a), \quad a = 0, 1, \ldots, N, \quad \sum_0^N \omega(a) = 1. \qquad (4.5.9)$$

$\omega(a)$s are known numbers. Clearly for any $(x_1, \ldots, x_n, x_{n+1}, \ldots, x_N)$, exchangeability implies that the joint pmf of $X_1, \ldots, X_n, X_{n+1}, \ldots, X_N$ is

$$f(x_1 \ldots, x_n, x_{n+1}, \ldots, x_N) = \binom{N}{\tilde{a}}^{-1} \omega(\tilde{a}), \quad \text{where } \tilde{a} = \sum_1^N x_i, \quad (4.5.10)$$

and $x_i = 0, 1, i = 1, \ldots, N$. Hence for any $1 \leq n < N$, the marginal pmf of $X_1, \ldots, X_n$ is

$$f(x_1, \ldots, x_n) = \sum_{a=\tilde{b}}^{N-n+\tilde{b}} \binom{N-n}{a-\tilde{b}} \binom{N}{a}^{-1} \omega(a), \quad \text{where } \tilde{b} = \sum_1^n x_i, \quad (4.5.11)$$

and $x_i = 0, 1, i = 1, \ldots, n$. From (4.5.10)–(4.5.11), given any $x_1, \ldots, x_n, x_i = 0, 1$ with $\sum_1^n x_i = b$, the conditional distribution of X_{n+1} has pmf

$$f(x_{n+1}|x_1, \ldots, x_n) = \frac{\sum_{a=b+x_{n+1}}^{N-n-1+b+x_{n+1}} \binom{N-n-1}{a-b-x_{n+1}} \binom{N}{a}^{-1} \omega(a)}{\sum_{a=b}^{N-n+b} \binom{N-n}{a-b} \binom{N}{a}^{-1} \omega(a)}, \quad (4.5.12)$$

where $x_{n+1} = 0, 1$ and the conditional distribution of $X_{n+1}, \ldots, X_N$ has pmf

$$f(x_{n+1}, \ldots, x_N|x_1, \ldots, x_n) = \frac{\binom{N}{b+\tilde{c}}^{-1} \omega(b+\tilde{c})}{\sum_{c=0}^{N-n} \binom{N-n}{c} \binom{N}{b+c}^{-1} \omega(b+c)}, \quad (4.5.13)$$

where $\tilde{c} = \sum_{n+1}^N x_i$, and $x_i = 0, 1, i = n+1, \ldots, N$. If Y stands for $\sum_{n+1}^N X_i$, the conditional distribution of Y given $x_1, \ldots, x_n$ has pmf

$$f(y|x_1, \ldots, x_n) = \frac{\binom{N-n}{y} \binom{N}{b+y}^{-1} \omega(b+y)}{\sum_{c=0}^{N-n} \binom{N-n}{c} \binom{N}{b+c}^{-1} \omega(b+c)}, \quad y = 0, 1, \ldots, N-n.$$

$$(4.5.14)$$

Particular choices of the $\omega(a)$s would give particular forms. For instance, if we assume

$$\omega(a) = \frac{1}{N+1}, \quad a = 0, 1, \ldots, N, \quad (4.5.15)$$

from (4.5.12), after some reduction we get

$$P(X_{n+1} = 1|x_1, \ldots, x_n) = \frac{b+1}{n+2}. \quad (4.5.16)$$

Example 4.5.2. Suppose successive random draws *with* replacement are made from a finite population in which the units are classified exhaustively into $k(\geq 2)$ mutually exclusive types $1, 2, \ldots, k$. The proportions corresponding to the different types in the population are unknown (subjectively random). We associate the random k-vector $X_i = (X_{i1}, \ldots, X_{ik})'$, $\sum_{j=1}^{k} X_{ij} = 1$ with the ith draw such that $X_{ij} = 1$ or 0 according as the ith draw produces or not a unit of type j, $i = 1, 2, \ldots$. Clearly the infinite sequence $X_1, X_2, \ldots$ is *infinitely exchangeable* in the sense that for any $n \geq 1$ the finite sequence $X_1, X_2, \ldots, X_n$ is exchangeable. (This is equivalent to the definition of infinite exchangeability which we gave in section 3.8.) Note that from the subjective point of view $X_1, X_2, \ldots$ is not an independent sequence; observing $X_1, X_2, \ldots, X_n$ one can get some idea about the composition of the population and hence modify one's opinion about X_{n+1}. Write

$$P\left(\sum_{i=1}^{n} X_{i1} = a_1, \ldots, \sum_{i=1}^{n} X_{ik} = a_k\right) = \omega(a_1, \ldots, a_k|n), \quad a_j \geq 0, \quad \sum_{1}^{k} a_j = n.$$

(4.5.17)

(Of course the values of $\omega(a_1, \ldots, a_k|n)$ would be subject to appropriate restrictions.)

Given any $(a_1, \ldots, a_k)$, if $x_1, x_2, \ldots, x_n$ are observations on $X_1, \ldots, X_n$ such that

$$x_i = (x_{i1}, \ldots, x_{ik})', \quad x_{ij} = 0, 1, \quad \sum_{j=1}^{k} x_{ij} = 1, \quad \sum_{i=1}^{n} x_i = (a_1, \ldots, a_k)', \quad (4.5.18)$$

as exchangeability implies that all the $n!/(\prod_{1}^{k} a_j!)$ ways of choosing the a_j draws producing units of type $j, j = 1, \ldots, k$ are equiprobable, we have the pmf of $X_1, \ldots, X_n$ in the form

$$f(x_1, \ldots, x_n) = \frac{\prod_{1}^{k} a_j!}{n!} \omega(a_1, \ldots, a_k|n). \qquad (4.5.19)$$

When $x_1, \ldots, x_n$ satisfy (4.5.18), for any $x_{n+1} = (x_{n+1,1}, \ldots, x_{n+1,k})'$ with $x_{n+1,j} = 1, 0, \sum_{j=1}^{k} x_{n+1,j} = 1$, the conditional pmf of x_{n+1} given X_i, $i = 1, \ldots, n$ is given by

$$f(x_{n+1}|x_1, \ldots, x_n) = \frac{a_{j*} + 1}{n + 1} \cdot \frac{\omega(a_1 + x_{n+1,1}, \ldots, a_k + x_{n+1,k}|n + 1)}{\omega(a_1, \ldots, a_k|n)}.$$

(4.5.20)

where $j* = j*(x_{n+1}) = \sum_{j=1}^{k} j x_{n+1,j}$ is that j for which $x_{n+1,j*} = 1$. From (4.5.20) we can readily write down the expression for the conditional joint pmf of $X_{n+1}, \ldots, X_{n+m}$ given $X_1, \ldots, X_n$ for any $m \geq 1$.

Simplifying postulates about $\omega(a_1, \ldots, a_k|n)$ will give particular expressions. A very simple postulate (Johnson (1924)) is that for every n all the $\binom{n+k-1}{k-1}$

ordered partitions $(a_1, \ldots, a_k)$: $\sum_1^k a_j = n$, $a_j \geq 0$ are equiprobable. This gives

$$\omega(a_1, \ldots, a_k | n) = \binom{n + k - 1}{k - 1}^{-1} \tag{4.5.21}$$

Using this in (4.5.20) we get

$$P(X_{n+1, j} = 1 | x_1, \ldots, x_n) = \frac{a_j + 1}{n + k} \tag{4.5.22}$$

Sufficiency

4.5.5   The concept of a sufficient statistic and the sufficiency principle which we introduced in section 4.3 in the parametric context have to be given a somewhat different interpretation under the purely subjective set-up. Thus any statistic $T = T(X_1, \ldots, X_n)$ based on the first n observations is called *sufficient* in the subjective set-up if the conditional distribution of any possible number of further observables $X_{n+1}, \ldots, X_{n+m}$ given $X_1 = x_1, \ldots, X_n = x_n$ depends on $x_1, \ldots, x_n$ only through $t = T(x_1, \ldots, x_n)$ (this is equivalent to saying that the conditional distribution is same as that given $T = t$). Thus in Example 4.5.1 from (4.5.13), $\sum_{i=1}^n X_i$ is a sufficient statistic. Similarly in Example 4.5.2 from (4.5.18) and (4.5.20), it follows that $\sum_{i=1}^n X_i$ is a($(k - 1)$-dimensional) sufficient statistic (cf. Lad (1996) pp. 193–194). The *sufficiency principle* here assumes the form: when a sufficient statistic $T = T(X_1, \ldots, X_n)$ exists, to induce about $X_{n+1}, \ldots, X_{n+m}$, whatever m, we may disregard the individual $X_1, \ldots, X_n$ and base our induction solely on the value of T.

4.5.6   The subjective approach to induction as described above with its disregard of natural laws and parameters and its exclusive adherence to the subjective view of probability was originally proposed by de Finetti (1937) (see also de Finetti (1975) Chapter 11) and was strengthened by the work of Savage (1954) (but with regard to Savage's contribution recall the remark made in footnote 38 of this chapter). Further details with applications based on diverse models have been worked out in Lad (1996). As we have emphasized earlier, the speciality of this approach is induction about unrealized observables. Stochastic independence of various observables, which is a key concept for model construction in a large part of the development in various objective approaches has no role to play in this approach; observables which are traditionally assumed independent in the former, typically influence each other in the probabilistic sense here (cf. Example 4.5.2). (It is interesting to note that Fisher, who was professedly an objectivist, conceded the practical importance of both these aspects of induction in the context of fiducial inference; see Fisher (1956) pp. 113–115.) In the subjective approach knowledge of analogy or symmetry among observables is incorporated into the model by making the model exchangeable in respect of analogous variables. The concept of exchangeability was introduced for generalized Bernoulli

trials by Johnson (1924). It was independently developed in the general case by
de Finetti (1937), who among other things proved his famous Representation
Theorem for infinitely exchangeable random variables. In section 3.8 we stated
and briefly discussed de Finetti's Representation Theorem. We now consider
how this theorem permits the interpretation of the problem of induction about
an infinitely exchangeable sequence as one about certain notional parameters.

| Infinite exchangeability | Consider an infinitely exchangeable sequence of discrete
or continuous random variables $X_1, X_2, \ldots$, all assum-
ing values in a common space $\mathcal{X}_1$. For any $n \geq 1$ the joint distribution of
$X_1, \ldots, X_n$ is represented by the pmf (or pdf) $f(x_1, \ldots, x_n)$. (Of course infinite
exchangeability entails some special restrictions on the sequence of distributions
$f(x_1, \ldots, x_n), n = 1, 2, \ldots$.) By de Finetti's theorem infinite exchangeability
implies that

$$f(x_1, \ldots, x_n) = \mathbb{E}_{f_1} \prod_{i=1}^{n} f_1(x_i), \quad n \geq 1, \tag{4.5.23}$$

where the common pmf (or pdf) f_1 appearing in all the factors in the product on
the right represents a 'random distribution' over $\mathcal{X}_1$ and the mean value $\mathbb{E}_{f_1}$ is
taken with respect to this 'random distribution'. If $\mathcal{F}_1$ is the space of all possible
pmf's (or pdf's) over $\mathcal{X}_1$, f_1 is a 'random distribution' means that it assumes
values in $\mathcal{F}_1$ subject to some probability distribution (mixing distribution) over
$\mathcal{F}_1$. In simple cases this would mean either f_1 has a countable set of realizations
and there is a discrete mixing distribution over this countable set or f_1 involves a
real- or vector-valued continuous parameter θ having a domain Θ and there is a
continuous mixing distribution over Θ. Generally the mixing distribution would
be a probability measure on a suitable class of subsets of $\mathcal{F}_1$. de Finetti (1937)
further established that the mixing distribution for an infinitely exchangeable
sequence of random variables is virtually determined by the sequence of joint
distributions $f(x_1, \ldots, x_n), n = 1, 2, \ldots$ (see also Savage (1954) p. 53).

| Notional 'objective–
pro-subjective' | The representation (4.5.23) suggests that we can regard
the sequence $X_1, X_2, \ldots$ as if it is generated by a
two-stage random process: in the first stage an f_1 is
probabilistically chosen out of $\mathcal{F}_1$ according to the mixing distribution; after f_1
is chosen, in the second stage a conditionally i.i.d. sequence $X_1, X_2 \ldots$ with common distribution f_1 is generated. The mixing distribution here has the role of a
prior distribution. This is more patent in the case when the different members
of $\mathcal{F}_1$ are distinguished by the values of some real- or vector-valued parameter θ.
The mixing distribution then can be looked upon as representing a (known) prior
distribution of θ. Hence induction in the purely subjective approach becomes
formally same as induction about the notional parameter θ according to the

objective–pro-subjective view-point. We illustrate all this with reference to the set-up of Example 4.5.2.

Example 4.5.2 (Contd.). If the infinite sequence $X_1, X_2, \ldots$ were i.i.d., the common distribution would be specified by a $(k-1)$-dimensional parameter $\theta = (\theta_1, \ldots, \theta_{k-1})'$, where using the Kronecker delta notation

$$P(X_1 = (\delta_{j1}, \delta_{j2}, \ldots, \delta_{j,k-1}, \delta_{jk})') = \theta_j, \quad \text{for } j = 1, \ldots, k-1,$$

$$P(X_1 = (0, 0, \ldots, 0, 1)') = 1 - \sum_1^{k-1} \theta_j. \tag{4.5.24}$$

θ belongs to the $(k-1)$-dimensional simplex

$$\Theta = \left\{ \theta : 0 \le \theta_j \le 1, j = 1, \ldots, k-1, \sum_1^{k-1} \theta_j \le 1 \right\}. \tag{4.5.25}$$

When $X_1, X_2, \ldots$ is an infinitely exchangeable sequence, by de Finetti's representation theorem there exists a $(k-1)$-dimensional random variable θ assuming values in Θ, such that given $\theta = \theta$, conditionally, $X_1, X_2, \ldots$ are i.i.d. with common distribution given by (4.5.24). From de Finetti (1937) we also get that there exists a $(k-1)$ dimensional cdf $M(\theta)$ such that

$$\lim_{n \to \infty} P\left(\frac{1}{n} \sum_1^n X_{i1} \le \theta_1, \ldots, \frac{1}{n} \sum_1^n X_{i,k-1} \le \theta_{k-1} \right) = M(\theta) \tag{4.5.26}$$

at all continuity points of $M(\theta)$. $M(\theta)$ represents the cdf of the distribution of θ (the mixing distribution) appropriate for the infinitely exchangeable sequence $X_1, X_2, \ldots$. Thus when $X_1, X_2, \ldots, X_n$ can be considered as part of an infinitely exchangeable sequence, the subjective set-up for $X_1, X_2, \ldots, X_n$ is equivalent to the objective–pro-subjective set-up in which we have a random sample of size n from a k-class generalized Bernoulli distribution with the notional parameter θ subject to the prior (4.5.26).

When $X_1, X_2, \ldots$ are infinitely exchangeable from (4.5.20) we get that given $X_1, X_2, \ldots, X_n, \sum_1^n X_i = (a_1, \ldots, a_k)'$ is a sufficient statistic for inducing about $X_{n+1}, \ldots, X_{n+m}$ for any $m \ge 1$. In particular, taking $m = 1$, from (4.5.20) we get, for any j

$$P(X_{n+1,j} = 1 | x_1, \ldots, x_n) = \frac{a_j + 1}{n + 1} \cdot \frac{\omega(a_1, \ldots, a_j + 1, \ldots, a_k | n + 1)}{\omega(a_1, \ldots, a_j, \ldots, a_k | n)}. \tag{4.5.27}$$

It can be shown (Johnson (1924, 1932); see also Good (1983) p. 101, and Skyrms (1996) section 4) that when $k \ge 3$ if we assume that (4.5.27) depends on $a_1, \ldots, a_k$ only through a_j, we must have for some $\lambda_1 > 0, \ldots, \lambda_k > 0$,

$$P(X_{n+1,j} = 1 | x_1, \ldots, x_n) = \frac{a_j + \lambda_j}{n + \sum_{\ell=1}^k \lambda_\ell}. \tag{4.5.28}$$

Further, when the form of dependence of (4.5.27) on a_j is same irrespective of j, we get

$$P(X_{n+1,j} = 1|x_1,\ldots,x_n) = \frac{a_j + \lambda}{n + \lambda k} \qquad (4.5.29)$$

for some $\lambda > 0$. When $\lambda = 1$, (4.5.29) reduces to (4.5.22). In the philosophical literature the rules of induction represented by (4.5.28)–(4.5.29) are known as G.C. (Generalized Carnapian) systems (see e.g. Festa (1993) pp. 57–59).

If we look at the infinitely exchangeable sequence $X_1, X_2, \ldots$ as generated by a two-stage random process with a notional parameter θ, then the assumption leading to (4.5.28) is equivalent to the condition that θ has a Dirichlet distribution (see Good *op. cit*) with pdf given by

$$\text{Const.} \prod_{j=1}^{k-1} \theta_j^{\lambda_j - 1} \left(1 - \sum_{1}^{k-1} \theta_j\right)^{\lambda_k - 1}. \qquad (4.5.30)$$

(4.5.29) corresponds to the case $\lambda_1 = \cdots = \lambda_k = \lambda$. Given θ the joint pmf of $X_1, \ldots, X_n$ is

$$\prod_{j=1}^{k-1} \theta_j^{a_j} \left(1 - \sum_{1}^{k-1} \theta_j\right)^{a_k} \qquad (4.5.31)$$

Combining, the conditional pmf of $X_{n+1} = (X_{n+1,1}, \ldots, X_{n+1,k})'$ given $X_1, \ldots, X_n$, is seen to be

$$\frac{\int_\Theta \prod_{j=1}^{k-1} \theta_j^{\lambda_j + a_j + x_{n+1,j} - 1} \left(1 - \sum_{1}^{k-1} \theta_j\right)^{\lambda_k + a_k + x_{n+1,k} - 1}}{\int_\Theta \prod_{j=1}^{k-1} \theta_j^{\lambda_j + a_j - 1} \left(1 - \sum_{1}^{k-1} \theta_j\right)^{\lambda_k + a_k - 1}} \qquad (4.5.32)$$

For any j, taking $x_{n+1,\ell} = \delta_{j\ell}, \ell = 1, \ldots, k$, (4.5.32) reduces to (4.5.28). Alternatively, when looked at from the objective–pro-subjective angle, (4.5.32) can also be interpreted as the posterior mean of θ_j (i.e the Bayes estimate of θ_j under squared error loss) given $x_1, x_2, \ldots, x_n$, under the prior (4.5.30). (Note that Dirichlet prior (4.5.30) is the conjugate prior for the present set-up.) From the subjective point of view, however, since θ is a notional parameter (possibly having different meanings for different persons) such an alternative interpretation has little practical significance.

From the above it is clear that not only does exchangeability allow the incorporation of the agent's perception about the similarity of certain characters,[40] but infinite exchangeability makes it possible for different agents to agree to some extent regarding the form of the underlying model; different persons accepting

---

[40]In our discussion, for simplicity we have considered only full-scale exchangeability under all permutations. It is also possible to have partial exchangeability where the joint distribution of a set of variables remains same under subgroups of permutations (see de Finetti (1975) pp. 212–213, Lad (1996) pp. 228–243).

infinite exchangeability can differ only in respect of the mixing distribution, i.e. the prior for the notional parameter.

| A matter of choice |

4.5.7 If one accepts only the personalistic view of probability and has the positivistic attitude which does not admit quantities other than those which can in principle be measured, it would be logical to go the whole hog and adopt the radical subjective approach to statistical induction described above. Those who regard probability as having some kind of objective or at least impersonal interpretation, however, consider the approach chimerical and unsuitable for serious scientific purposes (it may be suitable for short-term managerial decision-taking in business situations). The main criticism is that, doing without the conception of agent-independent parameters and associated probabilistic laws, this approach leaves little room for interpersonal communication and does not provide any impersonal guidelines for choosing an appropriate experiment. Ultimately, the choice depends on one's opinion. The rightness or wrongness of an opinion cannot be resolved through theoretical debate; the solution if there is any, has to be pragmatic, based on the experience of what works in various situations in practice.

4.6 Concluding discussion

In the preceding sections we have given a somewhat detailed resumé of the different approaches to statistical induction. As it has been noted in the various contexts, the differences in the approaches arise basically because of differing attitudes to certain fundamental questions relating to probability-based induction. In this concluding section we collect together these questions without going once again into the various answers to these that have been canvassed.

  (i) What is the nature of probability? Is it a characteristic of the external world or is it an index of our mental attitudes to phenomena in the external world? Are there two or more kinds of probability? More generally, what are laws of nature? Do these exist objectively or do these represent merely consensuses agreed upon by all concerned?

 (ii) Can all forms of uncertainty be graded or quantified in terms of probability obeying the standard probability calculus? Or, as Keynes (section 3.10) put it originally, are there propositions whose uncertainties being situated on different non-intersecting paths cannot be so graded?

(iii) What is the nature of statistical induction? Is it basically inferential or does it always end up with the recommendation of definite actions in practice?

 (iv) It is generally agreed that induction should follow the Principle of Total Evidence (subsection 4.3.4) according to which 'In any problem of induction, *all* relevant evidence available should be utilized for reaching the conclusion' (we follow here the version of the principle as stated by Wilkinson (1977) who calls it the Relevance Principle). But how should one decide which part of a given body of evidence is relevant and which spurious? In particular, is information about the design of the sampling experiment, after the observation has been realized, relevant? Also, if, before the experiment a knowledgeable experimenter has certain hunches about the value of the parameter of interest and these have been abstracted in the form of answers to various queries (maybe imaginary betting proposals), should that information (even though it cannot be subjected to empirical verification in any other way) be considered as part of the relevant evidence?

(v) It is generally conceded that subjective judgement in one form or other is an unavoidable ingredient of statistical induction. Judgement as to the model most appropriate in an experimental situation or the parameter most suitable for formalizing our inductive concern is usually agreed upon on the basis of experience, convention, or consensus. But there are other forms of judgement, for example about how important it is to avoid wrong conclusions of different types, regarding which there may be difference of opinion. In which stage and how should the latter be incorporated in the mode of induction? Should such judgements be entered in a detachable way so that their modification does not necessitate reworking the entire process?

(vi) What should be the guiding principle in choosing the mode of induction? Should we aim at some form of optimal performance for the mode with respect to some specific questions of interest? Or should we remain content if we are able to achieve self-consistency in answering all conceivable questions?

As we have seen, different attitudes and answers to the above questions have given rise to different lines of development to statistical induction. All such developments, one way or other, try to formalize modes of thinking we follow in everyday life to solve simple problems of induction and thus have some legitimacy. True, often it is possible to interpret the same rule of induction from more than one viewpoint. An admissible behavioural rule can be seen as a Bayes rule with respect to an implicit prior or a Bayes rule may be justified on the basis of its long-run performance under repeated application. As we examine the history of development of statistical thought in later chapters, we would see that different ways of thinking have gained ascendancy at different times and in different contexts. There have been attempts to subsume and interpret everything that is useful under one single paradigm to the exclusion of others. But each distinctive approach has its own natural setting in practice—some are appropriate in repetitive situations, some for one-time studies, for some personal hunches about the underlying state of things should be ruthlessly kept out of consideration, for others these constitute valid data, and so on. It would be artificial to try to cast all into one single mould. Induction is a kind of expert activity which works in various ways in various spheres, extending beyond the reach of normative logic. In our view, so far as statistical induction is concerned, eclecticism is a virtue.

PART II

HISTORY

PREHISTORY, BEGINNING OF HISTORY, AND THE TODDLING PERIOD

5.1 'Probability' before 'statistics'

'Which came first, probability or statistics?'—I. J. Good in an article raised this question (see Good (1983) Chapter 5). If we identify statistics, as we have done, with statistical induction, the answer should be: probability. For, as we have seen, probabilistic reasoning underlies all approaches to statistical induction and of course probability must have gained some footing in the thought-world of man before it could be used in induction.

| Genesis of probability |

However, the question of development of probability in this context can be formulated in at least three forms:

1. How did the concept of quantitative probability emerge?
2. How did the calculus of probability or the doctrine of chances originate?
3. What stimulated the development of the theory of probability?

Different authors have put varying emphasis on these forms and naturally come out with different answers. Thus Hacking (1975) in his eminently readable *The Emergence of Probability* has devoted considerable space to the first question; Kendall (1956), David (1962), and lately Hald (1990) have mainly addressed the second; and Maistrov (1974) has given importance to the third. In the next section where we will discuss the pre-history of probability and statistics we will deal with the first question. The other two questions, as also the interplay between probabilistic reasoning and statistical studies in the early periods of history, will be considered in the subsequent sections.

5.2 The long period of gestation

| Behaviour under uncertainty |

5.2.1 Before going for what traces of the idea of probability can be found in ancient and medieval records, we should keep in mind one thing. Ever since his cave days, man has been facing in everyday life situations of uncertainty where he has to act in the absence of definite knowledge and thus face the prospect or risk of pleasant or unpleasant consequences. In course of progressive adaptation to his environment through millenia, man has evolved certain *behavioural principles* such that in the face of uncertainty action taken in accordance with them is generally fruitful. Some of these principles involve ideas of probability implicitly. The principles may not have been clearly formulated in course of the evolution of civilization until much later—they have been rather part of the general lore of human experience and when they have been followed, it has often been at an unconscious level. Nevertheless, that human civilization has survived and progressed so far by generally acting upon them proves their soundness vis-à-vis the phenomenal world we live in. (As we noted in section 1.5, Bertrand Russell (1948) pp. 495–496, 507 put forward a similar thesis with reference to the principles of induction; we believe the same can be said of the principles of human behaviour in the face of uncertainty.) We list below six such principles.

(i) In repetitive situations of the same kind where an event sometimes occurs and sometimes does not, the relative frequency of its occurrence often becomes roughly stable in the long run so that we can judge whether such a frequency is close to 1 or 0 or one such frequency is definitely larger than another. Thus when dark thunder clouds appear, very often it rains; when the poles of a hut are made of *sal* timber usually they remain intact at the end of three seasons of rains, whereas bamboo poles become rotten at the end of the period more often; when in a case of infantile diarrhoea there is stoppage of urine, it turns out to be fatal more often than when there is no such complication; a certain physician succeeds in curing his patients more often than another; and so on and so forth.

(ii) The approximately stable long-term relative frequency of an event can be taken as a rational measure of trustworthiness of our expectation or degree of belief that it would take place. This degree of belief may alternatively be expressed as an odds, i.e. the number of instances in which the event occurs to the number in which it does not.

(iii) In determining the relative frequency of an event among similar situations, it is important to identify the important signs of the case at hand and define similar situations on their basis. Thus in a case of diarrhoea it is important to note whether there are any urinary complications. This is because a relative frequency among similar situations defined in this way is more trustworthy in practice.

(iv) It is rational to look upon the stable relative frequency of an event as representing a sort of causal tendency inherent in such situations. Thus it is generally agreed that irregular habits and indulgence in over-eating tend to reduce the longevity of a person: the proportion of such persons who survive beyond, say, 70 years, is less than that among others.

(v) When two incompatible events that can occur in an uncertain situation are distinguishable only by arbitrarily given names or labels and not on the basis of any essential characteristics of the events as natural phenomena, their relative frequencies in a long run of similar situations would be roughly equal. Thus since the marks one, two, etc. on the faces of a perfectly symmetric die can be regarded as arbitrary labels, the long run relative frequencies of getting two different numbers when the die is repeatedly tossed can be taken equal. Similarly since for a thoroughly shuffled pack of identical cards, the names of the cards vis-à-vis their order of occurrence in the pack are arbitrary, the relative frequency of the first card turning up to be the ace of spades or the queen of hearts would be same.

(vi) The last principle which may be called the *fair price principle* is less transparent than the earlier ones and possibly required longer experience to evolve. Suppose somebody is undertaking a risky venture and there is some uncertainty as to whether its outcome would be favourable or not. In such a situation the amount of resources (in terms of money, time or effort) it would be rational to expend (the fair price) should be large or small depending directly on, firstly, the strength of his belief that a favourable outcome would follow if the venture is undertaken in right earnest, and secondly, the magnitude of the benefit from a favourable outcome. To take an example from the legendary era, suppose an itinerant poet had to decide whether to undertake one month's journey to the court of an emperor to try his luck. For this he should weigh both the strength of his belief that he would be able to please the emperor by his versifying skill and the extent of munificence the emperor is known to show in such cases. He should not undertake the journey if the prospective reward is paltry compared to the cost of the journey or if he has little faith that the emperor would appreciate his skill. At a more practical level, suppose somebody living on the bank of a turbulent river fears that his house may collapse due to erosion of the bank. The river may be contained by piling a sufficient number of boulders at suitable places, but there is no certainty about it. To decide whether such a laborious measure should be undertaken, the person has to consider both the strength of his belief that the measure would be successful and the benefit of saving his house. He would definitely not carry out the measure if its cost is more than that of abandoning the house or if he is sure that the measure is bound to fail. Uncertain situations where the cost involves money but little labour or skill represent gambling. Suppose, a gambler plays a dice game in which he stakes the amount s_1 against an adversary whose stake is s_2. We can interpret it as an uncertain situation in which he spends the amount s_1 in any case and receives $s_1 + s_2$ in case of success. It would be rational for the gambler to assess both how sure he is about a win and how large the uncertain receipt $s_1 + s_2$ is in comparison to s_1 before deciding whether to play. In all such cases, ideally the strength of belief should be determined on the basis of the relative frequency of success observed in similar situations in the past. When such past records are not available or out of the question because the situation is not repetitive, it would have to be assessed subjectively by taking account of all available information. Sometimes, especially, in the latter case, it may be easier to think in terms of the present worth or expectation of the uncertain receipt and to undertake the venture only if the price or down expenditure (stake s_1) called for is less than that. Of course a high or low value of the expectation in relation to $s_1 + s_2$ would imply a high or low degree of belief in success.

We do not say that in ancient and medieval times, in situations of uncertainty, action-taking was always based on the above behavioural principles. Apart from aberrations due to individual passions and predilections, the visions of entire communities were sometimes befogged by superstitions. But these principles were generally accepted as representing ideals of rational behaviour, and, as civilizations evolved, were more and more adhered to in practice.

Looking at these principles today, we may feel that probability is ensouled in each one of them. Some of these even look like vague formulations of well-known concepts of probability theory. Thus (i) and (iv) look like the frequency and propensity interpretations of probability (subsections 3.2.1 and 3.2.5), (iii) like the principle of instantiation of Russell and Fisher (subsection 3.2.5) and (vi) like a vague statement of the principle of rational expectation (subsection 4.4.6)! But actually the above principles of rational behaviour were being tacitly followed in practice long before probability was recognized as an intellectual object. When probability theory developed, these were taken for granted and incorporated in it quietly. In fact, as we are going to show in the following, probability came into existence and the theory of probability started only when these principles were openly admitted in empirical philosophy and formalized.

5.2.2   The admission of the above behavioural principles into empirical philosophy and science was, however, a slow process. In fact it remained virtually a non-starter over many centuries. As mentioned in Chapter 1 (section 1.4), Greek philosophy dealt almost exclusively with demonstrative knowledge and uncertain events were generally not regarded as worthy of study.

Probabilistic ideas in Talmuds

One notable exception is the Jewish Talmudic literature which grew over several centuries upto the 5th century A.D. It deliberated on various juridical questions relating to rules of behaviour, distribution of rights, inheritance etc. in ancient Jewish society. The answers to the questions were often reached through elementary probabilistic reasoning. Such reasoning often required the assumption that various demographic ratios like the proportion of male births among all live births, proportion of miscarriages, or premature deliveries among all cases of pregnancy etc. were stable. Common sense values were taken for such ratios (like the value $\frac{1}{2}$ of the proportion of male births) and relative frequencies of more complex events (like 'a case of pregnancy will not miscarry and a male child will be born out of it') were arrived at through manipulation; these derived relative frequencies were interpreted as degrees of belief to settle questions like inheritance (Rabinovitch (1969)). Also drawing from urns containing indistinguishable balls or stone pieces was widely practised to reach decisions in situations of perplexity. Although sometimes this was merely for divination of God's will, often the purpose was to ensure a fair distribution of rewards or responsibilities (Hasover (1967)); the latter use clearly presumed principle (v). However, these early instances of probabilistic reasoning were somehow not followed up and did not influence general philosophical thinking. Does the Marxian thesis of the history of development of knowledge—a new science grows only when the practical needs of society requires it (see Maistrov (1974) p. 14)—explain this?

Renaissance—mutation of 'probable'

The Greek attitude to knowledge continued in the scholastic philosophy through the middle ages. Only conclusions which could be established demonstratively were admitted as knowledge. Empirical findings of practical sciences like medicine or agriculture were dubbed as 'opinion' which was regarded as inferior to demonstrative knowledge (see Hacking (1975) Chapter 3). Curiously, in the middle ages the adjective 'probable' in the sense of being 'approvable' came to be associated with opinion that was supported by external evidence, i.e. the testimony of respected people. In the pre-Renaissance period this 'probability' had no connection with relative frequency of occurrence in the past. According to Hacking, during the Renaissance, in the general climate of disregard for authority, the sense of the word 'probable' underwent a mutation and it came to signify something which was supported by the prevailing signs or intrinsic evidence, and naturally, relative frequency of occurrence came to be accepted as a measure of trustworthiness of such evidence.[1] Thus Fracastoro, a 16th century Italian physician averred that the probability of an impending contagion should be judged on the basis of signs in the environment and our experience of whether such signs in the past had proved trustworthy 'almost always' or 'often'. Hobbes, the English philosopher of the early 17th century, made the association of probability with frequency more explicit when he wrote, '... the signs are but conjectural; and according as they have often or seldom failed, so their assurance is more or less ... experience concluded nothing universally. If the signs hit twenty times for one missing, a man may lay a wager of twenty to one of the event ...' (see Hacking (1975) pp. 28, 48). All this indicates that the behavioural principles listed in the

---

[1]According to Hacking, because of this origin of the modern sense of the word probability, the concept of probability is ontologically Janus-faced—its two aspects 'degree of belief' and 'relative frequency' are inseparable (Hacking (1975) pp. 12, 15). This is fine if we confine ourselves only to repeatable situations in which the outcome is unpredictable. However, as we discussed in Chapters 2–3, probability with its subjective interpretation is also used in non-repeatable situations where uncertainty means just 'absence of knowledge'.

preceding subsection were now being accorded some place in the philosophical literature.

However, this did not still lead to the development of any comprehensive theory of probability. One reason might be that in all these cases, interest lay primarily in values of relative frequency close to 1 or 0—intermediate values hardly mattered in practice. Another certainly was that any sort of verification of putative long-term relative frequencies in such applications would require patient data collection over wide areas and a considerably long period—the frequency of onset of epidemics or cure-rate of a proposed treatment for a particular disease cannot be judged by locally conducted studies in one or two days.

| Aleatory contracts | On the other hand around the same time the same |

On the other hand around the same time the same behavioural principles gradually began to influence business practices as well in the form of aleatory contracts, i.e. contracts which took account of the risks involved in a venture. The first marine insurance companies were set up in the 14th century in Italy and Holland and soon the scope of insurance was enlarged to cover intra-continental deliveries. (Maistrov (1974).) The companies charged premiums at the rate of 12–15% of the value of goods for freight by sea and 6–8% for that by land. These premiums can be thought to have been set so as to provide sufficient profit margins over the expected losses due to non-delivery, in conformity with principle (vi) of the preceding subsection. The rates therefore represented upper bounds to the relative frequencies of non-delivery. However, since relevant large scale data were not available, these must have been set partly by judgment and partly by convention. Non-availability of data inhibited the development of any theory of probability through such business applications as well.

| Probability from gaming and gambling |

5.2.3 Thus around the 15th–16th centuries the behavioural principles mentioned earlier were being overtly and covertly applied more and more in scientific decisions and in business decisions involving uncertain situations. The concept of probability with both its objective and subjective interpretations was also gradually taking form. What finally led to the formalization of the concept and birth of the theory of probability was the invocation of those principles in certain concrete problems of betting in games of chance. Games of chance have certain advantages compared to the other areas of human activity discussed above, which made them particularly suitable as the prospective seed-bed for the theory of probability. Firstly, in games involving the tossing of coins or dice or drawing of cards, the outcome space (cf. section 2.3) is clearly defined, and in simple cases, finite. Secondly, the set-up in many games displays obvious symmetries. Thirdly—and this is rather crucial—various conclusions about long-term relative frequencies in the case of games can be empirically derived and verified provided one plays the game repeatedly an adequate number of times. We will consider in the next section how the theory of probability was born from consideration of games of chance and thus the history of probability and statistics started. (In the context of games, initially the term 'chance' was used; but as we will see, gradually 'probability' came to be used synonymously with it.) Before concluding the present subsection we recount some further prehistory of probability relating to gaming and gambling.

Gambling on the basis of games of chance like coin tossing, dice throwing, drawing of lots etc. had been in vogue in most human societies since earliest times. It had been a common pastime in ancient Egypt, India, Greece, Rome, and other countries for thousands of years. Yet there is no evidence that a beginning of a probabilistic theory of gambling had been made anywhere until about the 16th century, with *one notable exception*.

| A mysterious reference in an Indian epic | The exception occurred in the case of India, where gambling on the basis of dice-play used to be practised by kings and their courtiers. Passionate gambling con- |

tests used to be held between rival kings, in which challenges were thrown and various possessions including kingdoms and even wives were staked, sometimes with disastrous consequences. In fact, the story of the well-known Indian epic *Mahābhārata* (which, experts say, was compiled over a millenium starting not later than the 5th century B.C.; see Pusalker (1962)) revolves round one such gambling bout. The epic contains a sub-story, the story of *Nala* and *Damayantī* which has a reference to the science of dice play and a statement as to how the same science can be used for getting a quick estimate of the size of a finite population. This story[2] has been mentioned by a number of writers (Godambe (1976),

---

[2]The story runs as follows: King *Nala*, who had lost his kingdom in a gambling bout, was living incognito as the chariot-driver (he was an expert in equine science) of another king named *Ṛtuparṇa* (also called *Bhāngasuri*, which literally means son of *Bhangasur*). The latter was an expert gambler and Nala wanted to acquire that expertise from him. Once Ṛtuparṇa had to reach a distant place urgently and he asked Nala to drive him there. During the journey an incident occurred which prompted Nala to boast about his extraordinary charioteering skill. At this Ṛtuparṇa was piqued. Telling Nala not to boast so because there was nobody who knew everything, he showed off his own expertise by giving quick estimates for a way-side tree as to the numbers of leaves and fruit, both standing on the tree and fallen on the ground, and also of the same borne by two big branches (the number of fruit on the branches was estimated as 2095). Nala verified through direct enumeration the estimates given and was amazed. On his asking how such quick estimation was humanly possible, the king replied,

'Biddhi akṣahṛdayajñam mām saṃkhyāne ca viśāradam'

'Know that I am a knower of the secret of the dice and therefore adept in the art of enumeration' (*Mahābhārata* Vol. 3, *Āraṇyaka Parva* section 70, Verse 23 ed. V. S. Sukhtankar, Bhandarkar Oriental Research Institute, Poona, 1942.) The story goes on to describe how at Nala's importunity the king gave him the 'secret of the dice' in exchange for the former's 'secret of the horses' and how with his

Hacking (1975) pp. 6–8, Lad (1996) p. 21) in the context of the history of probability and statistics. No details of the science or of its use in estimation are
available. Yet, since the connection between the science of dice-play and population estimation is not something which occurs in the human mind naturally and
can be understood by somebody only after he has acquired sufficient knowledge
of the theory of probability and the theory of sampling, the statement linking the
two suggests that both theories must have been well-developed in ancient India.

| Conditions in the West |
| before the 16th century |

We can only speculate on what prevented the development of a theory of gambling based on probability in
the West before the 16th century. Existence of com-
mercially run gambling houses in any country would naturally give a fillip to
the development of such a theory. This is not only because such casinos would
have to keep regular accounts of the outcomes of games so as to ensure their
margin of profit (note that existence of such casinos itself confirms the fact of
frequential regularity in the case of games), but also because any habitual gambler playing again and again against the same adversary at fixed stakes would
readily notice any underlying regularity in the outcome frequencies. But commercial gambling establishments were discouraged both by the government and
the church in the various countries of Europe in the middle ages. Thus attempts
to suppress gambling on the basis of games of chance were made in the 13th century by Frederick II and Louis IX in continental Europe and sometime later by
Henry VIII in England. However, such attempts were generally unsuccessful and
gambling casinos periodically became part of life in many countries. It seems in
the 16th–17th centuries conditions in Italy and France were sufficiently conducive
for the systematic study of games of chance. We describe this development in
the next section.

5.3   The birth of probability theory

| Cardano—an |
| unrecognized pioneer |

5.3.1   Who started the theory of probability with its
distinctive calculus and when? Usually the names of
the well-known French scholars Pascal and Fermat and
the year 1654 are mentioned in this connection. But about a century earlier
there lived in Italy an unusual personality named Gerolamo Cardano (1501–
1576), who was a renowned physician, mathematician, astrologer, and an
inveterate gambler. He wrote a book entitled *Liber de Ludo Aleae* (The Book on
Games of Chance) around 1564 (David (1955) puts the year as 1526, but from
internal and circumstantial evidences 1564, the year accepted by Hacking (1975)
and Hald (1990), seems more likely). The book remained unpublished possibly
because of various misfortunes and tragedies that befell the author towards the

---

newly acquired skill Nala could ultimately recover his kingdom by winning a gambling
contest.

end of his life and saw the light of day only in 1663. In it Cardano made probability calculations for various games of chance and exemplified the following basic ideas and rules of probability theory.

1. The chance of an event in a random trial represents its long-run relative frequency. Thus, after discussing the chances of certain events defined for trials made with two dice, he observed, 'The argument is based upon the fact that such a succession is in conformity with a series of trials and would be inaccurate apart from such a series' (see Hacking (1975) p. 54). At another place (see Maistrov (1974) p. 19) he says that the frequency may deviate substantially from the chance in a small number of trials but the deviation would become negligible if the number of trials is large.
2. If a die is honest its different faces have equal chance of appearing. (In fact Cardano makes the statement, 'I am as able to throw 1, 3, or 5 as 2, 4, or 6' (see Hacking (1975) p. 54), which suggests that he had something like 'propensity' in mind.) From this he identifies the set of equally likely cases (the sets of all 36 or 216 permutations) when two or three honest dice are thrown. He uses the term 'circuit' for such a set.
3. When the 'circuit' for a trial is well-identified, the chance of an event is represented by the 'portion' of the whole circuit favourable to it. Cardano gives the rule that to obtain the odds we have to consider 'in how many ways the favourable result can occur and compare that number to the remainder of the circuit' (see Maistrov (1974) p. 22–23).
4. Cardano correctly uses the rule for addition of probabilities in terms of disjoint events. Thus in throwing two dice of 36 equally likely cases, 11 are favourable to the event 'at least one ace', 9 additional cases become favourable if we take the larger event 'at least one ace or deuce', 7 further cases come if we consider the still larger 'at least one ace, deuce, or trey' and so on. Similar computations are made for three dice (see Hald (1990) p. 39).
5. Cardano also correctly formulates the product rule for computing the chance of the simultaneous occurrence of events defined for independent trials. In terms of odds he says that if out of n equally likely cases just m are favourable to an event, then in r [independent] repetitions of the trial the odds that the event would occur every time are as $m^r/(n^r - m^r)$ which, writing p for m/n becomes $p^r/(1 - p^r)$. In particular, in throwing three dice, 91 out of 216 cases are favourable to the event 'at least one ace'. If the three dice are thrown thrice, Cardano correctly gets that the odds for getting the event at least once is a little less than 1 to 12 (see Hald (1990) pp. 39–40).

These instances of probabilistic computation by Cardano show that he recognized clearly the behavioural principles (i)–(v) listed in subsection 5.2.1 (except possibly principle (iii) which generally has little relevance for a game of chance) and correctly formulates these. What is most noteworthy, however, is that, as per available records, he was the first to sharpen and codify the fair price principle (vi). This he did as follows in the context of defining a fair game.

    6. In the case of throwing two dice the odds on getting at least one ace, deuce, or trey are 3 : 1. Cardano states that if the player who wants an ace, deuce, or trey 'wagers three ducats [a standard unit of currency at that time] and the other player one, then the former would win three times and would gain three ducats and the other once and would win three ducats; therefore in the circuit of four throws [impliedly in the long run] they would always be equal. So this is the rationale of contending on equal terms ...' (see Maistrov, p. 22). Note that this implies the precise rule that the expectation or present worth of an uncertain pay-off is equal to the product of the chance and the amount and a game is fair if the expected pay-offs are same for the two players, i.e. the stakes are in the ratio of the odds.

Cardano's motivation for the addition rule and the multiplication rule (for events in independent trials) in 4 and 5 above is undoubtedly based on the definition in terms of equally likely cases as given in 3. However, from his definition of a fair game it appears that the frequency interpretation of probability was very much at the back of his mind.

In view of the above we fail to see why Cardano is not generally regarded as the founder of the theory of probability. True, his pioneering book remained unpublished for too long, but he had lectured extensively on its contents. One reason may be that he treated problems of games of chance more as those of gambling than of serious mathematics; his book reads like a manual for gamblers in which particular problems are solved by direct enumeration. Another may be that because of the bad reputation he had earned on various counts he was looked upon as a charlatan and not given his dues even when those were deserved. However, many of his ideas on probability (like those of equally likely cases, the product rule for probabilities, the definition of a fair game) came to be considered as part of the general lore of the subject and were accepted as such by his successors.

Before finishing with Cardano, we cite two problems which he discussed but failed to solve correctly. They are of some importance, because their solution later heralded the advent of the theory of probability in the arena of mathematics.

    (i) Problem of minimum number of trials: What should be the minimum value of r, the number of throws of two dice, which would ensure at least an even chance for the appearance of one or more double sixes?

    (ii) Problem of division (or the problem of points): Two players start playing a series of independent identical games in each one of which one or other would win, with the agreement that whoever wins a pre-fixed number of games first would win the series and the total stake. The series is interrupted when the two players have respectively a and b games still to win. What division of the total stake between the players would be fair?

Obviously (i) means that we have to find r so that $1 - (\frac{35}{36})^r$ is just $\geq \frac{1}{2}$. Cardano came out with the solution $r = 18$, whereas the correct value is $r = 25$. (See Epstein (1967) p. 2.) Since Cardano correctly solved similar problems elsewhere, this was very likely a computational error (logarithms came into vogue only around 1620). (ii) was an outstanding problem of the time; others besides Cardano had unsuccessfully tried their hands at it.

Incidentally, Cardano as an astrologer had predicted the date of his death and committed suicide to prove himself. (Does this not provide a tragic illustration to bear out de Finetti's dictum that probabilists should previse but desist from prediction? See footnote 39 in Chapter 4.)

Galileo

5.3.2 After Cardano, a small piece of work was done, presumably around 1620 (see Hald (1990) p. 41), by the illustrious Galileo Galilei (1564–1642), who is regarded as one of the pioneers in introducing experimental methods in science. In it Galileo sought to resolve a puzzle about a dice game which somebody had asked him to explain. In throwing three dice, the numbers of unordered partitions producing the total scores 9 ($\{1,2,6\}$, $\{1,3,5\}$, $\{1,4,4\}$, $\{2,2,5\}$, $\{2,3,4\}$, $\{3,3,3\}$) and 10 ($\{1,3,6\}$, $\{1,4,5\}$, $\{2,2,6\}$, $\{2,3,5\}$, $\{2,4,4\}$, $\{3,3,4\}$) are both equal to 6. Yet, why is it 'that long observation has made dice-players consider 10 to be more advantageous than 9'? Galileo pointed out that there is a 'very simple explanation, namely that some numbers are more easily and more frequently made than others, which depends on their being able to be made up with more variety of numbers' (see Hacking (1975) p. 52). A variety of numbers making up a score here represents an ordered partition. There being 27 such ordered partitions for the score 10 and 25 for the score 9 and all ordered partitions or permutations being equally likely, the chance of getting a 10 is higher. Although no new ideas beyond those considered by Cardano were introduced by Galileo, it is to be noted that he also tacitly accepted the relative frequency interpretation of chance as represented by the behavioural principles (i)–(ii) and also principle (v) according to which for three symmetrical (identifiable) dice what marking on each defines a permutation will not affect its frequency and all permutations will be equally frequent. Also, speaking of the 'ease' of making a score, he seems to concede principle (iv) according to which chance can be looked upon as a sort of propensity. It may be further noted that (contrary to what Hacking (1975) p. 52 says) the problem considered by Galileo is not an inductive problem where some uncertainty attaches to the conclusion; it is rather a deductive problem of modelling where we presume an order relation among the chances of certain events on the basis of empirical experience and then try to hit upon a model from which the presumed order relation would follow.

Probability is officially born—Pascal and Fermat

5.3.3 Probability was 'registered as a new-born member' in the family of mathematics in the year 1654. The problems whose solution occasioned this again arose in the context of gambling. That year, Chevalier de Méré, a French nobleman with considerable experience in gambling and a nibbling interest in mathematics, brought to the notice of Blaise Pascal (1623–1662), the renowned Parisian mathematician, some gambling puzzles that he had been unable to resolve. A version of the 'problem of minimum number of throws' and the 'division problem' that we mentioned in connection with Cardano (subsection 5.3.1), were among those. As Pascal solved the problems he discussed them through correspondence with his friend, Pierre de Fermat (1601–1665) who was stationed at Toulouse. Although a jurist by profession, Fermat had become famous for his contributions to mathematics and the other branches of knowledge. Fermat threw light on some of the steps in the development and sometimes suggested alternative approaches to solution. Around the same time, Pascal wrote a *Treatise on Arithmetic Triangles* which contained a section devoted to the use of the arithmetic triangle for the solution of the division problem. Although the *Treatise* was published posthumously in 1665, it had been printed in 1654, and from one of his letters to Pascal it is clear that Fermat had received a copy at that time. Pascal's treatment of the division problem in the *Treatise* is more systematic and general than in his letters. A detailed account of the development in the Pascal–Fermat correspondence and also in Pascal's treatise is given by Hald (1990) pp. 54–61. It should be noted that Pascal and Fermat freely drew upon the general lore of probability concepts that had been part of the tradition, at least since the time of Cardano. In fact, concept-wise there was very little that was new in Pascal and Fermat, beyond what was considered by Cardano. However, algebra including the theory of permutations and combinations, had progressed a lot from the time of Cardano, and it is to the credit of Pascal and Fermat that they could formulate the gambling puzzles posed to them in formats that were mathematically interesting.

| Huygens |

The mathematics of probability is generally regarded to have begun with Pascal and Fermat in 1654. But almost at the same time a third luminary also appeared on the scene. This was the young Dutch scientist Christiaan Huygens (1629–1695). During his visit to Paris in 1655, Huygens came to know of the problems of games of chance being considered by Pascal and Fermat at that time, but remained in the dark about their methods and solutions. Huygens attacked the problems on his own and came out with solutions which, through some common friends, received the endorsement of Pascal and Fermat. Huygens wrote those and some other results in the form of a book entitled *De Ratiociniis in Ludo Aleae* (Computations in Games of Chance) which was published in 1657. This was the first published book on probability which remained the standard text for about half a century and influenced the development of the subject considerably. Accounts of Huygens's work on probability are given in Maistrov (1974), Hacking (1975) and Hald (1990).

What actually were the contributions of Pascal, Fermat and Huygens to the calculus of probability?

| Problem of minimum |
| number of trials |

As regards the problem of minimum number of trials, de Méré had posed it to Pascal in the form: throwing a single die 4 times is enough to ensure that one has a better-than-even chance of getting at least one six. Then why is there a disadvantage in undertaking to throw at least one double-six in throwing two dice 24 times although '24 is to 36 (which is the number of faces of two dice) as 4 to 6 (which is the number of faces of one die)'? Of course the presumption that the minimum number of throws would bear strictly a constant ratio to the number of equally likely cases is rather naive. Pascal resolved the puzzle by showing that for a single die in 4 throws, chance of at least one six is $1 - \left(\frac{5}{6}\right)^4 = \frac{671}{1296}$ so that the odds are $\frac{671}{625}$ whereas for two dice in 24 throws the chance of at least one double-six is $1 - \left(\frac{35}{36}\right)^{24}$ which gives approximately the odds $\frac{491}{509}$. Note that the argument requires:

(i) the idea of equally likely cases (all 36 cases for two dice are equally likely),
(ii) computation of chance as the proportion of equally likely cases that are favourable (the chance of not getting a double-six with two dice is $\frac{35}{36}$),
(iii) the product rule for independent events (the chance of not getting a double-six in all 24 throws of two dice is $(\frac{35}{36})^{24}$), and
(iv) the additivity of chance (the chance of getting at least one double-six in 24 throws is $1 - \left(\frac{35}{36}\right)^{24}$).

Apparently these ideas and rules had by that time become generally accepted and were taken as obvious. Further, the disadvantage mentioned by de Méré must have been observed in terms of empirical relative frequencies (if one could compute the odds $\frac{491}{509}$ as Pascal did, there would be no occasion for a puzzle).[3] Therefore the explanation based on computation of probabilities would hold water only if probability is interpreted as long-term relative frequency.

| Division problem |

Now consider the major achievement—solution of the hitherto unsolved division problem—by Pascal and

---

[3]It has been pointed out that to detect such a small discrepancy as between 0.491 and 0.500 of the probability of an event in a trial empirically, one would require to perform the trial almost an impossibly large number of times. (The same objection has been raised with regard to the discrepancy between the probabilities $\frac{27}{216}$ and $\frac{25}{216}$ in the problem considered by Galileo, subsection 5.3.2; see Maistrov (1974) pp. 30, 43.) But without such empirical conclusions, there is no *raison d'être* for the problem. Was it that Chevalier de Méré had access to the records of some gambling house? But even ruling that out, such games were being played over centuries by many many generations of gamblers. The experience accumulated, not as records, but in the form of hearsay and traditional wisdom might have led to such conclusions (cf. Hacking (1975) p. 53).

Fermat and a little later by Huygens. As mentioned earlier (subsection 5.3.1), in the division problem in its general form, two players A and B start a series of independent games in each of which of A and B have chances p and $q = 1 - p$ of winning, with the understanding that whoever wins a prefixed number of games first wins the series and the total stake. If the series is interrupted when A has still a games and B has b games to win, how should the total stake be divided between the two? All three of Pascal, Fermat, and Huygens discussed the division problem in the symmetric case $p = q = \frac{1}{2}$. Pascal and Huygens gave the solution by recursion on the expectation, whereas Fermat solved it combinatorially.

| Combinatorial solution | First consider Fermat's solution since it is more elementary, although historically Pascal's solution preceded it. Fermat argued that if play were continued for another $a + b - 1$ games, either A or B would certainly win within that. Since the games are symmetric all the 2^{a+b-1} sequences of outcomes for these additional games are equally likely. A can win if and only if anyone of the b mutually disjoint events $A_0, A_1, \ldots, A_{b-1}$ occurs, where

A_j: A wins in $a + b - 1 - j$ and B in j of the additional games,

$\quad j = 0, 1, \ldots, b - 1$.

Since, just $\binom{a+b-1}{j}$ equally likely cases are favourable to A_j, by the additive rule, we get that the chances of A and B winning the series, if it were continued until termination, are

$$P_A = 2^{-(a+b-1)} \sum_{j=0}^{b-1} \binom{a+b-1}{j}, \quad P_B = 1 - P_A. \qquad (5.3.1)$$

The prescription was to divide the total stake in the ratio P_A/P_B. No justification in support of the fairness of this was given; apparently division in the ratio of odds had been generally accepted by that time. (Recall that Cardano had sought to justify his definition of fair game in terms of frequencies in repeated play.) (If the games were asymmetric with $p \neq \frac{1}{2}$, the 2^{a+b-1} outcome sequences would be no longer equally likely. By invoking the product rule for independent events nowadays we would argue that the chance of every sequence constituting A_j is $p^{a+b-1-j}q^j$ and hence get a modified expression for P_A. But recall that Fermat would use the product rule only if it could be justified in terms of equally likely cases. Thus the development could be sustained through a circuitous argument for a rational-valued p but not for any $0 < p < 1$.)

| Recursive solution | In solving the division problem both Pascal and Huygens took the present value or expectation as their starting point. Each player's expectation was taken as the share of the total

stake to which he is entitled. As the winner here gets the entire stake and the loser gets 0, the fraction of the total stake representing his share is equal to his chance of winning. Thus the ratio of their chances is same as the ratio of their expectations.

If X is any random variable and H is an event with probability P, nowadays the expression

$$E(X) = P\, E(X|H) + (1 - P)E(X|\,\text{not-}H) \qquad (5.3.2)$$

for the unconditional expectation of X in term of its conditional expectations given H and not-H is quite familiar. Pascal and Huygens repeatedly made use of (5.3.2) in their derivation. Denoting by $e(a, b)$ the pay-off that player A can expect in the situation when A and B have still a and b games respectively to win if the series were played until termination and taking 'A wins in the next game' as the conditioning event H, from (5.3.2) we get the recursive relation

$$e(a, b) = \tfrac{1}{2}\{e(a - 1, b) + e(a, b - 1)\} \qquad (5.3.3)$$

If the total stake is 1, clearly we can take $e(0, n) = 1$, $e(n, 0) = 0$ for $n > 0$. Pascal used (5.3.3) to compute $e(a, b)$ numerically for low values of a, b, and with the help of his results for Arithmetic Triangles also deduced the general solution $e(a, b) = P_{\mathrm{A}}$, where P_{A} is given by (5.3.1). Huygens derived only numerical solutions.

The pioneers' world of probability

It may be noted that Pascal, Fermat, and Huygens in their development confined themselves only to trials with identifiable equally likely cases and (at least implicitly) used for (rational-valued) probabilities the notions of finite additivity and the product rule for independent trials (trials are independent if all combinations of their individual equally likely cases can be taken to be equally likely). In Huygens's book there are some problems involving a potentially unlimited sequence of games. For example, in throwing two dice, A or B wins according as the total score is 7 or 6, and B and A successively take turns until one of them wins. In another problem, known as that of the gambler's ruin, two players having finite numbers of counters go on playing a sequence of games for each of which the winner gets a counter from the loser, until one of them loses all his counters. (This problem later came to be known as 'random walk with absorbing barriers' and assumed importance in particular in physics and in connection with sequential experimentation.) Naturally questions involving countable additivity of probability can be asked in the context of these problems. However, Huygens solved the problems within the ambit of finite additivity through clever dodges based on expectation and avoided such questions.

Although Pascal and Huygens both worked in terms of expectation of a game, there seems to be some difference in their view-points. Thus Pascal seems to have taken the value or expectation of a game as a primitive concept; in his development the probability of winning a game whose pay-off is equal to the total stake in case of a win and 0 otherwise is just the fraction of the total stake represented by the expectation. This is very similar to the attitude taken by the present-day exponents of subjective probability (cf. subsections 3.3.2 and 3.3.3). (Pascal discussed expectation only in the context of a sequence of games which would be generally accepted as symmetric. Hence the probabilities implied can be taken to be impersonal subjective.) We would see that Pascal later used probabilistic arguments to discuss theological questions and only a probability of the subjective kind can have relevance there. Huygens, however, seems to have taken probability itself as the primitive concept. In fact for a game which to a player yields the pay-offs a and b with (rational valued) probabilities p and $q = 1 - p$ he defined $pa + qb$ as the expectation. (By implication he extended the definition also to the case of more than two pay-off values; see Hald (1990) p. 70.)

What is most interesting is that Huygens sought to give a justification for regarding the expectation as the fair price to be paid by a player for the oppor-tunity to play a game. We saw earlier (subsection 5.3.1) that Cardano argued for the expectation as the fair price in terms of the relative frequency interpretation of probability, assuming that the game would be played repeatedly. Huygens, however, considered the one-time situation. He defined a fair lottery as one in which if n tickets are to be sold to n persons at a fixed price x per ticket, the win-ner of the lottery would get the entire pool nx and all the ticket-holders would have the same probability of winning. (The lottery is fair because neither the house nor the community of ticket-holders would have a net gain.) For a game in which there are two pay-offs a and b with corresponding probabilities $p = m/n$ and $q = (n - m)/n$, where $m < n$ are positive integers, he presumed that the player can regard himself as participating along with $n - 1$ other persons in a fair lottery in which there are n tickets each priced at $\{ma + (n - m)b\}/n$. The player enters with $m - 1$ of the other participants into an arrangement under which he would pay each of them an amount a if he wins, provided they also pay him the same amount in case of winning. With each of the remaining $n - m$ participants he enters into a similar arrangement but the amount this time is b. To the player, participation in such a lottery (with such an internal arrangement with the other participants) is equivalent to the given game since his probability of getting the amounts a or b is p or q in both cases. Since the fair price for the lottery is $pa + qb$, the same is true of the game. If we adopt the relative frequency interpretation of probability, the argument really gives a reason for instantiation (cf. subsection 3.2.5) in the case when the player plays only once. If one takes a subjective view of probability, the argument is akin to that given these days by some protagonists of subjective probability (cf. Hacking (1975) p. 97).[4]

---

[4]Maistrov (1974) p. 51 puts forward the Marxian thesis that in the 17th century the arithmetic mean was in use as a tool for accounting in Holland which was commercially well-advanced, and Huygens imported the same idea in the form of expectation in studying the fair value of a game. But we see that in his book at least, Huygens

As noted above, Pascal and Fermat treated problems of games of chance merely as a new kind of mathematical problem and did not bother about the applicability of the ideas of chance in any other field. On the other hand Huygens, right from the beginning, seems to have a wider vision. In a letter written in 1657, which appeared as the foreword of his book, Huygens remarked, 'I would like to believe that in considering these matters closely, the reader will observe that we are dealing not only with games but rather with the foundations of a new theory, both deep and interesting' (see Maistrov (1974) p. 48). We will see that within a few years Huygens came across some practical problems which provided ready scope for the application of this new theory.

5.4 Beginning of data analysis

5.4.1 At the beginning of this chapter we had said that if statistics be identified with statistical induction, then it could not have emerged before the theory of probability. But statistical induction requires statistical data which would represent the empirical component in its premises. Hence, the question naturally arises: when did the systematic and purposeful collection of statistical data start?

| Ancient and medieval official statistics |

Information about the population, land, agriculture, and other economic activities in a country are required for military mobilization, revenue collection and the undertaking of welfare measures. From the earliest times, in every civilization, whenever a powerful government was at the helm of a country attempts were made to gather such information through administrative channels. Thus population, agricultural, and economic censuses were undertaken in ancient Egypt and Judea millenia before the birth of Christ, in the Roman empire for several centuries up to the 1st century A.D., and in India during the Mauryan period (4th–3rd centuries B.C.). In the medieval period collection of such data was organised for instance in England by William the Conqueror (11th century), in Russia by the Mongols after their conquest of the country (13th century), and in India by the Mughal emperor Akbar (16th century). Data on population and various trading activities used to be collected by the Italian city states during the 13th–15th centuries. Also parish churches in some European countries used to maintain registers of weddings, christenings, and burials in their parishes regularly (see Mahalanobis (1950), Maistrov (1974), Hacking (1975)). However, such information was mainly used to serve its immediate purpose; nobody thought of analysing the data in depth to derive interesting conclusions. In fact, the consciousness that new knowledge could be projected from such data through the superimposition of plausible assumptions was generally lacking.

---

justified the expectation solely in terms of fair lotteries and not as the idealization of the mean receipt from a number of repetitions of the game. Therefore such a thesis seems hardly sustainable.

London bills of mortality—John Graunt

5.4.2 The official data which occasioned such an exercise for the first time, as per available records, were those contained in the London *Bills of Mortality*; these were weekly gazettes of the numbers of christenings and burials in the different parishes of the city of London as registered in the established church. The person who undertook the study was John Graunt (1620–1674), a London businessman (draper) with keen intellectual interests, an innovative mind, and a strong common sense.

Throughout the middle ages right upto the 17th century, the scourge of plague used to recur in the cities of Europe periodically. In London, after a devastating plague epidemic occurred in the 16th century during the reign of Henry VIII, it was decided that weekly gazettes reporting plague mortality in the city would be brought out to give forewarning about an impending epidemic. Later on, the scope of these bills of mortality was enlarged to cover other information. The bills, which at the outset were brought out from time to time, started being regularly published from 1604 onwards.

When Graunt took it upon himself the task of analysing the information in the bills of mortality around the 1660s, he had at his disposal continuously published bills for some 60 years for the parishes of London. (Graunt had data from 123 city parishes; later he also got the figures for three country parishes.) These gave the weekly numbers of births (christenings) and deaths (burials), both classified by sex (from 1629 onwards). Deaths (including abortions and still births) were also classified according to the cause of death (Graunt could procure the data for 81 causes). As indicators of the current state of health of the city, the weekly bills had been of flitting interest to its residents. Nobody had thought that their collective information as accumulated over the years could be otherwise useful. In the mean time, however, people were coming to realize the importance of empirical studies and induction as tools for generating knowledge. Bacon published his *Novum Organum* in 1620 and made a strong plea for the use of inductive methods in scientific studies (see section 1.9). It is clear that Graunt was influenced by Bacon's ideas. He undertook the study of the volume of data represented by the bills of mortality and on the basis of that in 1662 published a book entitled *Natural and Political Observations Made upon the Bills of Mortality* (in one of the dedications in the book Graunt refers to Bacon). The book ran into five editions within the next fourteen years. In it he summarized the data and after critical appraisal followed by necessary adjustments superimposed plausible assumptions to derive various demographic conclusions. A detailed account of Graunt's work is given by Hald (1990). Here we consider only a few selected topics to illustrate how Graunt's study influenced the growth of statistics.

Graunt's estimation of population

No comprehensive census data on the size of the population were available in Graunt's time. (Decennial population censuses in England were started only in

1801.) People used to make wild guesses (sometimes one million, sometimes two) about the size of the London population. Graunt hit upon the idea that an objective estimate could be derived on the basis of the total number of births reported for the city. He argued like this. If the annual number of births is B, the number of child-bearing women (in the age group 16–40, according to Graunt) would be $2B$ for such women can bear 'scarce more than one child in two years' (this amounts to a general fertility rate of $\frac{1}{2}$). Assuming that the number of women in the age group 16–76 is twice this and that is the same as the number of households, the latter turns out to be $4B$. Graunt assumed that the average household size is 8 (man, wife, three children, and three servants or lodgers). This gave him $32B$ as the size of the population. To get B for a year, Graunt did not directly use the number of christenings. (Comparing with the number of deaths in non-plague years, he had found that after 1642, what with the civil war and other reasons, the practice of baptism had become less universal.) Instead he multiplied the number of burials, 13,000, by the factor $\frac{12}{13}$ (the ratio of the number of christenings to the number of burials over the non-plague years before 1642) to get the estimate 12,000 of B for the year 1661. Hence the size of the population would come out as 3,84,000. To account for the population in several suburban parishes whose number of burials was about $\frac{1}{5}$ of that in the city proper, he inflated the figure by 20% to get the estimate 460,000 of the population of greater London. Note that the formula $32B$ implies the assumption of the value $\frac{1}{32}$ for the Crude Birth Rate (CBR = ratio of number of births to total population).[5,6]

It is remarkable that Graunt sought to validate the above estimate in other ways. Thus on the basis of observations made in three parishes, he found that on an average there were 3 burials per 11 households in a year. Taking the average household size as 8, this gave an estimated Crude Death Rate (CDR = ratio of number of deaths to total population) of $\frac{3}{88}$. For an annual number of 13,000 of deaths this would give an estimate of population pretty close to the earlier figure. He also gave a third argument based on the area of the city and the density of resident families in it.

---

[5]Strangely, using Graunt's formula on the number of births in 1851, Greenwood (1941) found that the estimated London population (about 2,400,000) differed from the actual census figure for the year by only 1.7%! Greenwood conjectured that possibly Graunt's over-estimated value $\frac{1}{2}$ for the general fertility rate was offset by an over-estimation also of the average house-hold size, so that the CBR value of $\frac{1}{32}$ was close to the actual. Incidentally, in recent times there has been drastic fall in the CBR in most western countries (e.g. in U.K. it is around $\frac{1}{80}$) but in India in 1990 it was approximately $\frac{1}{33.5}$.

[6]Graunt's method of population estimation on the basis of the total number of births and the assumption that CBR remains constant over a period, was generally accepted as sound. In fact, as we will see in Chapter 7, even one and a half centuries later, Laplace used essentially the same method to estimate the population of France in 1802.

What did the assumptions made by Graunt amount to? Apart from partic-
ularities like the average household size, in the approach based on births he was
essentially assuming a stable population in which at any time the relative proportions of women of different ages remain the same and in any year the relative
frequency of women in the child-bearing period who give birth to a child is the
same. (This would hold if the life-histories of all female children born represent
independent and identical stochastic processes.) A similar assumption of stability of the population as a whole underlies the approach based on deaths. It is
to be noted that these assumptions really involve assuming, what in Chapters 2
and 3 we called frequential regularity, in respect of certain events. In the next
subsection we consider Graunt's treatment of another question where the idea
of frequential regularity comes out more openly.

| Graunt's life table | Graunt's intention also was to determine the age-wise
composition of the population. For that he asked him-
self the question: of 100 children born, how many die before attaining the age
of 6 years, between 6 and 16, between 16 and 26, and so on? At that time, age
at death was not reported in the London bills of mortality, and anybody would
have considered it baffling to answer the question on this basis. However, Graunt
ingeniously conceived that an estimate of the proportion dying before 6 years
could be obtained on the basis of the number of deaths due to causes specific to
children. Over the period of investigation (1604–1661) he considered the number
of deaths due to ten such causes (abortion and still births, thrush, convulsion,
rickets, teething, etc.) and inflated it by adding half the number of deaths due to
certain other causes (like small pox, measles etc.) which were common to both
children and adults. Relating the figure to the total number of non-plague deaths
over the period, he concluded that of 100 children born, 36 die before the age
6 (this involved the assumption that the population remained stationary over a
period). From the number of deaths due to old age reported in the bills, Graunt
inferred that only 1 among 100 children born die after attaining the age 76. He
took six 'mean proportional numbers' to determine the survivors at the intermediate ages $16, 26, \ldots, 66$. Presumably, he used the uniform decrement factor $\frac{5}{8}$
on the number at the beginning of each ten-yearly age interval to get the number
of survivors at the end, except towards the tail where he used a larger factor. In
this way he got the following ℓ_x column (as we would now call it) of an *abridged
life-table* with cohort 100.

Note that the idea of stability of mortality rates or relative frequencies of
persons dying in different age intervals (of course apart from plague) is inherent
in this development. In fact the ℓ_x column gives the 'more than cumulative'
relative frequency distribution of the variate 'age-at-death' for a new born child
in a tabular form. Following the lead given by Graunt, this mode of representing
the distribution became standard in demography.

Graunt's purpose in all this was to know the age composition in the stable
population (presumably, what would take shape under existing conditions, if

Table 5.1. *Graunt's Life Table (1662)*

Age x	Number of Survivors ℓ_x	Age x	Number of Survivors ℓ_x
0	100	46	10
6	64	56	6
16	40	66	3
26	25	76	1
36	16		

plague could be eliminated) and hence to estimate the proportion of 'fighting men', i.e. men who are fit to bear arms in the case of a war (males in the age-group 16–56). He broke up the figure $34(= \ell_{16} - \ell_{56})$ percent according to the appropriate sex ratio and took the corresponding male part as this estimate. (This would be correct if all births in a year occurred at the same instant and all deaths in an age-interval occurred just before reaching the upper boundary.)

| Graunt's other studies | Besides the above, on the basis of the bills of mortality Graunt also made a number of other investigations. These included studies on the temporal variation in the intensity of the plague epidemic, trend in the mortality due to certain other causes, extent of migration into the city from the country, correspondence between the rise in sickliness and fall in fertility over the years, and the near-constancy of the sex ratio at birth. Generally, his technique was to compute and compare relevant numbers and ratios at different times and places. To smooth out irregularities while studying the trend of a time series he often considered subtotals for successive non-overlapping groups of years. As regards the sex-ratio at birth, he found overall a slight preponderance of males over females. (Apparently, it was Graunt who noted this fact for the first time on objective grounds.) He remarked on the significance of this fact as providing an argument against polygamy: generally, there is a reduction in the number of men available for marriage due to emigration and higher risk of death, but the higher male birth rate maintains a balance between the sexes at the ages of marriage. We would see that later on this imbalance in the sex-ratio at birth became a standard topic of investigation which was thought to have theological significance.

| Speciality of Graunt's methods | Can we call Graunt's methods of data-analysis statistical induction? Of course through these, inferences representing new knowledge were sought to be drawn utilizing all available information, empirical and other. Graunt must have been clear in his mind about the uncertain nature of his conclusions, otherwise he would not have tried to validate them through alternative arguments as in the

case of his estimate of population. But at that stage, there was no question of formulating the assumptions and assessing the uncertainty in the conclusions in probabilistic terms. In view of our discussion in sections 1.7, 2.7, and 2.10, we can say that Graunt's methods fell short of but foreshadowed statistical induction. From another point of view we can say that before Graunt, empirical studies were generally based on qualitative descriptions (like in what Mill (1843) later called method of agreement and difference, i.e. co-presence and co-absence of cause and effect); Graunt emphasized the importance of comparing quantitative indices in the form of demographic ratios across various situations.

Impact of Graunt's work

5.4.3   Graunt's book had far-reaching impact on the thinking of people not only in England but also in many continental countries.

Firstly, in a general way, the value of numerical data as the basis for projecting interesting and useful conclusions under plausible assumptions was realized through his study. This paved the way for the development of statistical induction in a broad sense. Graunt's friend and confidant, William Petty visualized very well the potential of such data-based methods. He himself followed up with studies on economic data to estimate the national income of England by first computing the per-capita income (which he took to be same as consumption) and then multiplying that by the estimated population of the country. He examined the scope for further taxation by the Government in course of that study. Although his database was scanty and hardly convincing, his problems and approach were quite novel. For this, Petty is generally regarded as a pioneer in the field of political economy. Petty also made some demographic studies and urged that data along the lines of the London bills be gathered on a wider scale for comparing the degrees of salubrity of different localities and countries. The kind of work that Graunt and Petty did was for a time known as 'Political Arithmetic'. Note that Graunt only analysed available data for drawing various conclusions, whereas Petty further suggested that new data be purposefully collected to answer various pre-formulated questions.

Graunt's and Petty's investigations and the latter's pleadings for comprehensive data collection within a short period led to the introduction of bills of mortality similar to those of London in several other centres in Europe. Further, a general awareness about the indispensability of demographic and economic data for governmental decision-making gradually grew, albeit slowly. By the early 19th century, regular population censuses were started and statistical offices set up in many countries. The official recognition gained by statistics helped the development of certain methods of data collection and certain modes of statistical induction (theory and practice of sample surveys, time series analysis etc.). (The word 'Statistics' is etymologically related to 'State' and originally meant descriptive facts useful for statecraft. By a process of mutation, by the turn of the 18th century it came to mean numerical data relating to the State and later, any numerical data gathered through observation. Around the beginning of the 20th century, the further connotation of a branch of knowledge was added to the word.)

A second consequence of Graunt's work was realization (first by mathematicians, and later by authorities) of the importance of life tables in providing a rational basis for the computation of present values of life annuities. Christian Huygens's brother, Lodewijk, in a letter written to the former in 1669, suggested the computation of expected life at different ages from Graunt's table for this purpose and the two had some correspondence on the matter at the time. Around 1671 their compatriots Jan de Witt and Jan Hudde (who were mathematicians turned administrators) carried out similar exercises more thoroughly, on the basis of tables prepared from annuity data collected in the Netherlands. A little later, Edmond Halley, the well-known British astronomer, prepared a life table on the basis of five years' detailed records of births and deaths by sex and age, maintained at a German township called Breslau, and in a paper presented to the Royal Society in 1693 gave the table and various computations based on it. (The records had been procured by the Society through the courtesy of Leibniz and made available to Halley). Halley's life table came to be regarded as a standard for nearly a century after this. In course of time, apart from annuities for life, life insurance (which is a sort of reversed annuity[7]) came into practice. Of course all these studies were mainly deductive in nature—no element of uncertainty in the conclusions was kept in purview while making them.

The third consequence of Graunt's treatise was also related to his idea of a life table, but it had wider implications. It followed from Graunt's work that frequential regularity occurs in the case of various demographic phenomena just like in the case of games of chance. Hence, even though in the context of the former no equally likely cases can be identified, if one interprets probability objectively in terms of relative frequency, one can extend the scope of application of probability theory to such phenomena. We recall that Huygens, in the preface to his 1657 book, had visualized that probability theory would have deeper and more interesting applications than in problems of games of chance. In 1662, when Graunt's book came out, the Royal Society referred a copy of it to Huygens for his opinion. Looking through the book, Huygens must have immediately seen the probabilistic interpretation that could be given to Graunt's work. (He gave a favourable assessment and on that basis Graunt was admitted as a Fellow of the Royal Society.) Later, in course of his 1669 correspondence with his brother Lodewijk, Huygens clearly spelt out this probabilistic interpretation (see Hald (1990) pp. 108–110) by pointing out that Graunt's life table can be regarded as a

---

[7]The annuitant pays a lump-sum amount initially, in return for an assured annual remittance for life; the insurant pays annually a certain premium in return for a lump-sum amount in case of death.

lottery with 100 tickets, 36 of which have values in [0, 6), 24 have values in [6, 16) and so on. The ℓ_x column thus represents in percentage terms the right-hand cumulative probabilities of a random variable. He drew the corresponding curve (the 'more-than' *ogive* curve, to use a term introduced by Galton two centuries later) and showed how it can be used to find graphically the solution for t in the equation $\ell_{x+t} = \frac{1}{2}\ell_x$ for each x. Huygens said that a person aged x has an even chance of dying before or after a further t years and distinguished this 'median residual life' (as we would call it nowadays) from the mean residual life which is the excess of the expected life at age x over x. In the words of Hacking (1975) p. 109, by the 1670's 'problems about dicing and about mortality rate had been subsumed under one problem area'. The gates were now open for the application of probability theory in any practical situation where we can presume frequential regularity to hold.

5.5   Diffusion of probabilistic thinking—objective and subjective

| Pascal and Port Royal |
| Logic |

5.5.1   We mentioned earlier that Pascal, during his 1654 correspondence with Fermat, had treated problems of chance as bare mathematical problems and had not considered any wider applicability of the concept of chance. Pascal was a member of the Roman Catholic sect of Jansenites (who called themselves followers of St. Augustine and were staunch believers in predestination). After 1654 he left Paris and retired to the retreat of Jansenites at Port Royal. Pascal's associates there, among whom were leading French philosophers like Antoine Arnauld and Pierre Nicole, were busy writing a voluminous book on logic which came out in 1662 with the Latin title *Ars Cogitandi* ('The Art of Thinking'—but the book was generally known as the Port Royal *Logic*). There are four chapters in the fourth part of the book which discuss probabilistic reasoning from a broad perspective and it is believed that these, even if not authored by Pascal, reflected his thinking.

It is remarkable that in the Port Royal *Logic*, the term 'probability' was freely used in place of 'chance', which had been in vogue in the context of games. In the case of uncertain events or propositions like 'a person would be struck by lightning' or 'a notarized contract has been post-dated', it was agreed that probability should objectively be judged on the basis of past frequencies. But in instantiating to a particular case, the importance of utilizing all available information about the case and taking similar instances on that basis as the reference set was emphasized. Thus in judging a particular contract, knowledge about the lack of probity of the notary involved is important (cf. behavioural principle (iii), subsection 5.2.1 and also Russell and Fisher, subsection 3.2.5). Further, the rationality of taking account of both the gravity of an uncertain event and its probability in forming an attitude or taking an action was clearly pointed out. Thus it was argued that as death through lightning is very uncommon (reportedly less than 'one in two million') one should not be unduly concerned about it. 'Fear of harm ought to be proportional not merely to the gravity of the harm, but also the probability of the event' (see Hacking (1975) p. 77). Thus the principle of expectation was sought to be justified in a wider context.

Pascal's wager   The expectation principle was invoked explicitly, and indubitably by Pascal himself, in course of a theological argument for the maxim that every rational man must decide to lead a pious life. The argument, which came to be known as 'Pascal's wager' appeared in 1670 in a posthumous publication of Pascal's; but it was summarized in the concluding part of the *Logic*. From our modern viewpoint it can be regarded as the solution of a no-data decision problem with two hypotheses, namely, H_1: God exists (Pascal of course had the Roman Catholic faith in mind) and H_2: God does not exist, and two decisions (or actions) for wagering, namely, a_1: lead a pious life and a_2: lead a worldly life. Pascal constructed utilities (or negative losses) as follows.

	H_1	H_2
a_1:	∞ (eternal bliss)	finite negative (loss of worldly enjoyments)
a_2:	finite positive (no more than worldly enjoyments) or $-\infty$ (if possibility of damnation is admitted)	finite positive (worldly enjoyments)

Pascal argued that if to a person H_1 has a priori a positive probability, however small, the expected utility of a_1 would exceed that for a_2 for such a set of utilities. The argument is exactly like what would be offered nowadays by a Bayesian with no data at hand (see subsections 4.4.6 and 4.5.3). (See Hacking (1975) for a detailed analysis of Pascal's wager.) It is most noteworthy that in this argument Pascal is postulating subjective probabilities (which may be personal) of propositions which have no relation to any random experiment.[8] This is arguably the earliest occasion after the advent of the calculus of probability,

---

[8]Hacking (1975) p. 70 observes that Pascal did not speak of a quantitative measure of degree of belief but merely said 'that we are in the same *epistemological* position as someone who is gambling about a coin whose aleatory properties are unknown'. But this is exactly how some proponents interpret probabilities based on judgement (see e.g. Carnap (1950) p. 237; cf. also Fisher's view of fiducial probability in subsection 4.3.15).

when computations are notionally made with a probability which is intrinsically subjective.

Leibniz **5.5.2** The subjective nature of probability was openly brought out in the writings of the German philosopher–mathematician Gottfried W. Leibniz (1646–1716). Leibniz did not make any contribution of note to the calculus of probabilities, but it was he who first conceived the probability of an uncertain proposition A relative to a body of evidence B as a measure of the degree of certainty of A given B or partial implication of A by B. Thus according to Leibniz probability is primarily impersonal and subjective. (Probability is subjective because according to Leibniz's philosophy, in the ultimate analysis every proposition A is either true or false; see Russell (1961) pp. 573–574. Uncertainty about a proposition arises only because of the limitedness of the evidence B available to us.) Thus Leibniz anticipated the explication of probability offered by Keynes and Jeffreys two and a half centuries later (subsection 3.3.2).

Leibniz, who had a legal background, in his early writings conceived probability in the context of evaluating the strength of the conditional legal right to a property given a mixed body of evidence, parts of which established and parts negated the right. Later, while visiting Paris during 1670s, he became acquainted with the probability mathematics developed by Pascal, Fermat, and Huygens and visualized the applicability of the same to subjective probability. Since in the law of conditional rights it is possible to break up the evidence into a number of disjoint irreducible parts, some of which are favourable and the others unfavourable to the claim under study, probability could be defined as in games of chance provided symmetry of the break up allows the irreducible parts to be interpreted as equally likely cases. To justify this interpretation he invoked what later came to be known as the *principle of insufficient reason* or *principle of indifference* (cf. non-informative priors in subsection 4.4.4). This means, when there is no reason to 'fear or hope' that one case is true rather than another, we can take them as equally likely, or in other words, 'equal suppositions deserve equal consideration' (see Hacking (1975) p. 136). (As we will see, this principle was later worked upon to the point of absurdity by some later writers and that led to a lot of controversy.) Leibniz looked upon probabilistic reasoning as an extension of classical two-valued logic. As early as in that toddling period of probability, he even visualized that a sound logic based on probability (he called it 'natural jurisprudence' because of its similarity with legal reasoning) could be developed to provide a unified impersonal approach to problems of induction according to which one should choose the proposition which has the highest probability *ex datis*. But of course he could not foresee what efforts and conflicts, construction, demolition and reconstruction would be enacted during the next three and half centuries in the process of working out the details of the programme and still a universally accepted final solution would remain elusive?

(An account of Leibniz's work on probability is given in Chapters 10, 14, and 15 of Hacking (1975)).

5.6 An attempted synthesis of various concepts of probability

James Bernoulli—*Ars Conjectandi*

5.6.1 From the account given in the preceding sections, it is apparent that as one approached the end of the 17th century, although there was general agreement about how probability is to be mathematically manipulated, the inner scene in the conceptual field of probability was one of messy confusion. On the one hand, there was the concept of probability of a contingent event for a game of chance defined as the proportion of equally likely cases favourable to it. This we may call *a priori* probability. On the other, for repeatable natural phenomena like demographic events whose outcomes are unpredictable but for which no equally likely cases are identifiable, probability was conceived as some sort of physical tendency or propensity which got reflected in long-term relative frequency. Then there was the question about the nature of probability: does probability exist objectively for repeatable phenomena as the tendency of a contingent outcome to be realized, or is probability purely subjective as in the context of Pascal's wager or Leibnitz's 'natural jurisprudence', being attributable to *any* uncertain proposition relative to the evidence available? (A priori probability is objective or subjective according as cases are considered equally likely because they have equal physical tendencies or because they are subjectively judged to be so.)

The first person who sought to clear the confusion and attempted a synthesis of the various concepts was the Swiss mathematician James (Jacob or Jacques) Bernoulli (1654–1705). He toiled with the issues involved for twenty years and came out with a partial resolution. Apparently, for one reason or other, he was not fully satisfied. For although he wrote a book containing the fruits of his toils, entitled *Ars Conjectandi* or the Art of Conjecture (the similarity with the title of the Port Royal book of logic *Ars Cogitandi* suggests that Bernoulli looked upon the process of probabilistic conjecture as an extension of logical thinking), he left it unpublished. The book was published by his nephew and former student Nicholas Bernoulli in 1713, eight years after his death. The publication of the book marked the beginning of a broader theory of probability, not confined solely to games of chance.

Bernoulli's book consisted of four parts. The first three, which covered traditional grounds, contained respectively a critical review of Huygens's book, a systematic development of the theory of permutations and combinations, and the solution of a number of problems of diverse types on games of chance. Bernoulli was the first to derive the general form of binomial distribution for the frequency of successes in a number of independent and identical trials each of which may result in either a success or a failure. Such trials came to be subsequently known after him as Bernoulli trials. (Recall that a particular case of the binomial distribution had been considered by Fermat in connection with the division problem; subsection 5.3.3.) Also, it is noteworthy that Bernoulli explicitly introduced and handled the property of countable additivity of probability in course of solving some problems involving a possibly unlimited number of trials (in subsection 5.3.3 we saw that Huygens earlier had avoided the question of countable additivity in such contexts by clever dodges).

| Bernoulli's limit theorem |

5.6.2 In the fourth part of his book, Bernoulli broke new grounds. His object was to mathematically validate the practice of determining the probability of an event for a repeatable trial, for which no equally likely cases can be enumerated and as such no a priori probability can be computed, by its relative frequency in a large number of repetitions. He realized that for this one would require to establish that when an a priori probability exists and the trial is repeated a sufficiently large number of times, the relative frequency approximates the probability closely in some reasonable sense. Then only can one explain an observed discrepancy between the relative frequencies of two events by showing that a comparable discrepancy exists between their a priori probabilities. (Recall that Cardano, Galileo, Pascal, and Fermat all advanced this kind of explanation for the empirical findings of gamblers.) Also given such a result, when the a priori probability of an event cannot be computed, *we can postulate its existence* and substitute an observed long-term relative frequency as an estimate for it under the presumption 'that a particular thing will occur or not occur in the future as many times as it has been observed, in similar circumstances, to have occurred or not occurred in the past' (Bernoulli (1713), quoted by Stigler (1986) p. 65; note how the assumption of 'Uniformity of Nature' is invoked here, cf. section 2.2). Bernoulli visualized that the establishment of such a result would allow one to apply the art of conjecture based on probability empirically in games of skill, natural phenomena, and events relating to 'civil, moral, and economic affairs'.

The result Bernoulli proved is known as Bernoulli's limit theorem. Its modern version is: if S_n is the random number of successes in n Bernoulli trials with probability of success p, then given any $\epsilon > 0$ and $0 < \delta < 1$, we would have

$$P\left(\left|\frac{S_n}{n} - p\right| < \epsilon\right) > 1 - \delta \tag{5.6.1}$$

provided the number of trials n is not less than a threshold value $n_0(\epsilon, \delta; p)$ depending on ϵ, δ, and p. In other words, as $n \to \infty$, for any fixed ϵ the probability on the left of (5.6.1) tends to one. Of course we can express this in terms of convergence in probability (section 3.10) and say that as $n \to \infty$, $S_n/n \overset{P}{\to} p$. Bernoulli considered the situation when $p = r/t$ is an a priori probability based on t equally likely cases of which $r(0 < r < t)$ are favourable to success and $\epsilon = 1/t$; however, since we can always replace r and t by rk and tk where $k \geq 1$ is any integer keeping p the same, the restriction on ϵ is unnecessary.

In (5.6.1) S_n follows a binomial distribution with parameters (n, p). To prove the theorem, Bernoulli derived a lower bound depending on ϵ, p, and n to the probability on the left by directly examining the relative values of the various binomial terms involved and then showed that this lower bound can be taken arbitrarily close to 1 by making n large. For our purpose it is more convenient to consider the much shorter proof given by Tshebyshev about one and a half centuries later; this allows p to be any real number in $(0, 1)$ and the merits and limitations of Bernoulli's theorem can be brought out equally well on its basis. However, it requires the notion of variance of a random variable which was unknown in Bernoulli's time. Writing q for $1 - p$

$$\mathrm{Var}\left(\frac{S_n}{n}\right) = \frac{pq}{n} = \sum_{s=0}^{n}\left(\frac{s}{n} - p\right)^2 P(S_n = s) > \sum_{s:|s/n-p|\geq\epsilon}\left(\frac{s}{n} - p\right)^2 P(S_n = s)$$

$$\geq \epsilon^2 P\left(\left|\frac{S_n}{n} - p\right| \geq \epsilon\right).$$

Hence

$$P\left(\left|\frac{S_n}{n} - p\right| \geq \epsilon\right) \leq \frac{pq}{n\epsilon^2} \tag{5.6.2}$$

so that

$$P\left(\left|\frac{S_n}{n} - p\right| < \epsilon\right) \geq 1 - \frac{pq}{n\epsilon^2} \tag{5.6.3}$$

Therefore if we take $n_0(\epsilon, \delta; p)$ to be the least integer for which $(pq)/n\epsilon^2 < \delta$ i.e.

$$n_0(\epsilon, \delta; p) = \text{smallest integer} > \frac{pq}{\delta\epsilon^2}, \tag{5.6.4}$$

(5.6.1) would hold for $n \geq n_0(\epsilon, \delta; p)$. Bernoulli suggested that for a small δ (of the order of $1/100$ or $1/1000$) an event with probability less than δ may be regarded as 'morally impossible' and dually an event with probability greater than $1 - \delta$ may be regarded as 'morally certain'. He thought that since it is rarely possible to achieve total certainty, in practice an event which is morally certain may be regarded as absolutely certain—such an event 'cannot be perceived as not to happen'. Thus for small δ and $n \geq n_0(\epsilon, \delta, p), S_n/n$ in (5.6.1) can be interpreted as providing an estimate of p upto a margin of approximation ϵ. Of course $n_0(\epsilon, \delta, p)$ depends on p; but since $\max_p n_0(\epsilon, \delta, p)$ is finite (in (5.6.4) this maximum corresponds to $p = \frac{1}{2}$), we can make n large enough so that (5.6.1) holds, whatever p. But there are other problems which arise when we try to apply the theorem in practice. Firstly, the probability in (5.6.1) being derived from the a priori probability p is itself a priori. To assume that an event with a priori probability as high as, say 0.99 or 0.999 would occur practically always, involves an element of circularity, virtually begging our original question why the relative frequency of an event would lie close to its a priori probability (at least when the latter is high). To take a morally certain event for granted amounts to postulating an interpretation clause as in the present-day formalistic approach to probability (subsection 3.2.4). But even under such a postulate, (5.6.1) is really a pre-experimental assertion in which S_n is a yet-to-be-observed random variable. After S_n has been observed no direct significance can be attached to the absolute level of the probability in (5.6.1) (see the discussion of the behavioural view point in section 4.2 and of backward instantiation in section 4.3). However all these issues were not clearly understood in Bernoulli's time.

It is natural to identify in Bernoulli's idea of moral impossibility and moral certainty the seeds of later concepts like significance and confidence levels (sections 4.2 and 4.3). But perhaps it would not be proper to give Bernoulli sole credit for these. The ideas must have been in the air for a long time; others besides Bernoulli invoked these independently (cf. Arbuthnott in the next chapter).

| Bernoulli's probability | **5.6.3**   There is some confusion about Bernoulli's view about the nature of probability (see Hacking (1975) Chapter 16). On the one hand, being a believer in the Christian doctrine of predestination, Bernoulli professed that everything in this world is objectively certain: God being omniscient knows the truth-value of every proposition, and in particular, whether an event which appears contingent to us, will or will not occur. To us things are uncertain only because of the limitedness of our information. Thus, like Leibniz a little earlier, Bernoulli averred that probability is always subjective and relative to the evidence available. He laid down certain broad principles which should be followed at the time of assessment of a probability. Interesting among these are the maxims that 'one must search for all possible arguments or evidence concerning the case' and that 'for individual events ... special and individual arguments have to be taken into account'(see Hald (1990) p. 250). We can see that these essentially represent 'the principle of total evidence' (Chapter 1, footnote 5) and 'the principle of instantiation' of Russell and Fisher (subsection 3.2.5). But he did not give any concrete rule for the evaluation of a subjective probability in the general case.

Apart from games of chance, Bernoulli considered numerical probabilities only of events defined for random experiments for which independent and identical copies can be generated. He postulated that for such an experiment, knowing the initial conditions and the underlying physical laws, we can always presume that the outcome belongs to some conceptual set of equally likely cases, a certain proportion p of which is favourable to a given event. Although the set may be unidentifiable and the proportion may be unknown, he took the proportion as fixed over independent repetitions of the experiment and on the basis of his limit theorem proposed the long-term relative frequency as an estimate of it. Clearly here Bernoulli's position deviates from that of a true-blue subjectivist; to a subjectivist the uncertainty about the value of p also would be probabilizable by a prior distribution. The different repetitions would not be independent (unconditionally over variation of p) and the distribution of p would continually change in the light of the observations realized (sections 4.4 and 4.5). This is why subjectivists generally consider Bernoulli's Theorem as having very limited practical relevance (cf. sections 2.3 and 3.6; also see Keynes (1921) p. 341). Indeed, in the context of his limit theorem, Bernoulli seems to lean towards the objective viewpoint.

Bernoulli's work influenced the development of probability and statistical thought in a significant way. His limit theorem, despite its limitations, was generally supposed to validate the use of long-term relative frequencies for determining the probabilities of events in repeatable experiments. His emphasis on the use of probability in the art of conjecture 'in order to be able in our judgments and actions always to choose or follow the path which is found to be the best, most satisfactory, easy and reasonable' (see Maistrov (1974) p. 68) suggested a programme for statistical induction. The roots of later developments by De Moivre, Bayes, and even Laplace can be traced to the work of Bernoulli. Bernoulli's limit theorem was the first such theorem in the theory of probability and was the precursor of all those results which later came to be known as Laws of Large Numbers (section 3.10; the name was tagged to Bernoulli's theorem by Poisson more than a century after its appearance). Because of this, some authors go so far as to call James Bernoulli 'the real founder of mathematical probability' (Keynes (1921) p. 41) and 'father of quantification of uncertainty' (Stigler (1986) p. 63).

5.7   A historical puzzle

Before concluding this chapter we address a historical question which has drawn the attention of and puzzled a number of authors : why did probability theory and statistics both emerge with prominence suddenly and almost synchronously around the middle of the 17th century and not earlier? Several explanations, none very convincing, have been put forward (see e.g. Maistrov (1974), Hacking (1975)).

Thus David (1955) thinks perfect dice were not available before the 16th century and this prevented the discovery of the equiproportionality property of the fall of the die in gambling. (But reasonably perfect cubes were there even in ancient times.) Kendall (1956) speculates that among other things, possibly the existence of moral and religious barriers discouraged the development of ideas of randomness or chance—whatever happened, happened because it was ordained by God. (But even after the emergence of probability, as we would see, facts such as the constancy of the sex-ratio were advanced as evidence of Divine Providence.) Kendall (1956) also gives the absence of a sufficiently well-developed combinatorial algebra as a possible reason. (But, to solve simpler problems enumeration was enough, and in fact, the needs of probability theory provided an impetus for the development of combinatorics.)

Then there are teleological explanations, one economic and the other philosophical. The economic explanation put forward by Maistrov (1974) pp. 7, 14–15 says that probability and statistics emerged to meet the needs of the economy after the development of the capitalistic system with its monetary exchange, competition, and profit motive. (But the emergence of probability did not find any large-scale application in economic and commercial practice—not even in annuities and insurance—for a fairly long time.) According to the philosophical explanation probability emerged, in the words of Mahalanobis (1986) pp. 227–228, because with 'the emergence of the scientific view of an objective world of physical reality in which events were regulated by laws of nature ... it became necessary for the human mind to find some order or regularity in the occurrence of chance of events.' (But the mechanistic determinism referred to here developed in the 17th–18th centuries after the emergence of probability; also such an explanation fits well only with an objective interpretation of probability, not accepted by many, including Laplace, one of the principal protagonists of such determinism.)

To us it seems the most tangible cause behind the emergence of probability and statistics is the rise of empiricism. Beginning with the 17th century, empirical methods became gradually ascendant in science and rationalization of the process of induction became the need of the day. Probabilistic arguments were found handy for this and naturally this led to the development of probability theory and methods of statistical induction based on that theory.

6

NEW CONCEPTS AND METHODS—PRE-BAYESIAN ERA

The century after Pascal–Fermat correspondence

6.1   In the preceding chapter we traced the course of development of probability and statistics upto the time of James Bernoulli, i.e. roughly upto the end of the first half-century after the year of Pascal–Fermat correspondence. It is remarkable that so many basic ideas (e.g. those of finite and countable additivity, independence and multiplication rule, objective and subjective probability, the expectation principle, and convergence in probability) all germinated during these fifty years. One limitation of this development that has sometimes been commented upon (see Hacking (1975) p. 97) is that since this early development was rooted in the concept of proportion of equally likely cases, it was restricted to rational-valued probabilities only. (One could not get rid of this restriction even by appealing to Bernoulli's theorem, as the theorem as then proved, postulated a rational-valued limiting probability.) However, in practice this restriction was quietly ignored as more and more results with proofs valid for real-valued probabilities came to be available. (In fact, as we would see in the present and the next chapter, as soon as one brings in the idea of a random variable whose cdf increases continuously from 0 to 1 or a continuous prior distribution for some unknown probability, one perforce has to break free from this restriction.) Nevertheless, the question about how to introduce an irrational-valued probability subject to the usual rules of operation was left hanging for a long time.[1,2]

In the present chapter we review briefly some new concepts and methods related to the mathematics of probability and also its application to induction that came up more or less in the half-century after Bernoulli. As we will see later, so far as the application of probability to induction is concerned, a sort of reorientation of viewpoint was suggested by Thomas Bayes just after the end of this period. All the work that we consider in this chapter belongs to the pre-Bayesian era at least in spirit, if not strictly according to chronology.

---

[1]In this connection see Bayes's definition of probability in the next chapter.

[2]However, Koopman (1940*b*), has shown that one can reach real-valued probabilities starting from rational-valued probabilities, provided qualitative comparison of probabilities is admissible.

Arbuthnott—use of
probability for
hypothesis testing

6.2   We first discuss an isolated piece of work by an amateur probabilist of the name of John Arbuthnott (1667–1735). Arbuthnott was by profession a physician (in the court of Queen Anne), but was also a literary wit (a collaborator of Jonathan Swift) and a mathematician in his spare time. In his younger days (1692) he had authored an English translation of Huygens's book on probability. In 1710 he read before the Royal Society a paper entitled 'An argument for Divine Providence, taken from the constant regularity observed in the births of both sexes'. The paper, which was published in the *Philosophical Transactions* about a year later, contains 'the first published test of a statistical hypothesis' (Hacking (1965) p. 75) and was one of the earliest instances of the application of probability for statistical induction.

Arbuthnott considered data on the annual number of christenings in London classified by sex for the 82 years 1629–1710. Equating christenings with live births, his intention was to test the hypothesis H_0: births represent independent trials with a constant probability $\frac{1}{2}$ of a male child being born. If H_0 holds, he would say that 'chance' governs the birth of a child. He used the original data only to check whether each year was a 'male year' (i.e. having a majority of male births) or not. As it turned out, all the 82 years considered by Arbuthnott were male years.

If S denotes the number of successes in a sequence of n Bernoulli trials with probability of success $p = 1 - q$,

$$P(S > n - S) = \sum_{s:s>n-s} \binom{n}{s} p^s q^{n-s} = P(p,n) \quad \text{(say)}. \qquad (6.2.1)$$

Putting $p = \frac{1}{2}$, clearly, whatever n, $P(\frac{1}{2}, n)$ is exactly equal to or slightly less than $\frac{1}{2}$ according as n is odd or even. Thus, whatever the number of births in the different years, under H_0, the probability of all the 82 years considered being male years does not exceed $\frac{1}{2^{82}} \simeq 0.207 \times 10^{-24}$. As this is exceedingly small, Arbuthnott rejected H_0 and concluded that 'it is Art, not chance that governs' the birth of a child. The implication was that the sex of a child was determined willfully by a wise Creator who arranged it so that always an adequate excess of males was born and thus the higher mortality among males due to their more risky way of living got compensated. (Recall that Graunt, who had also noted the slight preponderance of male over female births, advanced the phenomenon as an argument against polygamy; subsection 5.4.5.)

How can we look at Arbuthnott's test from a modern viewpoint? Obviously he replaced the original H_0 by the wider hypothesis that the probability of any year being a male year is no more than $\frac{1}{2}$ and thus avoided the slight complication due to the number of births varying over the years. Although his actual mode of reasoning was somewhat obfuscated by the fact that all the 82 years considered turned out to be male years, it seems that he took 'the number of male years' as his test statistic and followed what we have called extended *modus tollens* under the instantial approach (subsection 4.3.11). Such a line of reasoning here is intuitively appealing because we can conceive alternatives under which the observed occurrence would have probability close to 1 (cf. subsection 4.3.12). (Apparently, Arbuthnott admitted alternatives to H_0 under which the genders of children born were non-stochastically controlled by Divine Providence. But even for an alternative under which births are independent Bernoulli trials with $p > \frac{1}{2}$, when n is large, $P(p, n) \simeq 1$, so that the probability of all the years being male years would be close to 1.) Arbuthnott stated his conclusion by noting that if H_0 were true someone betting on all the 82 years being male years would have a negligible chance of winning (see Hald (1990) p. 278). From his language it appears that Arbuthnott was interested in the instance at hand and did not adopt a behavioural attitude with repeated occasions in mind.

The conclusion of Arbuthnott's test was seized upon by some 18th century theologians as providing 'argument from design' in support of the existence of God: if things occurred in nature not haphazardly but according to some definite order and purpose then that bespoke the presence of Divine Providence. (Such arguments had become very popular among proponents of a natural religion (i.e. religion not dependent on revelation) during the post-Newtonian era.) Mathematically, there were some attempts to improve upon Arbuthnott's reasoning and to see what would happen to the argument if more information from the records of births was utilized. Further, Nicholas Bernoulli thought that Arbuthnott's use of the word 'chance' as meaning only 'equitable chance' was not proper. He sought to show that if one regarded births as Bernoulli trials with $p = \frac{18}{35}$, i.e. as throws of a dice with 35 faces, 18 of which favour the birth of a male child, one could get the observed data with a substantial probability. But then, as De Moivre pointed out, from the theological point of view the operation of a chance mechanism involving such an unusual die would be no less an evidence of the presence of art in the matter of determination of the sex of a child. (See Hacking (1975) pp. 169–175 for theological discussions and Hald (1990) pp. 279–284 for the mathematical developments originating from Arbuthnott's work.)

| Abraham De Moivre | **6.3** During the period under discussion a fair amount of work on probability mathematics was done by a number of workers. But generally it was oriented towards solving more complex problems in games of chance and had very little bearing on the development of statistical theory and practice. However, the same cannot be said of the work of De Moivre, who in course of tackling various problems of games of chance, among other things, introduced a technique and derived a result which had far-reaching consequences. In the following, we take note of only these two contributions of De Moivre.

Abraham De Moivre (1667–1754) was a Huguenot (French Protestant) who emigrated to England in the prime of his life to escape religious persecution. He had to struggle hard in his adopted country, earning his living as a visiting private tutor of mathematics in wealthy homes. But his mathematical genius was irrepressible and he soon started making research contributions to various branches of mathematics. He had already done important work in other areas before he published his first treatise on probability *De Mensura Sortis* (Measurement of Lots) in 1712. This was followed by the first editions of *The Doctrine of Chances* (1718) and *Annuities upon Lives* (1725) and also a mathematical tract *Miscellanea Analytica* (1730). The second edition of *The Doctrine* appeared in 1738 and also a posthumous third edition which was seen through by the author in 1756. It was De Moivre who initiated in England work in the mathematics of probability (as distinct from applications of probability on the basis of empirical data) in a significant way.

| Generating functions | De Moivre, first in his *Miscellanea Analytica* and later in the second edition of *The Doctrine of Chances*, developed an ingenious technique for finding the probability that when n symmetric dice each with f faces marked $1, 2, \ldots, f$ are thrown the total score would be $s(n \leq s \leq fn)$. In essence the technique involved the representation of the distribution of a non-negative-integer-valued random variable by what we now call the (probability) generating function (the name was introduced by Laplace about fifty years later). Generally, if X is such a random variable with $P(X = x) = p_x, x = 0, 1, 2, \ldots$, the generating function of X is given by

$$\psi_X(t) = E(t^X) = \sum_{x=0}^{\infty} p_x t^x. \tag{6.3.1}$$

(The power series (6.3.1) always converges at least for $|t| \leq 1$.) One particular advantage of the generating function is that if $X_1, \ldots, X_n$ are independently distributed, the generating function of the sum $\Sigma_1^n X_j$ is given by

$$\psi_{\Sigma X_j}(t) = E(t^{\Sigma X_j}) = \prod_{j=1}^{n} \psi_{X_j}(t). \tag{6.3.2}$$

When $X_1, \ldots, X_n$ are further identically distributed, this reduces to

$$\psi_{\Sigma X_j}(t) = \{\psi_{X_1}(t)\}^n. \tag{6.3.3}$$

Expanding the right-hand member of (6.3.2) or (6.3.3), as the case may be, in powers of t and identifying the successive coefficients, we can easily get the distribution of $\sum X_j$.[3] In the problem involving n f-faced dice considered by De Moivre, the generating function (6.3.1) of the score on a single die is a finite

---

[3]In recent years (6.3.3) has been generalized to cover a sum $\sum_1^N X_j$ where $X_1, X_2, \ldots$ are independently and identically distributed independently of N, which is a positive-integer-valued random variable. If the generating function of N is $\zeta(t)$, we have $\zeta(\psi_{X_1}(t))$ as the generating function of $\sum_1^N X_j$ (see Feller (1950) pp. 222–223).

series the only positive coefficients being $p_x = 1/f, x = 1, 2, \ldots, f$. Hence the distribution of the total score readily follows via (6.3.3).

The generating function is also useful for the derivation of moments and for the solution of difference equations. De Moivre and other workers extensively used the generating function as a handy general tool for working out discrete distributions in various ways. To anticipate some future developments, about forty years later after the concept of a continuous random variable had become established in statistical theory, Lagrange suggested that one could use $E(t^X)$ as the definition of the generating function of any random variable (provided the mean value exists) and use (6.3.3) to determine the distribution of a sum of independently and identically distributed copies of a random variable. Laplace around 1810 proposed the replacement of t by e^{it} and using $Ee^{itX} = \phi_X(t)$ (which exists for all real t since $|e^{itx}| \equiv 1$) to represent the distribution of X. The function $\phi_X(t)$, which later came to be called the *characteristic function*, turned out to be an extremely useful tool for the derivation of the distribution, especially the large sample distribution of various statistics. In particular, the Central Limit Theorem (section 3.9) giving the limiting distribution of the normalized partial sum of a sequence of random variables could be neatly proved in terms of it. However, it required well over a century and the contributions of a number of workers (culminating in the appearance in 1925 of Lévy's book *Calcul des Probabilités*) for the inversion and continuity properties of the characteristic function, which form the basis of all such results to be established in sufficient generality (see Cramér (1946) Chapter 10).

| Normal approximation to binomial |

The second contribution of De Moivre which we would note had its root in Bernoulli's limit theorem. As we saw in subsection 5.6.2, for a random variable S_n following the binomial distribution with parameters $(n, p = 1 - q)$, Bernoulli considered a lower bound to the probability

$$P\left(\left|\frac{S_n}{n} - p\right| < \epsilon\right) = \sum_{s:|s/n-p|<\epsilon} b(s; n, p), \qquad (6.3.4)$$

where

$$b(s; n, p) = \frac{n!}{s!(n-s)!} p^s q^{n-s}. \qquad (6.3.5)$$

He showed that this lower bound can be taken arbitrarily close to 1 by making n large. De Moivre conceived the idea that if an easy way of computing (6.3.5) to a close approximation could be found then the probability (6.3.4) itself could be closely evaluated. He grappled with the problem for many years. A satisfactory solution was finally reached when the approximation formula $n! \simeq \sqrt{(2\pi n)}e^{-n}n^n$ evolved through the sustained efforts of Stirling and of De Moivre himself. Employing this formula to the factorials in (6.3.5), De Moivre derived

$$b(s; n, p) \simeq \frac{1}{\sqrt{(2\pi npq)}} \exp -\frac{(s - np)^2}{2npq}. \tag{6.3.6}$$

Substitution of this in (6.3.4) led to

$$P\left(\left|\frac{S_n}{n} - p\right| < \epsilon\right) \simeq \frac{1}{\sqrt{(2\pi)}} \int_{-\epsilon(pq/n)^{-1/2}}^{\epsilon(pq/n)^{-1/2}} e^{-z^2/2} dz. \tag{6.3.7}$$

De Moivre gave the derivation for $p = \frac{1}{2}$; for general p he stated (6.3.6) explicitly and (6.3.7) by implication. (Details of the stages in the derivation are given in Hald (1990), Chapter 24.)

Nowadays (6.3.6) is called a local limit theorem which can be rigorously stated as

$$\lim_{n \to \infty} \sqrt{(npq)} b(np + z_n \sqrt{(npq)}; n, p) = \frac{1}{\sqrt{(2\pi)}} e^{-z^2/2} \tag{6.3.8}$$

for any sequence z_n such that $s_n = np + z_n \sqrt{(npq)}$ is an integer, $0 \leq s_n \leq n$, and $z_n \to z$ for some $-\infty < z < \infty$. (6.3.7) is stated in the form

$$\lim_{n \to \infty} P\left(z' < \left(\frac{S_n}{n} - p\right) \Big/ \left(\frac{pq}{n}\right)^{1/2} \leq z''\right)$$

$$= \frac{1}{\sqrt{(2\pi)}} \int_{z'}^{z''} e^{-1/2z^2} dz, \quad -\infty \leq z' < z'' \leq \infty, \tag{6.3.9}$$

which is called the integral form of the limit theorem.[4]

De Moivre published his approximation formulae in the form of a short Latin tract which was printed and circulated in 1733. Later the result was incorporated in the second edition of *The Doctrine of Chances*. But somehow it did not attract much notice from his contemporaries. Later Laplace worked on the same problem and published his results in 1812. Unlike De Moivre, who just like James Bernoulli earlier assumed that p is rational-valued, Laplace in his derivation allowed p to be any positive proper fraction. Also, Laplace gave an improved version of the approximation formula (6.3.7) by bringing in a correction term of order $n^{-1/2}$ (see Hald (1990) pp. 496–497). (6.3.8)–(6.3.9) are nowadays generally referred to as De Moivre–Laplace Limit Theorem.

Since $e^{-z^2/2}/\sqrt{(2\pi)}, -\infty < z < \infty$ is the density function of the standard normal distribution $\mathcal{N}(0, 1)$, (6.3.9) may be concisely expressed as

$$\left(\frac{S_n}{n} - p\right) \Big/ \left(\frac{pq}{n}\right)^{1/2} \xrightarrow{\mathcal{L}} \mathcal{N}(0, 1) \tag{6.3.10}$$

---

[4]There is a general result (see Billingsley (1968) pp. 49–50) which implies that (6.3.9) follows from (6.3.8).

(see section 3.10). Because De Moivre's Latin tract in which the approximation formulae (6.3.6)–(6.3.7) first appeared was dated 12 November 1733, sometimes it is said that the normal distribution was born on that date. In the same vein De Moivre is given credit as the originator of the normal curve (see Stigler (1986) p. 76). However, as we will see, the idea of a continuous random variable and a continuous distribution appeared explicitly in the work of Simpson just after De Moivre died. De Moivre looked upon (6.3.6) and (6.3.7) simply as approximation formulae. Although he used integrals as in (6.3.7) to evaluate probabilities (by expansion of the exponential and term-by-term integration for small values of $\epsilon\sqrt{n}$, and by quadrature for large values), he never explicitly noted that the total area under the curve $e^{-z^2/2}/\sqrt{(2\pi)}$ over $(-\infty, \infty)$ is 1 (this was proved by Laplace about 40 years later; see subsection 7.3.6) and did not look at it as the representation of a probability distribution.

The result (6.3.9) can be regarded as the precursor of the Central Limit Theorem. For this, S_n has to be represented as the sum of n i.i.d. Bernoulli random variables. De Moivre, however, did not look at the problem from this angle and naturally did not visualize the scope for any possible generalization.

As noted earlier, De Moivre's primary purpose in establishing (6.3.6)–(6.3.7) was to evaluate the probability $P(|S_n/n - p| < \epsilon)$ in Bernoulli's Theorem to a close approximation. Using his formula, when p is known one could estimate the probability with which the experimental ratio S_n/n would lie within a given interval around p, or as De Moivre put it in the second edition of *The Doctrine*, one could state 'the Degree of Assent which is to be given to Experiments' (see Hald (1990) p. 490). (Recall that, about fifty years earlier Locke (section 1.9) had laid down that 'degree of assent we give to any proposition should depend upon the grounds of probability in its favour'.) De Moivre's result could be put to advantage for testing any hypothesis specifying the value of p, but De Moivre did not look at it in that way. Stigler (1986) p. 86 comments that De Moivre's contemporaries were more concerned about the inverse of the problem considered by De Moivre—the problem of getting an interval estimate for an unknown p given the observed ratio S_n/n—and because of this, they failed to recognize the importance of the result.

Besides his work on probability theory proper, De Moivre did a lot of work on the application of probability to annuity mathematics. For this he used Halley's life table (subsection 5.4.6) and suggested a smoothing formula for its survival probabilities which gave simplified solution to various problems. His contributions in this regard were contained in his *Annuities upon Lives* (see Hald (1990) Chapter 25). But all that study was deductive in nature and had little implication for statistical induction.

| De Moivre's view of probability | De Moivre was a thorough-going mathematician who rarely took the liberty of making philosophical comments. Nevertheless, from some remarks made in the |

third edition of *The Doctrine* (see Hald (1990) pp. 27, 490–491), it seems that he
was inclined to the objective view of probability. Thus he states, 'there are, in the
constitution of things certain Laws according to which Events happen...' and
'altho' Chance produces irregularities, still the Odds will be infinitely great, that
in the process of Time those Irregularities will bear no proportion to the recur-
rence of that order which naturally results from original design.' Thus to him
probability was a kind of propensity applying to repeatable situations which got
reflected in the long-term relative frequency. But laws of nature or the propensities of things, De Moivre believed, 'must all be from *without*; the *Inertia* of matter,
and the nature of all created Beings, rendering it impossible that anything should
modify its own essence...'. These propensities, according to De Moivre, were
decided by Divine Providence. In this way, it seems, De Moivre steered clear of
full-scale determinism (which would have forced him to accept a subjective view
of probability) and at the same time was able to repudiate the doctrine of free
will which was foreign to his professed faith.

Behaviour of observational errors

6.4   In tracing the genesis of probabilistic reasoning
for analysing statistical data, so far we have come across
almost exclusively with data originating in games of
chance or demography. From ancient times, however, astronomers had made
extensive observations on heavenly bodies. In course of time, it was realized
that all such observations were affected by observational errors, which it was
impossible to eliminate altogether or control. Astronomical data thus provided
a ready field for the application of probabilistic reasoning. However, this possibility was not explored until the development of probability theory had got well
under way.

Generally, part of the error in an astronomical observation is systematic and
part unpredictable or random. In the second half of the 16th century, the Danish
astronomer Tycho Brahe introduced in astronomy the practice of combining
repeated observations on a phenomenon to make a more accurate determination
(see Hald (1990) pp. 146–148). Although Tycho proposed taking the arithmetic
mean of observations mainly to eliminate a particular kind of systematic error
(which were made to counterbalance each other in duplicated observations by
suitable planning), it seems gradually a line of thinking developed that the arithmetic mean of repeated observations is more reliable than one single observation,
however carefully taken. In the mean time in the 17th century, Galileo, while
discussing the (random) errors affecting the measurements of stellar distances,
made some qualitative observations on the behaviour of such errors. Specifically,
he noted that: (i) errors are inescapable in making instrumental observations,
(ii) smaller errors occur more frequently than larger ones, and (iii) observers
are equally prone to err in one direction or the other. (See Maistrov (1974)
pp. 30–34.) (We can now express these features by saying that the errors follow
a symmetric bell-shaped probability distribution centred at 0.) After probability theory had made some progress, the behaviour of errors was sought to be

studied in terms of probability models; the observations of Galileo influenced the building of such models to a large extent.

| Thomas Simpson | The first person who made any serious attempt to study observational errors probabilistically and through such

a study to develop a reasonable mode of estimation of a true value was an Englishman named Thomas Simpson (1710–1761). Simpson was a junior contemporary of De Moivre who was greatly influenced by and freely drew upon the latter's ideas (this was something which De Moivre did not always appreciate). Simpson was a prolific writer on mathematical subjects. But, apart from his work on actuarial science and the one important paper on probability and statistical estimation which we are going to discuss, he did not make much of an original contribution. The paper mentioned appeared in its full form as part of his *Miscellaneous Tracts on some curious and very interesting subjects in Mechanics, Physical Astronomy, and Speculative Mathematics* (1757) and was entitled 'On the Advantage of Taking the Mean of a Number of Observations, in practical Astronomy'.

Simpson addressed himself to the problem of determination of a true value of μ on the basis of a number of measurements made independently. The practice of taking the arithmetic mean was then being fairly widely followed by astronomers and geodesists. Simpson wanted to justify the practice theoretically. For this he started with the assumption that the error $Z = X - \mu$ in an observation X has no systematic part and is discrete with possible values

$$-h, -h+1, \ldots, -1, 0, 1, \ldots, h-1, h \qquad (6.4.1)$$

where h is some positive integer. He further assumed that Z follows a probability distribution over the set and this of course meant that the errors in repeated observations obey frequential regularity (section 2.7). His preferred choice for the model was the discrete symmetric triangular distribution with pmf

$$\frac{h+1-|z|}{(h+1)^2}, \quad z = 0, \pm 1, \ldots, \pm h. \qquad (6.4.2)$$

Thus like Galileo earlier, Simpson also supposed that the probability of an error decreases as its magnitude increases and for an error of any magnitude both signs are equally probable. Using De Moivre's technique he then deduced that the generating function of Z is given by

$$\psi_Z(t) = \frac{t^{-h}(1 - t^{h+1})^2}{(h+1)^2(1-t)^2}, \qquad (6.4.3)$$

and hence, that of the sum $\Sigma_1^n Z_i$ of n i.i.d. errors by

$$\psi_{\Sigma Z_i}(t) = \{\psi_Z(t)\}^n. \qquad (6.4.4)$$

Noting that when h is even (6.4.3) is same as the generating function of the sum of two independent random variables each following the uniform distribution over $0, \pm 1, \ldots, \pm \frac{1}{2}h$, Simpson could interpret (6.4.4) as the generating function of the sum of $2n$ i.i.d. random variables each following that same distribution. Thus $\sum_1^n Z_i$ could be interpreted as the total score obtained when $2n$ symmetric dice each having $h+1$ faces marked $0, \pm 1, \ldots, \pm \frac{1}{2}h$, were thrown. A simple adjustment of a by then well-known result (cf. De Moivre, section 6.3) would then give the probability distribution of $\sum Z_i = n\bar{Z}$. Hence Simpson could obtain $P(|\bar{Z}| > k)$ in the form of a sum (for details, see Hald (1998) section 3.3) and numerically compare its values with those of $P(|Z| > k)$ for different k for particular choices of h and n. In particular, he made such comparison for $h = 5$ and $n = 6$ and showed that $P(|\bar{Z}| > k)$ is much smaller than $P(|Z| > k)$—very much so for large values of k.

Up to this technically Simpson's analysis closely followed that of De Moivre (for the problem of total score on a number of generalized dice), although there was originality in his conception of a location model for observations under which the distribution of the error $Z = X - \mu$ does not depend on the parameter μ. After this, however, Simpson made a conceptual break-through. He considered the situation when, in his own words, 'the error admits of any value whatever, whole or broken, within the proposed limits' (see Stigler (1986) p. 95), i.e. he introduced the idea of a continuous probability distribution. Simpson's line of reasoning here can be described as follows: let the unit in which the error Z is expressed be changed from 1 to $\delta(>0)$ so that (6.4.1) is replaced by

$$-h\delta, -(h-1)\delta, \ldots, -\delta, 0, \delta, \ldots, (h-1)\delta, h\delta \qquad (6.4.5)$$

and the pmf becomes

$$\frac{h+1-|z|/\delta}{(h+1)^2}, \quad z = 0, \pm\delta, \ldots, \pm h\delta. \qquad (6.4.6)$$

Let us now make $h \to \infty$ and $\delta \to 0$ so that $h\delta = a$ remains fixed. Without loss of generality taking $a = 1$, this means that the probability of Z lying in any interval would approach the corresponding area under the continuous curve

$$1 - |z|, \quad -1 < z < 1. \qquad (6.4.7)$$

We nowadays describe this by saying that as $h \to \infty$, the sequence of discrete triangular distributions (6.4.6) converges in law to the continuous distribution (6.4.7). If we apply the same limiting process to $P(|\bar{Z}| > k\delta)$ by keeping $k\delta = d$ (say) fixed, we get the limiting value

$$P(|\bar{Z}| > d) = \frac{2}{(2n)!} \sum_{j=0}^{[n-nd]} (-1)^j \binom{2n}{j} (n - nd - j)^{2n}, \quad 0 < d < 1. \qquad (6.4.8)$$

We now say that (6.4.8) represents the two-sided tail probability of the distribution of the mean $\bar{Z}$ of a sample of size n from the continuous distribution (6.4.8). Simpson gave graphical representations of the density (6.4.7) and the one corresponding to (6.4.8) in the same figure; both are symmetric curves centred at 0 but the latter is very much more peaked than the former. The figure confirms Simpson's conclusion that '... it appears that the taking of the Mean of a number of observations, greatly diminishes the chances for all the smaller errors and cuts off almost all possibility of any great ones'. (See Hald (1998) pp. 37–39. Hald observes that Simpson's drawing of the second curve is somewhat erroneous, but that does not affect the conclusion.)

Notable features of Simpson's work

Of course Simpson's conclusion hinged on the particular triangular model he assumed. Still his contribution is remarkable on several counts:

(i) As noted earlier, Simpson's was the earliest recorded treatment of the errors in astronomical observations (or for that matter, in any kind of physical measurements) in terms of probability models.

(ii) Simpson introduced for the first time in statistical practice the location model under which the distribution of the error $Z = X - \mu$ does not depend on the unknown μ. This model later proved to be very fertile in generating various ideas. Thus, as we will see, Gauss started from this model to give his development of the normal distribution. Much later, assumption of particular forms of this model made it easy to develop certain approaches to the problem of interval estimation (like the confidence interval and fiducial interval of subsections 4.2.4 and 4.3.15), at least initially.

(iii) Simpson's argument for using the sample mean $\bar{X}$ in preference to a single observation X for estimating μ rested on showing that $P(|\bar{X} - \mu| > d) < P(|X - \mu| > d)$ for different choices of d. This essentially means he was comparing the alternative estimators $\bar{X}$ and X in terms of their sampling distributions. This can be regarded as the beginning of the behavioural approach to the problem of point estimation.

(iv) Simpson did not show that the sampling distribution of $\bar{X}$ is more concentrated around μ than that of X by arguing in terms of criteria like unbiasedness and smaller variance (in fact variance as a standard measure of dispersion came up about fifty years later). Instead, he made the comparison in terms of the two-sided tail-probabilities, or equivalently, the central humps over intervals like $[\mu - d, \mu + d]$. Interestingly, in modern times we find similar ideas of comparison in the formulation of 'maximum probability estimators' by Weiss and Wolfowitz (1967).

(v) Simpson's introduction of the concept of a continuous probability distribution had far-reaching consequences for the development of statistical theory. Admittedly, in practice all measurements are expressed as multiples of some

discrete unit and continuous random variables and distributions are nothing but idealizations which give satisfactory approximation (cf. section 2.4).
Yet the assumption of continuous models simplifies theoretical treatment
immensely and it is doubtful whether the edifice of statistics, as it stands,
could have been built without such simplification.

After Simpson considered his continuous triangular distribution, it must have
been generally realized that any curve above the horizontal axis with total area 1
under it could be interpreted as representing a continuous probability distribution. De Moivre's approximation formula could be recognized as a continuous
distribution (the name 'normal', however, came much later). Various other continuous distributions including the beta, the double-exponential, and the Cauchy
distribution were proposed as models from time to time. As regards Simpson's
justification for using the sample mean as an estimator of a true value under
the triangular model, similar studies were made for other continuous models by
some later workers who developed various analytical techniques for deriving the
distribution of the mean of a sample from a continuous population directly (see
Hald (1998) pp. 46–63). However, from the point of view of conceptual novelty,
it cannot be said that these studies made much of an addition.

| Daniel Bernoulli | **6.5**   We will next review in this chapter the contribu-
tions to statistical thought of Daniel Bernoulli (1700–
1782). He was a member of the famous Swiss family of Bernoullis—a nephew of
James and a junior cousin of Nicholas Bernoulli.[5]

We have considered Thomas Simpson ahead of Daniel Bernoulli, although the
latter was born and started contributing to the field of probability and statistics
much earlier. It was expedient to do so, partly because Simpson's mathematics
were in a sense a continuation of De Moivre's, and partly because, as we would
see, Bernoulli's range of contributions is so wide and varied that he is best
taken up at the end. Bernoulli was a versatile scholar who had a long life and
left his mark in many branches of knowledge. Unlike De Moivre, who made
sustained investigations over a broad spectrum covering interconnected problems
of probability, and Arbuthnott and Simpson who are remembered mainly for
single pieces of work, Bernoulli sporadically worked on challenging but disjointed
problems of probability and its applications, which were often posed by others.
In discussing the contributions of Bernoulli, it would be convenient to proceed

---

[5]The genealogical tree of the Bernoullli family reads as follows:

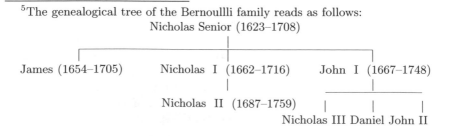

by broad groups of topics, although that would mean we would have to skip over decades back and forth more than once.

| Testing a cosmogonical hypothesis |

First consider Daniel Bernoulli's contributions to inference. In section 6.2 we saw that Arbuthnott used probabilistic reasoning to test whether the genders of children born are determined by 'chance' (i.e. equi-probabilistically). This was a hypothesis about an ongoing biological phenomenon, but as we noted at that time, Arbuthnott's test seems to be a test of significance along the lines of the instantial approach. In 1735 Daniel Bernoulli used similar reasoning to test a hypothesis about a one-time astronomical phenomenon which took place in the distant past—at the time of the creation of the solar universe—and his approach was even more patently of the instantial variety. Astronomers had noted that the angles of inclination of the orbital planes of the planets with respect to each other as well as the equatorial plane of the sun are all remarkably small. Newton in 1730 had observed that the smallness of the mutual inclinations must be the effect of 'choice' and not of 'blind fate'. Daniel Bernoulli wanted to explain the phenomenon, and as the first step towards this, to dispose of the possibility that the orbital planes were set quite at random.

Bernoulli's approach can be explained conveniently in terms of the test he applied on the inclinations of the orbital planes of the six planets Mercury, Venus, Earth, Mars, Jupiter, and Saturn to the equatorial plane of the sun. Denoting the observed values (in degrees) of these by $x_i, i = 1, 2, \ldots, 6$ his object can be described as testing the hypothesis H_0: $x_i, i = 1, 2, \ldots, 6$ represent observations on a random sample from $\mathcal{U}(0, 90)$ (the uniform distribution over $(0, 90)$). He wanted a test for H_0 that is sensitive against departures which tend to make all the inclinations low. Writing $x_{(6)}$, for the largest value among $x_i, i = 1, \ldots, 6$ and $X_{(6)}$ and $X_i, i = 1, \ldots, 6$ for the corresponding random variables, we can say Bernoulli applied the left-tailed test based on the test criterion $X_{(6)}$. (Note that $X_{(6)}$ meets the requirements of a test criterion as laid down in subsection 4.3.11— the smaller the value of $X_{(6)}$, the greater the discrepancy between H_0 and the sample.) As

$$P = P_{H_0}(X_{(6)} \leq x_{(6)}) = P_{H_0}(X_i \leq x_{(6)}, \ i = 1, \ldots, 6) = \left(\frac{x_{(6)}}{90}\right)^6, \quad (6.5.1)$$

in Bernoulli's case, the value $x_{(6)} = 7°30'$ (the largest inclination of the orbital plane to the sun's equatorial plane occurs in the case of Earth) gives $P = \frac{1}{12^6} \simeq 0.335 \times 10^{-6}$. This probability being extremely small, a discrepancy between H_0 and the sample as much as or more extreme than that observed can be regarded as 'a moral impossibility' under H_0. Thus following the form of argument which we have called 'extended *modus tollens*', H_0 is to be rejected. Just like Arbuthnott earlier, after rejecting H_0 Bernoulli did not seek to explain the observations in terms of any alternative probability model. Rather, he attempted a physical explanation and suggested that the orbital planes which had been originally wider apart, had been brought closer by the pull of the sun. (See Hald (1998) section 4.2 for further details of Bernoulli's test.)

The type of probabilistic reasoning used by Daniel Bernoulli in the above testing problem was extended by others to test various hypotheses involving unusual astronomical occurrences. The probability of an occurrence at least as unusual as that observed was computed under an hypothesis postulating an unmotivated 'random' creation of the universe. This probability being too low, the hypothesis was rejected. Taking the alternative to the hypothesis as manipulation by a conscious Creator, theologians utilized such studies for building up their 'arguments from design'. In particular, about thirty years after Daniel Bernoulli, John Michell in England considered the hypothesis that stars up to any given magnitude are distributed randomly over the celestial sphere. He examined the angular distances among the six brightest stars in the constellation Pleiades. Since the probability of a formation as close as these under the hypothesis that the 1500 stars of at least the same magnitude of brightness are randomly distributed is too low, Michell concluded against the hypothesis. (Michell's computation and argument were rather obscure; Fisher (1956) pp. 38–39 in course of discussing test of significance of a null hypothesis, re-examined Michell's problem and came out with an upper bound to the relevant P-value.) We will see later that an extended version of Daniel Bernoulli's problem related to inclinations of planetary orbits was also studied by Laplace. What is noteworthy is that in all these cases tests were performed in the instantial spirit without bothering about possible repetitions and, as Fisher (1956) p. 21 points out, 'without reference to or consideration of, any alternative hypothesis which might be actually or conceivably brought forward'.

| The maximum |
| likelihood principle |

Going back to Daniel Bernoulli, his was one of the earliest applications of what later came to be known as the maximum likelihood method for the estimation of a true value. Denoting the true value by μ and a random determination by X, just like Simpson earlier, he assumed that the distribution of the error $Z = X - \mu$ does not depend on μ, i.e. the distribution of X follows a location model with a density of the form $f(x - \mu)$. Also just like Simpson, he investigated about the practice of taking the arithmetic mean of a number of independent determinations for the estimation of μ. But whereas Simpson showed that the practice can be justified in terms of the sampling behaviour of the estimator if one assumes a particular form (the triangular) for $f(.)$, Bernoulli assumed another form of $f(.)$ and showed that if one chose among all possible values of μ the value which gives 'the highest degree of probability for the complex of observations as a whole' one may be led to an estimator quite different from the sample mean. Specifically Bernoulli took the semi-circular form

$$f(x - \mu) = \text{const.} \left\{ r^2 - (x - \mu)^2 \right\}^{1/2}, \quad 0 < |x - \mu| < r, \tag{6.5.2}$$

where r, the maximum numerical error possible, is taken as known. (This of course conforms to Galileo's requirements of symmetry and unimodality; section 6.4.) If the sample observations are $x_1, x_2, \ldots, x_n$, the probability element at the observed sample is proportional to

$$L = \prod_{i=1}^{n} \left\{ r^2 - (x_i - \mu)^2 \right\}^{1/2}. \tag{6.5.3}$$

Bernoulli suggested that we take the value of μ which maximizes (6.5.3) as an estimate. Setting up the equation

$$\frac{dL^2}{d\mu} = 0, \tag{6.5.4}$$

the determination of the estimate required the solution of a polynomial equation of degree $2n - 1$. Bernoulli noted that this would be troublesome for large n; nevertheless for $n = 3$ he showed that the solution would differ from the sample mean $\bar{x}$ unless the second order-statistic $x_{(2)}$ coincides with $\bar{x}$.

Bernoulli did not consider the sampling behaviour of his estimate on repeated sampling, but instead thought that the principle of choosing μ so as to maximize the probability of the realized sample is in itself intuitively appealing. (We can now recognize that he was suggesting here what Fisher one and a half centurie later called the maximum likelihood method of estimation.) He drew a parallel between this and the choice of the most probable outcome in a game of chance in which one is forced to choose beforehand what would happen. Although this betrays a confusion between what we have called problems of forward and backward instantiation (subsection 4.3.3), it is clear that Bernoulli's approach to the problem of point estimation is instantial rather than behavioural in spirit. This impression is further strengthened by his remark that the estimate of μ would remain unaffected if the semicircular law (6.5.2) is replaced by the parabolic law

$$f(x - \mu) = \text{const.} \left\{ r^2 - (x - \mu)^2 \right\}, \quad 0 < |x - \mu| < r, \tag{6.5.5}$$

which shows that the sampling behaviour of the estimate remained farthest from his mind. (Although the above investigation of Bernoulli was reported finally in a paper in 1778, he had been working on the problem since much earlier. The maximum likelihood principle for point estimation had been enunciated by Lambert also a few years earlier; but Lambert's treatment here was rather slipshod.)

| The normal model | Apart from statistical induction, Daniel Bernoulli made some interesting contributions to the construction of probability models. One characteristic feature of his work here was the extensive use he made of the tool of differential equations. An instance was the alternative derivation he gave of the normal approximation formula for the binomial.

Examining the differences of the binomial pmf $y_x = b(np + x; n, p)$ as given by
(6.3.5) and replacing $(\Delta y_x / \Delta x)$ by (dy_x / dx), he set up essentially the differential
equation

$$\frac{d \log y_x}{dx} = -\frac{(x/q) + 1}{np + x + 1} \simeq -\frac{x}{npq} \tag{6.5.6}$$

for $x = O(\sqrt{(n)})$. From this equation, Bernoulli was able to derive an approximation formula (see Hald (1990) pp. 502–503) which was equivalent to (6.3.6).
(We will see in Chapter 10 that more than a century later, Karl Pearson made a
similar manipulation on the hypergeometric pmf to set up a differential equation
which was a generalized version of (6.5.6) and the solution of that led to the
different curves of the Pearsonian system.) Daniel Bernoulli also tabulated the
function $e^{-x^2/100}$ over the range $x = 1(1)5(5)30$ (this essentially gave the normal
density function with $\sigma^2 = 50$, covering a range upto a little beyond 4σ) and
used the table for the computation of binomial probabilities.

Inoculation problem—
a survival model

A more novel instance of a probability model was
derived by Daniel Bernoulli in the context of his treatment of the inoculation problem. Around the middle of
the 18th century, a type of question that exercised the minds of demographers
greatly was: what gain in longevity would result if some particular killer disease like smallpox or tuberculosis could be totally eradicated? The question
was especially topical with regard to smallpox which was one of the principal
scourges of the time and against which only a risky form of immunization through
inoculation (administration of actual human smallpox virus) was available. Such
questions could not be answered on the basis of available demographic data
as records of death by both age and cause of death (which would have made
computation of age-specific death rates excluding particular diseases possible)
were generally not maintained. The ℓ_x column of a life table represents the
survival function (complement of cdf) of the distribution of age at death and
forms the basis for computation of expectation of life at birth. To determine the
gain in longevity, i.e. the increase in expectation of life at birth, one needs the
modified ℓ_x (say ℓ'_x) values that would occur if smallpox could be completely
eradicated. Daniel Bernoulli (1760) derived the latter by making two sweeping
assumptions: (i) the probability that a person aged x who had not previously
contracted smallpox will do so within a year is a constant (say, π) for all x,
(ii) the probability that a person aged x contracting smallpox will die of the
disease is a constant (say, τ) for all x. His line of argument was as follows.

Let s_x denote the number among the ℓ_x persons aged x who have not been
previously attacked by smallpox: only such persons are exposed to the risk of
an attack ($s_x \leq \ell_x, s_0 = \ell_0$). Of the ℓ_x persons aged $x, -d\ell_x$ die before crossing
age $x + dx$ and out of these the number dying from smallpox is $\pi \tau s_x \, dx$. Hence
the probability of death in the age-interval $(x, x + dx]$ due to causes other than

smallpox is

$$\frac{1}{\ell_x}(-d\ell_x - \pi\tau s_x \, dx).\tag{6.5.7}$$

Since the decrement $-ds_x$ in s_x over the interval $(x, x+dx]$ is partly due to some of these s_x persons contracting smallpox (whether they die or not) and partly due to death from other causes, (6.5.7) gives

$$-ds_x = \pi s_x \, dx - \frac{s_x}{\ell_x}(d\ell_x + \pi\tau s_x \, dx).\tag{6.5.8}$$

Writing $\ell_x/s_x = r_x$, we can re-cast (6.5.8) in the form

$$\frac{d\log r_x}{dx} = \pi - \frac{1}{r_x}\pi\tau,$$

solving which under the initial condition $r_0 = 1$, we get r_x, and hence

$$s_x = \frac{\ell_x}{\tau + (1-\tau)e^{\pi x}}.\tag{6.5.9}$$

When smallpox is eradicated, the decrement $-d\ell'_x$ in ℓ'_x over the age interval $(x, x + dx]$ would be solely due to causes other than smallpox. Hence by (6.5.7)

$$-d\ell'_x = -\frac{\ell'_x}{\ell_x}(d\ell_x + \pi\tau s_x \, dx),$$

from which using (6.5.9) we get

$$\ell'_x = \ell_x \frac{e^{\pi x}}{\tau + (1-\tau)e^{\pi x}}.\tag{6.5.10}$$

Note that the factor appearing after ℓ_x on the right is greater than 1 for $x > 0$. This means eradication of smallpox would increase the survival function of the distribution at all x, or in other words, would stochastically increase the distribution of life, which is as it should be.

Bernoulli took $\pi = \tau = \frac{1}{8}$ and worked on the ℓ_x column of Halley's life table. The modified ℓ_x column, obtained gave him an increase in longevity from 26 years 7 months to 29 years 9 months. (Interestingly, at that time the risk of death due to smallpox inoculation itself was about 1 in 200. This would mean Bernoulli's mathematics provided little inducement for getting oneself inoculated against smallpox: a gain in expected longevity of about 3 years was poor compensation for undergoing a risk of immediate death as high as $\frac{1}{200}$. Towards the end of the 18th century, however, Jenner discovered smallpox vaccination (which involved the administration of cow-pox virus) which had negligible risk and that laid the way for the eventual elimination of smallpox.)

Daniel Bernoulli has been criticized for the sweeping assumptions he introduced in making the derivation. But the ingenuity of his approach cannot be denied. In fact, his study of the smallpox problem can be looked upon as the beginning of the theory of what later came to be known as multiple decrements tables and competing risks which are of importance in actuarial science and in survival analysis.

| St. Petersburg paradox and moral expectation |

We have reserved Daniel Bernoulli's major intellectual contribution—that related to his idea of moral expectation—to the last, although chronologically it was one of the earliest. It appeared in a paper entitled '*Exposition of a new theory on the measurements of risk*' in the year 1738 and generated a lot of controversy. We can now see that conceptually it broke new grounds and had a far-reaching impact.

Earlier Daniel's cousin Nicholas Bernoulli II had posed a probabilistic teaser in a letter to Montmort, which had appeared in the second edition (1713) of the latter's probability book. It concerned a game between two players Peter and Paul, in which a fair coin was tossed repeatedly until the first head appeared. If the first head appeared at the jth toss, Paul would receive an amount 2^{j-1} from Peter, $j = 1, 2, \ldots$. The question was: how much initial amount or entry fee should Paul pay Peter to make the game equitable?

By that time, the principle that the fair entry fee for a game should be equal to the expected gain of the player had become well-established in the theory of probability. (This principle could be justified in the long run if the game were played repeatedly, in terms of the frequency interpretation of probability (cf. Cardano, subsection 5.3.1.) True, a rigorous justification along these lines would require the application of the Law of Large Numbers (section 3.10), which, as we will see was proved in a general form only around the middle of the 19th century; yet the Law implicitly follows from Bernoulli's Theorem, at least when the gain can assume only a finite number of values.[6] When the game is played only once, as we saw in subsection 5.3.3, Huygens gave a justification for the principle in terms of fair lotteries.) What caused trouble in the case of Nicholas Bernoulli's problem was that the expected gain

$$\sum_{j=1}^{\infty} 2^{j-1} \frac{1}{2^j} = \sum_{j=1}^{\infty} \frac{1}{2}$$

of Paul for the proposed game was ∞. Yet common sense dictates that it would be foolish on the part of Paul to give an exorbitant amount to Peter to get an opportunity to play the game. The questions that arose were: is something really wrong with the principle setting the fair entry fee for a game as equal to its expected gain, or has its application to be limited to situations where the expectation is finite; if so, how should one determine the fair entry fee for a game

---

[6] According to Keynes (1921) p. 314 such a justification of the expectation rule was explicitly advanced by Condorcet in a piece of writing that was reported in 1781.

with infinite expected gain? It should be noted that equating the entry fee to the expected monetary gain is really a particularization of the general principle ('the crude expectation principle') according to which in an uncertain situation an action is to be preferred to another if the expected monetary gain for the former is more. Thus Nicholas Bernoulli's problem questioned the validity of this principle which so far had been regarded as basic to all rational decision making under situations of risk.

Daniel Bernoulli in his 1738 paper proposed a resolution of the problem and since the paper was published in the annals of the St. Petersburg Academy, the problem itself came to be known as the St. Petersburg paradox. (In the paper, Bernoulli quoted some excerpts from an unpublished letter of Gabriel Cramer to Nicholas Bernoulli and acknowledged that, unknown to him, some of the ideas in the paper had been anticipated by Cramer.) Bernoulli's basic argument was that the moral worth or utility of an absolute increment in the wealth of a person depended inversely on his initial wealth: a small addition to wealth may mean a lot to a pauper but practically nothing to a millionaire. Specifically, denoting nominal wealth by x and its utility by y he assumed

$$dy = c\frac{dx}{x}, \tag{6.5.11}$$

which by integration showed that when a person's nominal wealth increased from α to $\alpha + \Delta x$ the utility of the wealth increased by

$$\Delta y = c\log\frac{\alpha + \Delta x}{\alpha} \tag{6.5.12}$$

This means when Paul's initial wealth is α, his monetary gain being 2^{j-1}, the gain in utility would be

$$c\log\frac{\alpha + 2^{j-1}}{\alpha}, \quad j = 1, 2, \ldots,$$

so that his expected utility gain, or as it was called, the *moral expectation* from playing the game with Peter would be

$$c\sum_{j=1}^{\infty}\frac{1}{2^j}\log\frac{\alpha + 2^{j-1}}{\alpha} = c\log\left\{\frac{\prod_{j=1}^{\infty}\left(\alpha + 2^{j-1}\right)^{1/2^j}}{\alpha}\right\}. \tag{6.5.13}$$

By (6.5.11) this means in terms of moral expectation the upshot of playing the game with Peter would be equivalent to Paul gaining an increment of monetary possession equal to

$$e(\alpha) = \prod_{j=1}^{\infty}\left(\alpha + 2^{j-1}\right)^{1/2^j} - \alpha \tag{6.5.14}$$

Therefore $e(\alpha)$ would be a reasonable entry fee for the game. Thus the entry fee for Paul depends on his initial wealth α: $e(0) = 2, e(10) \simeq 3, e(100) \simeq 4$, $e(1000) \simeq 6$, and so on. Because of this, Daniel Bernoulli's solution of the St. Petersburg Paradox appeared somewhat bizarre to many probabilists and occasionally he was ridiculed for this. (But strangely, Laplace a few decades later accepted and worked on it!)[7]

The rational expectation principle

The assumption (6.5.11) leading to the formula (6.5.12) implied *diminishing marginal utility* of money, i.e. a concave utility function. Economists give Daniel Bernoulli great credit for pioneering this idea (the English translation of Bernoulli's 1738 paper was reprinted in the *Econometrika* in 1954). But from the title of Bernoulli's paper also from the other examples discussed in it, it becomes clear that although the St. Petersburg paradox provided the initial inspiration for writing the paper, this and even the idea of concavity of the utility function were only incidental to its central theme. The main thrust of the paper was towards replacing the crude expectation principle by the *rational expectation principle* (cf. subsection 4.4.20) for determining the fair entry fee or price. According to the latter, in a situation involving risk when we have to choose between two actions, we should choose the one for which the expected utility gain (or moral expectation) is larger. Thus in the St. Petersburg game the act of playing the game for Paul was regarded as equivalent to the act which resulted in a certain increase of his wealth by the amount (6.5.14). Bernoulli discussed also the example of a pauper who is in possession of a lottery ticket which may give him (in some units of money) either an amount 0 or 20,000, each with probability $\frac{1}{2}$. The pauper may understandably prefer to sell the ticket for 9,000—a decision which cannot be justified in terms of crude expectation, but would follow from the rational expectation principle if utility is taken as some suitable concave function of the amount of money. As another instance, Bernoulli cited the example of a prudent merchant who decides to insure his ship against

---

[7]Solutions which took into account Peter's limited capacity to pay any sum won by Paul, were derived by others (see Epstein (1967), pp. 111–112). Another solution, applicable when the game is played repeatedly was suggested by Bertrand (see Feller (1950) pp. 200–201). According to it, a game in which the player's expected gain is $\mu < \infty$, is fair when the entry fee for each play is taken as μ because denoting the total gain when the game is played n times by S_n, we have $S_n/n\mu \xrightarrow{P} 1$ as $n \to \infty$. It can be shown that if S_n is Paul's total gain from playing the St. Petersburg game n times, then $S_n/(n \log_2 n) \xrightarrow{P} 1$. This implies if we lay down rules for entry so that the accumulated entry fee for n plays is $n \log_2 n$, then the arrangement can be regarded as fair. Of course, this would mean that the entry fee for successive plays would be variable, being 0 for the first play, 2 for the second, $3 \log_2 3 - 2 \simeq 2.8$ for the third, $4 \log_2 4 - 3 \log_2 3 \simeq 3.2$ for the fourth and so on!

loss at sea, even though he knows perfectly well that the premium charged by the insurance company exceeds the expected loss.

In modern times there has been a renewed interest in the rational expectation principle following the work of Ramsey (1926) and of von Neumann and Morgenstern (1947) (see also Savage (1954)). But one distinction is noteworthy. Daniel Bernoulli assumed a utility function given a priori. Considering decision taking under a situation of risk where the possible states have known objective probabilities, he ordered the actions in terms of their expected utilities. On the other hand, in the modern development one starts from a complete ordering of the different possible consequences or gains and their randomized mixtures and derives the utility function therefrom; the probabilities that come into the picture are also generally personal subjective probabilities. Nevertheless it must be conceded that it was Bernoulli who first took the revolutionary step of breaking away from the crude expectation principle and comparing actions in terms of their mean utilities in order to explain the behaviour of rational persons in situations involving risk. The oft-quoted saying, 'Probability is the very guide of life'[8] can be justified only if one brings in the concept of utility.

In the next chapter we will introduce the reorientation of viewpoint to problems of statistical induction suggested by Thomas Bayes which we mentioned earlier. Although Daniel Bernoulli outlived Bayes by more than twenty years, we do not find any trace of Bayesian thinking in his work. After this, however, whatever solution to any problem of statistical induction came up, it either conformed to or was in opposition to Bayes's line of thinking. Daniel Bernoulli was arguably the last of the stalwarts making significant contributions on such problems, who belonged to the pre-Bayesian era.

---

[8]There seems to be some confusion about the origin of this adage. Keynes (1921) p. 309 and Russell (1948) p. 340 attribute it to the 18th century English moral philosopher Bishop Joseph Butler. But Good (1950) p. v traces it to Cicero, the Roman philosopher–statesman of 1st century B.C. Of course one can question the sense of the term 'probability' in the statement.

7

BEGINNING OF THE PRO-SUBJECTIVE APPROACH

Parameter as random variable

7.1 In the preceding chapter we considered certain early instances of application of probability for statistical induction. In some of these induction was about a parameter with a known domain. There the value of the parameter, though unknown, was supposed to be fixed. In some others (e.g. in the case of Arbuthnott's and Daniel Bernoulli's testing problem in section 6.2 and at the beginning of section 6.5) certain null hypotheses were examined without bothering to specify the entire range of alternative hypotheses possible—the set-up was not fully parametrized, i.e. described in terms of a parameter and its domain. Nevertheless it was implicitly assumed that the process generating the random variables observed, so far as it was specified, was something fixed, though not fully known. We now come to a reorientation of viewpoint that was proposed in the second half of the 18th century. According to it, statistical induction is to be considered in a scenario in which the set-up is fully parametrized but the unknown parameter is regarded not as something fixed once and for all, but as the realization of a random variable subject to a probability distribution. The person from whom this revolutionary idea, which has remained contentious even to this day, originated was an unpretentious English clergyman of the name of Rev. Thomas Bayes (1702–1761). But as we would see, a few years after Bayes, and apparently independently of him, the great French mathematician–astronomer Pierre Simon Laplace also adopted the same viewpoint in those parts of his work on statistical induction where he took the posteriorist stance, i.e. regarded the observations as already taken, and hence, fixed. These works of Bayes and Laplace marked the beginning of what in section 4.4 we called the objective pro-subjective approach to induction in statistics.

7.2 Thomas Bayes

Bayes's life and his *Essay*

7.2.1 Thomas Bayes, who was born the son of, and was himself a non-conformist (i.e. non-believer in the Trinity) minister had an uneventful life. In his youth he was privately educated. (Barnard (1958) has speculated that he might have been tutored by De Moivre himself, but there does not seem to be any solid evidence in support of this.) Apparently, he gained considerable reputation among his contemporaries for his scholarship in mathematics, for he became Fellow of the Royal Society at the age of 40, even though at that time he had only one mathematical publication—a tract on the Theory of Fluxions (as calculus used to be called then)—to his credit. After his death two other such publications came out, one on the asymptotic series related to the De Moivre–Stirling approximation formula for $n!$ and the other his principal contribution—a paper on probability which has had enormous influence on the development of statistical thought and which has given Bayes a lasting place in the annals of the subject. All these papers bear testimony to the penetrative insight and depth of understanding of their author.

Bayes's paper on probability was entitled '*An Essay towards solving a Problem in the Doctrine of Chances*'. From an unpublished notebook of the author, which has been discovered some years back (see Hald (1998) p. 134), it appears that the work was largely completed between 1746 and 1749. But for some reasons Bayes was hesitant about its publication. More than two years after his death, his literary executor and friend, Rev. Richard Price (who was 21 years his junior) read it before the Royal Society on 23 December 1763. Price himself evinced interest in probability and its applications and was the author of the Northampton Life Table. He edited Bayes's *Essay*, replacing its original (presumably shorter) introduction by an introductory letter and adding an appendix containing some further developments. The *Essay* was published as a two-part paper in the *Philosophical Transactions of the Royal Society* (1964, 1965). (It has been reprinted, in particular, in *Biometrika* (1958) with a biographical note by Barnard.)

In his introductory letter to Bayes's *Essay*, Price stated that as per the author's own introduction, the purpose of the paper was to solve a problem which can be regarded as inverse of the one considered by De Moivre earlier. The latter had been concerned with the evaluation of the probability that in a number of future Bernoulli trials with a given constant probability of success p, the relative frequency of success would lie within a specified neighbourhood of p. Given the relative frequency of successes in a number of past trials with p unknown, Bayes wanted to find with what probability we can assert that p happens to lie within specified limits.

Bayes's conception of probability

7.2.2 Bayes's programme was to assume an initial probability distribution of the unknown success probability p, to combine that with the probability of the observed outcome in the trials performed for a given p, and then manipulate according to the usual rules of probability calculus to derive the required probability. It speaks volumes about the perspicacity of Bayes that he realized clearly that such a programme could be solidly implemented only if one started with a wider definition of probability which would cover statements about the value of an unknown parameter. What is more, he saw the logical necessity of showing that the usual probability calculus would be applicable to the wider definition. In this respect Bayes thought far ahead of almost all his contemporaries and successors of the next one and a half centuries, who nonchalantly operated with all sorts of probabilities according to the usual rules without bothering to ponder about the justification of such operations.

Bayes started with the definition: 'The probability of any event is the ratio between the value at which an expectation depending on the happening of the event *ought to* [italics ours] be computed and the value of thing expected upon its happening.' Several things are noteworthy about this definition. Just like Pascal and Huygens (subsection 5.3.3) a century earlier, Bayes took 'expectation' as his basic concept and defined probability in terms of it. From Price's introduction it appears that Bayes considered the meaning of expectation as self-evident, whether it refers to a contingent event relating to the future performance of a repeatable trial or to an uncertain proposition which may be true or false. Thus Bayes's definition applies whether probability is interpreted objectively or subjectively.[1] Also use of the words 'ought to' suggests that when subjectively interpreted Bayes would mean probability to be of the impersonal subjective kind (subsection 3.3.2). However, operationally the definition implies that in the subjective context, to determine the probability of an event (proposition) we have to agree about the value of the prize that we would receive if actually it happens (is true) and also about its expectation or present worth. This is just like the case of personal subjective probability (subsection 3.3.3), except that the feasibility of interpersonal agreement is presumed. Also, as Hacking (1965) p. 192 points out, 'Bayes evidently assumes that for given information the ratio between the stake and the prize is independent of the size of the prize' which would be realistic only if the size of the prize is small.

| Probability rules— |
| Bayes's derivation |

7.2.3 Starting from his definition of probability, Bayes methodically deduced the usual rules of operation of probability calculus. In this regard the resemblance between Bayes's reasoning and that based on the principle of coherence as put forward in the modern personalistic subjective theory of probability (subsection 3.5.4) is remarkable. (In fact, as would be seen in the following, self-consistency or coherence is implicit in Bayes's argument.) Thus to prove the addition law for two mutually exclusive events A_1, A_2, Bayes would have us suppose that for each of these the prize is the same amount N. If a_1, a_2 denote the respective expectations from these, according to Bayes the expectation 'from the happening of one or other of them' is $a_1 + a_2$. Since the prize for this is also N, the addition law follows from Bayes's definition of probability. If we interpret the prize as the total stake and the 'expectation from an event' as the amount we consider as the fair bet on it, this looks like a rewording of the standard proof of the law for subjective probability (subsection 3.5.4). Of course one can establish only finite additivity in this way; we will see that Bayes did not require countable additivity in the proof of his main result.

---

[1]On this point we disagree with Fisher who expressed the opinion that Bayes meant only the objective frequency sense through his definition (Fisher (1956) p. 14).

Bayes stated the multiplicative law as 'The probability that two subsequent [i.e. successive] events will both happen is a ratio compounded of the probability of the 1st and the probability of the 2nd on supposition the first happens.' Denoting the two events by C and A and assuming that the prize for each of the three events $C, C \cap A$, and $A|C$ (A on supposition that C happens) is the same amount N, let the expectations for the latter be $c = pN, d' = p'N$, and $d'' = p''N$ respectively. Bayes's argument was that the original expectation d' on $C \cap A$ increases by $d'' - d'$ as soon as we know that C has happened. Interpreting expectation as the amount which gives a fair bet when the total stake is N, this means if someone bets the amount d' on $C \cap A$, he would lose the amount if C fails to happen (as $C \cap A$ can no longer happen) and notionally gain $d'' - d'$ if C happens. This implies the odds $d'/(d'' - d')$ on C (Bayes did not use the term 'odds' but considered the ratio of the contingent loss to the contingent gain). He argued that this must equal the original odds $c/(N - c)$ on C and that led to the multiplicative law $p' = p \cdot p''$. Referring to our derivation of the multiplicative law for subjective probabilities in subsection 3.5.4 when $P(A|C)$ is defined in terms of the conditional bet $A|C$, we see that Bayes's idea of a notional gain turns into a concrete gain of $d'' - d'$ when C happens if one simultaneously bets d' *on* C and $N - d''$ *against* A under C. Thus the reasoning effectively turns out to be same. (In his proof as given above, Bayes assumed that the event C is observed prior to A and gave an alternative proof for the opposite case. But clearly the assumption is not essential and the same proof applies in both cases.)

After proving the multiplicative law, Bayes reformulated it like this: *Given two successive events C and A, if it is first discovered that the 2nd event A has happened and from this it is guessed that the 1st event C also has happened, then the probability that the guess is right is $P(C \cap A)/P(A)$.* As we would presently see, Bayes's entire programme of solving the inverse of De Moivre's problem rested on this proposition.

| Bayes's problem | **7.2.4**   Bayes stated his problem in the following terms: 'Given the number of times in which an unknown event |

has happened and failed: required the chance that the probability of its happening in a single trial lies somewhere between any two degrees of probability that can be named.'

This means if in n performances of a Bernoulli trial with an unknown probability of success p (which we regard as a random variable, at least in the subjective sense), s successes and $n - s$ failures have been observed, we want to find the probability that subject to such data p lies in an interval (ℓ, u) whose limits $0 \leq \ell < u \leq 1$ are arbitrary.

Price in his introduction states that Bayes perceived that the required probability could be found provided some rule according to which 'we ought to estimate the chance' of p lying between any two limits prior to any experiment made about it could be presumed. Price also states that since we know *nothing* about p prior to the experiment performed, it appeared to Bayes that the appropriate rule should be such that *the probability of p lying between any two 'equi-different' limits should be the same*. A modern Bayesian would immediately interpret this as amounting to the assumption of a uniform prior for p over the interval (0, 1) and then routinely proceed to derive the posterior distribution of p given the experimental data. But for Bayes every step in the derivation had to be justified. Obviously, in general, this would necessitate that the prior probability distribution be interpreted in the subjective sense. Although Bayes started his paper by carefully setting up the axiomatic apparatus required for this, apparently he thought, and Price confirms this, that such a conceptual reorientation would be too revolutionary to be accepted by others. Therefore he first considered a two-stage random experiment (of the type described in subsection 4.4.3), whose first stage determines the value of p, and after this, the second stage, the value of S (the random number of occurrences of the event). The experiment was so conceived that for it (i) the prior probability distribution of p can readily be interpreted in objective frequency terms and (ii) there can be no dispute about this distribution being uniform over (0, 1). Bayes solved his problem for this two-stage model and then by analogy extended the result to the general case. We will see that such a strategy also allowed Bayes to attempt a justification of the assumption of uniform prior in terms of a postulate which he considered to be more acceptable.

Solution for a two-stage model

7.2.5   Bayes considered a physical two-stage experiment which involved the throwing of balls on a square table, but we would present his argument in an equivalent simpler form. Choosing notations according to convenience, let AB be a line segment of unit length. In the first stage of the experiment a point O is randomly located on AB. Let p denote the length of the sub-segment AO. Clearly the requirement that the probability of p lying between any two equi-different limits should be same is met here. Without measuring p, in the second stage n further points are independently located on AB at random. Let S stand for the random variable representing the number of points that happen to be situated on the sub-segment AO and s for its observed value. Bayes considered the conditional probability of p lying in (ℓ, u) given $S = s$ where $0 \leq \ell < u \leq 1$.

By the proposition stated at the end of subsection 7.2.3 this is given by

$$P(\ell < p < u | S = s) = \frac{P(\ell < p < u, S = s)}{P(S = s)}. \tag{7.2.1}$$

Nowadays we would note that the set-up implies marginally a uniform distribution of p over (0, 1) and given $p = p$ a binomial (n, p) distribution of S, and

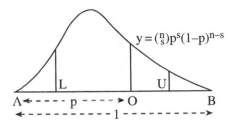

FIG. 7.1. Showing the geometric representation of $P(\ell < p < u, S = s)$.

hence, write down

$$P(\ell < p < u, S = s) = \int_{\ell}^{u} \binom{n}{s} p^s (1 - p)^{n-s}\, dp, \qquad (7.2.2)$$

$$P(S = s) = \int_{0}^{1} \binom{n}{s} p^s (1 - p)^{n-s}\, dp. \qquad (7.2.3)$$

From this we would get

$$P(\ell < p < u | S = s) = \frac{1}{B(s + 1, n - s + 1)} \int_{\ell}^{u} p^s (1 - p)^{n-s}\, dp. \qquad (7.2.4)$$

However, Bayes derived (7.2.4) by a geometric argument within the framework of finitely additive probability. His argument can be described by drawing over AB the curve whose ordinate at O is $y = P(S = s|p) = \binom{n}{s} p^s (1 - p)^{n-s}$ (see Figure 7.1). If L and U are points on AB such that the lengths of AL and AU are ℓ and u, the integral in (7.2.2) represents the area under the curve between the ordinates at L and U. Bayes showed that the probability on the left of (7.2.2) can neither be more nor less than this area since otherwise we can circumscribe or inscribe suitable histograms on the curve to get contradictions. Taking $\ell = 0, u = 1$, i.e. A for L and B for U, we get (7.2.3) and hence (7.2.4) follows. (See Bayes's *Essay* as reprinted in Biometrika (1958).)

Note that in Bayes's model p can assume any real value in $(0, 1)$ and the prior (uniform) distribution assumed and the posterior (beta) distribution implied by (7.2.4) are both continuous. We have seen in section 6.4 that the concept of a continuous random variable and distribution was first clearly articulated by Simpson in 1757. Bayes's quiet application of the same concept at about the same time without special mention suggests that possibly it was already part of the accepted lore of probability theory.

For the above two-stage experiment the prior, and hence, the posterior probability distribution of p can be interpreted in frequency terms. But the way Bayes enunciated the result (7.2.4) (his proposition 9) shows that he considered it legitimate to instantiate (subsection 3.2.5) a posterior probability statement about p to the case at hand. Thus what he stated amounted to: given the data, if one guesses that the value of p 'was' somewhere between ℓ and u, the probability that one 'is in the right' is given by (7.2.4). Such instantiation renders a sort of subjective interpretation to the frequency probability and provides the stepping stone to the conceptual platform of subjective probability where the general problem has to be tackled.

| Solution of the general |
| problem |

7.2.6 Bayes presented the solution to the general problem stated at the beginning of subsection 7.2.4, in the form of a *Scholium* (explanatory note). Evaluation of the integral in (7.2.3) gave

$$P(S = s) = \frac{1}{n+1}, \quad s = 0, 1, \ldots, n. \tag{7.2.5}$$

Thus for the above two-stage model, before the point O is chosen or its position discovered, the probability that S will assume the value s is same whatever s. Bayes thought that the result (7.2.4) was a consequence of this fact, implicitly assuming that (7.2.2) holds whenever (7.2.5) holds. With reference to the general problem he defined an 'unknown event' as one 'concerning the probability of which nothing at all is known antecedently to any trials made or observed concerning it.' He reasoned that for such an event we 'have no reason to think that, *in a certain number* [italics ours] of trials, it should rather happen any one possible number of times than another', i.e. if S denotes the random number of occurrences of the event in n trials, $P(S = s)$ would be same for $s = 0, 1, \ldots, n$. Hence denoting the unknown probability of the event in a single trial by p (which is generally random in the subjective sense) and drawing an analogy with the two-stage model, Bayes concluded that (7.2.4) holds for p.

From the modern point of view, if p has in general a prior density $\pi(p), 0 < p < 1$, in place of (7.2.3) and (7.2.4) we would have (cf. (4.4.1) and (4.4.2))

$$P(S = s) = \int_0^1 \binom{n}{s} p^s (1 - p)^{n-s} \pi(p) \, dp, \tag{7.2.6}$$

$$P(\ell < p < u | S = s) = \frac{\int_\ell^u p^s (1 - p)^{n-s} \pi(p) \, dp}{\int_0^1 p^s (1 - p)^{n-s} \pi(p) \, dp}. \tag{7.2.7}$$

(7.2.7) would reduce to (7.2.4) if and only if the prior is uniform i.e.

$$\pi(p) = 1, \quad 0 < p < 1. \tag{7.2.8}$$

(The 'only if' part follows by equating the expressions in (7.2.4) and (7.2.7) and differentiating with respect to u keeping ℓ fixed.) The assumption made in Bayes's scholium therefore really was that the condition

$$\int_0^1 \binom{n}{s} p^s (1 - p)^{n-s} \pi(p) \, dp = \frac{1}{n+1}, \quad s = 0, 1, \ldots, n, \tag{7.2.9}$$

is *not only necessary but sufficient* for (7.2.8). It has been pointed out (see Stigler (1986) p. 129, and also Hald (1998) p. 142 where other references are given) that mathematically the assumption is valid if we suppose (7.2.9) holds not for a particular n but for *all* $n \geq 1$. For then, taking $s = n$ we would have $\int_0^1 p^n \pi(p)\, dp = 1/(n+1)$, $n = 1, 2, 3, \ldots$ and the uniform distribution (7.2.8) over the finite interval (0, 1) being uniquely determined by its moment sequence (see e.g. Rao (1965) p. 86), the result would follow. But conditions for a distribution to be uniquely determined by its moment sequence were studied only in the first quarter of the 20th century. Also, it is doubtful whether Bayes while asserting that (7.2.9) holds 'in a certain number of trials' really meant that it holds for all n.

Bayes has been criticized by Fisher and others (Fisher (1956) pp. 16–17, Hacking (1965) pp. 199–201) for postulating the uniform prior (7.2.8) for the probability p of an unknown event on the ground that we know nothing about p prior to any experiment made about it. It has been pointed out that the postulate involves an arbitrary element because if we consider some one-one transform $\tau = \tau(p)$ of p (say, $\tau(p) = \sqrt{p}$) as the parameter, on the same ground one should assume a uniform distribution for τ and that would give a distribution of p different from (7.2.8). But as our above discussion shows, if we allow for the gap in Bayes's reasoning, his postulate really was: *the prior distribution of p is such that (7.2.9) holds for all $n \geq 1$.* This postulate can withstand the above criticism because if $\tau = \tau(p)$ is our parameter, the prior distribution of τ for which the counterpart of (7.2.9) holds is the one obtained from (7.2.8) by the same transformation.[2] Actually, as we will see later, in the context of any problem of induction involving a parameter θ about which nothing is known, Laplace prescribed that we should proceed by postulating a uniform prior for θ. Such a blanket postulate would be vulnerable to criticism of the type made by Fisher.

In the latter part of his essay, Bayes considered the evaluation of the incomplete integral in (7.2.4). He noted that there is no problem if either $n - s$ or s is small. If $n - s$ is small one can expand $(1 - p)^{n-s}$ by binomial theorem and integrate term by term; if s is small one can first transform from p to $q = 1 - p$ and do the same after expanding $(1-q)^s$. If neither s nor $n-s$ is small, however, this will not work—the very large number of terms in the binomial expansion would make the computation intractable. Yet this is the case in practice when one would have the greatest interest in knowing with what assurance limits can be set for an unknown p. Bayes expended tremendous efforts on finding lower and upper bounds for the integral when both s and $n - s$ are large, but his derived bounds, besides being complicated, were rather crude. Price in his follow-up of the essay improved upon those bounds, but the improved bounds also were not

---

[2]Fisher (1956) p. 17, also criticizes Bayes on the ground that in situations most frequently encountered in scientific practice, a priori probabilities are inapplicable. Obviously the subjective interpretation of probability is *non grata* to Fisher.

sharp enough. We would see that a practically useful solution to the problem gradually took shape in the hands of Laplace who started working on the problem a few years later.[3]

| Bayes's reticence |

7.2.7   Some authors have called the result (7.2.4) as applied to the general case as *Bayes's Theorem* (see Hald (1998) p. 141) although nowadays we use that epithet to denote the more general proposition later enunciated by Laplace (to be be considered in the next section). In any case, Bayes's result marked the beginning of what in Chapter 4 we have called the objective pro-subjective approach to statistical induction. From Bayes's derivation of the result it is clear that quite generally he interpreted the probability p in the objective sense and in the general context the prior, and hence the posterior probability appearing in (7.2.4), in the subjective sense. The ultimate conclusion follows from the premises deductively without the need of importing any extra principle beyond those usually associated with probability. However, apparently Bayes was not fully satisfied with the result and that led him to withhold its publication. Two explanations for it have been advanced. Fisher ((1944) p. 20, (1956) pp. 9–10) always maintained that Bayes himself was not convinced about the validity of the argument through which he extended the conclusion from the two-stage experiment to the general case; this view is duplicated by, among others, Hacking (1965) p. 201. On the other hand, Stigler (1986) pp. 129–130 is emphatic that Bayes was quite sure of his reasoning and his inability to find a workable method for the evaluation of the incomplete beta integral was the cause of his unwillingness to publish. To us, this second view seems to be more plausible; otherwise there would be no reason to develop the calculus for subjective probability the way Bayes did it. In any case the dissatisfaction of Bayes with his own work seems to have been the reason for his not discussing any practical application of it.

| Price's commentary |

Price in his introductory letter and also in the appendix to the essay discussed the importance of Bayes's result and considered some applications. The one application of (7.2.4) for interval estimation that he considered related to drawings from a lottery in which blanks and prizes are mixed in an unknown ratio. Suppose one is interested in estimating the unknown ratio $p/(1-p)$ where p is the proportion of blanks. Let the number of blanks observed in n drawings be s_n, with S_n denoting the corresponding

---

[3]More than a century later, Karl Pearson organised a massive computational programme and prepared tables for the incomplete beta integral which plays a keyrole in the fitting of Pearsonian curves of certain types (see Pearson (1968)). The computation was based on successive use of recursion relations on p' (our $s+1$) and q' (our $n-s+1$) and the incomplete beta-function ratio (7.2.4) with $\ell = 0$ was tabulated over $0 < u < 1$ for $0 < p' < q' < 50$ including fractional values of p', q' (to avoid confusion we write p', q' instead of p, q which are the notations in Pearson's tables).

random variable. Then, if $s_n/n = \hat{p}$, and $\delta_1, \delta_2 > 0$ satisfy $\delta_1 < \hat{p}$ and $\hat{p} + \delta_2 < 1$, we would have

$$P\left(\frac{\hat{p} - \delta_1}{1 - \hat{p} + \delta_1} < \frac{p}{1 - p} < \frac{\hat{p} + \delta_2}{1 - \hat{p} - \delta_2}\middle| S_n = n\hat{p}\right)$$

$$= P(\hat{p} - \delta_1 < p < \hat{p} + \delta_2 | S_n = n\hat{p}), \qquad (7.2.10)$$

where the probability on the right can be obtained from (7.2.4). In Price's example $\hat{p} = \frac{10}{11}, \delta_1 = \frac{1}{110}, \delta_2 = \frac{1}{132}$, so that the left hand member of (7.2.10) reduced to

$$P\left(\frac{9}{10} < \frac{p}{1 - p} < \frac{11}{12}\middle| S_n = n\frac{10}{11}\right). \qquad (7.2.11)$$

Taking n successively equal to 11, 22, 44, 110, he got the values 0.07699, 0.10843, 0.1525, 0.2506 for the credibility coefficient (7.2.11). From these and some further calculations made for even larger n-values, Price concluded: 'it appears ... that the chance for being right in guessing the proportion of *blanks* to *prizes* to be nearly the same with that of the number of *blanks* drawn in a given time to the number of *prizes* drawn, is continually increasing as these numbers increase and that therefore when they are considerably large this conclusion may be looked upon as morally certain.' We will see in the next section that this follows quite generally from a proposition proved by Laplace.[4] Of course the assumption of a uniform prior for p here is hardly realistic as anyone familiar with lotteries would know that p is always chosen close to 1.

Bayes did not dwell on any philosophical implications of his result. But Price enthusiastically asserted that it provides a method for determining 'the strength of *analogical* or *inductive reasoning*' and hence, holds great interest for

---

[4]The English psychologist David Hartley in a 1749 publication mentioned that a friend of his had privately communicated to him a solution of the inverse of De Moivre's problem from which it followed 'what the Expectation is when an Event has happened p times, and failed q times, that the original Ratio of the Causes for the Happening or the Failing of an Event should deviate in any given Degree from that of p to q' and that it appeared from this solution 'that where the Number of Trials is very great, the Deviation must be inconsiderable.' (see Stigler (1986) p. 132). This remark of Hartley has proved a veritable red herring. It has been speculated that priority for Bayes's result should go to some other scholar (Stigler (1983)). Also wonder has been expressed that Hartley's notations p, q and mention of the case when $n \to \infty$ anticipate Laplace whose first publication on the problem was dated 1774. But as observed in subsection 7.2.1, recent evidence suggests that Bayes's work was largely completed between 1746 and 1749. Also, although the fact of increase in credibility with increasing n is commented upon by Price in the context of the above example, there is no reason to suppose that Bayes himself had not observed this. The notations p, q (for our s and $n - s$) were used by Bayes too; use of the word 'Expectation' by Hartley is also suggestive. In view of all this it seems reasonable to suppose that it was none other than Bayes who communicated the result to Hartley (see Hald (1998) p. 132).

'experimental philosophy'. It may be recalled (section 1.4) that Hume created 'the scandal of philosophy' by articulating his objection to the principle of induction in his 1739 book. The resulting controversy was very much in the air at the time of Bayes and Price. Without mentioning Hume by name, Price discussed how Bayes's result gives support to Hume's contention and at the same time shows a way out. For this he considered the case where no failures occur. Price's actual argument here is a bit confusing because he gave undue emphasis on the posterior probability of $p > \frac{1}{2}$ (the given event has 'more than an even chance'). But it seems that his intended meaning was somewhat like this: Putting $s = n$ and $u = 1$ in (7.24), we get

$$P(\ell < p < 1|S = n) = 1 - \ell^{n+1}, \tag{7.2.12}$$

$$P(0 < p \le \ell|S = n) = \ell^{n+1}. \tag{7.2.13}$$

Clearly for $\ell > 0$ (7.2.13) is never zero however large n, the number of trials in which success invariably occurs. Thus Hume's objection that however large the number of past instances in which a particular cause has been found to produce an effect we can never conclude that in future the same cause will *always* produce that effect, is upheld. On the other hand, taking ℓ close to 1, (7.2.12) gives a measure of the assurance that we can provide to the conclusion that the event will occur in future, thus providing a kind of pragmatic solution to Hume's problem. Also this assurance would increase as n increases. Price cited examples of natural phenomena like 'a piece of wood put into fire will burn', 'a stone detached from all contiguous objects will fall', and 'the Sun after setting one night will rise next morning'. The last example had been considered both by the 17th century philosopher Hobbes ('for though a man have always seen the day and night to follow one another hitherto; yet can he not thence conclude they shall do so, or they have done so eternally'—see Maistrov (1974) p. 6) and by Hume (see Hald (1998) p. 127). (We will see later that Laplace also gave his attention to this example.) Taking $\ell = \frac{1}{2}$, and denoting the probability of the sun rising on a day by p, Price used (7.2.12) to compute the probability of $p > \frac{1}{2}$, given that the sun has been found to rise without failure on $n = 1$ million past days.

In discussing all these examples involving natural phenomena, Price emphasized that one can use formulae such as (7.2.12) only if one is totally ignorant about nature previously; for otherwise, the assumption of uniform prior would be inapplicable. In this connection Price made a peculiar observation: 'After having observed for sometime the course of events it would be found that the operations of nature are in general regular, and that the powers and laws that prevail in it are stable and permanent. The consideration of this will cause one or a few experiments often to *produce a much stronger expectation of success in further experiments than would otherwise have been reasonable* [italics ours].' This can be explained if we think that at the outset one has two candidate priors $\pi_1(\cdot)$ and $\pi_2(\cdot)$ in mind; $\pi_1(\cdot)$ is the uniform prior and $\pi_2(\cdot)$ is a prior which is negatively skewed with a concentration of mass towards 1. As one goes on performing successive trials if some failures also occur along with successes, one sticks to the uniform prior $\pi_1(\cdot)$. But if only successes occur, knowing the regularity of natural phenomena, one can switch over to $\pi_2(\cdot)$ which would give a higher posterior probability to statements like $p > \ell$. In particular, taking the prior density in the form $\lambda p^{\lambda-1}$ where $\lambda > 0$, we can suppose $\pi_1(\cdot)$ corresponds to $\lambda = 1$ and $\pi_2(\cdot)$ to some suitably chosen high positive value of λ. Viewed from this angle, we can see that Price's recommendation represents a rudimentary form of what nowadays we call the empirical Bayes approach (see footnote 33 in Chapter 4).

Price also added a theological dimension to Bayes's result. He concurred with De Moivre's thesis (section 6.3) that the latter's result on the regularity in the behaviour of empirical relative frequencies provided reasons for 'believing that there are in the constitution of things fixt laws according to which events happen, and that, the frame of the world must be the effect of the wisdom and power of an intelligent cause.' But, over and above, he expressed the opinion that Bayes's result, as it showed 'in every case of any particular order or recurrency of events, what reason there is to think that such recurrency or order is derived from stable causes or regularities in nature,' provided even stronger arguments in support of the 'existence of the Deity.'

Although Bayes's paper was published during 1764–1765, it did not immediately attract the attention it deserved. Whether it was due to his considering a very special (though quite important) problem, or due to his result being geared to the assumption of total ignorance about p before the experiment, or his failure to find a workable method of evaluating the incomplete beta integral when neither s nor $n - s$ is small, is debatable. But it is a historical fact that Bayes's contribution gained widespread recognition only after Laplace had developed the same approach to serve broader purposes.

7.3   Pierre–Simon Laplace

Laplace's life and career   **7.3.1**   Pierre-Simon Laplace (1749–1827) was born in a village in Normandy and was educated at the nearby University of Caen. He showed his promise as a mathematician quite early. Moving to Paris at the age of 21 or 22, he impressed contemporary academics with his mathematical capabilities and secured an appointment as a Professor at the Paris Military School in 1771 and subsequently, within a few years, membership of the Paris Academy of Sciences. In 1794 he became Professor of Mathematics at the prestigious École Normale. He remained and worked in Paris before and during the French Revolution and thereafter into the Napoleonic and post-Napoleonic era. Apparently he had a knack for setting his sails obedient to the wind, for all through this period of turmoil punctuated by political upheavals, he managed to remain in the good books of successive ruling governments. He became member of various official commissions, procured support for his research projects, and received honours and recognition from different regimes. Some chroniclers of Laplace's life have suggested that all this bespoke his opportunism. Keynes (1921) p. 341 cites a concrete instance: how Laplace made an eloquent dedication, *A Napoléon-le-Grand* in the 1812 edition of one of his books on probability, only to replace it in the 1814 edition by an observation that 'Bernoulli's Theorem must always bring about the eventual downfall of a great power which, drunk with the love of conquest, aspires to a universal domination'. However, nobody can question the towering genius and achievements of Laplace as a scientist and mathematician.

| Scientific contributions | **7.3.2** Although Laplace evinced interest in many branches of science his prinicipal fields of research were theory of probability and its applications and astronomy. Besides, he made lasting contributions to various areas of mathematics. But, as Poisson said, Laplace was a 'natural philosopher, who sought to understand nature by drawing into service the highest mathematics'. Laplace adapted mathematical analysis to various uses but always subordinated the particular method to the special problem. (We will see that this pragmatic attitude showed through his approach to problems of statistical induction as well.) On probability theory and its application to the analysis of statistical data, he wrote about twenty memoirs capping them with the publication of two books—the highly mathematical *Théorie Analytique des Probabilités* and the less technical *Essai Philosophique sur les Probabilités*. The former was first published in 1812, followed by two more editions and a number of supplements during Laplace's life time; the latter, which was later included as an introduction to the former, was published in 1814, although an earlier version in the form of lecture notes came out in 1812.

There was a peculiar break in the continuity of Laplace's work on probability. He started working on probability early in life and published regularly on the subject until 1786. Then from 1786 to 1805 he was engrossed in his epoch-making work on astronomy, the fruits of which, in the form of his monumental *Traité de Mécanique Céleste*, came out in four volumes. Through it he earned wide acclaim by providing reassuring answers to certain knotty questions about the fate of the solar universe (will the moon crash into the Earth in the course of time? will Jupiter crash into the Sun and Saturn escape out of the system?), which seemed to be implied by Newton's laws of gravitation and which had earlier baffled his predecessors including Euler and Lagrange. But there is little evidence of his working on problems of probability during this period, although from 1795 onwards he lectured on probability at the École Normale. Karl Pearson (1978) p. 651 has made the interesting suggestion that Laplace fought shy of probability during the years of the revolution because probability which deals with mundane matters like voting and testimony was an unsafe subject at that time. (After all, Condorcet dabbling in such matters courted the displeasure of the Jacobins and was ultimately driven to death!) Whatever the reason, it is a fact that Laplace resumed working on probability only after relative internal stability set in with the assumption of powers by Napoleon.

Nature of Laplace's work on probability

7.3.3   In accordance with our general policy, we will skip the part of Laplace's work on probability dealing with the solution of intricate games of chance that have little bearing on statistical induction. The rest of his work in the area can be studied under three broad headings:

(a) Induction based on posterior distributions as in the pro-subjective approach.
(b) Derivation and study of probabilistic models.
(c) Induction following the sampling theory (behavioural or instantial) approach.

It is remarkable that Laplace at a quite early stage saw the possibility of two distinct approaches to problems of statistical induction: one from the pre-observational and the other from the post-observational position. In a memoir which he read at the Academy of Sciences in 1777 he considered 'the mean to be chosen among the results of many observations' and observed:

The problem we are concerned with may be regarded from two different points of view, depending on whether we consider the observations before or after they are made. In the first case the search for the mean we should take among the observations consists in determining a priori which function of the results of the observations is most advantageously taken for the mean result. In the second case the search for this mean consists in determining a similar function *a posteriori*, that is paying regard to the distances between the respective observations. We see easily that these two manners of regarding the problem must lead to different results, but it is at the same time clear that *the second is the only one that should be used.* [italics ours] (See Stigler (1986) p. 119.)

We see that in this paper Laplace expressed himself unequivocally in favour of the post-observational or posteriorist point of view. In fact, most of the work on statistical induction in the early (pre-revolution) phase of his career was carried out from this viewpoint. However, in the later (post-revolution) phase in a large part of his work he adopted the pre-observational or sampling theory point of view. Whether this was because he saw the difficulties of full specification and parametrization of the set-up in many situations and with the discovery of the CLT (section 3.10) realized that in large samples those can be largely sidetracked in the sampling theory approach, or he really changed his earlier opinion, we cannot say. In this chapter we mainly review the posteriorist part, i.e. the part coming under the heading (a) above, of Laplace's work. The parts coming under the headings (b) and (c) will be taken up in the next chapter.

In reviewing Laplace's work we face one difficulty. Many of his ideas evolved over time and were continually revised. As a result, the presentation in the early papers is sometimes at variance with that finally given in his two books. We would go as far as possible by the final version given in the books and would refer to earlier papers only as it would be necessary for historical reasons. But it would be expedient to first say a few words about Laplace's ideas about probability.

Laplace's conception of probability

7.3.4   Like most scientists in the post-Newtonian era, Laplace professed the view that all phenomena in nature happen according to deterministic laws and chance has no objective existence. But whereas to his predecessors like Pascal and James Bernoulli, nature was deterministic because it is controlled by an omniscient and omnipotent God, Laplace's determinism was mechanistic. In the first section of the *Essai Philosophique* he wrote:

An intelligence which for a given instant knew all the forces by which nature is animated and the respective situations of the existences which compose it; if further that intelligence were vast enough to submit these given quantities to analysis; it would embrace in the same formula the greatest bodies in the universe and the lightest atom: nothing would be uncertain to it and future as the past would be present to its eyes.... All men's efforts in the search for truth tend to carry him without halt towards such an intelligence *as we have conceived, but from which he will remain always remote*[5] [italics ours] (see Karl Pearson (1978) p. 656).

According to Laplace, limitations of the human mind and senses rendered most natural phenomena unpredictable to us and thus made probability relevant in practice. 'Probability', he said, 'has relation part to our ignorance and part to our knowledge'. (In fact, he went so far as to assert that, since induction and analogy, our principal means of reaching the truth, are based on probability, almost all our knowledge is only probable knowledge.) Such a stance naturally meant that probability to Laplace was essentially subjective.

But Laplace did not deny the existence of intersubjectively agreed (and hence, virtually 'objective') probabilities for random phenomena in the real world. According to him phenomena like the tosses of a coin are produced by a combination of constant causes (the same coin is tossed by the same device) and variable causes (other subtle conditions specific to each toss). In the *Essai* he observed that 'in a series of events indefinitely prolonged, the action of regular and constant causes ought to render them superior in the long run to the effects of irregular causes' (see Karl Pearson (1978) pp. 653, 664–665). Thus ignoring the variable causes we can associate with a combination of constant causes definite probabilities for random phenomena and generally definite values of parameters for random observables. (Laplace's language here seems to foreshadow the propensity interpretation of probability of Popper (subsection 3.2.5).) Because of this attitude, inspite of Laplace's subjective view of probability, his approaches

---

[5]It is generally thought that Laplace was a staunch believer in a completely deterministic model for the universe and that ideal has been undermined by developments like the uncertainty principle of modern sub-atomic physics. Thus Stephen Hawking (1988) p. 59 remarks, 'One certainly cannot predict future events exactly if one cannot measure the present state of the universe precisely!' But as the qualification at the end of the above excerpt shows, Laplace's determinism was conceptual and he conceded that operationally complete determinism would for ever remain out of reach to the human mind.

to problems of induction were objective in the wider sense of subsection 4.4.1. (In the next chapter we will see that while approaching problems of induction from the sampling theory viewpoint, Laplace regarded the unknown value of a parameter as fixed over repetitions of an experiment, which were considered independent. In such contexts Laplace accepted that the probability of an event gets reflected in its long-term relative frequency; see Hald (1998) p. 162.)

To define numerical probability Laplace took the urn model as his prototype:

The theory of chances consists in reducing all events of the same kind to a certain number of cases *equally possible that is to say that we are equally undecided as to their existence* [italics ours] and in determining the number of cases favourable to the event of which we seek the probability. The ratio of this number to that of all the cases possible is the measure of this probability (*ibid* p. 657).

The wording here further confirms that Laplace's probability was of the *impersonal* subjective kind (subsection 3.3.2).

Although, strictly speaking, the above definition of probability is appropriate only in situations where the total number of equally possible cases is finite, Laplace did not hesitate to use the notion of probability in the context of a random variable which can assume any value in a continuum. Implicitly, in such a situation he would suppose that the number of cases favourable to an event and the total number of cases both tend to infinity maintaining a constant ratio. In the finite case the usual rules of probability follow straightaway from the definition; Laplace took those for granted in the infinite case as well.

Laplace's form of Bayes's theorem

7.3.5 We have seen in subsections 7.2.5–7.2.6 how Bayes based his mode of inference about an unknown Bernoulli parameter on the posterior distribution of the parameter given the outcomes of a number of trials. To develop the same approach in the general case Laplace considered 'probabilities of causes'. Suppose an even E can occur if and only if one of the mutually exclusive contingent causes $C_1, C_2, \ldots, C_r$ obtains. Suppose further that E has been observed, but which particular cause produced E is not known. Then in modern notations, Laplace's expression for the probability that C_j produced E is

$$P(C_j|E) = \frac{P(C_j)P(E|C_j)}{\sum_{i=1}^{r}P(C_i)P(E|C_i)} \qquad (7.3.1)$$

In the 1814 edition of *Théorie Analytique*, Laplace derived (7.3.1) in the usual way starting from the product rule representation $P(C_j|E) = P(C_j \cap E)/P(E)$, just as Bayes had done earlier to derive his particular version (7.2.4). But this is not how Laplace arrived at the expression (7.3.1) initially. In a memoir published in 1774 he exogenously introduced the formula

$$P(C_j|E) = \frac{P(E|C_j)}{\sum_{i=1}^{r}P(E|C_i)} \qquad (7.3.2)$$

as representing the probability of the cause C_j given E. (7.3.1) reduces to (7.3.2) if $P(C_i)$ is same for $i = 1, \ldots, r$, but Laplace did not derive (7.3.2) that way. In fact there is no mention of the a priori probabilities $P(C_i)$ in Laplace's 1774 memoir; instead (7.3.2) is sought to be justified as an intuitive 'principle'—'the probabilities of the existence of these causes given the event are to each other as the probabilities of the event given the causes' (see Hald (1998) p. 160). (Recall that at about the same time, Daniel Bernoulli (section 6.5) proposed 'the maximum likelihood principle' on a similar intuitive basis. In fact from the point of view of some modern proponents of likelihood inference, (7.3.2) can be looked upon as standardized likelihood; cf. Basu (1975) Part II.) From his writings of that time it seems Laplace distinguished between two kinds of probability—'probability of events' and 'probability of causes'—the names 'direct' and 'inverse' probability also came to be associated with these concepts. Later it gradually dawned upon him that (7.3.2) can be interpreted as a conditional probability if we assume that the causes are all equally probable. He gave a circuitous proof of this in a memoir in 1781 and the direct proof based on the product rule in a later publication in 1786. As noted above, the proof of the general result (7.3.1) appeared only in 1814 (see Hald (1998) pp. 159–161).

Nowadays in standard textbooks on probability the result (7.3.1) is called Bayes's theorem. But as we saw in subsections 7.2.5–7.2.6, Bayes derived a version of (7.3.1) only in the context of a special problem. (However, in contrast with Laplace, Bayes from the beginning saw that the problem was one of derivation of a conditional probability.) Apparently, when Laplace wrote his 1774 memoir, he was not aware of Bayes's result as contained in the latter's 1764 *Essay*. But Condorect referred to Bayes in his announcement of Laplace's 1781 memoir and in lectures given in the 1780's interpreted (7.3.2) as representing a conditional probability derived from equal a priori probabilities (Hald (1998) p. 187). This suggests that Laplace's 1786 derivation of (7.3.2), and naturally therefore, his derivation of the general result (7.3.1) in 1814 might have been influenced by Bayes. However, although Laplace referred to Bayes at one place in the *Theorie Analytique* (see Fisher (1956) p. 18), he did not explicitly acknowledge any indebtedness to Bayes in this regard.

Posterior probability of a future event

Under the set-up of (7.3.1), Laplace also considered the probability of a future event E' given the event E. The general result for the situation when under each C_i the events E and E' are conditionally independent, as described in the *Essai* (see Karl Pearson (1978) p. 659), can be put as

$$P(E'|E) = \sum_{i=1}^{n} P(C_i|E)P(E'|C_i), \tag{7.3.3}$$

where $P(C_i|E)$ is given by (7.3.1), or when the C_is are equally probable, by (7.3.2). The proof, implicit in Laplace's work, follows by writing

$P(E'|E) = P(E \cap E')/P(E)$ and then breaking up the numerator and denominator on the basis of the partition $E = \cup_1^r (E \cap C_i)$.

Continuous versions
For purposes of statistical induction Laplace had to consider versions of (7.3.1) and (7.3.3) corresponding to a continuum of causes. These are best described by representing the cause as an unknown parameter θ which can assume values in a continuum Θ (if θ is real-valued, Θ is an interval). The a priori probabilities $P(C_i)$ get replaced by a prior density function $\pi(\theta)$ over Θ. In place of (7.3.1) we have the posterior density function given E (cf. (4.4.1)) as

$$\pi(\theta|E) = \frac{\pi(\theta)P(E|\theta)}{\int_\Theta \pi(\theta)P(E|\theta)\,d\theta}. \tag{7.3.4}$$

For a future event E' such that, given θ, E and E' are independent, we have (cf. (4.5.4))

$$P(E'|E) = \frac{\int_\Theta \pi(\theta)P(E|\theta)P(E'|\theta)\,d\theta}{\int_\Theta \pi(\theta)P(E|\theta)\,d\theta}, \tag{7.3.5}$$

which is the continuous version of (7.3.3). Laplace's posteriorist approach to induction was generally based on (7.3.4) and (7.3.5).

Posterior-based
estimation
7.3.6   For estimating a real parameter assuming values in an interval, depending on the nature of the problem, Laplace suggested the use of the mode, the median, or the mean of the posterior distribution (see Hald (1998) pp. 89–90, 175).

Thus to estimate the unknown probability of success $p \in (0, 1)$ in a Bernoulli trial on the basis of n independent (given p) performances of the trial, Laplace assumed a uniform prior distribution of p over (0, 1) and took the observed sequence of n outcomes for the event E. Then, writing s for the realized number of successes, (7.3.4) gives the posterior density

$$\pi(p|s) = \frac{1}{B(s+1,\, n-s+1)} p^s (1-p)^{n-s}, \quad 0 < p < 1. \tag{7.3.6}$$

This is the same as the density appearing in Bayes's formula (7.2.7). Differentiating and equating to zero one gets that the mode of this is $\hat{p} = s/n$, which was Laplace's preferred estimate here.

To judge the accuracy of this estimate Laplace considered the evaluation of

$$P(|p - \hat{p}| < \epsilon|s) = \int_{\hat{p}-\epsilon}^{\hat{p}+\epsilon} \pi(p|s)\,dp \tag{7.3.7}$$

He made a significant breakthrough in this regard and derived an asymptotic expansion for $\pi(p|s)$ by developing $\log \pi(p|s)$ in a Taylor series around $\hat{p}$.

The leading term of this expansion is the density function of the distribution $\mathcal{N}(\hat{p}, \hat{p}(1 - \hat{p})/n)$. No suitable tables for the cdf of the standard normal distribution were available when Laplace first worked on the problem; therefore Laplace developed workable expansions for the same to facilitate the computation of (7.3.7), *inter alia* proving rigorously that $\int_{-\infty}^{\infty} e^{-z^2/2}dz = \sqrt{(2\pi)}$. (These results were obtained in two papers appearing in 1774 and 1785; see Hald (1998) pp. 206–210. Hald notes that the first tables for probability integrals of the normal distribution were given by the French astronomer Kramp in 1799.) Further, using the asymptotic expansion for (7.3.7), Laplace showed that for any fixed ϵ if s and n tend to ∞ so that $s/n = \hat{p}$ remains fixed, the probability (7.3.7) tends to 1. Thus making n sufficiently large one can be practically certain that the unknown p is close to the realized $\hat{p}$ upto any specified margin of error ϵ.[6] This can be regarded as the inverse of Bernoulli's result.

The asymptotic expansion of $\pi(p|s)$ derived by Laplace implies that for large n the posterior distribution of $\sqrt{n}(p - \hat{p})$ is approximately $\mathcal{N}(0, \hat{p}(1 - \hat{p}))$. Here $\hat{p}(1 - \hat{p}) = \{(-1/n)d^2/dp^2 \log \pi(\hat{p}|s)\}^{-1}$. Actually, by a method which Laplace worked out in particular problems (see Hald (1998) p. 243), we get a more general result of this type. Consider the situation where $p = p(\theta)$ is a function of a real parameter θ which can be differentiated any number of times with $p'(\theta) > 0$ for all θ in an interval domain. Suppose, as θ varies in its domain, $p(\theta)$ assumes every value in (0, 1) (exactly once). Then as per Laplace's method, under a uniform prior for θ the solution $\hat{\theta}$ of the equation

$$\frac{d}{d\theta} \log \pi(p(\theta)|s) = 0 \qquad (7.3.8)$$

can be taken as the estimate of θ; for large n the posterior distribution of $\sqrt{n}(\theta - \hat{\theta})$ is approximately $\mathcal{N}(0, v^2)$, where

$$v^{-2} = -\frac{1}{n}\frac{d^2}{d\theta^2} \log \pi(p(\hat{\theta})|s). \qquad (7.3.9)$$

In fact by Laplace's method asymptotic normality of the posterior follows under mild assumptions even for a non-uniform prior. Further, similar results follow under broad conditions when instead of the outcome of Bernoulli trials we have a random sample from some distribution whose pmf or pdf involves the unknown real parameter θ (see Hald (1998) pp. 245–246). These results foreshadow what later came to be known as the *Bernstein–von Mises Theorem* (subsection 4.4.6).

---

[6]Stigler (1986) p. 133 refers to this result as *posterior consistency*, but in the current literature (see e.g. Diaconis and Freedman (1986) and Ghosal *et al* (1995)) the term is used to connote a somewhat different property. Thus according to current usage posterior consistency holds because under any *fixed true value* p_0 of p, as $n \to \infty$, $P(|p - p_0| < \epsilon|S) \to 1$ in probability (almost surely), S being the random counterpart of s. However, Laplace's result can be adapted to prove this by plugging in a suitable Law of Large Numbers.

Obviously here there is a formal similarity between (7.3.8) and the maximum likelihood equation, (7.3.9) and the estimated asymptotic variance of the maximum likelihood estimate, and the asymptotic form of the posterior distribution and the asymptotic sampling distribution of the MLE. This is a particular instance of the formal similarity in the general context of point estimation which we noted in subsection 4.4.6.[7]

For the estimation of the true value of θ of a real quantity on the basis of n independent measurements $x_i, i = 1, \ldots, n$ where the errors $Z_i = X_i - \theta$ are subject to the density $f(z)$, $-\infty < z < \infty$, Laplace assumed an improper uniform prior over the real line giving

$$\pi(\theta|x_1, \ldots, x_n) = \text{const.} \prod_{i=1}^{n} f(x_i - \theta), \quad -\infty < \theta < \infty. \qquad (7.3.10)$$

To get a point estimate of θ here, Laplace followed what nowadays we call a 'decision theoretic approach' with absolute error loss (subsection 4.4.6). Thus denoting the estimate by t, he suggested t should be determined so that

$$\int_{-\infty}^{\infty} |\theta - t| \pi(\theta|x_1, \ldots, x_n) \, d\theta \qquad (7.3.11)$$

be minimized. Laplace showed that this amounted to taking the median (which in this context he called the 'probability mean') of (7.3.10) as t. During 1774–1781 when Laplace made such investigations the normal law had not as yet been recognized as the standard (or even a possible) choice as the law of error. With the forms of $f(\cdot)$ that Laplace tried, he could not make much headway towards finding t for a general n. The limited results that he got showed that for the forms taken t would not generally be equal to the sample mean—something he apparently had hoped to establish through his investigations (see Stigler (1986) pp. 112–122, Hald (1998) pp. 171–181).

It appears that in his early investigations Laplace did not explore the possibility of using the posterior mean as a possible estimate. However, later in the context of the estimation of a parameter representing the precision of measurements he proposed and studied the posterior mean as the appropriate estimate (see Hald (1998) pp. 418, 421).

---

[7]Incidentally, the essence of Laplace's method of asymptotic expansion for the approximate evaluation of the probability integral of a bell-shaped function consists in noting that the major contribution to the value of the integral comes from a neighbourhood of the maximal point of the function. This method has been found useful in many contexts; for some recent applications, see e.g. Schwarz (1978), Ghosh and Samanta (2001).

| Posterior-based testing | **7.3.7**   Laplace utilized (7.3.6) also to develop tests for |

7.3.7   Laplace utilized (7.3.6) also to develop tests for one-sided hypotheses involving binomial probabilities. The problems considered related to sex ratios at birth which had remained a sort of *cause célèbre* to demographers and probabilists ever since the time of Graunt (subsection 5.4.5 and section 6.2). Denoting the probability of birth of a male child by p, Laplace first considered the birth records of Paris for the 26 years 1745–1770 (n = number of births = 493472, $\hat{p}$ = observed proportion of male births = 0.50971) to examine whether the data supported the conclusion $p > \frac{1}{2}$. For this, assuming a uniform prior over $(0,1)$, Laplace got the posterior distribution of p in the form (7.3.6) with $s = n\hat{p}$. Using an asymptotic expansion for the tail probability of (7.3.6), Laplace computed $P(p \le \frac{1}{2}|n\hat{p}) = \int_0^{1/2} \pi(p|n\hat{p})\,dp = 1.521 \times 10^{-42}$. Practically the same value is obtained here if one takes the leading term in Laplace's expansion for $\pi(p|n\hat{p})$ around $\hat{p}$ which corresponds to the distribution $\mathcal{N}(\hat{p}, \hat{p}(1-\hat{p})/n)$. Writing Z for a standard normal variable, from the latter we get

$$P\left(p \le \frac{1}{2}|n\hat{p}\right) = P\left(Z \le \frac{1/2 - \hat{p}}{\{\hat{p}(1-\hat{p})/n\}^{1/2}}\right) = P\left(Z \ge \frac{(\hat{p}-1/2)}{\{(\hat{p}(1-\hat{p}))/n\}^{1/2}}\right).$$
(7.3.12)

As $P(p \le \frac{1}{2}|n\hat{p})$ is negligibly small, Laplace concluded that 'one can regard it as equally certain as any other moral truth' that for Paris $p > \frac{1}{2}$, or in other words, nature favours more the birth of boys than of girls.

Laplace also considered the problem of comparison of two binomial probabilities on the basis of independent samples. Denote the proportion of male births for Paris and London by p_1 and p_2. For Paris during 1745–1770 the proportion $\hat{p}_1$ of male births out of a total number of n_1 births observed were as given above (we now write $\hat{p}_1, n_1$ for $\hat{p}, n$). For London during the 94 years 1664–1757, the corresponding figures were $\hat{p}_2 = 0.51346, n_2 = 1,436,587$. In view of $\hat{p}_2$ being appreciably larger than $\hat{p}_1$, one would like to know whether p_2 is really greater than p_1. Under the assumption that a priori p_1 and p_2 are independently distributed uniformly over $(0,1)$, given the data, a posteriori, these follow independent beta distributions of the form (7.3.6) with parameters $n_i, s_i = n_i\hat{p}_i, i = 1,2$. Using large sample approximations to the joint posterior distribution of p_1, p_2, after some manipulation Laplace obtained

$$P(p_2 - p_1 \le 0|n_i\hat{p}_i, i = 1,2) = P\left(Z \ge \frac{\hat{p}_2 - \hat{p}_1}{\{\hat{p}_2(1-\hat{p}_2)/n_2 + \hat{p}_1(1-\hat{p}_1)/n_1\}^{1/2}}\right)$$
(7.3.13)

where Z as before is distributed as $\mathcal{N}(0,1)$. In Laplace's case this probability came out as 2.43×10^{-6} and he concluded that one could bet at very high odds on the proposition that there is some cause which makes the probability of birth of a male child higher in London than in Paris. Laplace attempted a physical explanation of this phenomenon—he suggested that the higher proportion of girls among the foundlings registered in Paris pulled p_1 downwards.

Although Laplace applied his tests for binomial probabilities only on birth data, he noted that those could be used in other situations, e.g. for comparing the cure rates of different treatments. With these tests, Bayesian hypothesis testing—or as we called it in Chapter 4, the objective pro-subjective approach to hypothesis testing—really got under way. (Note that in conformity with what we said in subsection 4.4.1 the parameters involved in Laplace's hypotheses, e.g. the long-term proportions of male births have objective or at least inter-subjective interpretation.)

Regarding how low the posterior probability of a hypothesis should be in order that it may be rejected (and its complement accepted), Laplace seems to have set rather stringent norms. Thus for comparing the probability of birth of a boy for Paris ($\hat{p}_1 = 0.50971$ as above) and Naples ($\hat{p}_2 = 0.51162$), as the posterior probability (7.3.13) came out to be about $1/100$, he did not favour outright pronouncement that $p_2 > p_1$ (Stigler (1986) p. 135). Of course in these problems Laplace had very large sample sizes and therefore could afford to be finicky.

Looking at (7.3.12) and (7.3.13) we see that the posterior-based tests for $H_0: p \leq \frac{1}{2}$ and $H_0: p_2 \leq p_1$ are formally same as the large-sample tests one would apply for these in the sampling theory (behavioural or instantial) approach. Only the expressions (7.3.12) and (7.3.13), which represent posterior probabilities in the former, would be interpreted in the latter as P-values based on large-sample sampling distributions under the 'boundary hypotheses' $p = \frac{1}{2}$ and $p_2 = p_1$ in the two cases. This coincidence here arises because of the symmetry of the normal distribution and the formal similarity which exists (as we saw in the preceding subsection) between the posterior mode and the large sample posterior distribution on the one hand and the MLE and its large-sample sampling distribution on the other. However, on a point of detail, large-sample approximations are used in the case of (7.3.12) and (7.3.13) only to facilitate computation—the true values of the posterior probabilities here are fully known, no unknown parameters being involved. But the exact P-value in the sampling theory approach may involve nuisance parameters (e.g. the common value of $p_2 = p_1$ in the counterpart of (7.3.13)), and hence, may be indeterminable except in large samples.

As we saw in Chapter 6, the role of hypothesis-testing in an inductive study was implicit in the work of Arbuthnott and Daniel Bernoulli. This role was clearly perceived and spelt out by Laplace. In the *Essai* he remarked that given an apparently unusual phenomenon one should not indulge in 'vain speculations' but first test whether the unusuality is something real; only if this is confirmed should one proceed to search for a cause of the phenomenon. Thus although for a certain community the proportion of male births over a short period was found to be less than $\frac{1}{2}$, as the probability of the hypothesis $p \geq \frac{1}{2}$ was not small enough Laplace concluded that it was not worthwhile to investigate the cause (see Hald (1998) p. 241). It is to be noted that this idea of testing as a first step

in an inductive study has in modern times become the motto of statisticians
under the influence of R. A. Fisher.

<table>
<tr><td>Posterior-based
prediction</td></tr>
</table>

7.3.8   Laplace used (7.3.3) and (7.3.5) to evaluate the
probability of a future event given a set of data which
has bearing on it, generally assuming uniform a priori
probabilities. The most well-known application of this type concerned the prediction of the outcome of a future trial for a sequence of Bernoulli trials with
unknown success probability p, given the records of the first n trials. If the latter
have resulted in just s successes (in some order), writing $E_{n,s}$ for this observed
event and E' for the occurrence of success in the future trial, we have

$$P(E_{n,s}|p) = \binom{n}{s} p^s (1-p)^{n-s}, \quad P(E'|p) = p.$$

Substituting these in (7.3.5) (with $E_{n,s}$ for E) and assuming the prior distribution of p to be uniform over (0,1), we get

$$P(E'|E_{n,s}) = \frac{\int_0^1 p^{s+1}(1-p)^{n-s}\,dp}{\int_0^1 p^s(1-p)^{n-s}\,dp} = \frac{s+1}{n+2} \qquad (7.3.14)$$

as the probability of success in the future trial. In the special case $s = n$

$$P(E'|E_{n,n}) = \frac{n+1}{n+2}. \qquad (7.3.15)$$

This formula which gives the probability that an event which has occurred in
all of n successive trials will repeat itself in a future trial, became famous as
Laplace's *Rule of Succession*. This was supposed to provide a probabilistic route
for getting round Hume's problem of induction (section 1.4) and has been much
discussed about by philosophers (see e.g. Kneale (1949) p. 203, Cohen (1989)
p. 97). But it should be kept in mind that the validity of (7.3.14) and (7.3.15)
depends crucially on the assumption of a uniform prior for p.[8]

---

[8]There is an alternative discrete model which also leads to (7.3.14) and (7.3.15).
Suppose there are $N + 1$ urns each containing N balls, of which, for the ith urn i are
white and $N - i$ black, $i = 0, 1, \ldots, N$. One of the urns is chosen at random but its
composition is not noted. Out of $n(<N)$ balls randomly taken from it without replacement $s(0 \le s \le n)$ turn out to be white. Given this, probability that another ball taken
from the same urn would be white is given by (7.3.14) (cf. (4.5.16)). As this is independent of N, writing $p = i/N$ and making $N \to \infty$, 'without replacement' becomes
equivalent to 'with replacement' and the discrete model passes off to the continuous
one. Hald (1998) pp. 262–263 notes that this discrete model originated within a few
years of Laplace's derivation of (7.3.14) and was due to two Swiss mathematicians. It
is this discrete model which has been commented upon by, among others, Peirce (see
Buchler (1955) Chapter 13) and Russell (1948) p. 349.

Laplace illustrated the application of the rule of succession by noting that if a coin about which nothing is known showed a head in its first toss, the (subjective) probability of a head in the second toss should be computed as $\frac{2}{3}$ (the case $n = 1$ in (7.3.15)). This occasioned much criticism. Actually, as Hald (1998) pp. 258–259 points out, in the light of our existing knowledge, the assumption of a uniform prior is totally inappropriate here; a prior concentrated around $\frac{1}{2}$ would give a more realistic figure.

Laplace also examined from the point of view of the rule of succession, the age-old question of philosophers about the probability of 'the sun rising tomorrow'; as we have seen in subsection 7.2.7, Price had considered the same question in terms of posterior probabilities. Laplace reasoned that since the sun has risen every day in 5000 years $= 5000 \times 365.2426 = 1826213$ days of recorded history, the probability of its rising tomorrow is $\frac{1826214}{1826215}$. This observation has attracted much ridicule ever since it was made. To cite a recent instance, Polya (1968) (see Maistrov (1974) p. 131) has remarked that it is as silly as saying that the probability of a 10-year old boy surviving his 11th year is $\frac{11}{12}$, while that of his 70-year old grandfather surviving through his 71st year is $\frac{71}{72}$. Apparently, in this case Laplace turned a blind eye to the requirement of applicability of a uniform prior as also of the analogy of the trials with a Bernoulli sequence, only to illustrate his point.

It is clear that (7.3.14) can also be interpreted as the mean of the posterior distribution of p given that n trials have produced just s successes. In other words (7.3.14) represents what nowadays we call the Bayes estimate of p under squared error loss (subsection 4.4.6). Writing

$$\frac{s+1}{n+2} = \frac{n}{n+2} \cdot \frac{s}{n} + \frac{2}{n+2} \cdot \frac{1}{2},$$

it can be seen as a sort of shrinkage estimate of p, pulling the standard estimate $\frac{s}{n}$ towards the fixed value $\frac{1}{2}$ (first instance of a shrinkage estimate in the literature—according to Good (1983) p. 100).

Laplace extended (7.3.14) for the prediction of the number X of successes in m future Bernoulli trials. The posterior *predictive* distribution of X given s successes in first n trials is readily obtained by taking the event $X = x$ for E' in (7.3.5) with $P(X = x|p) = \binom{m}{x} p^x (1 - p)^{m-x}$. Laplace showed that when n, m and s are all large of the same order, the predictive distribution is approximated by $\mathcal{N}(m\hat{p}, m\hat{p}(1 - \hat{p}))$ where $\hat{p} = \frac{s}{n}$ (see Hald (1998) pp. 250–252 for details).

Laplace also considered the extension of (7.3.14) to the case where each trial can result in one of $k(\geq 3)$ mutually exclusive events with unknown probabilities $p_1, \ldots, p_k, \sum_1^k p_i = 1$. Given that in n initial trials the frequencies of the events are $a_i, i = 1, \ldots, k, \sum_1^n a_i = n$, the problem considered was to predict that a particular event, say, the jth would take place in a future trial. The prior that Laplace took in this case amounted to assuming that $(p_1, \ldots, p_{k-1})$ has a uniform distribution inside the simplex $0 < \sum_1^{k-1} p_i < 1$. Writing $E_{n;a_1,\ldots,a_k}$ for the outcome of the first n trials and E'_j for the occurrence of the jth event in a future trial, and arguing as before, from (7.3.5) we get

$$P(E'_j|E_{n;a_1,\ldots,a_k}) = \frac{a_j + 1}{n + k} \tag{7.3.16}$$

Note that $\sum_{i=1}^{k}((a_i + 1)/(n + k)) = 1$. (In the philosophical literature one sometimes finds unnecessary quibbling on the point that when each trial can result in 3 or more mutually exclusive events, application of (7.3.14) to each may be anomalous. For example Cohen (1989) p. 98 (see also Kneale (1949) p. 204) argues that if an urn contains only white, black, and yellow balls in unknown proportions and if three draws with replacement have produced balls of each colour just once, the probability of drawing a ball of each of the three colours in a 4th draw is $\frac{2}{5}$, which is clearly impossible. Actually (7.3.14) is applicable only if each trial can result in one of two complementary events. A prior distribution of $p_1, \ldots, p_{k-1}$ with $p_k = 1 - \sum_{1}^{k-1} p_i$ for which marginally $p_i, i = 1, \ldots, k-1, k$ each has a uniform distribution over (0, 1) does not exist for $k > 2$).

The result (7.3.16) was studied and extended by subsequent workers. In particular, as we mentioned in subsection 4.5.6, in recent times Carnap has generalized it to propose his G–C system for induction. Further, although Laplace's deduction was parameter-based, as we saw in the same subsection, (7.3.16) and its extensions have relevance even in the purely subjective parameter-free approach to induction.

| Laplace's sample survey | **7.3.9** As we observed in subsection 5.4.3, estimation of the total population of a city or country was a topical problem before regular population censuses were instituted in various countries generally around the beginning of the 19th century. We also saw how in the 17th century John Graunt sought to estimate the population of London on the basis of the annual total number of births under the assumption that this number bears a stable ratio (the CBR) to the size of the population. In 1802 Laplace used the same idea to estimate the population of France. But whereas Graunt's value $(=\frac{1}{32})$ of CBR was a mere guess, Laplace, with the support of the then Republican Government of France, organized a country-wide sample survey to get an objective estimate of the CBR. Further, Laplace utilized his theory of inverse probability to get a credible interval for the estimate of the size of the population. This was the earliest instance of a large-scale sample survey in which an attempt was made to assess the uncertainty in the estimate.

Using modern terminology we can characterize the sample survey conducted by Laplace as two-stage purposive. In the first stage, 30 departments (large administrative districts) spread over the country were selected, and in the second, a number of communes were purposively chosen in each selected department. For each sample commune the population was enumerated on a particular day (22 September 1802, the Republican New Year's day) and the average annual number of births over the preceding three years was ascertained from available records (the latter was taken as the smoothed value of the number of births in the year ending on the date of the survey). Writing n for the total population and b for the total annual number of births of the sample communes, the CBR was estimated to be b/n. Writing B for the total annual number of births in France, the estimate of the population N of France was $\hat{N} = B \cdot n/b$. In Laplace's case, b/n came out to be $1/28.352845$. Strangely, Laplace did not find the value of B from official records, but assuming $B = 1$ million (actually the annual numbers of births during 1801–1811 were nearly constant with values around 920,000) gave the estimated population size as 28,352,845.

Considering the communes as our ultimate sampling units and taking for each commune the annual number of births as the auxiliary variable and the population enumerated as the study variable, clearly b and B are the totals of the former variable over the sample and population of communes and n is the total of the latter variable over the sample. Thus the estimate $\hat{N}$ has the form of a typical ratio estimate (Cochran (1977) p. 66). However Laplace did not look at the problem from that angle. Instead he assumed that the collection of citizens on the day of the survey including those born in the preceding year, in the selected communes as also in the whole of France, represented *independent* random samples of sizes n (supposed to be pre-fixed) and N from an infinite super-population with an unknown proportion p of children born in the preceding year. By implication he also assumed that the number of births registered in the preceding year was equal to the number of children born in the preceding year who were alive on the date of the survey. Thus the data can be looked upon as giving independent binomial samples of sizes n, N with a common unknown p and with N as an additional unknown parameter. Using b and B to denote random variables, we then have

$$P(b = b, B = B|p, n, N) = \binom{n}{b}\binom{N}{B}p^{b+B}(1 - p)^{n+N-b-B}. \qquad (7.3.17)$$

Assuming p follows a uniform prior over (0, 1) and independently N follows an improper uniform prior over its domain, from (7.3.5) we can get the corresponding posterior joint distribution of p and N. Integrating out p from the same, we obtain the posterior pmf of N as

$$\pi(N|n, b, B) = \text{const.} \frac{N!}{(N - B)!} \frac{(N + n - b - B)!}{(N + n + 1)!}, \quad N = B, B + 1, \ldots$$

$$(7.3.18)$$

Writing $N = \hat{N} + Z$, by the method of asymptotic expansion, Laplace derived from (7.3.18) that for large n, b and B, Z is approximately distributed as

$$\mathcal{N}(-B(n - b)/b^2, B(B + b)n(n - b)/b^3).$$

In Laplace's case $E(Z)(= \text{'the Bayesian bias' in } \hat{N})$ being negligible relative to $\hat{N}$, could be taken as 0. By actual computation he found that the posterior probability of N lying between $\hat{N} \pm 500,000$ is about $1/300,000$ (see Cochran (1978)).[9]

Admittedly, Laplace assumed an oversimplified model to assess the uncertainty in the estimate. Leaving aside the question of propriety of the priors assumed, he ignored the fact that the selected communes represent a sample from the country as a whole and took the two samples from the superpopulation as independent. But it should be remembered that Laplace's study was made more than a century before the theory of sample surveys took shape. The superpopulation idea mooted by Laplace was not pursued by later workers for a long time; it has been re-introduced in the theory of sample surveys only around the 1970's (see e.g. Royall (1970)).

| Choice of prior— |
| Laplace's prescription |

7.3.10   From the above account it is clear that Bayes initiated and Laplace laid a comprehensive foundation of the objective pro-subjective approach (section 4.4) to statistical induction. As regards the choice of the prior, Bayes indirectly, and Laplace directly, postulated a uniform distribution for the unknown parameter over its domain. Laplace laid down such a postulate 'when there is nothing to make us believe that one case should occur rather than any other, so that these cases are, for us, equally possible' (see Hacking (1975) p. 132). The roots of this postulate were traced to the philosophy of Leibniz and in contrast with the latter's 'principle of sufficient reason', it was called the 'principle of non-sufficient reason', presumably meaning the absence of knowledge about any sufficient reason; see Hacking (1975) Chapter 14. (Keynes (1921) introduced the simpler term 'principle of indifference' for it.) But it seems that Laplace's considerations for introducing the principle were more pragmatic than metaphysical. In fact, in some early papers Laplace admitted the possibility that non-uniform priors might be appropriate some times. But later, when solving problems, he invariably took the prior to be uniform, presumably because this often led to a neat result and did not require the introduction of any personal element in the inductive process. (Stigler (1986) p. 103 suggests that while assuming equally likely 'causes', Laplace meant that the causes be enumerated and defined in such a way that they are equally likely, not that any specification or list of causes may be taken a priori equally likely. But, apart from the fact that this would be impracticable in most situations, it would bring an element of personal judgment in the process—something which Laplace seems to have been keen to avoid.)

---

[9]It is interesting to note that if we take into account the fact that the sample of size n is a part of the population of size N and leave aside finite population correction, from the sampling theory point of view the estimated variance of $\hat{N}$ comes out as $B^2 n(n-b)/b^3$, which is same as Laplace's expression for posterior variance apart from B appearing in place of $B + b$.

Pro-subjective
approach—later
developments
7.4 Laplace's influence over his contemporaries was overwhelming. (It was comparable to that of R. A. Fisher in the first half of the 20th century. Even today we can see traces of that influence in the thinking of some authors—see e.g. Jaynes (1976).) But dissenting voices against the sweeping assumption of a uniform prior and its derivative, the rule of succession, started being raised towards the middle of the 19th century. As we will see in the next chapter, even Gauss, who initially followed Laplace, later on distanced himself from the pro-subjective approach, privately expressing dissatisfaction with its 'metaphysics' (see Hald (1998) p. 467).

The objections raised were of various types. First, there were many, particularly representatives of the English frequentist school like Ellis and Venn and their sympathizers like C. S. Peirce who could not accept the subjective view of probability (see Keynes (1921) Chapter 4). Then there were the actuaries and others who pointed out that when there were several simple events and a complex event was derived from these by disjunction or conjunction, application of the rule of succession to the simple events and the complex event would give conflicting results. (The example we gave in subsection 7.3.8 of drawing from an urn containing white, black, or yellow balls is of this type; see Hald (1998) pp. 269–276 for further examples.) But careful listing of all the possibilities conceivable in a problem situation and assumption of a uniform prior encompassing all the parameters involved, generally preclude such anomalies. More serious objections relate to the arbitrariness and hazard involved in assuming a uniform prior for a parameter in the absence of knowledge about the underlying process generating the set-up. Thus consider the discrete model leading to the rule of succession (see footnote 7) in which we draw without replacement first n balls and then an additional ball from an urn containing N balls, an unknown number i of which are white and the rest black. Boole (see Keynes (1921) p. 50) pointed out that, instead of assuming that a priori i has a uniform distribution over the values $0, 1, \ldots, N$, we can equally well assume that all the 2^N possible constitutions of the urn resulting from each ball being either white or black are equiprobable, i.e. i has the symmetric binomial distribution $\binom{N}{i} 1/2^N$. This would give a probability $\frac{1}{2}$ of a white ball in the next draw irrespective of the outcome of the first n draws.

Apart from criticism on the above counts, frivolous applications without regard to the validity of the underlying assumptions of Laplace's posteriorist methods, and in particular, of the rule of succession, induced many mathematicians and philosophers to shy away from pursuing them. As we saw in subsection 7.3.8, Laplace himself was sometimes guilty of such misapplication. Keynes (1921) p. 383 has cited the instance of a German scholar who carried Laplace's computation of 'the probability of the sun rising tomorrow' to the extreme, by working out that the probability of the sun rising everyday for the next 4000 years is no more than $\frac{2}{3}$. As John Stuart Mill remarked such applications made the calculus of probability 'the real opprobrium of mathematics'.

In revulsion, Chrystal amputated the entire material on inverse probability from his text book on Algebra in the revised second edition.

Despite such rumblings, Laplace's posteriorist reasoning continued to be employed from time to time until the beginning of the 20th century, even by those like Edgeworth and Karl Pearson who were conscious of its limitations, arguably because it often came handy for solving a problem. Things changed in the second decade of the 20th century with the advent of R. A. Fisher on the scene. Fisher articulated the objections against the Laplacean approach forcefully with telling examples. Thus to emphasize the hazard involved in assuming arbitrarily a uniform prior he gave the real-life example of a genetic experiment in which a black mouse of unknown genotype is mated with a brown mouse to produce offspring that are all black. Here there are two possible hypotheses: the test mouse is homozygotic (BB) or heterozygotic (Bb). To assign equal probabilities to these would amount to the assumption that the test mouse is the offspring of one homozygotic and one heterozygotic parent and unwarranted exclusion of the possibility that both its parents were heterozygotic. (In the latter case the two hypotheses would have had respective probabilities $\frac{1}{3}$, $\frac{2}{3}$; see e.g. Fisher (1956) pp. 19–20.) Another kind of arbitrariness inherent in extraneously assuming that an unknown parameter is uniformly distributed, had been mentioned by others earlier, but was brought sharply into focus by Fisher who pointed out that for the same problem the implication of the principle of indifference depends on the form of the parametrization chosen. We have already discussed this point in the context of Bayes (subsection 7.2.6).

Along with demolishing Laplace's system based on the indifference principle, Fisher vigorously developed his alternative system of induction based on sampling theory and likelihood. A little later Neyman and Pearson built up their behavioural system also based on sampling theory (sections 4.2 and 4.3). The effect of all this was that in the first half of the 20th century, the pro-subjective approach to induction proposed by Bayes and Laplace was almost completely pushed to the background. One of the few who persisted with it and sought to remove its dependence on the principle of indifference by introducing more rational priors was Jeffreys (subsection 4.4.4). As interest in the pro-subjective approach revived in the latter half of the 20th century, Jeffreys's work was taken as the starting point. In the meantime the personalistic subjective interpretation of probability had been rationalized (subsections 3.3.3, 3.5.4). This added a new dimension to the objective pro-subjective approach to statistical induction, as in it the experimenters and statisticians could utilize their own hunches for choosing the prior. But inevitably this led to further pluralism in statistical induction; some of that we have recounted in section 4.4.

8

PRO-SUBJECTIVE APPROACH LOSES AS
SAMPLING THEORY GAINS GROUND

First half of 19th
century

8.1 In the preceding three chapters we traced the historical development of statistical thought roughly upto the end of the 18th century. We saw how the rudiments of the sampling theory approach towards statistical induction made their appearance in the work of some researchers like Arbuthnott, Simpson, and Daniel Bernoulli in the context of isolated problems. However, no full-scale development of the approach occurred during the period. Rather, as we saw in the last chapter, in the second half of the 18th century, Laplace vigorously pursued the objective pro-subjective approach begun by Bayes and himself to derive pro-subjective (posteriorist or Bayesian) solutions to various inductive problems. Towards the end of that chapter, we made brief anticipatory mention of the expansion in the sampling theory approach that occurred later and the tussle between that and the pro-subjective approach which began in the first quarter of the 20th century (and has been continuing to the present time). But these happenings had as their antecedent certain developments which took place in the first half of the 19th century. In the present chapter we direct our attention to the latter.

The first half of the 19th century is a difficult period to describe: there were so many different thought currents running in parallel and so much interplay among them. Laplace and Gauss were the principal personalities on the stage in that period, but there were many others (like Legendre, Poisson, and Bessel) who had more or less important roles to play. Laplace and Gauss often worked on similar problems along parallel tracks and it is not always easy to see who floated which idea first and to what extent the work of one drew upon the other's. Also, as we will see in course of the present chapter, both Laplace and Gauss changed their tracks, initially starting from the pro-subjective position and later veering round to the sampling theory point of view, but not always irrevocably (at least in the case of Laplace). It is difficult to say to what extent one was influenced by the other. Another fact which occasionally tends to confuse us is that both Laplace and Gauss at the beginning started from particular parametric models and both were instrumental in popularizing the normal model. Yet, for different reasons, both later opted to work under the 'model-free' set-up in which no particular form is assumed for the parent distribution. It was this period which saw the emergence and the beginning of the extensive use of the normal distribution both as a model and for approximating large-sample distributions of statistics and also the germination of seminal concepts like relative efficiency of estimators.

Further, the foundations of the theory of linear models, which later proved to be one of the most fruitful topics from the point of view of application, were laid during this period. All this makes the first half of the 19th century one of the most exciting periods to study in the history of statistical thought.

8.2   Quest for a model for the estimation of a true value

| Background |

8.2.1   We saw in Chapter 6 how in 1773 the formula of the normal density function was first derived by De Moivre as an approximation to the binomial mass function. At that time, however, it was like an uncut diamond. Even the fact that the total area under the curve is 1 and therefore it represents a proper continuous probability distribution was established by Laplace about half a century later (see Hald (1998) pp. 206–207). In Chapter 7 we saw that the normal distribution appeared time and again in the course of investigations carried out by Laplace in the earlier phase of his career, as asymptotic approximation to various distributions— mostly posterior distributions under the binomial set-up. In retrospect it seems to us strange today that, in spite of all this, Laplace at that time did not envisage the possibility of using the normal law for representing the distribution of error under the measurement error model.

Not that Laplace did not feel the need of a well-founded and analytically tractable law of error in that phase. As we observed in subsection 7.3.6, during the 1770s and 1780s Laplace, besides solving various problems of induction involving binomial proportions in the pro-subjective way, considered also the estimation of the true value of a location parameter under the measurement error model. The additive measurement error or location model for observables, it may be recalled, had been introduced by Simpson (section 6.4). Such models are natural to assume in the context of astronomical, geodetic, and physical experiments where repeated measurements on the same quantity differ mainly due to observational errors. The defining feature of this type of model is that the distribution of the error (= observation − true value) in the same experimental situation is supposed to be same irrespective of the true value of the quantity measured. In other words, considering the continuous case, the observable X has a density function of the form $f(x - \theta)$, where $f(\cdot)$ does not depend on the value of θ. For a long time it had been generally agreed upon that the form of the curve representing $f(z)$ should be bell-shaped and usually symmetric around $z = 0$. (Recall that these properties had been enunciated by Galileo qualitatively even before the theory of probability took shape; section 6.4.) But what form? If mathematical simplicity is the key consideration here, many different forms can be suggested. As we saw in sections 6.4 and 6.5, Simpson and Daniel Bernoulli worked respectively with the triangular and semicircular forms; other forms involving polynomial and trigonometric functions were also proposed. But these were all ad hoc proposals. Besides it was found that apparent simplicity of form did not necessarily lead to ease of analytical manipulation, particularly with regard to the study of the posterior distribution of θ and the sampling distribution of the sample mean (see Hald (1998) Chapter 3). Laplace in the course of his search for a suitable form of error distribution for the measurement error model on different occasions came out with certain more sophisticated proposals, such as the double exponential distribution (later also called the Laplace distribution)

$$f(z) = \tfrac{1}{2}me^{-m|z|}, \quad -\infty < z < \infty \ (m > 0), \tag{8.2.1}$$

which allows the error to assume any real value, and the finite-range distribution

$$f(z) = \frac{1}{2a}\log_e \frac{a}{|z|}, \quad -a < z < a \ (a > 0). \tag{8.2.2}$$

Laplace sought to justify these models starting from certain postulates but was hardly convincing in his arguments. Besides, as mentioned in subsection 7.3.6, he got bogged down in trying to solve the problem of point estimation of θ in the posteriorist way on the basis of these models. Analytically they were not easy to handle—the density (8.2.1) is not differentiable at 0 and (8.2.2) is not even continuous at the point. Thus, it seems, towards the end of the 18th century the search for an ideal error distribution had reached a dead end—the problem remained as wide open as ever.

| Carl Friedrich Gauss | **8.2.2** It was at this stage that the German mathematician–scientist Gauss stepped into the arena. Carl Friedrich Gauss (1777–1855) was junior to Laplace by 28 years and died 28 years after him. He was born in Brunswick, Germany and had his education at the University at Gottingen, where he later settled down and worked as Professor and Director of Observatory. He had wide-ranging interests and made lasting contributions to many areas of pure and applied mathematics and science, such as algebra, number theory, geometry, astronomy, probability and statistics (theory of errors), geodesy, and various branches of physics. Unlike in the case of Laplace, probability and statistics did not represent a focal point of Gauss's broad interests. Also, he came to this field not, as Laplace and many others did, through the solution of problems of games of chance, but by way of analysing sets of data collected in course of astronomical studies. Further, he made contributions in the field well after he had gained wide-spread recognition for his researches in pure mathematics and astronomy.

| A.M. Postulate and the normal law | **8.2.3** Up to that time all investigators including Laplace had tried to solve the problem of estimation of a true value under the measurement error model, starting from some particular form of the error density $f(\cdot)$. Experimental tradition had long established that usually the arithmetic mean of a sample of independent measurements on a quantity gives a reliable estimate for it (cf. section 6.4). Earlier attempts to justify this theoretically at a sufficient level of

generality had floundered because of the intractability of the error models chosen (see section 6.5 and subsection 7.3.6). Gauss decided to attack the problem from the opposite angle. He accepted the empirical fact that the sample mean is a good estimate of the true value and enquired what model for the error would make the sample mean the best estimate in some reasonable sense. Specifically, he followed Laplace in assuming that θ in the model $f(y - \theta)$ is a priori subject to an (improper) uniform distribution over the real line but in contradistinction with Laplace (subsection 7.3.6) took the mode of the posterior distribution

$$\pi(\theta|y_1, \ldots, y_n) = \text{const.} \prod_{\ell=1}^{n} f(y_\ell - \theta), \quad -\infty < \theta < \infty, \qquad (8.2.3)$$

given the sample $y_1, \ldots, y_n$, as the best estimate of θ. Gauss asked himself the question: what form of $f(\cdot)$ made the posterior mode based on (8.2.3) coincide with the sample mean? In other words, his starting point was what came to be called the Arithmetic Mean Postulate: *the form of $f(\cdot)$ must be such that for any n and $y_1, \ldots, y_n$, the posterior density attains its maximum at $\bar{y} = (1/n)\sum_1^n y_i$.*

If we assume that $f(z)$ is everywhere differentiable and write $f'(z)/f(z) = g(z)$, then equating the derivative of the logarithm of (8.2.3) with respect to θ to zero, from this postulate we get that $\theta = \bar{y}$ must satisfy

$$\sum_{\ell=1}^{n} g(y_\ell - \bar{y}) = 0, \qquad (8.2.4)$$

whatever n and $y_1, \ldots, y_n$. Putting $y_\ell - \bar{y} = u_\ell, \ell = 1, \ldots, n$, this means

$$\sum_{\ell=1}^{n} g(u_\ell) = 0, \qquad (8.2.5)$$

for all $u_1, \ldots, u_n$ subject to $\sum_1^n u_\ell = 0$. Gauss tacitly assumed the function $g(\cdot)$ to be continuous. He considered particular configurations of $u_1, \ldots, u_n$ for all possible n (see Hald (1998) p. 354). However, it is enough if (8.2.5) holds for $n \leq 3$. Thus taking $n = 1, 2, 3$ successively, we get $g(0) = 0, g(z_1) = -g(-z_1)$ and $g(z_1) + g(z_2) = g(z_1 + z_2)$ for all positive z_1, z_2. Hence through recursion we derive that

$$\sum_{\ell=1}^{n} g(z_\ell) = g\left(\sum_1^n z_\ell\right), \qquad (8.2.6)$$

for all n and $z_1, \ldots, z_n > 0$. This implies that for a positive rational z and therefore by continuity and symmetry for any real z

$$g(z) = zg(1),$$

whence since $f(\cdot)$ is a density,

$$f(z) = \frac{h}{\sqrt{\pi}} e^{-h^2 z^2}, \quad -\infty < z < \infty, \qquad (8.2.7)$$

for some $h > 0$. This is the form in which Gauss derived the normal law, or as it was called at the time, 'the law of error'. (It also came to be called the Gaussian law. We will see in the next chapter that towards the end of the 19th century Galton, among others, thought that the law had a normative status and vigorously canvassed for it; the name 'normal' became standard after that.) Clearly if we put

$$h = \frac{1}{\sigma\sqrt{2}} \qquad (8.2.8)$$

in (8.2.7), we get the normal density in its familiar form. Gauss observed that h directly measures the precision of measurement—as h increases, for the random error Z the probability $P(|Z| < a) = \frac{1}{\sqrt{\pi}} \int_{-ah}^{ah} e^{-t^2} dt$ increases for every $a > 0$. Because of this, the parameter h came to be called 'the modulus of precision'. For many years, the direct measure of precision h continued to be used side by side with the inverse measure σ representing the standard deviation.

Gauss arrived at the normal law while searching for a suitable model for the error distribution in the course of his investigations into the orbits of planets on the basis of astronomical observations. The result appeared for the first time in his 1809 book *Theoria Motus Corporum Coelestium* (The Theory of Motion of Celestial Bodies) which was written in Latin. (But it had an earlier unpublished German version, completed in 1806; see Hald (1998) p. 394.)

Since the sample likelihood is proportional to the posterior density (8.2.3) under uniform prior, clearly the A.M. postulate can also be stated as: the form of $f(\cdot)$ should be such that for any n and $y_1, \ldots, y_n$ the MLE (subsection 4.3.6) of θ is equal to $\bar{y}$. Indeed, the postulate can be looked upon as representing an exclusive property, or what nowadays we call, a 'characterization' of the normal distribution.[1] Some authors (see Stigler (1986) pp. 141–143) do not regard it as a real postulate at all. However, we should remember that it was an empirically observed fact that the sample mean is a satisfactory estimate of the true value in most experimental situations. Gauss laid down the principle that the error distribution must be such that the posterior density (or likelihood) is as high as possible at the sample mean so that this fact is vindicated.

Gauss's estimate of a true value

8.2.4   After justifying the normal model for measurement errors as above, Gauss proceeded to apply it. The major application was in the context of linear observational equations and least squares; we will take up these in a later section. For the present we consider application to the special case of estimation of a true value. Assuming, as Gauss did, that the precision h is known, from (8.2.3)

---

[1]See Teicher (1961) who shows that to derive normality of $f(\cdot)$ from this condition one need not assume differentiability—lower semi-continuity of $f(z)$ at $z = 0$ and validity of the condition for $n = 2, 3$, are enough.

and (8.2.7) we get the posterior distribution of θ as

$$\pi(\theta|y_1, \ldots, y_n) = \text{const. } e^{-h^2 \sum_1^n (y_\ell - \theta)^2}$$

$$= \text{const. } e^{-nh^2 (\theta - \bar{y})^2}. \tag{8.2.9}$$

Of course the break-up

$$\sum_1^n (y_\ell - \theta)^2 = \sum_1^n (y_\ell - \bar{y})^2 + n(\bar{y} - \theta)^2 \tag{8.2.10}$$

and redefinition of the const. by absorption of a factor free from θ is crucial for this. (8.2.9) means that the posterior distribution of θ is normal with mean $\bar{y}$ and modulus of precision $\sqrt{(n)}h$. That the mode (= median = mean) of the posterior is $\bar{x}$ is immediate. Thus $\bar{y}$ comes out as the natural estimate of θ.

We can use (8.2.9) to set up an interval estimate (credible interval) for θ. Of course for this we require an estimate of h. Gauss did not consider this in the *Theoria Motus*. Later in a treatise in 1816 he obtained a posteriorist solution to this problem assuming θ to be known. Taking the (improper) uniform prior for h over $(0, \infty)$, he got the posterior density of h as

$$\text{const. } h^n e^{-h^2 \sum_\ell (y_\ell - \theta)^2}, \quad 0 < h < \infty, \tag{8.2.11}$$

and proposed the posterior *mode*

$$\hat{h} = \left\{ n / \left(2 \sum (y_\ell - \theta)^2 \right) \right\}^{1/2} \tag{8.2.12}$$

of this as the estimate of h. From this we get

$$\hat{\sigma} = \left\{ \sum (y_\ell - \theta)^2 / n \right\}^{1/2} \tag{8.2.13}$$

as the estimate of $\sigma = 1/(\sqrt{(2)}h)$. Gauss also showed along Laplacian lines that in large samples the posterior distribution of $\hat{h}$ is approximately $\mathcal{N}(\hat{h}, \hat{h}^2/2n)$ from which credible intervals for h and σ can be set up. In practice one of course has to substitute $\bar{y}$ for θ when θ is unknown.

In this connection we may mention that Laplace considered the same problem of estimation of h and σ when θ is known in 1818, while considering a geodetic application in which normality of the observations can be taken to hold. But in contrast with Gauss, he took h^2 to be uniformly distributed over $(0, \infty)$. This would result in the appearance of $n + 1$ in place of n in the expression (8.2.11) of the posterior density and the estimates of h and σ would also undergo corresponding modifications. (This provides another illustration of how the conclusion in the pro-subjective approach based on an indifference prior depends on the parametrization used; cf. subsection 7.2.6 and section 7.4.) Laplace, however, in this case, proposed the *mean* $(n+2)/2 \sum(y_\ell - \theta)^2$ of the resulting posterior distribution of h^2 as the estimate of the latter (cf. subsection 7.3.6). Of course for large n we may replace $n+2$ by n and get (8.2.12) as the estimate of h. In another study relating to the tides of the sea, Laplace considered both θ and h^2 as unknown and examined their joint posterior under the uniform prior over $(-\infty, \infty) \times (0, \infty)$. Hence he derived the marginal posterior of h^2 whose mean is $(n+1)/2 \sum(y_\ell - \bar{y})^2$ and the conditional posterior of θ given h^2 which is $\mathcal{N}(\bar{y}, 1/(2nh^2))$. Combining, he arrived at an interval estimate for θ (see Hald (1998) § 20.8).

Going back to Gauss, in the *Theoria Motus* he also considered the two more general situations (i) the different measurements y_ℓ of θ have different precisions $h_\ell, \ell = 1, \ldots, n$ and (ii) the precision h is same for all the measurements but y_ℓ measures a known multiple $a_\ell \theta$ of $\theta, \ell = 1, \ldots, n$. (Of course each of the two cases is reducible to the other.) On the basis of break-ups similar to (8.2.10), he showed that the posterior distribution of θ is normal in both the cases with mean and precision $\sum h_\ell^2 y_\ell / \sum h_\ell^2$ and $\sqrt{\sum h_\ell^2}$ in case (i) and $\sum a_\ell y_\ell / \sum a_\ell^2$ and $h\sqrt{\sum a_\ell^2}$ in case (ii).

Implication of Gauss's derivation

8.2.5 Gauss opened the eyes of one and all to the tractability of the normal model under linear transformations. As we now know, in general, when the joint distribution of a set of observables involves only one unknown real parameter θ, the posterior distribution of θ can be neatly factored out if there exists a real-valued sufficient statistic. In particular, this holds if there is a one–one transformation of the observations such that (i) one of the transformed variables is distributed independently of the others and (ii) θ is involved *only* in the distribution of the former (obviously then, the first transformed variable represents a sufficient statistic). Implicit in Gauss's derivation of the posterior distribution of θ when the variables are independently normally distributed as above, is the existence of a non-singular linear transformation satisfying (i) and (ii)[2] (the transformation can be taken to be orthogonal when all the observations have same precision). The posterior distribution of θ here becomes normal because the first transformed variable is normally distributed with mean a known multiple of θ.

Not everybody was taken in by Gauss's A.M. postulate argument for the assumption of normality of errors. (However, attempts were made to deduce the A.M. postulate starting from simpler ones; see Whittaker and Robinson (1952) p. 216. Also later Keynes and others investigated what models are generated if in the postulate the sample mean is replaced by other averages; see Hald (1998) pp. 377–378.) Soon other less convoluted postulates implying the normal model

---

[2]From characterization theorems discovered in recent times (see e.g. Rao (1965) pp. 443–444), it follows that for a set of independent random variables such a linear transformation exists only if they are normally distributed.

were put forward (we will consider some of these later in this section). But the advantage that accrues from normality when linear transformations come into the picture became plain to all. As we will see in this as well as in later chapters, this was cashed in upon both in small-sample developments made under the assumption of a normal model and in large-sample ones made without such an assumption.

Laplace's CLT   **8.2.6**   Laplace in the 1780s, while investigating the distributions of the sum of i.i.d. random variables in simple particular cases, had obtained some partial results which suggested that these might be approximated by the normal distribution. Around the time when Gauss's *Theoria Motus* was published, Laplace was busy generalizing those results. In 1810, i.e. a year after the publication of Gauss's book, Laplace came out with an epoch-making paper. In it he derived the result that if $X_1, \ldots, X_n$ are any i.i.d random variables with (say) mean μ_1 and variance σ_1^2, then for large n, $\sum_1^n X_\ell$ is approximately normally distributed with mean $n\mu_1$ and variance $n\sigma_1^2$. (Laplace toned up the proof of the result further in subsequent papers and in his 1812 book *Theorie Analytique*.) This was the first instance in which a *general* theorem on the approximate normality of the sampling distribution of a sum made appearance in the literature. A group of results, of which Laplace's theorem was a special case, later came to be known as Central Limit Theorem (CLT; see section 3.9). This has had far-reaching implication for the development of both theory of probability (where it has been generalized further and further to reach Gaussian distributions in various abstract spaces) and theory of statistical inference (where it forms the backbone of large-sample methods under the sampling theory approach).

In view of the importance of Laplace's result we describe very briefly the essence of the heuristic technique employed by Laplace in its derivation. The technique itself developed into a powerful tool indispensable to probabilists and statisticians.

Laplace's starting point was the case when the common distribution of the independent random variable $X_1, \ldots, X_n$ is discrete with a finite set of consecutive integers, say, $a, a+1, \ldots, b-1, b$ as mass points. As we saw in section 6.3, almost a century ago De Moivre had introduced the probability generating function (pgf) as a handy tool for finding the probability distribution of a sum in such a case. Writing $E(t^{X_1}) = \psi_1(t)$ for the common pgf of $X_1, \ldots, X_n$, the pgf of $S_n = \sum_1^n X_\ell$ is given by $\{\psi_1(t)\}^n$. Hence for any integer $s(na \le s \le nb)$, $P(S_n = s)$ turns out to be same as the term free from t in the expansion of $t^{-s}\{\psi_1(t)\}^n$.

Laplace conceived the brilliant idea of replacing t by e^{it} in $\psi_1(t)$. Let $\psi_1(e^{it}) = Ee^{itX_1} = \phi_1(t)$ (towards the end of the 19th century Poincaré called it the *characteristic function* of X_1 and this name has since become standard). Then $P(S_n = s)$ is the term free from t in the expansion of $\bar{e}^{its}\{\phi_1(t)\}^n$ in powers of e^{it}. The advantage in this stems from our knowledge that

$$\frac{1}{2\pi} \int_{-\pi}^{\pi} e^{itr} \, dt = 1 \quad \text{if } r = 0,$$

$$= 0 \quad \text{if } r \text{ is a non-zero integer.} \qquad (8.2.14)$$

Because of this we can write

$$P(S_n = s) = \frac{1}{2\pi} \int_{-\pi}^{\pi} e^{-its} \{\phi_1(t)\}^n \, dt. \qquad (8.2.15)$$

Let us utilize modern terminology and notations. From the definition of $\phi_1(t)$ we have

$$\log \phi_1(t) = (it)\mu_1 + \frac{(it)^2}{2!}\sigma_1^2 + \text{terms involving higher powers of } (it)$$
$$\text{and higher cumulants.} \qquad (8.2.16)$$

We now take $s = n\mu_1 + \sqrt{n}\sigma_1 z$ (assuming z is such that this is an integer), make use of (8.2.15) in (8.2.14), and transform $\sqrt{n}\sigma_1 t = u$. We then get

$$P\left(\frac{S_n - n\mu_1}{\sqrt{n}\sigma_1} = z\right) = \frac{1}{2\pi\sqrt{n}\sigma_1} \int_{-\sqrt{n}\sigma_1\pi}^{\sqrt{n}\sigma_1\pi} e^{-iuz - u^2/2 + 0(n^{-1/2})} \, du. \qquad (8.2.17)$$

For large n, neglecting terms of order $n^{-1/2}$, the integral on the right of (8.2.17) may be approximated by

$$\frac{1}{2\pi\sqrt{n}\sigma_1} \int_{-\infty}^{\infty} e^{-iuz - u^2/2} \, du = \frac{1}{\sqrt{(2\pi n)}\sigma_1} e^{-z^2/2}. \qquad (8.2.18)$$

(Laplace proved (8.2.18) using the result that if $g(z) = \int_{-\infty}^{\infty} e^{-u^2/2} \cos uz \, du$, then $g'(z) = -zg(z)$ with $g(0) = \sqrt{(2\pi)}$; see Hald (1998) p. 370.) Hence, by the obvious replacement of a sum in which the argument varied in steps of $1/(\sqrt{n}\sigma_1)$ by the appropriate integral, it may be concluded that for large n,

$$P\left(z' < \frac{S_n - n\mu_1}{\sqrt{n}\sigma_1} \leq z''\right) \simeq \frac{1}{\sqrt{(2\pi)}} \int_{z'}^{z''} e^{-z^2/2} \, dz, \quad -\infty \leq z' < z'' \leq \infty.$$
$$(8.2.19)$$

Laplace stated that when X_1 can assume any positive or negative integer as its value we can make $a \to -\infty$ and $b \to +\infty$ and the same result as above would hold provided μ_1 and σ_1^2 exist. The result can of course be adapted to the case when X_1 assumes integral multiples of an arbitrary quantity δ as its values. When X_1 has a continuous distribution, Laplace discretized it by approximating the density curve by a histogram with interval span, say δ, and then replacing the histogram by a polygon in an obvious way. For the discretized random variable the normal approximation formula would hold. As δ is made smaller and smaller, the histogram approaches the continuous density curve. Therefore Laplace argued that the result (8.2.19) would hold in all cases.

Laplace derived the CLT along the above lines first in particular cases like those of a uniform distribution and a symmetric distribution and then passed on to the general case (a detailed account of all this can be found in Hald (1998) Chapter 17). However, Laplace's proof in the general case does not come up to the standards of modern rigour. As soon as one goes beyond the simplest case of a discrete random variable assuming a finite number of values, one encounters in it various dubious limiting operations whose validity Laplace took for granted. Attempts were made by a host of probabilists over more than a century to extend and establish the CLT rigorously under suitably formulated conditions covering the case of non-identically distributed X_ℓs and to obtain valid correction terms to the normal approximation formula. Notable contributors to this development were Poisson, Tshebyshev, Markov, Liapounov, Lindeberg, Levy, Feller, Berry, and Essen (see Gnedenko and Kolmogorov (1954)). The standard proof of the CLT that is usually given in text books nowadays (see e.g. Cramér (1946) pp. 214–218), just like Laplace's proof, proceeds through the characteristic function and invokes the continuity theorem and implicitly the inversion theorem for characteristic functions. However, if one examines the proof and remembers the representation of a definite integral as the limit of a sum, one can see that the modern highway is never laid too far way from the path hewn out by Laplace. (Laplace's formula (8.2.15) contains the seed of the inversion theorem for the sum in the case of an integer-valued random variable. The corresponding formula in the continuous case follows from Fourier's well-known integral theorem. However, Fourier's results were derived only in 1811, i.e. subsequent to Laplace's derivation of the CLT.)

In this context we anticipate some results derived later. Poisson, Bessel, and later Tshebyshev in their derivation explicitly considered certain further terms in the expansion of the log characteristic function. To describe that development (see Hald (1998) Chapter 17 and Gnedenko and Kolmogorov (1954) Chapter 8) in terms of the above heuristic approach, let k_{1r} be the cumulant of order r of X_ℓ ($r \geq 3$). When the terms involving higher powers of (it) in (8.2.16) are written explicitly, the integrand on the left of (8.2.18) becomes

$$\exp\left\{ -iuz - \frac{1}{2}u^2 + \sum_{r=3}^{\infty} \frac{k_{1r}}{n^{(r/2)-1}\sigma_1^r} \frac{(iu)^r}{r!} \right\}.$$

Repeated differentiation of the identity

$$\int_{-\infty}^{\infty} e^{-iuz-u^2/2}\,du = \frac{1}{\sqrt{(2\pi)}}e^{-z^2/2}$$

with respect to z shows that

$$\int_{-\infty}^{\infty} (-iu)^r e^{-iuz-u^2/2}\, du = D^r \left(\frac{1}{\sqrt{(2\pi)}} e^{-z^2/2} \right), \quad r = 1, 2, 3, \ldots, \quad (8.2.20)$$

where $D = d/dz$. Hence, if we explicitly introduce the additional terms in (8.2.16), (8.2.19) would be replaced by the formal expression

$$P \left(z' < \frac{S_n - n\mu_1}{\sqrt{n}\sigma_1} \le z'' \right)$$

$$\simeq \frac{1}{\sqrt{(2\pi)}} \int_{z'}^{z''} \exp \left\{ \sum_{r=3}^{\infty} \frac{k_{1r}}{n^{(r/2)-1}\sigma_1^r} \frac{(-D)^r}{r!} \right\} \left(e^{-z^2/2}\, dz \right). \quad (8.2.21)$$

Expanding the operator inside the integral on the right we get successive correction terms to the Central Limit approximation. (The validity of the expansion was investigated much later by Cramér; see Cramér(1946), Chapter 17.)

Laplace's response to Gauss's proposal

8.2.7   Apparently Laplace came to know of Gauss's derivation and application of the normal model for errors in *Theoria Motus* only after completing his 1810 paper on the CLT. For he prepared a supplement to the paper and got it published in the same volume of the *Memoires* of the Paris Academy in which the paper appeared. Laplace took cognizance of Gauss's work only in the supplement.

We saw that about 25 years earlier Laplace had striven to derive an ideal law of error but had not met with much success. Gauss's normal law seemed to provide a neat solution to the problem. What is more, Laplace's own CLT suggested a more convincing justification of the normal law of error. It must have been plain to Laplace that if the error in an observation could be regarded as the resultant of a large number of independent random components an approximate normal law for it would follow. But apparently Laplace was not very enthusiastic about this line of reasoning[3] (in the next subsection, we will see that soon others pursued this very line with avidity). It may be that he was already visualizing the possibility of developing a model-free large-sample theory of inference. His response to Gauss's work as contained in the supplement was as follows.

As is common in experimental practice (particularly in fields like astronomy, geodesy, etc.), suppose in the measurement error model each observation y_ℓ is itself the mean of a large number n_ℓ of i.i.d. primary observations with precision, say, h_ℓ. Then by CLT we can take y_ℓ to be approximately normal with mean θ and precision $\sqrt{(n_\ell)}h_\ell$ and just as in Gauss (1809) derive that under uniform prior the posterior distribution of θ is approximately normal with mean $\sum n_\ell h_\ell^2 y_\ell / \sum n_\ell h_\ell^2$ and precision $\sum n_\ell h_\ell^2$. The only point of novelty in Laplace's

---

[3]There seems to be a conflict here between the accounts given by Stigler (1986) pp. 143, 146 and Hald (1998) pp. 357–358. We have followed Hald's interpretation.

supplement, was his stress on the fact that $\sum n_\ell h_\ell^2 y_\ell / \sum n_\ell h_\ell^2$ should be taken as an estimate of θ because, being the posterior median it minimized the mean posterior absolute error and not because, as Gauss had argued, it was the posterior mode; cf. subsection 7.3.6. (This observation was, however, quite superfluous as for a symmetric bell-shaped distribution the median and the mode coincide. Besides, as we saw in subsection 7.3.6, Laplace himself had used the posterior mode in other contexts of point estimation.)

Postulate of elementary errors

8.2.8   A justification for assuming the normality of observational errors based on the CLT was advanced and developed comprehensively by W. F. Bessel in 1838. Earlier Bessel, who was an expert in observational astronomy, had examined sets of repeated observations y in the context of determination of various quantities θ. Comparing the grouped empirical distribution of the residuals $y - \bar{y}$ (which can be considered as the surrogate for the error in y) with the fitted normal distribution, he had found satisfactory agreement in every case. To explain this he postulated that the error $y - \theta = z$ in an observation y can be represented as

$$\epsilon_1 + \epsilon_2 + \cdots + \epsilon_n, \qquad\qquad (8.2.22)$$

where $\epsilon_1, \epsilon_2, \ldots, \epsilon_n$ are 'elementary errors' contributed by independent causes. With his vast experience as a practical astronomer, Bessel could identify for any experiment a large number of such causes depending on the conditions of the observer, the instrument and the environment at the time of observation. He postulated that the random elementary errors ϵ_ℓ follow independent (but not necessarily identical) distributions symmetric around 0. He further assumed that the different ϵ_ℓs are of comparable magnitude (earlier Poisson had produced some examples which showed that if a few terms in the sum dominate the others the CLT may not hold in the non-identical case), and although n is very large xthe second moment of the random error Z remains finite. Then discretizing the distribution of each ϵ_ℓ and arguing essentially like Laplace, Bessel derived an inversion formula for the density of Z in terms of its characteristic function. Writing $\sigma^2 = \text{Var}(Z)$, it followed that neglecting terms of order $n^{-1/2}$ and higher, Z is approximately distributed as $\mathcal{N}(0, \sigma^2)$.

Empirical studies based on residuals of the type made by Bessel were replicated by many subsequent 19th century workers for different types of observations and satisfactory agreement with the normal model was found. (Actually as later analyses showed, the agreement with normality in some of these cases was not as close as was claimed. But, of course, with no tests for goodness of fit at their disposal, workers at that time were somewhat handicapped in this regard.) The postulate made by Bessel came to be known as the 'postulate (or hypothesis) of elementary errors' and was widely accepted as providing a plausible explanation of the phenomenon of normality. (The postulate is sometimes attributed to Hagen who considered a very special case at about the same time.) As we will

see in the next chapter, soon Quetelet and others sought to extend the postulate
to random variables other than errors of observation.

| Other postulates | **8.2.9** Side by side, other postulates characterizing |

the normal model for errors were also proposed and
studied. At about the same time when Gauss's *Theoria Motus* appeared, an
American mathematician of the name of Adrain gave two characterizations of
the normal distribution (see Hald (1998) pp. 368–372). The first one was similar
to that of Gauss—it showed that if the errors are normally distributed then a
standard empirical practice of surveyors can be justified in terms of a normative
principle. (But Adrain did not consider A.M.—the empirical rule considered by
him was more specialized. Also he did not bring in posterior probability; his principle can be stated in terms of likelihood.) The other characterization given was
somewhat artificial. In any case, Adrain's paper being published in an obscure
journal did not attract notice until much later and had very little impact.

Around 1850 an interesting set of postulates leading to the normal law of
error was suggested by the British astronomer–mathematician John Herschel
and developed fully by Boole. Although Herschel's postulates are applicable
more generally, these may be conveniently described in terms of a prototype
experiment in which a point target O situated on a vertical plane in front is
being fired at by an experimenter. The random point of hit on the plane may be
represented by (X_1, X_2) with reference to two rectangular axes of co-ordinates
drawn through O. It is postulated that

(i) (X_1, X_2) are independently distributed with density functions, say $f(x_1)$
  and $g(x_2)$;
(ii) $f(x_1)$ and $g(x_2)$ are continuous; and
(iii) the probability of the actual point of hit lying in an infinitesimal neighbourhood of a given point $P(x_1, x_2)$ depends only on the distance
  $OP = \sqrt{(x_1^2 + x_2^2)}$

(iii) of course is equivalent to saying that the probability at P is unaffected by
rotation of the coordinate axes through O (i.e. under orthogonal transformation).
It means we have for some function $\alpha(\cdot)$

$$f(x_1)g(x_2) = \alpha \left(\sqrt{(x_1^2 + x_2^2)} \right). \tag{8.2.23}$$

Taking in turn $y = 0$ and $x = 0$, from this we get

$$\frac{f(x)}{f(0)} = \frac{g(x)}{g(0)} = \frac{\alpha(|x|)}{\alpha(0)}. \tag{8.2.24}$$

Now writing

$$\log \frac{\alpha(|x|)}{\alpha(0)} = p(x^2), \tag{8.2.25}$$

from (8.2.23)–(8.2.24)

$$p(x_1^2) + p(x_2^2) = p(x_1^2 + x_2^2),$$

and hence by recursion, for any real $x_1, x_2, \ldots, x_n$

$$\sum p(x_i^2) = p\left(\sum x_i^2\right). \qquad (8.2.26)$$

This functional equation is exactly like the equation (8.2.6) considered in the context of Gauss's derivation. Hence just as there using continuity of $p(\cdot)$ we get that $f(x) = \text{const}.e^{-h^2 x^2}$. (For a modern proof see Feller (1966) pp. 77–78 where it is shown that condition (ii) can be dispensed with and condition (iii) can be considerably relaxed.)

Herschel's postulates are important in another respect. Circumstantial evidence suggests (see Porter (1986) pp. 118–119, 123) that Herschel's ideas influenced the physicist Maxwell in his 1860 derivation of the probability distribution of the space velocity of the molecules in an enclosed mass of gas. Maxwell assumed that when the velocity of a molecule in space is resolved in three orthogonal directions the components are distributed independently and identically and the joint density of the components depends on the magnitude of the velocity alone. This was just an extension of Herschel's postulates from two to three dimensions and just like in the case of Herschel the normality of each component and hence the fact that the resultant space velocity has the distribution of a χ_3-variable followed. From this it is immediate that the probability distribution of the kinetic energy ($= \frac{1}{2} \times \text{mass} \times \text{velocity}^2$) of gas molecules obeys the distribution of a χ_3^2 apart from a scalar factor—a fact which was made use of by Boltzmann a few years later. These were the earliest instances of application of probability in physics and it is noteworthy that in such applications probability was interpreted concretely in the sense of relative frequency. Curiously, neither Herschel's nor Maxwell's derivation was inspired directly by the earlier work of Gauss and Laplace; rather both of them drew upon the application of the law of error to variation in human populations which we will consider in the next chapter (see Porter (1986) Chapter 5).

Gauss's main purpose in proposing the normal law of error was to rationalize the fitting of *linear parametric curves* (i.e. curves representing functions which are linear in the parameters) on the basis of observed data. In the next section, we consider successively the general problem of fitting such curves, the nonprobabilistic methods of fitting developed prior to Gauss, Gauss's initial breakthrough based on the normal model for errors, and Laplace's and Gauss's later model-free developments.

8.3   Fitting of linear parametric curves

General problem of curve-fitting

8.3.1   Suppose $y, x_1, x_2, \ldots, x_m$ are observable quantities which obey an approximate relation of the form

$$y \simeq \beta_1 x_1 + \beta_2 x_2 + \cdots + \beta_m x_m, \qquad (8.3.1)$$

where the values of the parameters $\beta_1, \beta_2, \ldots, \beta_m$ are unknown. $n \geq m$ tuplets of observations

$$(y_\ell, x_{1\ell}, x_{2\ell}, \ldots, x_{m\ell}), \quad \ell = 1, \ldots, n, \tag{8.3.2}$$

on $(y, x_1, x_2, \ldots, x_m)$ have been made. To 'fit' the equation (8.3.1) to these data we have to find values of $\beta_1, \beta_2, \ldots, \beta_m$ for which all the observations satisfy (8.3.1) closely. This means, writing

$$y_\ell = \beta_1 x_{1\ell} + \beta_2 x_{2\ell} + \cdots + \beta_m x_{m\ell} + z_\ell, \quad \ell = 1, \ldots, n, \tag{8.3.3}$$

the discrepancies, $z_1, \ldots, z_n$ should have values which are on the whole close to 0.

The problem of fitting a relation such as (8.3.1) may arise when some of the parameters have important physical significance and have intrinsic interest and we are primarily interested in 'estimating' them. The variable y in such cases does not have any special status except that it is known to enter the relation with a non-zero coefficient, which, without loss of generality, we can take as 1. Alternatively, our primary purpose may be to predict the value of y corresponding to various observed or chosen values of $(x_1, \ldots, x_m)$. This occurs when $x_1, \ldots, x_m$ are easier to observe or control and y is difficult to observe, uncontrollable, or its values of interest are future contingents. There y is called the dependent (or predicted) and $x_1, \ldots, x_m$ independent (or predicting) variables.

The relation (8.3.1) is linear in the unknown β_is. In terms of the variables it is much more general than it appears. Thus x_is may be subject to known subsidiary (nonlinear) relations. It may happen that some or all of the x_i's are known nonlinear functions, say, $x_1 = \varphi_1(u, v, \ldots)$, $x_2 = \varphi_2(u, v, \ldots) \ldots$ of a few primary variables $u, v, \ldots$. As a special case one x_i, say, x_1 may be a known constant which we may be taken as 1. Also, although the original relation among $y, x_1, \ldots, x_m$ may be nonlinear, it may be reduced to the form (8.3.1) by prior transformation of the variables. Further, a relation like (8.3.1) may hold locally under broad conditions. Thus if there is a true relation of the form $f(u, v, \ldots) = 0$ where $u, v, \ldots$ are primary variables and $f(\cdot)$ is some suitably smooth function, then in the vicinity of a point $(u_0, v_0, \ldots)$ which satisfies the true relation, we have the usual Taylor expansion

$$f(u, v, \ldots) = (u - u_0) f_u(u_0, v_0, \ldots) + (v - v_0) f_v(u_0, v_0, \ldots)$$
$$+ \tfrac{1}{2}(u - u_0)^2 f_{uu}(u_0, v_0, \ldots) + \cdots .$$

Defining the x_is and β_is suitably, this can clearly be recast in the form (8.3.1).

Curve-fitting problems of the above type became increasingly important in scientific investigations in the post-Newtonian era, particularly in astronomy (for studying the movements of heavenly bodies) and in geodesy (for determining the shape of the earth on the basis of terrestrial and stellar observations). In such applications the relation (8.3.1) was generally deducible from accepted theories (like the theory of gravitation, the laws of dynamics etc.). The variation in $y, x_1, \ldots, x_p$ was partly systematic and, at least for some of the variables, partly unpredictable.

If all the z_ℓs in (8.3.3) are taken as zero, the problem of determination of the β_is can be looked upon as one of finding an approximate solution of n generally inconsistent equations in $\beta_1, \ldots, \beta_m$. The naive approach for this would be 'the method of selected points' in which the solution is found from a subset of m selected equations ignoring the rest. But this is arbitrary, wasteful and the solution often gives large discrepancies for the excluded equations. Hence attempts were made to develop better methods. In the next subsection we describe some more broad-based methods that were proposed from around the middle of the 18th century. These methods are non-probabilistic in that they regard the z_ℓs only as unknown numbers and treat the problem of curve-fitting as one of numerical analysis. In these, no assumptions about the origin and nature of the z_ℓs are involved—these may arise due to errors of measurement in one or more of the variables or due to omission of additional terms involving further variables in the equation, or due to both.

Non-probabilistic methods

8.3.2 (i) *Method of Averages*: This was proposed and used by the German astronomer Tobias Mayer in 1750 in the context of his study of the movements of the moon. The idea was to divide the set of n equations (8.3.3) into m non-overlapping subsets of (nearly) equal size and to replace each subset by a single equation obtained by averaging over it and assuming that the subset-averages of the z_ℓ-terms are all zero. From the resulting m equations the β_is could be solved out. In Mayer's main application there were $n = 27$ equations and $m = 3$ parameters of which one, say, β_1 was of primary interest. Mayer ordered the equations according to the coefficients of β_1 and took disjoint groups of 9 consecutive equations as three subsets. This gave a good 'estimate' of β_1 and the problem of lunar movements could be resolved satisfactorily. (This paved the way for the construction of lunar tables which were very important in those days for the determination of longitude at sea.)

The subset-wise division of the equations would be problematic in Mayer's method unless there is one β_i of primary interest. Laplace, in course of his epoch-making study on the movements of the outer planets in 1787 suggested a modification. Mayer's method can be looked upon as taking linear compounds of the n original *observational equations* (obtained by omitting the z_ℓs in (8.3.3)) with compounding coefficients 1 or 0. Laplace's modification consisted in allowing the coefficients to be ± 1 or 0 and the same equation to appear in more than one compound according to a set rule based on the $x_{i\ell}$s (see Stigler (1986) pp. 33–37). Laplace used his method successfully to estimate 4 parameters from 24 equations and was able to settle long-standing questions about the movements of Jupiter and Saturn.

In the 20th century there was a revival of interest in the method of averages. Wald (1940) considered Mayer's method from the probabilistic viewpoint for solving the problem of straight-line-fitting 'when both variables are subject of error'. Nair and Shrivastava (1942) and Bartlett (1949) noted that for fitting a straight line it is more advantageous to order the equations according to, say, the values of x and divide them into *three* (nearly) equal subsets of contiguous equations. We can then take the two extreme subsets omitting the middle third. (As the two points of support based on subset centres of gravity are thereby pushed further apart, this makes the fitted straight line more stable.) These versions of the method of averages as well as Laplace's modification were later generalized into what came to be called as 'the method of instrumental variables' in econometrics (see Johnston (1984) p. 363).

(ii) *Method of Least Absolute Deviations*: The method of averages is heuristic and is not based on any criterion of 'goodness of fit'. A method which starts from such a criterion was proposed for fitting a straight line by a Dalmatian Jesuit priest named Roger Boscovich in 1760, in course of a geodetic study of the ellipticity (extent of flatness at the poles) of the earth. In Boscovich's case m was 2 and $x_{1\ell}$ was 1 for $\ell = 1, \ldots, n$. Writing for the present x and x_ℓ for x_2 and $x_{2\ell}$, the problem was to fit the straight line $y = \beta_1 + \beta_2 x$ to n pairs of observations $(y_\ell, x_\ell), \ell = 1, \ldots, n$. The criterion proposed was the sum

$$\sum_{\ell=1}^{n} |y_\ell - \beta_1 - \beta_2 x_\ell| \tag{8.3.4}$$

of the absolute deviations. Boscovich laid down that this should be minimized subject to

$$\sum_{\ell=1}^{n} (y_\ell - \beta_1 - \beta_2 x_\ell) = 0 \quad \text{i.e. } \beta_1 = \bar{y} - \beta_2 \bar{x}. \tag{8.3.5}$$

Eliminating β_1, this means β_2 should be determined so that

$$\sum_{\ell=1}^{n} |(y_\ell - \bar{y}) - \beta_2 (x_\ell - \bar{x})| \tag{8.3.6}$$

becomes minimum. Boscovich developed an algorithm for this through a geometric argument. Laplace who, as we have seen (subsection 7.3.6), always held the absolute deviation as the correct measure of 'loss' in point estimation, considered the method and gave an algebraic formulation of the algorithm about 40 years later.

Assume without loss of generality that $x_\ell - \bar{x} \neq 0$, $\ell = 1, \ldots, n$ and denoting $(y_\ell - \bar{y})/(x_\ell - \bar{x}) = b_{(\ell)}$, suppose the observations are numbered so that $b_{(1)} \leq b_{(2)} \leq \cdots \leq b_{(n)}$. Writing

$$\frac{|x_\ell - \bar{x}|}{\sum |x_\ell - \bar{x}|} = p_{(\ell)},$$

Laplace's argument consisted essentially in interpreting the problem as the minimization of

$$\sum_{\ell=1}^{n} |b_{(\ell)} - \beta_2| p_{(\ell)}$$

with respect to β_2. Clearly this is achieved when β_2 equals a median of the discrete distribution with mass points $b_{(1)}, \ldots, b_{(n)}$ for which the mass at $b_{(\ell)}$ is $p_{(\ell)}, \ell = 1, \ldots, n$.

Boscovich and Laplace both used the above method ('the method of situation' according to Laplace) for studying the relation between the lengths of meridian arcs at different locations on the surface of the earth to the corresponding latitudes. (Laplace also considered an alternative method of fitting in which the sum (8.3.4) was replaced by the maximum of the absolute deviations, but here the algorithm developed was cumbersome and practically unworkable for large n.)

Although unlike the method of averages the method of least absolute deviations is criterion-based, it is 'nonlinear': the fitted coefficients do not come out as solutions of linear equations and are not linear compounds of the y_ℓs. Also the difficulty of extending Boscovich's algorithm to the case of more than two β_is is obvious. The method went into oblivion once the more tractable least squares method described below became available, although from time to time it has been pointed out that the former is less affected by grossly erroneous outlying observations than the latter (see e.g. Edgeworth's observations as quoted by Eisenhart (1961)). In recent years with the advent of computers there has been a renewal of interest in the method and extensions of the algorithm to more complex cases have been developed (see Arthanari and Dodge (1981)).

(iii) *Method of Least Squares*: This method which is both 'linear' and criterion-based was first proposed in print in the appendix of an 1805 memoir on cometary orbits by the French mathematician Adrian Marie Legendre. Considering the general case (8.3.1), Legendre took the sum of *squares* of deviations

$$\sum_{\ell=1}^{n} (y_\ell - \beta_1 x_{1\ell} - \cdots - \beta_m x_{m\ell})^2 \tag{8.3.7}$$

as his criterion of goodness of fit and sought to determine $\beta_1, \ldots, \beta_m$ so that this is minimized. Equating the partial derivatives of (8.3.7) with respect to β_i to zero, he was led to the equations (later called *normal equations* by Gauss)

$$\beta_1 \sum_{\ell=1}^{n} x_{i\ell} x_{1\ell} + \cdots + \beta_m \sum_{\ell=1}^{n} x_{i\ell} x_{m\ell} = \sum_{\ell=1}^{n} x_{i\ell} y_\ell, \quad i = 1, \ldots, m, \tag{8.3.8}$$

which are clearly linear compounds of the observational equations. Nowadays it is customary to use the matrix notations

$$\mathcal{X}^{n \times m} = (x_{i\ell})_{\substack{\ell = 1, \ldots, n \\ i = 1, \ldots, m}}, \quad y = (y_1, \ldots, y_n)', \quad \beta = (\beta_1, \ldots, \beta_m)', \tag{8.3.9}$$

to write (8.3.8) in the compact form

$$\mathcal{X}'\mathcal{X}\beta = \mathcal{X}'y. \tag{8.3.10}$$

(The column vector y in the present context may not be confused with the variable y appearing in (8.3.1).) When rank $\mathcal{X} = m$, the case is said to be one of *full rank* (we can, without loss of generality, assume this since, otherwise, we can reduce the number of parameters and modify $\mathcal{X}$ to ensure the condition). In this case the matrix $\mathcal{X}'\mathcal{X}$ is nonsingular and (8.2.10) has a unique solution $\hat{\beta} = (\hat{\beta}_1, \hat{\beta}_2, \ldots, \hat{\beta}_m)' = (\mathcal{X}'\mathcal{X})^{-1}\mathcal{X}'y$ which is called the *Least Squares* (LS) *estimate* of $\beta \cdot \hat{\beta}_i, i = 1, \ldots, m$ are clearly linear compounds of $y_\ell, \ell = 1, \ldots, n$.

Legendre took the criterion (8.3.6) apparently from considerations of mathematical tractability. He claimed that through its use 'a kind of equilibrium is established among the errors' (see Stigler (1986) p. 13) and pointed out that (i) when $n = m$ and the observational equations (8.3.3) with $z_\ell = 0$ are consistent, the method leads to the correct values of the β_is and (ii) for the problem of estimation of a true value, i.e. the case $m = 1, x_{1\ell} = 1, \ell = 1, \ldots, m$, it gives $\hat{\beta}_1 = \bar{y}$. But what is remarkable is that unlike his predecessors he described the method quite generally in algebraic terms and not with reference to particular numerical problems.

After proposing the method Legendre applied it to the geodetic problem of determining the relation of the length of a meridian arc to the latitude at which it is measured. The data used were those collected under the aegis of the French Academy of Sciences in connection with their project for the standardization of the meter (the Academy proposed to define the meter as 10^{-7} times the length of the quadrant arc from the equator to the North Pole through Paris, and for this the latter had to be ascertained). The same data had been analysed earlier by Laplace by the method of minimization of the largest absolute derivation. In their reduced form the observational equations related to the differences $L_\ell - L_{\ell+1}, \ell = 1, \ldots, 4$, in the latitudes at 5 chosen locations to the corresponding arc-lengths linearly via certain nonlinear functions. Thus formally it was the case $n = 4$, $m = 2$ in the general model. But since the variables related arose as differences of original observations, following Laplace, Legendre took the discrepancies in the form $z_\ell - z_{\ell+1}, \ell = 1, \ldots, 4$, and fixing one of the z_ℓs, treated it as an extra parameter.

The methods of curve-fitting described above were undoubtedly tools of induction, but they did not represent statistical induction. No probability models were explicitly assumed and naturally there was no scope for assessing the uncertainty in the conclusion probabilistically. However, probabilistic ideas seem to be implicitly present at various places in their development. Thus in the method of averages, the assumption that the average of the discrepancies taken over several equations is negligible, and hence, can be equated to zero, can be justified if we assume that the discrepancies are realizations of random variables with expectation zero. Also Laplace's and Legendre's trick of treating one of the z_ℓs as fixed in the geodetic problem described above, assumes meaning if we take the z_ℓs as values of independent random variables, since such conditioning would render the remaining components independent.[4]

The curve-fitting problem can be regarded as an extension of the problem of estimation of a true value from repeated observations. However, although for the latter the use of the A.M. had become an established practice by the middle of the 18th century, the realization that for the estimation of multiple parameters it would be profitable to combine observations taken under diverse conditions, grew rather slowly (see Stigler (1986) pp. 28–30). It may be asked: why, even as late as 1805, was not the curve-fitting problem treated explicitly in terms of probability, regarding the discrepancies as random errors? After all, for the estimation of a true value probabilistic measurement error models had been assumed by Simpson, Daniel Bernoulli, and Laplace (sections 6.4, 6.5, and 7.3). The plain answer seems to be that people, having burnt their fingers with the simpler true-value problem, did not venture to pursue that course in the more complex case.

| Gauss's first approach |

8.3.3   The situation however changed with the entry of Gauss, who was the first person to approach the problem of curve-fitting starting from a probability model. He interpreted the problem as one of estimation of the β_is and gave a posteriorist (i.e. Bayesian) justification for the least squares method of estimation. The development was contained in his 1809 book *Theoria Motus* where, as we have seen in subsection 8.2.3, he gave his derivation of the normal law of error. In fact, Gauss's derivation of the normal law was plainly motivated by his desire to use it in this development.

In Gauss's model it is assumed that the z_ℓ in (8.3.3) are realizations of unobservable random variables Z_ℓ and the $x_{i\ell}$ are known numbers which are values of either non-stochastic variables or conditionally fixed stochastic variables. This means if Y_ℓ denotes the random variable corresponding to y_ℓ, the probability model underlying (8.3.3) is

$$Y_\ell = \beta_1 x_{1\ell} + \cdots + \beta_m x_{m\ell} + Z_\ell, \quad \ell = 1, \ldots, n. \tag{8.3.11}$$

The Z_ℓ are supposed to be i.i.d. $\mathcal{N}(0, \sigma^2)$[5] (for conditionally fixed $x_{i\ell}$ when these are observations on stochastic variables). (In conformity with current practice we use σ instead of Gauss's modulus of precision $h = 1/(\sqrt{2}\sigma)$.) Such a model

---

[4]This is just like one of the ways of handling a mixed model, say for a two-way classification in analysis of variance. When we fix the random effects conditionally, we can treat the model like a fixed-effects one; cf. Scheffé (1959) p. 171.

[5]Actually there is no loss of generality in assuming $E(Z_\ell) = 0$. If $E(Z_\ell) = \mu$, we can introduce μ as an extra parameter and take $Z_\ell - \mu$ for Z_ℓ.

presumes that only the observations y_ℓ may be subject to random measurement errors.[6]

Gauss's model implies that given the observations (8.2.2), the likelihood, and hence, the posterior density of $\beta_1, \ldots, \beta_m$ under an (improper) uniform prior is given by

$$\text{const.} e^{-1/(2\sigma^2)\sum_1^n (y_\ell - \beta_1 x_{1\ell} - \cdots - \beta_m x_{m\ell})^2}, \quad -\infty < \beta < \infty, \quad i = 1, \ldots, m. \tag{8.3.12}$$

(It is implicitly assumed that σ^2 is a known fixed number, and when the $x_{i\ell}$s are conditionally fixed values of certain random variables their marginal distribution is free from the β_is.) Gauss's argument was that the *most probable values* of $\beta_1, \ldots, \beta_m$ corresponding to the maximum of the posterior density would be their best estimates. This meant that the estimates were to be determined by minimizing (8.3.7), i.e. from the normal equations. Assuming that the case is one of full rank, Gauss was thus led to the LS estimates $\hat{\beta}_i, i = 1, \ldots, m$. Gauss's argument here is along the lines of what is now regarded as a standard principle of point estimation under the pro-subjective approach (subsection 4.4.16). Obviously the estimates here are same as the MLEs.

If we introduce matrix notations as in (8.3.9) and write $\hat{\beta} = (\hat{\beta}_1, \ldots, \hat{\beta}_m)'$, then as

$$y - \mathcal{X}\beta = (y - \mathcal{X}\hat{\beta}) - \mathcal{X}(\beta - \hat{\beta}),$$

we have the break-up

$$\sum_{\ell=1}^n (y_\ell - \beta_1 x_{1\ell} - \cdots - \beta_m x_{m\ell})^2 = (y - \mathcal{X}\beta)'(y - \mathcal{X}\beta)$$

$$= (\beta - \hat{\beta})'(\mathcal{X}'\mathcal{X})(\beta - \hat{\beta})$$

$$+ \text{ a term free from } \beta_1, \ldots, \beta_m. \tag{8.3.13}$$

Substituting (8.3.13) in (8.3.12) and absorbing the factor free from the β_is in the 'const.', we can readily see that the posterior distribution of β is the m-variate normal distribution $\mathcal{N}_m(\hat{\beta}, \sigma^2(\mathcal{X}'\mathcal{X})^{-1})$ (see Lindley (1965) Ch. 8 for this modern version of Gauss's result). From this result we can draw various types of conclusions about β.

But in Gauss's time the multivariate normal distribution was not a familiar entity (although in a sense, it had already appeared in the work of Lagrange and was soon to appear in the work of Laplace, its properties were not generally known; see Hald (1998) pp. 44, 403–404). Also, Gauss did not have the advantage of matrix notations. So instead of characterizing the joint posterior

---

[6]Any random measurement errors in the variables would clearly contribute to the Z_ℓs. If the $x_{i\ell}$s are subject to error when these are conditionally fixed, the parts in Z_ℓs contributed by them would be affected and the Z_ℓs may no longer have conditional means 0.

distribution as above, he reduced it to derive the univariate marginal posteriors of the individual β_is. He followed the sweep-out method for the solution of the normal equations and applied a matching triangular transformation on the β_is to express the quadratic form in the exponent of (8.3.11) as a sum of squares. Integrating out the other variables, from this he could deduce that marginally each β_i follows a normal distribution with mean $\hat{\beta}_i$; the marginal variance of β_i, which, we know, is the ith diagonal element of $\sigma^2(\mathcal{X}'\mathcal{X})^{-1}$, could be found as an appropriate entry in the tableau representing the sweep-out solution of the normal equations (see Hald (1998) pp. 383–389 for details).

The above conclusion holds whatever the fixed value of σ. However, to get actual interval estimates for the β_is we require to know σ. Gauss did not consider this problem in 1809. However, if we consider the β_is as known and assume that $h = 1/(\sqrt{2}\sigma)$ is subject to a uniform prior over $(0, \infty)$, then arguing as Gauss later did in the case of estimation of a true value (subsection 8.2.4), we get the estimate

$$\hat{\sigma} = \left\{ \sum_{\ell=1}^{n} (y_\ell - \beta_1 x_{1\ell} - \cdots - \beta_m x_{m\ell})^2 / n \right\}^{1/2}.$$

The observations made in subsection 8.2.5 on the implications of Gauss's derivation of the posterior distribution of θ extend themselves to the present case. Derivation of the posterior as above is possible because there exists a non-singular transformation of the independent normal variables $y_1, \ldots, y_n$ under which m of the new variables are distributed independently of the remaining $n - m$ and the β_is, which appear in the means of the former, are not involved in the distribution of the latter. The first m variables then represent a sufficient statistic for $\beta_1, \ldots, \beta_m$. Hence the joint distribution of $\hat{\beta}_i - \beta_i, i = 1, \ldots, m$ turns out to be multinormal with means 0 and from this the posterior distribution of the β_is under uniform prior is immediate. All this points to the malleability of the multivariate normal distribution under linear transformation of the variables.

From Gauss's justification of the LS method, a simple generalization of the method follows and this was pointed out by him. If the Z_ℓ have different variances σ_ℓ^2, the appropriate course would be to use weighted LS, i.e. to minimize

$$\sum_{\ell=1}^{n} \frac{1}{\sigma_\ell^2} (y_\ell - \beta_1 x_{1\ell} - \cdots - \beta_m x_{m\ell})^2 \tag{8.3.14}$$

for estimating the β_is.

Gauss's above derivation of the LS method was an instance of the pro-subjective approach to statistical induction (section 4.4). Strangely, after this Gauss gradually distanced himself from this kind of argument and during 1823–1828 advanced an alternative justification of LS based on sampling theory. But it drew its ideas from a large sample approach to the problem which had been chalked out by Laplace in the meantime. For the sake of historical justice we will first consider Laplace's development and then take up Gauss's second approach.

The presentation of the LS method by Gauss in the *Theoria Motus* in 1809 triggered a long-drawn priority dispute. Gauss in the context of his discussion of the method claimed that he had been using it since 1795. Legendre, whose memoir containing the LS method came out only in 1805, took exception to this unsubstantiated claim of priority. Since then, from time to time various archival investigations have been made to settle the question (see Stigler (1986) pp. 145–146 and Hald (1998) pp. 394–396), but it cannot be said that the mist surrounding it has been fully cleared.

Laplace's large-sample approach

8.3.4   It is remarkable that Laplace, who in his early life had been an unequivocal votary of the post-experimental stand-point and the pro-subjective approach to inference based on it (subsection 7.3.3), later enthusiastically took to the pre-experimental position and developed the sampling theory approach to linear models, albeit in large samples. The reason was two-fold: perception of the difficulty of specifying an appropriate model as soon as one moved out of the ambit of the binomial set-up and his discovery around 1810 of the CLT. The latter opened his eyes to the possibility of developing a model-free (by this, we of course mean not depending on the form of the parent distribution) approach to inference based on the approximate normality of statistics in large samples. Also, Gauss's posteriorist use of the normal model must have made him see how amenable the normal distribution was to linear transformations. In the following, as earlier, we freely draw upon matrix theory and the properties of the multivariate normal distribution to reach the mathematical results quickly with a view to highlighting the conceptual break-throughs involved.

In the following, given a sequence of random m-vectors $u_{(n)}, n = 1, 2, \ldots$, conventionally we say that $u_{(n)}$ is asymptotically distributed as $\mathcal{N}_m(\mu_{(n)}, \sum_{(n)})$ where $\mu_{(n)}$ and $\sum_{(n)}$ are sequences of m-vectors and $m \times m$ p.d. matrices, when for *every* fixed non-null m-vector l, $l'u_{(n)}$ is asymptotically $\mathcal{N}(t'\mu_{(n)}, t' \sum_{(n)} t)$ in the sense of section 3.10.

Let $Y_\ell, \ell = 1, \ldots, n, \ldots$ be a sequence of random variables such that for each ℓ an equation of the form (8.3.11) holds, $(x_{1\ell}, \ldots, x_{m\ell})$ representing given coefficients and $Z_\ell, \ell = 1, \ldots, n, \ldots$ being i.i.d. random variables *not necessarily normal* with mean 0 and variance σ^2 (Laplace assumed the distribution of Z_ℓ to be symmetric but that is unnecessary). Using matrix notations we can write (8.3.11) in the form

$$Y_{(n)}^{n \times 1} = \mathcal{X}_{(n)}^{n \times m} \beta^{m \times 1} + Z_{(n)}^{n \times 1}, \tag{8.3.15}$$

where the subscript indicates dependence on n, and where we assume rank $\mathcal{X}_{(n)} = m$ for every $n(\geq m)$.

Let for each $n \geq m$, $C_{(n)}^{n \times m} = (c_{i\ell})$ (cf. (8.3.9)) be a matrix of rank m such that

$$\text{rank } C'_{(n)} \mathcal{X}_{(n)} = m. \tag{8.3.16}$$

For any $t^{m \times 1}$, from CLT it follows that as $n \to \infty$ under mild regularity conditions on the sequence $C_{(n)}$,[7] $t'C'_{(n)}(Y_{(n)} - \mathcal{X}_{(n)}\beta) = t'C'_{(n)}Z_{(n)}$ is asymptotically $\mathcal{N}_m(0, t'C'_{(n)}C_{(n)}t)$. This means that the set of linear functions of the Y_ℓs represented by $C'_{(n)}Y_{(n)}$ is asymptotically $\mathcal{N}_m(C'_{(n)}\mathcal{X}_{(n)}\beta, C'_{(n)}C_{(n)})$ so that

$$(C'_{(n)}\mathcal{X}_{(n)})^{-1}C'_{(n)}Y_{(n)} = \tilde{\beta}_{(n)} = (\tilde{\beta}_{1(n)}, \ldots, \tilde{\beta}_{m(n)})'. \tag{8.3.17}$$

is asymptotically distributed as

$$\mathcal{N}_m(\beta, \sigma^2(C'_{(n)}\mathcal{X}_{(n)})^{-1}(C'_{(n)}C_{(n)})(\mathcal{X}'_{(n)}C_{(n)})^{-1}).\text{[8]} \tag{8.3.18}$$

From this it follows that marginally each linear function $\tilde{\beta}_i$ in (8.3.17) is for large n approximately normally distributed with mean β_i. Thus $\tilde{\beta}_i$ can be taken as an estimate of β_i.

Laplace derived the joint distribution of $\tilde{\beta}_1, \ldots, \tilde{\beta}_m$ in the form (8.3.17) only for $m = 1, 2$. For $m = 1$ the asymptotic variance of $\tilde{\beta}_1 = \sum_1^n c_{1\ell}Y_\ell/\sum_1^n c_{1\ell}x_{1\ell}$ is $\sum_1^n c_{1\ell}^2/(\sum_1^n c_{1\ell}x_{1\ell})^2$. Differentiating it with respect to $c_{1\ell}$ and equating to 0, Laplace showed that the minimum occurred for the choice $c_{1\ell} = x_{1\ell}, \ell = 1, \ldots, n$, i.e. when $\tilde{\beta}_1$ equals the LS estimate $\hat{\beta}_1$. Similarly in the case $m = 2$ he showed that the choice $c_{i\ell} = x_{i\ell}, \ell = 1, \ldots, n, i = 1, 2$, which gives $\tilde{\beta}_i = \hat{\beta}_i, i = 1, 2$, minimizes the two marginal asymptotic variances simultaneously. Laplace took it for granted that $C_{(n)} = \mathcal{X}_{(n)}$, which makes $\tilde{\beta} = \hat{\beta}$, and simultaneously minimizes all the m asymptotic variances, whatever m.

In general Laplace held that in estimating β_i if we confine ourselves to linear estimates of the form $\tilde{\beta}_i$, then our aim should be to minimize the mean absolute error $E|\tilde{\beta}_i - \beta_i|$. As for large n, $\tilde{\beta}_i$ is approximately normal with mean β_i, $E|\tilde{\beta}_i - \beta_i|$ can be taken as proportional to the asymptotic S. E. of $\tilde{\beta}_i$. Thus, Laplace argued that our aim would be achieved upto large sample approximations, if we minimized the marginal asymptotic variance of $\tilde{\beta}_i$, i.e. took $\tilde{\beta}_i = \hat{\beta}_i$. Laplace called the LS method of estimation which led to the $\hat{\beta}_i$s as 'the most advantageous method'. In conformity with modern usage we can say that Laplace proved that $\hat{\beta}_i$ is the *best linear asymptotically normal* (BLAN) estimator of β_i.

---

[7]Laplace sought to prove the result from first principles by proceeding as in his proof of CLT for the i.i.d. case, but since the terms of $t'C'_{(n)}Z_{(n)}$ are *not* in general i.i.d. some restriction on the $C_{(n)}$ would be required. Writing $\nu_{(n)}$ for the smallest characteristic root of $C'_{(n)}C_{(n)}$ it can be shown (cf. Hájek *et al* (1999) p. 184) that it is enough if, as $n \to \infty$, $\nu_{(n)}^{-1}\{\max_{1 \le \ell \le n}\sum_{i=1}^m c_{i\ell}^2\} \to 0$. Incidentally, Laplace required for his proof that the coefficients $c_{i\ell}$ be integer-valued, but this restriction is superfluous.

[8]As $(C'_{(n)}\mathcal{X}_{(n)})^{-1}$ depends on n, for this we actually require that the convergence of the distribution of $t'C'_{(n)}Z_{(n)}/(t'C'_{(n)}C_{(n)}t)^{1/2}$ to $\mathcal{N}(0, 1)$ be uniform in t. A sufficient condition for this, which implies the condition in footnote 7 above, is that $\nu_{(n)}^{-1}\{n \max_{1 \le \ell \le n}\sum_{i=1}^m c_{i\ell}^2\}$ be bounded; see Chatterjee (1972) pp. 263–265.

The above is clearly an instance of the sampling-theory approach (subsection 4.3.1)—the β_is are supposed to have fixed unknown values and the performance of an *estimator* is judged in terms of its sampling distribution.

It is remarkable that Laplace emphasized that it is not enough to estimate a parameter but we must give probabilistic bounds to the margin of error. Thus he did not stop with the point estimate $\hat{\beta}_i$ of β_i but suggested interval estimates of the form $\hat{\beta}_i \pm k$ (as. S. E. of $\ddot{\beta}_i$) where k should be chosen so that $P = \text{Prob}\,(|\hat{\beta}_i - \beta_i| < k$ (as. S. E. of $\hat{\beta}_i$)) as determined from the normal approximation is suitably high. Of course for this one has to tackle the question of estimation of σ. Laplace's prescription here was to substitute

$$\hat{\sigma} = \sqrt{\left\{\frac{1}{n}\sum_{\ell=1}^{n}(Y_\ell - \hat{\beta}_1 x_{1\ell} - \cdots - \hat{\beta}_m x_{m\ell})^2\right\}}$$

$$= \sqrt{\left\{\frac{1}{n}(Y - \mathcal{X}\hat{\beta})'(Y - \mathcal{X}\hat{\beta})\right\}} \qquad (8.3.19)$$

for σ. In a later paper (1816) he sought to justify this step by considering the joint asymptotic distribution of $\hat{\beta}$ and $\hat{\sigma}^2$ under the assumption that the Z_ℓs possess a finite 4th moment.

In the context of interval estimation of β_i, Laplace even used the term 'le degré de confiance' (the degree of confidence) for P. Was he thinking of the interval here as a (large sample) confidence interval as in the behavioural approach (section 4.2) or as a fiducial interval as in the instantial approach (section 4.3)? The facts that Laplace was basically a believer in impersonal subjective probability (subsection 7.3.4) and he interpreted the P above by saying that one can bet P against $1 - P$ that the statement holds (see Hald (1998) p. 399), seem to point to the latter. (Incidentally, the idea of large-sample fiducial limits predates Laplace. Hald (1998) pp. 23, 43 points out that around 1776 Lagrange proposed such limits for binomial and multinomial proportions.)

Going back to the problem of estimation of σ in large samples, we mention that in an 1816 paper Gauss considered it in the simplest case when we have observable i.i.d. $N(0, \sigma^2)$ random variables $Y_\ell, \ell = 1, \ldots, n$, and examined some estimators of σ other than $\sqrt{\{1/n\sum Y_\ell^2\}}$. These we will take up in the next section while discussing the asymptotic relative efficiency of estimators.

| Gauss's second approach |

8.3.5   We have seen that Gauss in his 1809 book followed the Bayes–Laplace posteriorist approach and proposed that the maximal point of the posterior distribution of β be taken as the estimate of β. When Laplace in his 1811 paper introduced the sampling theory approach leading to the BLAN estimatior of β_i, Gauss was attracted to the new point of view not only because, as Laplace had pointed out, it did not involve any hard specification of the form of the error distribution, but apparently also because it did not require any 'metaphysical' assumption of a subjective prior. (It seems Gauss was less of a subjectivist than Laplace in his conception of probability; see section 7.4.) In 1816, in a paper dealing with the estimation of σ (which we will consider in a later subsection), he followed Laplace to compare alternative estimators in terms of their large sample S.E.s. But gradually he felt the need of providing some justification for the LS method which would be convincing even when n is small. He came out with such a justification in his treatise *Theoria Combinationis Observationum Erroribus Minimis Obnoxiae* (Theory of Combination of Observations in Error Leading to Minimum Loss) which was published in instalments during 1823–1828. In this, he followed Laplace in basing the choice of the estimator on the principle of minimization of the expected loss. But instead of taking the absolute error, as Laplace did, he took the squared error as the loss function.

Gauss, just like Laplace, started from the model (8.3.11) where $Z_1, \ldots, Z_n$ are i.i.d. with $E(Z_\ell) = 0$, Var $(Z_\ell) = \sigma^2$. (To get Gauss's result, we could as well replace 'independence' by 'uncorrelation', i.e. assume $\text{cov}(Z_\ell, Z_{\ell'}) = 0$ for $\ell \neq \ell'$.) As earlier, we freely make use of matrix notations to express Gauss's arguments in a compact form. The assumptions made about the Z_ℓs mean that the representation (8.3.15) holds with $E(Z) = $ o, disp $(Z) = \sigma^2 I_n$, where as earlier, rank $\mathcal{X}$ is assumed to be m (as n is held fixed, we now dispense with the subscript (n)). Gauss directly derived the expressions for the variances and covariances of the LS estimates $\hat{\beta}_1, \ldots, \hat{\beta}_m$. From the matrix representation $\hat{\beta} = (\mathcal{X}'\mathcal{X})^{-1}\mathcal{X}'Y$ we can readily put his findings as

$$\text{disp } (\hat{\beta}) = \sigma^2(\mathcal{X}'\mathcal{X})^{-1}. \tag{8.3.20}$$

From any set of m linear functions $C'Y$, where $C^{n \times m}$ satisfies (8.3.16), we can construct an unbiased linear estimator (ULE) $\tilde{\beta}$ of β as in (8.3.17). (Gauss here did not bring in unbiasedness directly but required C to be such that when the Z_ℓs equal 0, i.e. $Y = X'\beta$ exactly, $\tilde{\beta}$ equals β. In the case of linear functions of the Y_ℓs, this requirement of course is equivalent to unbiasedness. However, as we will see in the following, in the context of estimation of σ^2 Gauss introduced unbiasedness more explicitly.) For such an ULE we have

$$E(\hat{\beta} - \beta)(\tilde{\beta} - \hat{\beta})' = \sigma^2(\mathcal{X}'\mathcal{X})^{-1}\mathcal{X}'\{C(\mathcal{X}'C)^{-1} - \mathcal{X}(\mathcal{X}'\mathcal{X})^{-1}\} = 0,$$

which means that $\hat{\beta}_i$ is uncorrelated with $\tilde{\beta}_j - \hat{\beta}_j, i, j = 1, \ldots, m$. Hence writing

$$\tilde{\beta} = \hat{\beta} + (\tilde{\beta} - \hat{\beta}),$$

$$\text{disp } (\tilde{\beta}) = \text{disp } (\hat{\beta}) + \text{disp } (\tilde{\beta} - \hat{\beta}),$$

so that for any $\lambda^{m \times 1}$,

$$\text{Var } (\lambda'\tilde{\beta}) \geq \text{Var } (\lambda'\hat{\beta}), \tag{8.3.21}$$

the equality holding if and only if $\lambda' \tilde{\beta} = \lambda'\hat{\beta}$ (with probability 1). We thus reach the well-known *Gauss–Markov Theorem: among all ULE's of a linear function*

$\lambda'\beta$, there exists a unique estimate which has minimum variance and it is given by $\lambda'\hat{\beta}$, $\hat{\beta}$ being the LS estimate of β. In particular $\hat{\beta}_i$ has minimum variance among all ULE's of $\beta_i, i = 1, \ldots, m$.

In contrast with his first approach where he called $\hat{\beta}_i$ 'the most *probable* value', Gauss now termed it 'the most *plausible* value' of $\beta_i, i = 1, \ldots, m$. Nowadays $\lambda'\hat{\beta}$ is generally called the *best linear unbiased* (BLU) estimate of $\lambda'\beta$.

In the context of the present model Gauss also considered the problem of estimation of σ^2. He showed that the residual variance (in the case of full rank) given by

$$s^2 = \frac{1}{n-m}\sum_{\ell=1}^n (Y_\ell - \hat{\beta}_1 x_{1\ell} - \cdots - \hat{\beta}_m x_{m\ell})^2 = \frac{1}{n-m}(Y - \mathcal{X}\hat{\beta})'(Y - \mathcal{X}\hat{\beta})$$

(8.3.22)

gives an unbiased estimate of σ^2 and derived the expression for $\mathrm{Var}(s^2)$ (see Hald (1998) p. 477).

Going back to the Gauss–Markov Theorem, Gauss also noted that when $\mathrm{Var}\,(Z_\ell) = \sigma_\ell^2, \ell = 1, \ldots, n$ are different, the BLU estimates may be obtained by the method of weighted LS, i.e. by minimizing (8.3.14). For this of course one would require to know at least the mutual ratios of the σ_ℓ^2s.

The Gauss–Markov Theorem has been one of the main pillars on which the application of the statistical theory has stood and developed for almost two centuries now. After Gauss, a host of authors dealt with various versions of the result.[9] Specifically, as we will see in the next chapter, towards the end of the 19th century in the context of a sample from the multivariate normal population, Karl Pearson, unaware of the details of Gauss's result, arrived at the LS estimates of the β_is, by preceding along a different route. Later, R. A. Fisher considered *in extenso* the case when the independent variables are qualitative factors and built up his theory of analysis of variance and design of experiments on the basis of Gauss's result. To anticipate some future developments, Aitken in 1935 gave the matrix formulation of the problem as considered above and extended the method of weighted LS to the correlated case when $\mathrm{disp}\,(Z) = \sigma^2 W, W^{n \times n}$ being a known p.d. matrix; see Seal (1967). R. C. Bose (1940) clarified the situation when $\mathrm{rank}\,\mathcal{X} < m$ by introducing the concept of a linearly *estimable* function (i.e. a function $\lambda'\beta$ for which there exists at least one ULE). Specifically, he showed that when $\mathrm{rank}\,\mathcal{X} < m$ for such a function $\lambda'\hat{\beta}$ is unique (although $\hat{\beta}$ may not be so) and the conclusion (8.3.21) continues to hold.

---

[9]In 1912 Markov gave a proof of the result, and Neyman (1934), wrongly thinking it to be originally due to him, called it 'Markov Theorem'. That is how the name of Markov came to be associated with it (see Seal (1967)). Actually, from subsection 8.3.4 it follows that for the estimation of individual β_is, the proof of the result for $m = 1, 2$ is essentially contained in Laplace's derivation of the BLAN estimators. Gauss in his treatise acknowledged that Laplace's idea was the inspiration behind his result. Thus a more appropriate name for the theorem would be 'Laplace–Gauss Theorem'.

It looks somewhat odd that in the *Theoria Combinationis* Gauss confined himself solely to the problem of point estimation. If one added the assumption of normality of Z_ℓs to the model, then knowing σ^2, one could easily set up an interval estimate for $\lambda'\beta$ using Gauss's result. Gauss was silent on this point possibly because he considered it too obvious, and having once discarded the assumption of normality of errors as too restrictive, he did not want to re-induct it purely for the sake of expediency. Another reason might be his discovery of a probability inequality (see Seal (1967)). In the same treatise he proved that for any random variable ξ with mean μ and S.D. σ such that the distribution of $\xi - \mu$ is symmetric and unimodal around 0, we must have

$$P(|\xi - \mu| \leq a\sigma) \geq 1 - \frac{4}{9a^2}, \quad \text{for } a > \frac{2}{\sqrt{3}}. \tag{8.3.23}$$

If the distribution of $\lambda'\hat{\beta} - \lambda'\beta$ satisfies the requirement,[10] from the inequality we can get reasonably sharp lower bounds to the probability of inclusion for a sufficiently wide interval centred at $\lambda'\hat{\beta}$. Thus taking $a = 2$ and 3, the lower bounds would be 0.8989 and 0.9506 respectively. Gauss must have though that since these bounds are high enough one need not bring in questionable assumptions about the distribution of errors to get nominally high exact probabilities. (Incidentally, Tshebyshev's inequality, which applies generally whenever $E(\xi) = \mu$ and $\text{Var}(\xi) = \sigma^2$ exist but gives the less sharp bound $1 - (1/a^2)$ for the probability in (8.3.23), was introduced in the literature independently by Bienaymé and Tshebyshev much later—during the 1850s and 1860s; see Hald (1998).)

8.4 Comparison of estimators

ARE—Gauss

8.4.1 In the preceding section we saw how in the first quarter of the 19th century Gauss reinterpreted the problem of curve-fitting as one of the estimation of parameters of a probability model, and how first Laplace and then Gauss veered from the posteriorist to the sampling theory point of view for its solution. In the sampling theory approach, performances of alternative estimators are compared in terms of the characteristics of their sampling distributions. If asymptotic normality around the true value of the parameter holds, the relative values of the asymptotic variances will be useful for such comparison. Laplace in his seminal treatise of 1811 did not have any necessity to consider this, as he was confining himself to linear estimators of the β_is and within the class of such estimators a unique one minimizing the asymptotic variance exists. Gauss in his 1816 paper while approaching the problem of comparison of alternative non-linear estimators of σ for a normal population with known mean from the sampling theory angle, made the point explicit.

---

[10]However, something more than mere symmetry and unimodality around 0 of the independent errors Z_ℓ may be required to ensure the unimodality of any linear compound of the Z_ℓs, and in particular of $\lambda'\hat{\beta} - \lambda'\beta$; see Ibragimov (1956).

If $Y_1, \ldots, Y_n$ are i.i.d. as $N(0, \sigma^2)$, let

$$g_r = \frac{1}{n} \sum_1^n |Y_\ell|^r, \gamma_r = \frac{1}{\sqrt{(2\pi)}} \int_{-\infty}^{\infty} |y|^r e^{-y^2/2} \, dy, \quad r = 1, 2, \ldots. \qquad (8.4.1)$$

As $E(g_r) = \sigma^r \gamma_r$, a reasonable estimator of σ is

$$\tilde{\sigma}_r = (g_r/\gamma_r)^{1/r}. \qquad (8.4.2)$$

(We can now see that these estimators are not only consistent but actually Fisher-consistent in the sense of subsection 4.2.2.)

By Laplace's CLT for large n, g_r is approximately distributed as $N(\sigma^r \gamma_r, (1/n)\sigma^{2r}(\gamma_{2r} - \gamma_r^2))$. From this, Gauss derived an interval estimate for σ. This interval estimate follows directly from the fact that $\tilde{\sigma}_{(r)}$ is approximately distributed as $N(\sigma, \sigma^2 V_r/n)$, where

$$V_r = \frac{1}{r^2} \left(\frac{\gamma_{2r}}{\gamma_r^2} - 1 \right), \qquad (8.4.3)$$

and has the form $\tilde{\sigma}_r(1 \pm k n^{-1/2} V_r^{1/2})$, the probability of inclusion for any k being determined from normal tables. (According to modern terminology the interval estimate would be called a large-sample confidence or fiducial interval.) For any k the relative span of the interval estimate for different r is directly proportional to $n^{-1/2} V_r^{1/2}$.

Gauss numerically computed V_r for $r = 1, 2, 3 \ldots$ and found that its smallest value $\frac{1}{2}$ corresponding to the relatively narrowest interval estimate (see Hald (1998) p. 457) was attained for $r = 2$. The values of V_r/V_2 for different r were as follows:

r	1	2	3	4
V_r/V_2	1.14	1.00	1.09	1.33

He then gave the intuitively appealing interpretation that (thinking in terms of the relative span of the interval estimate or equivalently in terms of the asymptotic variance) $\tilde{\sigma}_2$ based on a sample of 100 observations would be as reliable as $\tilde{\sigma}_1$ based on one of 114, $\tilde{\sigma}_3$ based on one of 109, $\tilde{\sigma}_4$ based on one of 133 observations, and so on. Using modern terminology we can express this by saying that the asymptotic relative efficiency (ARE) of $\tilde{\sigma}_r$ versus $\tilde{\sigma}_2$ is V_2/V_r.

Gauss in the same study also considered an estimate of σ based on the sample median $g_{(0.5)}$ based on $|Z_\ell|, \ell = 1, \ldots, n$. Since the population median of $|Z_1|$ is 0.67449σ, it followed that a reasonable estimate of σ would be $\tilde{\sigma}_{(0.5)} = g_{(0.5)}/0.67449$. Gauss gave the value of the asymptotic variance of $\tilde{\sigma}_{(0.5)}$ as $2.72 \, \sigma^2 V_2$. Thus the ARE of $\tilde{\sigma}_{(0.5)}$ versus $\tilde{\sigma}_2$ would be $100/272$. (Among estimates based on the quantiles of the $|Z_\ell|$, maximum efficiency, $100/124$ relative to $\tilde{\sigma}_2$ is attained for the quantile of order 0.86; see Whittaker and Robinson (1944) p. 204.)

Gauss's interpretation of ARE as the inverse ratio of sample sizes required to achieve the same asymptotic variance has withstood the passage of time and appears today in various forms in the study of relative performance of rival inference procedures.

ARE—Laplace   **8.4.2**   Two years after Gauss had introduced the concept of ARE of estimators, Laplace pursued it at a greater depth for the problem of estimation of β_1 in the simple linear model

$$Y_\ell = \beta_1 x_\ell + Z_\ell, \quad \ell = 1, \ldots, n \tag{8.4.4}$$

where $Z_\ell, \ell = 1, \ldots, n$ are assumed to be i.i.d. having mean = median = 0, variance σ^2, and density function $f(z)$ continuous in the neighbourhood of 0 with $f(0) > 0$. (Laplace imposed some further restrictions on $f(z)$ like symmetry and the existence of the second derivative, which are unnecessary.) His object was an asymptotic comparison of the LS estimator $\hat{\beta}_1$ with the estimator $\tilde{\beta}_1$ obtained by Boskovich's method (subsection 8.3.2) of Least Absolute Deviations (LAD; 'method of situation' according to Laplace) for an $f(z)$ which is otherwise arbitrary.

From the CLT it readily followed that as $n \to \infty$,

$$\hat{\beta}_1 - \beta_1 = \frac{\sum_1^n x_\ell Y_\ell}{\sum_1^n x_\ell^2} - \beta_1 = \frac{\sum_1^n x_\ell Z_\ell}{\sum_1^n x_\ell^2} \tag{8.4.5}$$

is asymptotically $\mathcal{N}(0, \sigma^2 / \sum x_\ell^2)$. Although Laplace did not consider it, for this, one has to assume that the sequence $x_\ell, \ell = 1, 2, \ldots,$ is such that (see footnote 7) as $n \to \infty$,

$$\max_{1 \le \ell \le n} x_\ell^2 \Big/ \sum_1^n x_\ell^2 \to 0. \tag{8.4.6}$$

Here $\tilde{\beta}_1$ is the median of the discrete distribution which assigns the mass $|x_\ell|/\sum_1^n |x_\ell|$ at the mass point

$$\frac{Y_\ell}{x_\ell} = \beta_1 + \frac{Z_\ell}{x_\ell}, \quad \ell = 1, \ldots, n. \tag{8.4.7}$$

We assume, without loss of generality, that in (8.4.4), $x_\ell > 0$, $\ell = 1, \ldots, n$ and write $\sum_1^n x_\ell = S_n$. From (8.4.7) it then follows that $\tilde{\beta}_1 - \beta_1$ is the median of the distribution represented by the 'weighted' empirical cdf

$$G_n(t) = \sum_{\ell=1}^n \frac{x_\ell}{S_n} I_{(-\infty, x_\ell t]}(Z_\ell), \tag{8.4.8}$$

when $I_A(\cdot)$ denotes the indication function of the set A.

Laplace's own derivation of the asymptotic distribution of $\tilde{\beta}_1$ (see Hald (1998) pp. 447–448) is defective. (Laplace failed to take account of the fact that the ordering of $Z_\ell / x_\ell, \ell = 1, \ldots, n$ would determine the jumps at the successive order statistics, and hence, the rank of the particular order statistic representing the median.) However, we can easily prove his result from the following extension of the Bahadur representation for sample quantiles which may be derived along the lines of Ghosh (1971): as $n \to \infty$, under (8.4.6) we have

$$\sqrt{\left(\sum_1^n x_\ell^2 \right)} (\tilde{\beta}_1 - \beta_1) = \frac{S_n(1/2 - G_n(0))}{f(0)\sqrt{\sum_1^n x_\ell^2}} + o_P(1)$$

$$= -\frac{1}{f(0)\sqrt{\sum_1^n x_\ell^2}} \sum_{\ell=1}^n x_\ell \left\{ I_{(-\infty,0]}(Z_\ell) - \frac{1}{2} \right\} + o_P(1).$$
(8.4.9)

From this it readily follows that under (8.4.6) $(\tilde{\beta}_1 - \beta_1)$ is asymptotically $\mathcal{N}\left(0, \{4f^2(0)\sum_1^n x_\ell^2\}^{-1} \right)$.

From the forms of the asymptotic distributions of $\hat{\beta}_1$ and $\tilde{\beta}_1$ we get that the asymptotic efficiency of $\tilde{\beta}_1$ relative to $\hat{\beta}_1$ is given by $4\sigma^2 f^2(0)$. Hence Laplace concluded that the LS method is preferable to the LAD method unless $\sigma^2 > 1/(4f^2(0))$. In particular, when $f(\cdot)$ is the density of $\mathcal{N}(0, \sigma^2)$, $f(0) = 1/(\sigma\sqrt{(2\pi)})$ and $\sigma^2 < 1/(4f^2(0)) = (\pi/2)\,\sigma^2$, so that the LS estimator is definitely superior to the LAD estimator, the ARE of the latter versus the former being $2/\pi$.

After this, Laplace investigated whether knowing the form of $f(\cdot)$ it is possible to find a linear compound of $\hat{\beta}_1$ and $\tilde{\beta}_1$ which would be an improvement over both. Generally, such a linear compound for estimating β_1 may be taken in the form

$$\hat{\beta}_1 - c(\hat{\beta}_1 - \tilde{\beta}_1) = (1 - c)\hat{\beta}_1 + c\tilde{\beta}_1$$
(8.4.10)

where the coefficient c is at our choice. Laplace realized that to determine c optimally we have to consider asymptotic *joint* distribution of $\hat{\beta}_1$ and $\tilde{\beta}_1$. By employing the same kind of reasoning as in the case of $\tilde{\beta}_1$ and manipulating on the joint characteristic function, Laplace reached the conclusion that $u_n = \sqrt{(\sum_1^n x_\ell^2)}(\hat{\beta}_1 - \beta_1)$ and $v_n = \sqrt{(\sum_1^n x_\ell^2)}(\tilde{\beta}_1 - \beta_1)$ have asymptotically a bivariate normal distribution. We may sidetrack all that by considering the representations (8.4.5) and (8.4.9), employing the familiar Cramér–Wold device, and then invoking the CLT under (8.4.6). Writing

$$\nu_1 = \int_{-\infty}^{\infty} |z| f(z)\, dz, \quad \rho = \frac{\nu_1}{\sigma},$$
(8.4.11)

and using modern notations, we get that (u_n, v_n) is asymptotically distributed as

$$\mathcal{N}_2\left(0, 0, \sigma^2, \frac{1}{4f^2(0)}, \rho \right).$$
(8.4.12)

For (u, v) distributed as $\mathcal{N}_2(0, 0, \sigma_1^2, \sigma_2^2, \rho)$, the choice of c minimizing the variance of $(1 - c)u + cv$ is $(\sigma_1^2 - \rho\sigma_1\sigma_2)/(\sigma_1^2 - 2\rho\sigma_1\sigma_2 + \sigma_2^2)$. Hence, by (8.4.11)–(8.4.12), the choice of c which minimizes the asymptotic variance is

$$c_0 = \frac{2f(0)(2f(0)\sigma^2 - \nu_1)}{4f^2(0)\sigma^2 - 4\nu_1 f(0) + 1}. \tag{8.4.13}$$

Laplace noted that by taking (8.4.10) with $c = c_0$, it is possible to improve over $\hat{\beta}_1$ provided $c_0 \neq 0$ (and similarly over $\tilde{\beta}_1$ provided $c_0 \neq 1$). In particular, he observed that for $\mathcal{N}(0, \sigma^2)$, $f(0) = 1/\sigma\sqrt{(2\pi)}$, $\nu_1 = \sqrt{(2/\pi)}\sigma$ which gives $c_0 = 0$ and thus $\hat{\beta}_1$ has asymptotic variance less than that of every other linear compound (8.4.10). In fact, implicit in Laplace's derivation is the result that in the normal case, in the asymptotic joint distribution of $\hat{\beta}_1\tilde{\beta}_1$, the conditional distribution of $\tilde{\beta}_1$ given $\hat{\beta}_1$ does not involve β_1 (the conditional mean is just $-\hat{\beta}_1$), and hence, $\tilde{\beta}_1$ cannot supply any additional information about β_1 beyond that carried by $\hat{\beta}_1$.

| Excursus on sufficiency |

8.4.3 In subsection 8.4.1 we discussed Gauss's finding that when the observables Y_ℓ are i.i.d. $N(0, \sigma^2)$, the estimator $\tilde{\sigma}_2$ of σ based on their sample second moment is superior in terms of asymptotic variance to estimators based on the other sample absolute moments and the sample median absolute value. In subsection 8.4.2, for the model (8.4.4) with the errors Z_ℓ similarly distributed, we considered Laplace's result that LS estimator $\hat{\beta}_1$ is superior to the LAD estimator $\tilde{\beta}_1$ in the same sense. We now know that under normality $\tilde{\sigma}_2$ for the first problem and $\hat{\beta}_1$ for the second problem (assuming σ^2 is given) represent minimal sufficient statistics (subsection 4.3.3). From standard results on minimal sufficient statistics (see e.g. Lehmann (1983), Chapter 2), it follows that in these problems even for a fixed n, $\tilde{\sigma}_2^2$ and $\hat{\beta}_1$ have minimum variances among *all* unbiased estimators of σ^2 and β_1 respectively. In fact around 1920, R. A. Fisher hit upon the concept of sufficiency while investigating almost the same estimation problem as that considered by Gauss through a line of reasoning which resembles that of Laplace. Because of this at this point we will jump over a century and make an anticipatory excursus to discuss the discovery of sufficiency by Fisher.

But before that it will be fair to note that an American statistician of the name of Simon Newcomb had actually recognized and clearly enunciated sufficiency in the context of a particular problem in 1860—long before Fisher abstracted and formulated the general principle. Newcomb considered the estimation of N, the unknown number of tickets in a bag in which the tickets are numbered $1, 2, \ldots, N$, on the basis of the readings of n random draws made with replacement. If m be the largest number drawn, he noted that knowing m, we need not know any of the other numbers drawn, since given m, 'every combination of smaller numbers will be equally probable on every admissible hypothesis' about N and 'will therefore be of no assistance in judging of those hypotheses' (see Stigler (1978)). This was admittedly a very clear statement of the sufficiency principle in a particular context.

Fisher considered the estimation of σ for the population $N(\mu, \sigma^2)$ on the basis of a random sample $Y_1, \ldots, Y_n$, when both μ and σ are unknown. The occasion for his study was a claim made by the astronomer–physicist Eddington in a 1914 book on astronomy. With reference to the above problem, Eddington had in effect observed that it can be shown that the estimator $\hat{\sigma}_1 = \sqrt{(\pi/2)}\sum_1^n |Y_\ell - \bar{Y}|/n$ based on the sample mean deviation has better performance than the sample S. D. $\hat{\sigma}_2 = \sqrt{\{\sum_1^n (Y_\ell - \bar{Y})^2/n\}}$. Fisher, in a note which appeared in the *Monthly Notices of Royal Astronomical Society* (1920) (see Fisher (1950)) challenged the claim. He first derived the individual sampling distributions and hence the exact sampling variances of $\hat{\sigma}_1$ and $\hat{\sigma}_2$ and showed that the former sampling variance exceeds the latter by 14%.[11] He then found also the large sample variances of estimators based on higher order sample absolute moments and showed that all of them were larger than the sampling variance of $\hat{\sigma}_2$.

The above results essentially duplicated the 1816 findings of Gauss, but Fisher did not know it at that time. However, Fisher proceeded further. In the same note he derived the *exact* joint distribution of $\hat{\sigma}_1$ and $\hat{\sigma}_2$ in the particular case $n = 4$ and found that the conditional distribution of $\hat{\sigma}_2$ given $\hat{\sigma}_1$ still involved σ, but that of $\hat{\sigma}_1$ given $\hat{\sigma}_2$ is entirely free from σ. He stated further that the latter property of $\hat{\sigma}_2$ holds generally even when we replace $\hat{\sigma}_1$ by any other estimator of σ. (Strictly, this holds if by an 'estimator' of σ we mean a translation invariant estimator; see Basu (1978).) This, he concluded, implied, 'The whole of the information respecting σ, which a sample provides, is summed up in the value of σ_2 [Fisher's notation for $\hat{\sigma}_2$]'. Fisher immediately visualized the importance and scope of the property for statistical inference. In 1922 in a seminal paper he formulated the concept and investigated some of its consequences in the general parametric set-up. The term 'sufficient statistic' was introduced for the first time in that paper. (According to modern convention, under the set-up of Fisher's note, $(\bar{y}, \hat{\sigma}_2)$ would be called the minimal sufficient statistic. $\hat{\sigma}_2$ alone would be an 'invariably sufficient statistic' for σ with respect to the group of all translations in the sense of Stein; see Basu (1978).)

Just as Laplace had done a century earlier in the context of his own problem, Fisher also considered the joint distribution of the two competing estimators in his study. But unlike Laplace, he considered the *exact* joint distribution for a fixed n (the assumption of a normal population as Fisher himself admitted, was crucial for this) and this helped him to abstract the principle of sufficiency. Fisher had no idea of discovering a general principle when he set out to contend the untenable claim of Eddington. Thus the discovery of the sufficiency principle in the history of statistics can be said to be truly a serendipitious event.

---

[11]However, Eddington, while conceding Fisher's point, still maintained that in practice $\hat{\sigma}_1$ is more dependable than $\hat{\sigma}_2$ because it is less affected by doubtful observations, or as we say it now, it is more robust against outliers.

8.5   In Chapter 6 we saw how Arbuthnott (section 6.2)
and Daniel Bernoulli (section 6.5) employed sampling
theory argument (instantial form) to solve particular problems of hypothesis
testing. Of course that was before Bayes initiated and Laplace developed the
pro-subjective (i.e. Bayesian) approach to inference. In Chapter 7 we noted how
Laplace, in the earlier phase of his research, declared his preference for the pro-
subjective over the sampling theory approach (subsection 7.3.3) and followed the
former track to hypotheses involving proportions (subsection 7.3.7). It is remarkable, however, that at no time did Laplace hesitate to follow the sampling theory
approach to testing when it was convenient to do so. In the following we describe
two instances of this. The first relates to a problem which Laplace pursued off
and on right from 1776 until in 1812 he discussed it in the *Théórie Analytique*.
The second relates to a study undertaken in 1823. This shows that Laplace was
at no stage dogmatic in his approach to inference.

Daniel Bernoulli's
problem   The first problem was really an extension of Daniel
Bernoulli's problem of testing the cosmogonical hypothesis H_0: 'the planetary universe was created by
chance' on the basis of observations on the orbits of planetary bodies. As the
full parameter-space was unspecified, the pro-subjective line would be incon-
venient here. In the *Théórie Analytique* Laplace considered the inclinations
$x_i, i = 1, \ldots, 10$ to the equatorial plane of the sun of the orbital planes of 10 planetary bodies (6 planets and 4 satellites). Just like Bernoulli earlier, Laplace took
H_0 to imply that the corresponding random variables (measured in radians) represent a random sample from the uniform distribution $\mathcal{U}[0, \pi/2]$. Laplace took
the sample mean $\bar{x}$ as his test statistic. Since $\bar{x}$ was too low (much less than
$\pi/4$), to perform the test Laplace required the value of $P_{H_0}(\bar{X} \leq \bar{x})$. In the
case of a sample of size n, first considering sampling from a discrete uniform
distribution over equispaced points spanning $[0, \pi/2]$ and then making the step
between successive points tend to zero as in Simpson's study (section 6.4), it can
be shown (cf. Karl Pearson (1978) p. 715, Hald (1998) p. 52) that

$$P_{H_0}(\bar{X} \leq t) = \frac{1}{n!} \sum_{i=0}^{[2t/\pi]} (-1)^i \binom{n}{i} \left(\frac{2t}{\pi} - i \right)^n.$$

In Laplace's case he had $n = 10, 2\bar{x}/\pi = 0.914187 < 1$ so that only the term $i = 0$
was relevant for computing $P = P_{H_0}(\bar{X} \leq \bar{x})$. P came out to be $10^{-7} \times 1.1235$.
As this was extremely small, Laplace rejected H_0. (Actually, Laplace also took
account of the directions of motion of the planetary bodies relative to the earth.
He presumed that under H_0 the directions would be determined as in symmetric Bernoulli trials independently of the inclinations. As all the directions were
same, he reduced the above P further by multiplying by 2^{-10}. But as this point
is unimportant for our discussion, we ignore it.) It may be noted that unlike

Daniel Bernoulli who employed a left-tailed test based on the largest order statistic, Laplace used a similar test based on the sample mean. The two tests are sensitive against somewhat different alternatives, but of course this point could not be perceived at the time.

The language in which Laplace couched the conclusion of such tests is revealing. Thus in his 1776 study (see Hald (1998) p. 74) he observed that 'the probability of the actual disposition of our planetary system would be infinitely small if it were due to chance' and therefore we may be nearly certain that it is due to 'a regular cause'. This mode of thinking is clearly akin to the 'disjunction argument' of the instantial mode of testing as a form of extended *modus tollens* (subsection 4.3.6). In this context we note that Karl Pearson (1978) p. 716 took exception to Laplace's interpretation of untenability of the hypothesis of chance creation as indicating the existence of a regular cause. Undoubtedly Pearson's objection was prompted by his positivist persuasion according to which causation is *non sequitur*. However, judging by Laplace's pronouncements in other similar contexts, it seems by 'regular cause' he meant 'the operation of some natural law'. In the present case even if the inclinations of the planetary bodies followed a distribution $\mathcal{U}[0, a]$ with $a < \pi/2$, it would be suggestive of an underlying law.

Laplace also made a similar study of the orbital inclinations of 100 comets. In this case since the observed mean inclination $\bar{x}$ exceeded the mean $\pi/4$ of the distribution $\mathcal{U}[0, \pi/2]$, for testing Laplace required the probability $P_{H_0}(\bar{X} > \bar{x})$. As the sample size was 100, Laplace assumed approximate normality of $\bar{X}$ under H_0 to compute this right-tail probability. As the value 0.263 of this was not very small, Laplace's concluded that rejection of H_0 was not called for in the case of the cometary data (see Hald (1998) p. 76).

| Tides of the atmosphere | The second instance of Laplace's adoption of the sampling theory approach to testing that we cite relates to his 1823 study of 'the tides of the atmosphere' (Stigler (1986) pp. 148–153, Hald (1998), pp. 436–439). By that time, as we have seen, Laplace had more or less veered to the sampling theory view-point. In any case, the pro-subjective approach would have been troublesome here as the problem was one of testing a point null hypothesis (subsection 4.4.6). To investigate whether the daily variation in barometric pressure in Paris between 9 a.m. and 3 p.m. was affected by the phase of the moon, Laplace took two sets of data on daily variation each covering a period of $n = 792$ days spanning a period of 8 years. The first set related to the four days around each new moon and full moon and the second to the four days around each of the two days of quadrature (when the sun, earth, and moon form a right angle) in each of the 99 lunar months constituting the eight-year period. If the two sets are regarded as samples from two populations with means μ_1 and μ_2 respectively, the absence of lunar effect on atmospheric pressure would amount to the hypothesis $H_0: \mu_1 - \mu_2 = 0$. In Laplace's case the sample mean $\bar{x}_1$ was greater than $\bar{x}_2$. So assuming the two populations had the same S.D. σ, to test H_0 Laplace required the value of $P =$ the tail probability

to the right of $(\bar{x}_1 - \bar{x}_2)/(2\sigma^2/n)^{1/2}$. To compute P, Laplace used large sample normal approximation and substituted for σ^2 the estimate based on the total sum of squares of the combined sample. (As Stigler points out this would generally make the test less sensitive than if σ^2 were based on the within-sample sum of squares; but in this particular application it did not affect the conclusion.) Laplace got the value $P = 0.180$ and since this was not small enough, concluded that H_0 cannot be rejected, i.e. the observed difference may be due to chance alone.

| Choice of the rejection rule |

The above shows that the present-day sampling theory practice of computing the P-value of a test as the probability under H_0 of getting a value of the test criterion at least as extreme as that observed, crystallized, even if it did not originate, with Laplace. Rejecting if and only if P is too low, of course is equivalent to taking the appropriate tail with a small size as the rejection region. In this connection the question naturally arises: why should we take the tail as representing the rejection region? After all a region with as small a probability under H_0 can be chosen in innumerable ways (at least in the continuous case). As we noted in subsection 4.3.12, the question cannot be satisfactorily answered without bringing in the ideas of alternative hypothesis and likelihood ratio. The following passage from Laplace's *Essai Philosophique* (quoted in Hacking (1965) p. 79) shows that Laplace was at least vaguely aware of this:

On a table we see letters arranged in this order, CONSTANTINOPLE, and we judge that this arrangement is not the result of chance, not because it is less possible than the others, for if this word were not employed in any language we should not suspect it came from any particular cause, but this word being in use amongst us, *it is incomparably more probable that some person has thus arranged the aforesaid letters than that this arrangement is due to chance* [italics ours].

To anticipate the future development of sampling theory hypothesis testing, as we know, the idea of alternative hypothesis and of principles for choosing an optimum rejection region by consideration of power and likelihood ratio (subsection 4.2.2) were formulated precisely by J. Neyman and E. S. Pearson around 1930. Lehmann (1993) has recently analysed available evidence to trace the inspiration behind that work. Neyman's thinking seems to have been triggered by some remarks in a 1914 book of Borel where it was suggested that the region of rejection must represent some remarkable feature of the data decided upon beforehand. On the other hand, E. S. Pearson's source seems to be 'Student's' (W. S. Gosset's) mention of 'an alternative hypothesis which will explain the occurrence of the sample with a more reasonable probability' in a personal letter. The above excerpt shows that the same line of thinking had been adumbrated by Laplace, no less clearly, more than hundred years earlier.

Before concluding this subsection we note one curious fact. Laplace (just like Fisher a century later) was alive to the importance of hypothesis-testing in scientific investigations both while performing posteriorist tests of hypotheses involving demographic proportions (subsection 7.3.7) and sampling theory tests of hypotheses of 'no effects' in his study of tides of the atmosphere. In both contexts he emphasized that without performing a test one ran the risk of vainly searching for causes or laws behind chance occurrences. But strangely Gauss, who had contributed so much to the theory of estimation, does not seem to have applied his mind to hypothesis-testing at all. We can find no explanation for this except possibly that Gauss confined his studies to areas where the underlying physical laws were well-established and led to definitive models for the observables with only the values of certain parameters unspecified. If these parameters were (nearly) correctly ascertained, direct verification of the model through successful prediction of new observations was possible.

8.6   Reign of errors

| Theory of Errors |

8.6.1   We now come to the end of the saga of the Laplace–Gauss era. As we have seen, in the first century after the inception of probability and statistics around the 1650s and 1660s, the problems studied belonged almost exclusively to games of chance and demography. In the second century, which can be said to have begun with the work of Thomas Simpson in 1757 and extended roughly up to the middle of the 19th century, covering the Laplace–Gauss era, problems from the fields of astronomy and geodesy came to occupy the centre court. A large part of Laplace's and almost all of Gauss's work was concerned with the treatment of observational errors in investigations made in these fields. As a result, the group of probabilistic methods of induction developed during this period came to be known as the 'Theory of Errors'. Text books discussing statistical methods at that time discussed those under this head. (As for example, the popular text for such methods by the astronomer G. B. Airy (published 1861) from which pioneers like 'Student' and Fisher learnt their 'statistics', was entitled *'On the Algebraical and Numerical Theory of Errors of Observations and the Combination of Observations.'*[12]) Before concluding this chapter, in this section we briefly

---

[12]In fact the name 'Statistics' originally coined by the Germans in the 17th century was meant to signify a descriptive account of the constitution of states. Sir John Sinclair, a Scotsman, in his 1793 book *Statistical Account of Scotland* used it to denote numerical accounts. Through the major part of 19th century, 'Statistical theory' at least in the continent, was supposed to be concerned mainly with the estimation of probabilities of mortality, invalidity, etc., topics with which we nowadays associate the actuaries. These were the central problems discussed in the two 19th century text books on 'mathematical statistics' by Wittstein and Zeuner published in the 1860s (see Seal (1967)). It seems the use of the term 'Statistics' to denote statistical discipline in its modern sense with the theory of errors incorporated in it, started towards the end of the 19th century and the semantic mutation took place on English soil.

review the main contributions of the Theory of Errors to the growth of statistical theory.

Standardization of measures

8.6.2   The standard measures of location and variability of a distribution became well-established through various theoretical and applied studies based on the theory of errors. The former arose as standard estimates of a true value on the basis of repeated observations. The idea of loss and the principle of minimization of mean loss, when the loss was assessed in terms of squared error, led to the mean, and in terms of absolute error, to the median. Variability in such cases was taken as an indicator of the extent of measurement error. Both the mean deviation and standard deviation were proposed as measures of variability. However, instead of the S. D. σ, often the modulus of precision $1/\sqrt{(2)}\sigma$ was used. In fact, for a random variable X measuring a true value μ, Bessel proposed the median τ of $|X - \mu|$ as an index of measurement error and gave it the name 'probable error'. In the sequel, it was used only in the context of the normal distribution where $\tau = 0.6745\sigma$ (approx.). In the 19th century literature the probable error was a very frequently used measure. For a symmetric distribution τ is equal to the semi-inter-quartile range $\frac{1}{2}(Q_3 - Q_1)$ and, as we will see in the next chapter, in late 19th century in the case of empirical distributions based on a large body of data the probable error was often computed in this form. As regards nomenclature it took sometime to become standardized: Laplace used 'mean error' for the mean deviation about the mean, whereas Gauss used the same term for the S. D. (Incidentally, the term 'standard deviation' was coined by Karl Pearson towards the end of the 19th century. The term 'variance' for σ^2 came into vogue even later; it was introduced by Fisher in a 1918 paper dealing with genetic correlation.)

Development of approaches

8.6.3   As we saw in this as well as the preceding chapter, Laplace was instrumental in developing both the pro-subjective and the sampling-theory approaches to statistical induction, although with regard to the latter he confined himself only to large sample applications. As Laplace was avowedly a subjectivist in his view of probability, it seems that in setting large-sample interval estimates and tests he had what we have called 'instantial' interpretations (section 4.3) in mind. (But it would be too much to expect the distinction between the behavioural and instantial viewpoints to have been perceived in those early times.) As regards Gauss, he initially followed the pro-subjective approach for the study of linear models but later moved away from it and opted for the sampling-theory one. Although unlike Laplace he did not use sampling theory only in large sample situations, his investigations were limited almost exclusively to point estimation problems here. As regards the basic ideas of sampling-theory point estimation, we saw that concepts like consistency, unbiasedness, and ARE of estimators germinated in the work of Laplace and Gauss. An area which remained virtually unexplored was that covering interval estimation and testing in small samples from the sampling theory viewpoint. As we will see in later chapters, substantial progress here could be achieved only after the initial foray of 'Student' and the tremendous onslaught launched by Fisher in early 20th century. However, that would require the onset of that passing phase in the growth of statistical theory, when the normal model dominated the statistical scene in its role as an adequate representation of the population distribution.

It is noteworthy that the theory of errors was not totally impervious to the need of planning experiments to get sufficient accuracy for the ultimate results. This is illustrated implicitly by many of the applied investigations undertaken by Laplace. More concretely, in the book mentioned above, Airy considers the following 'variance components model' (according to modern terminology) for the jth observation y_{ij} in the ith night on an astronomical phenomenon:

$$Y_{ij} = \mu + c_i + e_{ij}, \quad i = 1, \ldots, I, \ j = 1, \ldots, J,$$

where μ is the true value, c_i is the random 'night-effect' and e_{ij} is the error. Assuming independence and homoscedasticity of the random components, he describes a crude test for the absence of night effects. Considering a similar model with a common S.D. σ_c for the c_is and σ_e for the e_{ij}s, a contemporary astronomer Chauvenet discusses how, when night effects are present, for the estimation of μ by the grand mean $\bar{y}_{..}$ (with Var $(\bar{y}_{..}) = (\sigma_c^2 + \sigma_e^2/J)/I$) it is pointless to increase J keeping the number I of nights small (see Scheffé (1956)).

Various models

8.6.4   Coming to the question of representation of the distribution of random variables encountered in practice, many particular models made their appearance during this time. As we have seen, the uniform, the triangular, the semicircular, the double exponential, the double logarithmic, and of course the normal model were proposed for representing the distribution of observational errors by various people in various contexts. Then there were distributions which were not initially formulated as models but were derived as combination of other distributions, or as limit or approximation to other distributions, or merely constructed to illustrate particular points. We saw how the beta distribution arose as the posterior distribution for the probability of an event in Bayes's work. The gamma distribution similarly arose as the posterior distribution of normal precision in the course of investigations by Gauss and Laplace. The utility of these distributions as models for graduation was, however, realized only towards the end of the 19th century mainly through the work of Karl Pearson and that we will consider in Chapter 10.

The most illustrious instance of a distribution arising as approximation to other distributions was of course the normal distribution itself. As we have seen, it first turned up as the limiting form of the binomial in the work of De Moivre, then as the limiting form of various posterior distributions in the work of Laplace, and finally in a big way in the CLT. Furthermore we have seen in subsection 8.2.6 that successive correction terms to the Central Limit approximation were obtained by expanding the operator within the integral in (8.2.21). The normal distribution together with these correction terms were later proposed as the basis of non-normal models for representing empirical distributions arising in practice. We will consider these developments in Chapter 10.

In another direction, seeking to show that the normal approximation to the binomial (n, p) distribution for large n may not hold if p varies with n, Poisson in 1830 considered the case when as $n \to \infty, p \to 0$ so that $np \to \lambda (0 < \lambda < \infty)$. In this way he derived the Poisson distribution for the number of rare events. (But very strangely he reached the distribution via what we now call the negative binomial distribution; see Stigler (1982).) This distribution, however, appeared implicitly in the work of De Moivre more than a century earlier as an approximation to the binomial in the context of a particular problem (see Hald (1998) pp. 214–215). Poisson did not consider any application of the distribution. It seems its first application as a model for the distribution of the number of stars was done in 1860 by the American statistician Simon Newcomb; see Stigler (1985). A more comprehensive application of the model was done later in 1898 by the Polish statistician Bortkiewicz to represent numbers of such rare events as death by horsekick in the Prussian army, birth of triplets etc; see Maistrov (1974) p. 160. (Incidentally, it was Bortkiewicz, who rather inappropriately called the Poisson distribution as 'the law of small numbers' and the name stuck for a long time.) Later in 1907 W. S. Gosset ('student') fitted the Poisson model to the distribution of the number of counts of yeast cells with a haemacytometer and this brought out the importance of the model in biological research; see Fisher (1944) p. 56. Poisson also considered the Cauchy distribution by way of deriving a counter example to Laplace's CLT (Cauchy himself considered the distribution as an illustration of a symmetric stable law about 30 years later).

| Families of random variables |

8.6.5   Coming to the treatment of the joint distribution of a family of random variables, we have already noticed that the bivariate and multivariate normal distributions appeared as the posterior joint distribution of the parameters of a linear model in the work of Gauss and as the limiting joint distribution of more than one statistic in the work of Laplace. (A particular case of it had appeared as the limit of the multinomial distribution even earlier in the work of Lagrange; see Hald (1998) p. 44.) But none of these authors seem to have realized the potential of the multivariate normal distribution as a model for representing the joint distribution of several random variables. They were mostly content to study the properties of univariate marginals derived from the joint distribution, although Laplace in exploring the scope for improving the LS estimator did obtain the distribution of a linear compound of two variables following asymptotically a bivariate normal distribution (subsection 8.4.2).

A more frontal investigation of the bivariate normal was done by Bravais in 1846; see Seal (1967). Considering n random variables ϵ_ℓ which independently follow the distribution $\mathcal{N}(0, \sigma_\ell^2), \ell = 1, \ldots, n$, Bravais considered the joint distribution of two linear compounds

$$X_1 = \sum_{\ell=1}^{n} a_{1\ell}\epsilon_\ell, \quad X_2 = \sum_{\ell=1}^{n} a_{2\ell}\epsilon_\ell, \tag{8.6.1}$$

where the vectors $(a_{i1}, \ldots, a_{in}), i = 1, 2$ are not linearly dependent. Introducing $n-2$ additional linear compounds of the ϵ_ℓs so that the whole set formed a nonsingular linear transformation, Bravais found the joint distribution of X_1 and X_2 in the familiar bivariate normal form by integrating out the additional variables in the joint distribution of the transformed variables. He studied the equidensity contours of the distribution and also found an expression for the probability outside a given such contour (as we now say, as an integral of the χ_2^2 distribution). He also partially extended the results to the case of trivariate normal. A similar study, with a fuller treatment of the three-dimensional case was done by Schols in 1875. But in none of these studies was the dependence among the variables examined in terms of measures like the correlation coefficient. In fact, these studies appear to have been prompted by the urge for theoretical generalization with no potential application prominently in view (although, as Seal (1967) points out, Schols did mention the possibility of using the bivariate normal for modelling the joint distribution of deviations in horizontal and vertical directions in the experiment of hitting a point target situated on a plane). As so often happens in the case of inventions that are born not out of the womb of necessity, these results did not attract much attention at the time.

It is remarkable that Laplace, in the same 1811 paper where he considered the asymptotic distribution of linear estimators of two or more coefficients in the linear model, also considered an uncountably infinite family of dependent random variables, or what now a days we call a *stochastic process in continuous time*. He started from the model of a continuing game of chance in which in each round of play a random interchange of single balls was made between two urns each containing n balls, half of all the balls in the two urns being white and the rest black. If $z_{x,r}$ be the probability that there will be exactly x white balls in the first urn after r rounds, one gets the partial difference equation

$$z_{x,n+1} = \left(\frac{x+1}{n}\right)^2 z_{x+1,r} + 2\frac{x}{n}\left(1 - \frac{x}{n}\right) z_{x,r} + \left(1 - \frac{x-1}{n}\right)^2 z_{x-1,r}. \tag{8.6.2}$$

Now setting $y = x/n, t = r/n, z_{x,r} = u(y,t), w = \sqrt{(4n)}(y - \frac{1}{2})$ and making $n \to \infty$, by the usual limiting process, (8.6.2) leads to the partial differential equation

$$\frac{\partial u}{\partial t} = 2u + 2w\frac{\partial u}{\partial w} + \frac{\partial^2 u}{\partial w^2}. \tag{8.6.3}$$

Here $\frac{1}{2}\sqrt{(n)}u\left(\frac{1}{2}+(w/\sqrt{(4n)}),t\right)$ is the density of w at 'time' t. Laplace derived the solution of (8.6.3) in the form of a series expansion in orthogonal polynomials, the leading term of which is the normal density irrespective of the distribution of w at time $t=0$. The implications of Laplace's result could be understood only after the physicists developed the theory of diffusion processes in the 20th century for their own purpose. (8.6.3) was then recognized as representing a Markovian diffusion process (see Hald (1998) pp. 337–343 for details).

| True value syndrome | **8.6.6** The era of the Theory of Errors which we have discussed in the present as well as the two preceding

chapters is marked by two inter-related features, one internal and the other external. Firstly, the methods of the theory of errors were all influenced by what we call the 'true value syndrome'. In astronomy and geodesy where these methods originated, generally the observations were measurements on certain true values. The primary object was to determine those true values and all the methods were geared to that. Variability arose mainly because of errors of measurement and although it had to be assessed, it was only because it was an unavoidable nuisance. Seldom was the study of variability undertaken for its own sake. Secondly, although the methods of the theory of errors within a short time became very popular with astronomers and geodesists all the world over, they were rarely applied outside these narrow fields. As Stigler (1986) p. 158 points out, the realization that these methods could be profitably used in other fields of study required the overcoming of a conceptual barrier. As we will see in the next chapter, this process started a little before the middle of the 19th century, and as it progressed the true value syndrome, although it persisted in the initial stages, was ultimately left behind.

9

BREAKING THE BARRIER:
OUT INTO A BROADER DOMAIN

Nature of conceptual barrier

9.1   At the end of the preceding chapter we mentioned that although by the first quarter of the 19th century a fairly wide repertoire of techniques for statistical induction had been developed under the name of Theory of Errors, a conceptual barrier seemed to limit their use only to a narrow area comprising the fields of astronomy, geodesy, etc. What exactly was the nature of this conceptual barrier and how was it broken?

In Chapter 1, in course of discussing the distinctive features of deductive and inductive inference, we observed that in deduction the reasoning is always formal depending on the core of the set of premises and we are concerned only with the formal validity of the conclusion—no question of its material validity beyond that of the premises is relevant. In the case of induction, however, apart from formal validity, material validity of the conclusion is of paramount importance. Furthermore, there is no core part of the set of premises, particularly of its empirical component representing the evidence. If information is subtracted from or added to it, the conclusion and its reliability may change. Nevertheless, if two problem situations are formally similar, generally it is agreed that the same methods of induction would be applicable to both. But the difficulty is that whether or not two inductive problems are formally similar often depends upon subjective judgement.

To elaborate the point further, recall 'the principle of total evidence' (section 1.4, footnote 5 in Chapter 1, subsection 4.3.4 and section 4.6) which, as we said, is of utmost importance in the context of induction. According to it, all *relevant* information available should be utilized in induction. But parts of the evidence which represent irrelevant details must be jettisoned—otherwise generally no process of induction can be conceived. (As the saying goes, 'If the map of England is as large as England, it is useless'!) Two problems of induction can be seen to be formally similar only after such irrelevant details are ignored. But which parts of the evidence are irrelevant has to be decided subjectively. There is not much difficulty in taking a decision when we have two problems of induction based on, say, data from two astronomical experiments. In such cases the different sets of observations often differ only in respect of their spatio-temporal and other like parameters which we are generally prepared to ignore (because of our innate belief in the uniformity of nature; see section 1.5). But when we try to match sets of data from, say, an astronomical experiment and an anthropo-metric study, or from two different anthropometric studies, it is quite a different proposition. The human beings who are observed in an anthropometric study have distinct identities of their own as represented by their names, places of res-idence, professions, social status, and so on and so forth. Unless these identities are obliterated and the observations are looked upon as pertaining to faceless members of some abstract population, no formal similarity can be established in such cases. Such an obliteration of identities requires a sort of conceptual *tour de force*. Historically, this was not undertaken before the 1820s.

We mention at this point that because such obliteration of identities to estab-lish formal similarity between different problems is subjective, there cannot be any final word about it. Thus, it may be that we would reach one solution to a problem of induction if all of certain identifying labels of the units of observation are ignored and another, if some of these are taken into account. In the theory of sample surveys, principles like post-stratification (cf. Example 4.2.2) and iden-tifiability of sampling units, really try to remedy the overdoing of the process of obliteration of labels (see e.g. Cassel *et al* (1977) Chapter 1).

In this chapter we are going to discuss roughly the developments in the period from around the 1830s to almost the end of the 19th century. As we will see, except towards the end of the period, so far as the mathematical techniques of statistical induction are concerned, very little was added that can be called really novel. However, during this period the conceptual frontier of the domain of stat-istics moved forward to a tremendous extent. What is more, the developments in statistics that took place at this time, created a sort of commotion in the thought-world of man, whose impact in various spheres has been a lasting one.

| 'Avalanche of numbers' | **9.2** What specific historical conditions led to the breaking of the conceptual barrier mentioned above |

after 1820? Ian Hacking (1990) has attributed this to the 'avalanche of printed numbers' that descended on many Western countries with the onset of the 19th century and then spread to other parts of the world. Let us examine the nature of this avalanche and the way it led to the expansion of statistical thought.

In Chapter 5 (subsection 5.4.3) we recounted how in the second half of the 17th century John Graunt in England superimposed certain plausible surmises on the data provided by the London bills of mortality to derive various conclusions relating to the underlying population. We also noted how Christiaan Huygens saw the underplay of probabilistic ideas in such studies and William Petty attempted an estimation of national wealth using similar data-based methods. After that there was a general recognition of the utility of such data and a movement, in which Petty and Leibniz (who evinced a lively interest in stat-istical studies relating to population) took prominent roles, to set up statistical offices and institute such studies in different countries. But the movement had only limited success. While data along the lines of London bills started being collected in many centres in Europe and statistical agencies to facilitate trade, taxation, and military recruitment were started in a number of countries in the 18th century, governments were often secretive and unwilling to divulge to the public information other than that relating to births, deaths, and marriages. The actuaries made studies to achieve their limited goals often by bringing in extraneous assumptions. Isolated comprehensive studies like those of Süssmilch in Germany and Sinclair in Scotland were outcomes of private enterprise.

The situation began to change towards the end of the 18th and the beginning of the 19th century. Regular periodic population censuses were started in many Western countries about that time and in some of the colonies some time later (Sweden 1750; United States 1790; Great Britain 1801; Prussia 1810; France 1836; Italy 1861; India 1881; Russia 1897). What is significant is that in the post-Napoleonic period of relative peace the inhibition about the publication of official statistics gradually disappeared. Various statistical reports giving numbers in classified form were regularly brought out in print by census offices and other government departments. Notable among these were the annual *Recherches Statistiques* of Paris edited by Joseph Fourier which, apart from giving the usual numbers of births, deaths, and marriages, also gave classified data relating to the number of cases of insanity, suicides, etc. Similar classified data on judicial cases and convictions for France were published by the Ministry of Justice. A little later data relating to births, deaths, and sickness for England and Wales started being published from the Registrar General's Office under the guidance of William Farr.

The process of classification is a great equalizer. When within a group we classify an individual, we regard him as just a member of the class to which he belongs and forget his other identities. (For this reason there was some hesitation about classifying the bourgeois citizens along with the ordinary people in a census carried out even in Napoleonic France and in the early censuses in the United States a black person was counted as $\frac{3}{5}$ of a white person; see Gigerenzer *et al* (1989) p. 252 and Hacking (1990) p. 17. But with the democratization of society such prejudices vanished in due course.) Thus, presentation of the outcomes of enumeration in a classified form with a frequency table replacing the original series itself foreshadowed the breaking of the conceptual barrier mentioned above. From another view point, this exemplified one of the features characterizing statistical induction which we emphasized in Chapter 2 (section 2.2), namely, regarding classes with their frequencies as the basic data.

What is significant is that in the 19th century the profusion of tabulated numbers became available for different countries more or less with regular periodicity—the 'avalanche' was not a one-time or sporadic phenomenon but a sustained sequence. This brought, what in Chapter 2 we called 'frequential regularity', with regard to various events relating to human affairs to the notice of people. The stability of demographic indices like the sex-ratio at birth or the mortality rates at different ages had been generally known since a long time. But that knowledge related to a few demographic events and did not have any wider impact. Now it was found that such stability ruled the day with regard to marriages, divorces, causes of insanity and causes of suicides, types of crimes, and even the number of dead letters in the Paris postal system. The constancy of the corresponding ratios and proportions (and numbers, since over a short span of time, the denominators usually did not vary much) were virtually rubbed into the eyes of anyone who cared to observe. Of those who made significant studies of such regularities around the 30s and 40s of the 19th century, the names of two—Poisson and Quetelet—deserve special mention, keeping in mind the impact their studies had on the course of statistical thought.

| Poisson |

9.3   We first take up the studies undertaken by Poisson, mainly because these can be considered as direct application of the theory developed by Laplace. Poisson's study of the stability of relative frequencies, as contained in an 1830 paper, related to the sex-ratios at birth for the years 1817–1826, as available for the whole of, and also for the different administrative districts of France. It was extended in a book published in 1837. Using Laplacian posteriorist methods, Poisson reached the conclusion that the proportion of male births for the whole of France remained stable over the ten years, but the proportion varied across the districts for a year and over the years for a particular district. If each child birth is regarded as a Bernoulli trial, this meant that the probability of a male birth for the years and districts could not be taken to be same and yet there was no instability in the overall proportion of male births. To resolve this dilemma, Poisson proved a result which he called 'the Law of Large Numbers' (cf. section 3.9).

Consider a sequence of independent Bernoulli trials with success probabilities $p_\ell = 1 - q_\ell, \ell = 1, 2, \ldots, n \ldots$ If $\hat{p}_n$ denotes the proportion of successes in the first n trials and $\bar{p}_n = \sum_{\ell=1}^{n} p_\ell / n$, Poisson showed that as $n \to \infty$,

$$\hat{p}_n - \bar{p}_n \xrightarrow{P} 0. \tag{9.3.1}$$

(Poisson used the CLT for non-identically distributed random variables to derive the result, but we know (9.3.1) follows without any restriction on the p_ℓs as a consequence of Tshebyshev's inequality (cf. subsection 5.6.2).) Of course Poisson's intention was to show that $\hat{p}_n$ becomes stable for large n. For this he assumed in effect (see Hald (1998) p. 576) a two-stage model under which the trials were themselves randomly selected from an underlying population of trials. (Poisson took the distribution of p_ℓs in the population to be discrete with a finite support, but this is inessential.) Of course as it was pointed out even at that time, this amounts to taking the sequence of trials to be unconditionally identical with the common success probability $E(p_\ell)$ so that Bernoulli's Theorem (section 5.6.2) would apply straightaway. But Poisson's model allowed him to explain why, say, the proportions of male births in the whole of France and in a particular district differed from each other. If the population of p_ℓs is divided into a number of strata and the overall population mean is different from the mean of a particular stratum, the relative frequency for a sample of trials taken from the whole population would tend to differ from that for the subsample of trials belonging to the stratum. Of course conceptually the assumption of a fixed distribution of p_ℓs amounts to assuming implicitly 'the uniformity of nature' (section 1.5).

Actually, although Poisson did not put it that way, the above model provided a sort of answer to the objection against sampling vis-à-vis complete enumeration that had been raised and canvassed vigorously since a few years earlier and that had resulted in stalling the introduction of sample surveys for many a decade. The objection was clearly articulated by Baron de Keverberg, an advisor to the Belgian Government in a letter to Quetelet sent around 1825 (see Stigler (1986) pp. 164–165), in which he questioned the validity of Laplace's sampling method for population estimation (subsection 7.3.9). The burden of the criticism was that while it would be legitimate to estimate the crude birth (or death) rate in a homogeneous stratum of the population on the basis of a (random) sample, in actual practice the population of a country is the union of a very large number of such strata. It would be impossible to take account of all these strata separately and preposterous to assume a stable rate for the aggregate population. Obviously, Poisson's model allows one to get round this difficulty.

In his 1837 book Poisson also analysed the judicial statistics for the years 1825–1833 published by the Ministry of Justice of France along similar lines and reached similar conclusions. Thus he found that the proportion convicted among those accused (conviction rate) remained stable for the whole of France during the period 1825–1830 in which the same system of trial by jury was followed, although there were differences in the rates between different districts and between 'crimes against person' and 'crimes against property'. As in the case of birth statistics, this type of situation could be explained by presuming that the individual rates, although heterogeneous, followed a stable pattern represented by a probability distribution.

After 1830 there was a change in the system of trial by jury: whereas upto 1830 essentially the verdict was by a simple majority among 12 jurors, in 1831 a majority of at least 8 to 4 was required. Apparently due to this there was a significant drop in the conviction rate. Poisson assumed that over the entire period the probability of an accused person being guilty had a fixed value κ and the jurors, independently of whether an accused person was guilty or innocent and independently of each other, gave the correct verdict with a fixed probability θ. (Actually here Poisson considered the unconditional expectations of the underlying distributions of probabilities as κ and θ.) This means when for an accused person to be convicted at least r out of n jurors were required to give the guilty verdict, the probability of conviction of such a person was

$$P_r(\kappa, \theta) = \kappa \sum_{i=r}^{n} \binom{n}{i} \theta^i (1-\theta)^{n-i} + (1-\kappa) \sum_{i=r}^{n} \binom{n}{i} (1-\theta)^i \theta^{n-i}. \qquad (9.3.2)$$

This model gave two different expressions for the theoretical probability of an accused person being guilty under the 1825–30 and 1831 systems. Equating these to the corresponding empirical conviction rates gave two equations in κ and θ solving which Poisson could estimate κ and θ (see Hald (1998) pp. 584–585). Hence he could estimate the probability of an erroneous verdict under each of the two systems and judge their reliability. It should be noted that in spirit this part of Poisson's study followed the sampling theory approach with κ and θ regarded as fixed unknowns.

Although Poisson conceived the Law of Large Numbers originally in the context of independent Bernoulli random variables, he visualized the more general result applicable to any sequence of independent random variables X_ℓ with $E(X_\ell) = \mu_\ell, \ell = 1, 2, \ldots, n \ldots$ (cf. section 3.9). Writing, $\bar{X}_n = 1/n\sum_1^n X_\ell, \bar{\mu}_n = 1/n\sum_1^n \mu_\ell$, here one has (generally under suitable regularity conditions)

$$\bar{X}_n - \bar{\mu}_n \xrightarrow{P} 0.$$

Stability of $\bar{\mu}_n$ would be assured if one assumes as before the μ_ℓs are randomly selected from an underlying population. Specifically, Poisson mentioned that such stability on the average would hold for gain from maritime insurance, length of life, income from taxes and lotteries, level of the sea, distance between molecules, etc. He thought that there would be a value specific in every such case to which $\bar{\mu}_n$ would approach as the number of observations increased.

9.4   Adolphe Quetelet

Quetelet's objective | **9.4.1**   It was Adolphe Quetelet, a Belgian scientist, who was in the main instrumental in breaking the conceptual barrier mentioned earlier. By training and profession, he was a mathematician–cum–astronomer. But he had a life-long interest in the arts and humanities. During a brief visit to Paris in 1823 with a view to learning the techniques of practical astronomy, he came under the influence of Fourier and Laplace and had some exposure to the basics of probability and the theory of errors. He then conceived the idea of employing the methods of theory of errors for the study of human populations.

Quetelet studied data relating to different human groups gleaned from various official and other publications, medical reports, military records, and also from a number of correspondents in several countries. Broadly speaking, his studies covered three kinds of data describing aspects of human populations: (a) data about natural events like births, deaths, sickness etc., (b) data about manifestations of human behaviour like marriages, drunkenness, duels, suicides, different types of crimes and so on, and (c) anthropometric measurements like those on height, weight, strength, pulse rate, etc. Starting from 1827 he published a series of memoirs dealing with these topics and consolidated his findings in two books one of which first came out in 1835 and the other in 1846.

Although Poisson and Quetelet dealt with somewhat similar sets and sometimes (as in the case of French judicial statistics) the same set of data, their attitudes and approaches were markedly different. As noted above, Poisson approached the problem of data-analysis from the point of view of a probabilist. To him each observation (e.g. about the gender of a newborn child or the judicial verdict on an accused person) was the outcome of a random experiment (subsection 3.2.1). When he tested homogeneity, he tested whether different kinds of trials could be taken as identical. He set up a model to explain why, even when a set of trials is palpably heterogeneous, the overall relative frequencies and averages based on them are stable. While studying judicial data, he postulated a model generating stable conviction rates and on its basis attempted an evaluation of the efficacy of the judicial system.

Quetelet, on the other hand, approached data-analysis from the point of view of, what we would now call a social and physical anthropologist. He was interested primarily in the study of society and invoked probabilistic ideas and tools solely for that purpose. To him a stable relative frequency or average relating to a social group represented a characteristic of the group and variations in such characteristics threw light on the laws in accordance with which inner movements in society took place.

| The 'Average Man' | **9.4.2** In the case of data on attributes like death, birth, marriage, drunkenness, suicide, crime, etc., |

Quetelet concentrated on the relative frequency of occurrence of the event in the human group studied. He presumed that for a homogeneous group such a relative frequency was stable and represented a characteristic of the group. In the case of variables like height, weight, etc., he similarly considered the average and took it as representing a stable group characteristic. Thus for any homogeneous group there was a set of relative frequencies and averages corresponding to various attributes and variates. Quetelet postulated that the set represented an imaginary typical member whom he called the 'average man' ('*l'homme moyen*') for the group considered. All this was introduced in his 1835 book.

In his 1846 book Quetelet bolstered the concept of the 'average man' further by studying the full relative frequency distribution in the case of certain variates. Specifically, he had data on the chest measurements of nearly 6000 Scottish soldiers collected by a military tailoring contractor. The relative frequency distribution based on the data had a form resembling the normal law of errors. Quetelet fitted a normal curve to the empirical distribution by using terms of the symmetric binomial distribution with $n = 499$.[1] (Arguably, this was the first instance of fitting a normal curve to an empirical distribution not related to errors of observations. As mentioned in the preceding chapter (subsection 8.2.8), Bessel had fitted the normal curve to residuals based on repeated observations on certain astronomical quantities, some years earlier.)[2] Inspection

---

[1]Tables of the normal distribution were already available in the published literature, but Quetelet was initially ignorant of their existence.

[2]Unlike Bessel whose method of fitting was straightforward, Quetelet fitted the normal curve indirectly by a sort of 'modified probit analysis'. Corresponding to each

of the observed and fitted relative frequencies showed the fit to be satisfactory. Quetelet made a number of other similar studies, notably one on the frequency distribution of heights of 100,000 French conscripts. He found that in almost all cases where the group was homogeneous, the normal curve fitted the empirical distribution well.

Quetelet interpreted the fact of approximate normality of variates such as chest measurements and heights of the men in a group as follows: if there were a *real* 'average man' and one were repeatedly measuring him, the different measurements would follow approximately a normal distribution—this follows from the 'postulate of elementary errors' (subsection 8.2.8) (Quetelet supposed the 'elementary errors' were i.i.d. 0–1 variables, so that approximate normality followed via the binomial model). In that case the mean of the observations would be an estimate of the true value for the real 'average man'. Since, when measurements were made on the different members of the group, normality still held, Quetelet postulated the existence of an underlying true value estimated by the observed mean and of a *conceptual* 'average man' to whom that true value belonged. The variation in the observations on the individuals of the group, Quetelet attributed to 'accidental causes'. The stable mean was supposed to be the effect of 'constant causes'. Nature, as it were, was trying to attain a targeted true value but was baffled in her objective due to accidental causes.

Quetelet supposed that even non-measurable normal characters such as propensity ('penchant') for acts like marriage, drunkenness, suicide, or crime were similarly distributed over the group around a true value representing the 'average man' in the group. However, he did not pursue this idea much further (we will see later that Galton used this idea for scaling different grades of non-measurable traits using their frequencies). Instead, Quetelet simply interpreted the rate or relative frequency with which an act was executed or perpetrated by members of a group as representing the propensity for that act of an 'average man' of the group. (He made statements like 'This probability (0.0884) can be considered as a measure of the apparent inclination to marriage of a city-dwelling Belgian'; see Maistrov (1974) p. 161.)

| Breaking the barrier | **9.4.3** From the above it is clear that Quetelet decisively imposed the process of 'obliteration of labels' |

mentioned earlier for studying social groups, or as they came to be called later, *populations*. His interest primarily was the study of the underlying population—the observations on different individuals were looked upon as interchangeable readings on the same population which was supposed to be a physical reality.

---

class he determined the value of a normal variable (the 'modified probit') with mean 0.5 and S. D. $\frac{1}{2}\sqrt{499}$ having the same cumulative relative frequency and then by trial fitted a linear formula expressing the modified probits in terms of the class-values. Thus he was able to tackle the problem of open end-classes. (The usual probit, it may be recalled, corresponds to mean 5 and S. D. 1; see Finney (1952).)

As it happened, within a few decades, the concept of population developed into a cardinal concept of statistics. (It became so very natural for the statistical psyche that Fisher about a century later supposed that there exists a conceptual or hypothetical population associated with every repeatable random experiment; see subsection 3.2.3.)

On the other hand, obviously Quetelet still remained mentally bound by what in the preceding chapter (subsection 8.6.6) we called 'the true-value syndrome'. The population was identified with the set of true values of the characters under study, representing 'the average man'. Just as in the theory of errors, variation of the characters within the population was regarded as of secondary importance. (As we will see later, the crucial step of breaking the syndrome was taken by Galton and his followers in the last quarter of the 19th century.)

The grotesqueness of Quetelet's idea of taking arithmetic means and proportions to characterize the 'average man' was criticized even at that time. Thus Cournot remarked that if one took the means of the sides of a number of right-angled triangles one would not in general get a right-angled triangle. Even stranger was Quetelet's suggestion that the 'average man' being the target of nature, represented the 'ideal man' in all his aspects, both physical and moral ('all our qualities', he wrote, 'in their greatest deviations from the mean produce only vices'; see Porter (1986) p. 103) and the progress of civilization meant bringing all members of society closer and closer to that ideal. As we will see later, directly in opposition to this idea was the thesis of the eugenists led by Galton, who put forth that the ideal of society should be to achieve all-round excellence and not to gravitate towards mediocrity.

| Social physics | 9.4.4 Quetelet's objective in defining an 'average man' was to unravel the laws according to which the |

profile of the 'average man' changed with different factors. Laws describing changes in the rates of birth and death with age had been of interest since the middle of the 17th century (subsection 5.4.2). In the late 18th century, the philosopher Kant had spoken about historical laws governing the long-term movement of birth, death, and marriage rates over time; see Hacking (1990) p. 15. But it was Quetelet's astronomical background which led him to visualize the existence of deterministic social laws connecting the variation in the profile of the average man in all its aspects—both physical and moral—with that in different factors. He asserted 'facts of the moral order are subject like those of the physical order to invariable laws' (see Hacking (1990) p. 73). He called this 'social physics'. In fact his 1835 book was subtitled 'Essai de Physique Sociale'. For identifying the laws he classified the population into different homogeneous groups or subpopulations according to the levels of the factor of interest and taking the rate or average corresponding to the 'average man' of each sub-population, studied how these varied with the factor-level.

For a measurable factor like age, temperature, height, etc., Quetelet generally sought to represent the variation in terms of a mathematical formula (cf. Hald (1998) p. 589). Thus when the factor was age and the characteristic of the average man studied was height, he found (separately for males and females) the average height $y = y_x$ corresponding to each age x, starting from $x = 0$ (birth) to $x = 30$ (maturity). He then determined empirically the equation

$$y + \frac{y}{1000(T - y)} = ax + \frac{t + x}{1 + \frac{4}{3}x}, \qquad (9.4.1)$$

where t and T are the average heights at birth and maturity and the appropriate value of the parameter a was found by studying the data. Another example is the relation between average weight w and average height h for which he found relations of the type

$$\begin{aligned}
w &= \text{const. } h^{5/2} \quad \text{in the growing period} \\
&= \text{const. } h^2 \quad \text{after maturity.}
\end{aligned} \qquad (9.4.2)$$

When the classifying factor was qualitative, like locality, gender, profession, social group, etc., Quetelet generally considered the order and relative values of the rates or averages corresponding to the different levels of the factor. Thus considering the data on crimes, he assumed that in any group the number of crimes committed bears a constant ratio to the number of accused so that the ratio of the number of accused to the size of the group would represent the propensity to crime. Generally taking the denominators as fixed he compared this propensity over different genders, level of education, etc. in terms of the absolute numbers (see Hald (1998) pp. 589–599). He also made more detailed analyses of the influence of various qualitative factors on the conviction rates (proportions of the accused convicted) based on French judicial statistics for the years 1825–1830 (recall that these also formed part of the data studied by Poisson at about the same time). In this case he examined qualitatively the stability of the rate for a particular group (say, females) over the different years, as also the homogeneity of the rates for different groups (say, males vs. females). He also suggested a 'measure of influence' of a state on the conviction rate. Thus if r_f denotes the rate for females and $\bar{r}$ the overall rate, the influence of gender was measured by the relative discrepancy $|r_f - \bar{r}|/\bar{r}$. Similar measures were computed for classes based on other factors such as level of education, type of crime, etc. (see Stigler (1986) pp. 176–177). Of course these conviction rates no longer represented the propensity to crime but were indicative of the efficacies of the agencies responsible for enforcing law and administering justice—'repression of crime' according to Quetelet.

The variation in a characteristic of the 'average man' over different groups was attributed by Quetelet to 'variable' causes (varying with the factors) acting on 'constant' ones. (Variation around the average value within a group, it may be recalled, was attributed to 'accidental' causes.)

Nature of Quetelet's studies

9.4.5 Can the kind of study performed by Quetelet be called statistical induction? Quetelet, it may be mentioned, was thoroughly convinced about the argument against sampling advanced by Baron de Keverberg to which we referred in the context of Poisson. He always sought to base his studies, as far as possible, on complete censuses and insisted that 'we must not make a selection, we must take all the men of the nation as they are' (Hald (1998) p. 597). But on the other hand, Quetelet argued that taking rates (relative frequencies) and averages eliminated accidental causes and interpreted rates as probabilities (Stigler (1986) p. 175). This bespoke the idea of an underlying superpopulation: otherwise his assertion that 'the precision of the results increases as the square root of the number of observation' (see Stigler (1986) p. 180) becomes meaningless. Thus Quetelet's investigations were in essence instances of covert induction (cf. section 1.2).

Quetelet—the organizer

9.4.6 Quetelet contributed to the development of statistics in other ways also apart from through his research investigations. He participated in 1833 in the meeting of the British Association for the Advancement of Science to present the results of his studies and took a leading role in the formation of a separate Statistics section in the Association. Shortly after that and immediately as a result of this campaign, the Statistical Society of London, progenitor of the present Royal Statistical Society, came into existence. In 1841 Quetelet organized the Belgian Central Statistical Commission of which he became the President. He conceived the idea of holding International Statistical Congresses regularly to standardize the terminology and methods of collection of official statistics in different countries and make them internationally comparable. The first such Congress was held in Brussels in 1853 under the leadership of Quetelet; the fourth in which also Quetelet was a prominent participant, took place in London in 1860. These Congresses were the forerunners of the permanent body known as the International Statistical Institute which was established in 1885.

The opening up which followed the breaking of the conceptual barrier and the statistical study of human populations by Quetelet and others, not only had implications for the development of statistics but influenced the thinking of people as regards social sciences, and at a more subtle level, philosophy and science in general. We describe these in the next section.

9.5   Aftermath of the opening up

Implications for statistics

9.5.1 Application of the methods of statistics for the analysis of data relating to human populations did not immediately lead to the development of any novel techniques, but it gave rise to certain new kinds inductive problems. Most prominent among these was the problem of judging the homogeneity of a group which is potentially divisible into more than one component subgroup, often in more than one way. In the case of data forming the grist for classical theory of errors, say astronomical data, a group of observations relates to some particular physical phenomenon and different groups correspond to different such phenomena. Generally there is unanimity about the grouping system. In the case of social data on the other hand, as noted above, groups are formed by ignoring various labels characterizing individuals; whether a collection represents a single group or the aggregate of several groups depends on what characteristics we choose to ignore. It is legitimate to ignore characteristics like gender, locality, time period, etc. as long as the homogeneity of the resulting group holds (and of course when homogeneity holds these factors cannot hold any interest for us). There is no underlying physical reality that determines the homogeneity of a group of observations; rather, a conceptual reality like an 'average man' underlying a group of observations is postulated when the group can be taken as homogeneous. Thus it is of paramount importance to judge whether a group of observations is or is not homogeneous on the basis of the internal evidence of the observations themselves.

In the case of a binary *attribute* the simplest interpretation of homogeneity of a collection of groups is that each group represents a set of identical Bernoulli trials and the probabilities of success are same for all the groups. As we saw earlier, Poisson looked at the problem from this angle (in the case of attributes like the gender of a newborn child or the verdict given on an accused) and sought to judge homogeneity across different years and provinces by Laplacian posteriorist methods. Essentially, he found a posteriorist interval estimate for the common probability under the assumption of homogeneity and checked how far the individual group proportions were contained in that interval. Apparently this was not found very convincing, for there were few instances of others using the same kind of reasoning. From Quetelet's writings it seems he understood homogeneity in a broader sense and allowed the trials in each group to be dependent with a common marginal success probability (see Lazarsfeld (1961)). But he judged the homogeneity of groups crudely essentially by inspection of the data and did not advance any probabilistic argument for that.

The problem of homogeneity in the context of social data of course could not be swept away. Sporadic attempts to tackle the problem continued to be made in the second half of the 19th century. The most comprehensive of these was made by the German statistician Wilhelm Lexis. To compare the probabilities p_i, $i = 1, \ldots, k$ for k groups of internally homogeneous Bernoulli trials of same size n, Lexis' idea essentially was to find estimates of the variance of the corresponding observed proportions $\hat{p}_i$ in two different ways and take their ratio. Writing

$$V = \frac{1}{k} \sum_1^k (\hat{p}_i - \bar{p})^2, \quad \bar{p} = \frac{1}{k} \sum_1^k p_i, \quad \bar{q} = 1 - \bar{p}, \quad (9.5.1)$$

$$Q = \frac{V}{(\bar{p}\bar{q}/n)}, \quad (9.5.2)$$

it is readily seen that

$$EQ = 1 + \frac{(n-1)}{k\bar{p}\bar{q}} \sum_1^k (p_i - \bar{p})^2. \tag{9.5.3}$$

Lexis proposed the use of $\hat{Q}$ obtained by substituting $\hat{\bar{p}} = \sum \hat{p}_i/k$ in (9.5.1)–(9.5.2) in place of $\bar{p}$ in Q.

He argued that in view of (9.5.3) if $\hat{Q}$ is close to 1 we may conclude that the groups are homogeneous, whereas a value of $\hat{Q}$ much larger than 1 (here Lexis suggested the ad hoc threshold 2) would indicate that the p_is are not all same. If $\hat{Q}$ is much smaller than 1, it would suggest that the binomial model is not valid for the individual groups—possibly the trials within each group are dependent. (But there are schemes of dependent trials also for which EQ is 1; see Uspensky (1937) p. 215. Stigler (1986) pp. 235–236 has remarked that over-dependence on the assumption of binomiality of the group proportions was the main weakness of Lexis' approach. But the same criticism applies to the χ^2-test for homogeneity for a $2 \times k$ table—the first more or less satisfactory test for homogeneity—which was originally proposed by Karl Pearson in 1911 (see Chapter 10). In fact, it seems resolution of the problem under some general model which covers all forms of attribute data would prove elusive even to a present-day statistician with his rich repertoire of techniques.)

When the observations were on a continuous *variate* like height, weight, etc., Quetelet suggested that the homogeneity of a group of observations may be judged in terms of the normality of the resulting distribution. The argument was similar to that advanced by Quetelet to support the idea of an 'average man'. For a homogeneous group such as that comprising repeated observations on the same astronomical phenomenon there is one constant cause and the variation is the outcome of the interplay of innumerable accidental causes. Because of this the observed frequency distribution is close to normal. Hence Quetelet postulated that whenever the latter condition holds for the observed distribution of a variate, we can presume that apart from accidental causes there is only one constant cause and therefore the group is homogeneous. Apart from the fallacy inherent in this kind of backward reasoning—misinterpreting a necessary condition as sufficient—the rule was patently self-contradictory. For, although it was not clearly perceived by Quetelet and his contemporaries, a normal distribution can be produced by mixing a large number of component normal distributions in appropriate proportions so that normality can hardly be considered as a determining feature of homogeneity. Most empirical distributions of continuous variates that Quetelet and others studied at that time (and quite a few such distributions including one on an exotic variate relating to the orientation of mountain ranges by a physicist named Spottiswoode were studied; see Stigler (1986) p. 219) turned out to be more or less close to normal. As a result it was falsely presumed that all the corresponding data related to homogeneous groups. Also, as normality apparently held in most cases encountered at that time and homogeneity was taken as virtually synonymous with normality, the myth grew that whenever there was no tangible source of heterogeneity and the variate was continuous, normality would rule the day. This presumption of universal normality was derisively called by skeptics *Quetelismus* or *Queteletismus* (see Stigler (1986) p. 203, Hacking (1990) pp. 113, 125–127). As we will see, the presumption held sway over statistical minds at least upto the time of Galton.

One very important problem of statistics that was implicit in Quetelet's criterion for judging homogeneity was that of testing goodness of fit of the normal distribution. The problem remained incipient for a few decades until, as we will see in the next chapter, towards the end of the 19th century it literally cried for solution. It was tackled in a more general context by Karl Pearson at the turn of the century.

Quetelet's method of studying the dependence of one variate on another as exemplified by (9.4.1) (height on age) and (9.4.2) (weight on height) in a population by classifying it by the values of the latter, was one of the earliest ones of its kind in statistics. Looking back, it seems that it might have influenced the later study of Galton (to be considered in section 9.7) of the dependence of measurements of offsprings on those of parents.

| Implications for social sciences |

9.5.2   Quetelet's kind of study generated a good deal of interest among social scientists and historians in the second half of the 19th century. Quetelet's thesis that society as distinct from the State had an identity of its own and the behaviour of society as a whole followed definite laws although the individual members of society might behave erratically, was widely accepted. An Englishman named Buckle carried Quetelet's ideas to the extreme and wrote a history of civilization in which it was sought to be established that the course of human societies was rigidly determined by conditions of climate and locality. There were also a fair number of social scientists who tried to draw sociological conclusions on the basis of the regularities revealed by statistics (see Hacking (1990) pp. 125–126, Porter (1986) pp. 68–69). Quetelet's limited use of probabilistic models to depict internal variation in human populations also opened the way for later development of more comprehensive stochastic models for human behaviour. Further, Ernst Engel's study of family budgets (made around 1860) to characterize the average behaviour of families as regards budgetary allocation in relation to income, which assumed importance for indexing cost of living, must have been influenced by Queteletian ideas. It is tempting to speculate that the revival of the inductive statistical approach for the development of macroeconomic theory in early 20th century (which after the pioneering work of people like Petty (subsection 5.4.6) was neglected for a long interregnum during which deductive studies dominated the field of economics) was also indirectly due to Quetelet's influence.

| Social activism |

The progress of human civilization is marked by collecting more and more information about nature, discovering laws about natural processes from such information and controlling nature on their basis. The presence of law-like regularities in social data, which were discovered by Quetelet, opened up the scope for improving society by making use of such knowledge . Hacking (1990) p. 2 makes the interesting point that dissemination of knowledge about the prevailing pattern in society tends by itself to change society in a subtle way, as those members who belong to the pathological or deficient classes strive to pull themselves to the 'normal' classes. Legistative and administrative intervention of course may expedite the process,but such interventions can be effective only if they draw upon the underlying laws to manipulate the causes. Quetelet himself emphasized this when in the context of reduction of criminality in society he wrote, 'since the crimes that are annually committed seem to be a necessary result of our social organization and since their number cannot be diminished without the causes inducing them first be modified, it is the task of the legislators to ascertain the causes and to remove them as far as possible...' (see Hald (1998) p. 592).

One of the most conspicuous 19th century converts to this kind of statistics-based social activism was Florence Nightingale. Starting with her crusade against the unhealthy conditions in hospitals to improve the lot of the sick and the wounded, she zealously pursued various programmes for reduction of morbidity, spread of education, removal of causes of famine, and generally for the welfare of society. And in every step in her course she masterfully garnered statistics to strengthen her case with administrators and achieve her objective.

It may be mentioned that the prevailing intellectual climate in Europe, and in particular in England, helped to foster this kind of social activism. In philosophy these were the days when positivism began to sprout. Along with that, utilitarianism set the objective of 'the greatest good of the greatest number' as an ideal to be followed. Statistical laws of society, so to say, formed the link between positivism and utilitarianism.

This phase of growth of statistical thought, in various forms, has lasted to this day. The role of statistics for planning and implementing welfare and development programmes (with objectives such as controlling unemployment, reducing poverty and other kinds of deprivation, and generally improving the quality of life of people) in both developed and developing countries is now well-established. The statistical data required for this are sometimes collected through large-scale sample surveys and sometimes through complete censuses, but in all cases induction, overt or covert, is involved—an element of uncertainty about the basis and success of such programmes is always present.

Implications for philosophy

9.5.3 Quetelet's work, within a few years started intense philosophical debates at two levels—we may say one of these related to the metaphysical and the other to the physical universe. And strangely, these challenged traditional beliefs and seemed to pull them in two diametrically opposite directions. We note these in this subsection briefly; for although of crucial importance from the point of view of philosophy, they had only tangential bearing on the growth of statistical thought. (Detailed discussions of these issues are given e.g., in Porter (1986) Chapters 5–7.)

| Deterministic social laws |

From the early times many philosophers believed that, although events in external nature might be subject to deterministic laws, human will enjoyed a degree of freedom. This was the view endorsed among others by the Platonists, Descartes, and Kant. True, the stoics and philosophers like Spinoza and Hegel were proponents of full-scale determinism encompassing even events in the mental world. But this view was not based on factual studies and there was a general belief that unless an individual had free will, he could not be held responsible for his actions and as such there could not be any basis for ethics and rule of law. Quetelet's studies revealed that at the macro-level, not only natural events like births and deaths, but even acts of volition like marriages, suicides, and crimes were subject to regular laws. This seemed to put curbs on individual free will. As Quetelet put it, 'experience proves... that it is society which prepares the crime and the guilty is only the instrument by which it is executed' (see Hald (1998) p. 592). Buckle, whom we have mentioned earlier, in course of popularizing Quetelet's ideas carried it to the extreme and negated the existence of any degree of freedom for individual will. He asserted that since there are deterministic laws charting the course of behaviour of society as a whole there must exist special laws determining individual behaviour in conformity with those. At this point we may note that in Christian theology too there have been two schools, one (generally represented by the Roman Catholics) preaching that faith of an individual is a deliberate and free meritorious act on his part and the other (generally represented by the Protestants, Jansenites, and Calvinists) that everything is predestined. But both the schools agree that God is free to grant an individual salvation or damnation. Buckle's 'mad fatalism' did not even leave room for Divine freedom.[3]

Broadly three kinds of resolutions of the dilemma were put forth (see Porter (1986) pp. 162–171 and Hacking (1990) pp. 127–132). One was to go with Buckle and deny the individual any kind of freedom. The second was what some Germans advanced at that time and presumably what would be endorsed by positivists like Karl Pearson some years later: the statistical regularities exhibited by society were mere uniformities and not 'causal laws' and therefore did not constrain the behaviour of individuals in any way. The third answer, which was vaguely articulated at that time but which superficially we can now look upon as an extension of Heisenberg's 'Uncertainty Principle' to philosophy, was that one cannot study the behaviour of the individual and the ensemble in one

---

[3]Interestingly, according to Indian Vedantic philosophy, the 'Absolute' which underlies the 'will' is free, but the 'will' is not free. The 'will' comes into existence, so to speak, when the 'Absolute' is viewed through the prism of time, space, and causation and therefore cannot but be subject to causal laws; see Swami Vivekananda, *Complete Works*, Vol. II, pp. 130–132, Advaita Ashrama, Calcutta, 14th edn. (1972).

breath—the two belong to different planes. (We can say that the 'gambler's fallacy' according to which a symmetric coin which has come up heads in the first three throws is more likely to show a tail in the fourth, is the outcome of a similar conflation.)

Physical indeterminism   The debate at the second level referred to the external physical world and began when first Maxwell and then Boltzmann developed their kinetic theory of gases. This was just after Quetelet's work had gained wide currency. At that time physicists had reached a dead-end trying to explain the behaviour of a body of gas under heat transmission. Maxwell became acquainted with Quetelet's work from an 1850 review by John Herschel and the exposition by Buckle. He seized upon Quetelet's idea that an ensemble can exhibit regularities in behaviour even though its individual members behave erratically. Regarding a body of gas as an ensemble of molecules and starting from postulates similar to those of Herschel (subsection 8.2.9), he derived the probability (i.e. relative frequency) distribution of their velocities and hence derived important conclusions. Boltzmann proceeded still further and interpreted entropy as a derivative of the distribution and explained the Second Law of Thermodynamics in terms of it. The basic idea was that systems tend to move from a less to a more probable state and this means that the entropy tends to increase (Maistrov (1974) pp. 227–228).

At first sight it looks a bit odd that Maxwell and Boltzmann reached their probability models for molecular behaviour via social sciences and not directly from the theory of errors which seems more congenial to physics. But in the theory of errors there are no separate physical entities corresponding to the individual observations, which constitute a collection; the parallel between a society of humans and a body of gas molecules in this respect is clear-cut. Also in the Laplacian thought world, the fundamental concept of probability was that of degree of belief; the identification of probability with relative frequency occurs naturally in the case of human beings and gas molecules.

The introduction of probability models for gas molecules marked the advent of an indeterministic world view in physics. Laplacian tradition, it may be recalled (subsection 7.3.4) considered phenomena in the world to be basically controlled by rigid deterministic laws—whether within or outside our range of comprehension. Gas molecules are too tiny and numerous to be observed individually. It would be hopeless to try to bind them under deterministic laws. But if mass-scale phenomena can be explained in terms of probabilistic models, why bring in deterministic laws at the individual level at all? So the question arose: is the physical world fundamentally deterministic or indeterministic? Proponents of full-scale indeterminism—the American philosopher–mathematician C. S. Peirce prominent among them—soon appeared. Further probabilistic models were also proposed in other areas of physics to represent phenomena like radioactive emission, random trajectories of moving particles, etc. (In fact a new branch of probability—the theory of stochastic processes—came up mainly to meet the

needs of physics, even before its application in statistics could be visualized.)
The determinism versus indeterminism debate has continued to the present day,
even though von Neumann's 1936 'no-hidden-variables' theorem seems to have
tilted the scale towards indeterminism at least with regard to phenomena in the
sub-atomic world (see Russell (1948) pp. 23–24, Hacking (1990) p. 116).

9.6   Developments in probability

| Advent of frequency probability |

9.6.1   In Chapters 7 and 8 we saw that the approach
to statistical induction which attracted more attention
in the early stages of development of statistics, was the
Bayesian prosubjective approach (Chapter 4). Even though the sampling theory
viewpoint was adopted at times, as in the case of the later work of Laplace
and Gauss, generally the conclusions were given an instantial interpretation
in a vague sort of way—the behavioural aspect of sampling theory was hardly
emphasized. The reason was that the behavioural viewpoint requires total com-
mitment to the objective frequency interpretation of probability (Chapter 3) and
this had remained undeveloped at that time.

Towards the end of the first half of the 19th century, however, there was a
change in the intellectual climate—the frequency interpretation of probability
was advanced by many as the only correct interpretation. Several reasons com-
pounded to induce this change. Firstly, as described in subsection 7.4, there was
widespread criticism of the way Laplace and his followers manipulated subjective
probability to cook up posterior probabilities for all sorts of uncertain hypotheses.
The second reason was the same as that which led Quetelet to propose his social
physics—the discovery of the stability of relative frequencies of demographic and
social events. Although this had been noticed earlier in some cases in the context
of demography and insurance, the 'avalanche of numbers' rolling down from the
start of the 19th century led people to think the stability of relative frequencies to
be a widely occurring phenomenon. Also unlike in the case of throws of coins or
dice, or natural events, one could hardly say that voluntary actions like suicides
or crimes occur according to deterministic laws and we ascribe probabilities to
them subjectively due to our ignorance (cf. Hacking (1990) p. 3). Furthermore,
social engineering for the control of an undesirable occurrence in future can be
directly justified only if its probability is supposed to have objective existence.
As noted above, in physics too, models of probability in the sense of relative
frequency started making their appearance during this period.

In the early 1840s, almost at the same time, Ellis and John Stuart Mill in
England and Fries in Germany proposed that the theory of probability may prop-
erly be based on the long-term stability of relative frequencies. A few years later
Boole endorsed the same idea. The first comprehensive exposition of the theory
was given by John Venn in his *Logic of Chance* whose first edition appeared in
1866. Although there were distinct nuances in the presentations of the different
authors—Mill and Venn emphasized the empirical aspect of the determination

of probability from a series whereas Ellis and Fries (who was a Kantian), the a priori faith of living beings in the (synthetic) principle of constancy of nature (cf. Porter (1986) pp. 80–88)—their general view about probability was effectively same. They all affirmed that it is pointless to posit a value of the probability of an event and then to prove mathematically via the law of large numbers that the relative frequency of the event approaches that value probabilistically in the long-run. For, in the words of Ellis, it is impossible 'to sever judgement that one event is more likely to happen than another from the belief that in the long run it will occur more frequently' (see Keynes (1921) p. 93). They agreed that probability of an event always relates to a series or group of instances and not to an individual instance and is to be measured by the long-run relative frequency of the event for the series. (Thus Venn argued that we cannot speak of the probability of death of a particular consumptive Englishman at Madeira; if too many individuals of that species are not available we may consider him as a member of the group of all Englishmen at Madeira or of all consumptive Englishmen and the probabilities computed in the two cases would generally be different; cf. subsection 3.2.5.) The question of how to determine the probability in the case of an unlimited sequence of instances, however, was left open and so was that of how to tackle subsequences. Apart from these limitations, we can say that the seed of the 20th century frequency probability theory of von Mises and Reichenbach (subsection 3.2.2) can be found in the work of Venn. (Incidentally, as we saw earlier in section 9.3 and subsection 9.4.4, Poisson and Quetelet inferred the stability of relative frequencies like conviction rates by comparing the same for groups of accused in different years and not by studying the progressive behaviour of the relative frequency in a sequence. In this respect their approach seems to be more akin to the formulation of Reichenbach.)

C. S. Peirce   After Venn, the chief spokesman for the frequency theory of probability in the 19th century was again C. S. Peirce. We will see later in this chapter that Peirce was arguably the first statistician who pleaded for a full-scale behavioural approach to statistical induction. But Peirce's conception of objective frequency probability (he called it the materialistic conception) had its special nuances. Thus he suggested that the probability of a die turning up a number divisible by three is $\frac{1}{3}$ means that the die has a certain '*would be*' for the event quite analogous to a certain habit that a man may have. In order that the full effect of a die's 'would be' may find expression it is necessary that the die should undergo a series of throws. According to Peirce the probability of a number divisible by 3 coming up is $\frac{1}{3}$ means that, as the number of throws increases, the relative frequency of the event oscillates above and below $\frac{1}{3}$ and there is no other number p for which this property holds. Because of this stand, Peirce was not altogether averse to using probability argument in the case of a one-time experiment. Thus he argued that if one has to choose between drawing a card from a pack containing twenty-five red and a black one and another containing twenty-five black and a red one *for*

a one-time draw and 'the drawing of a red card were destined to transport him
to eternal felicity and that of a black one to consign him to everlasting woe', it
would be a folly not to choose the first pack. (These arguments are contained in
an 1878 paper entitled 'The Doctrine of Chances' by Peirce; see Buchler (1955)
Chapter 12.) In all this, Peirce's theory can be regarded as the forerunner of
Popper's Propensity Theory of probability (subsection 3.2.5).

Mathematics of probability

9.6.2   In the second half of the 19th century, there
were also developments in the theory of probability of a
different kind, especially in Russia. The Russian development was spearheaded by the St. Petersburg school of which Tshebyshev was
the pioneer and Markov and Liapounov his principal successors. They were not so
much concerned about probabilistic reasoning for induction but occupied themselves with the development of probability mathematics and the sharpening and
rigorous derivation of standard results like the Law of Large Numbers and the
Central Limit Theorem with appropriate correction terms (subsection 8.2.6). The
object of probability theory according to this school, as stated by Tshebyshev
(see Maistrov (1974) p. 192), is 'the determination of probabilities of an event
based on the given connection of this event with events whose probabilities are
known'. Such an attitude led later workers of this school like Bernstein and
Kolmogorov to formulate the axiomatic or formalistic approach to probability
(subsection 3.2.4).

A characteristic of the history of statistics that has been verified repeatedly
is that whenever statistics has been called upon to play its role in a new field of
application, new ideas and techniques have come up. As the conceptual barrier
inhibiting the application of statistics for the study of human and other biological
populations got demolished in the first three quarters of the 19th century, the
stage was set for new developments. The principal role in this act belonged to
the versatile British scientist Sir Francis Galton. In the rest of this chapter,
we describe the contributions made by Galton and those who followed him—
contributions which in a sense changed the face of statistics.

9.7   Francis Galton

Galton's conceptual departure

9.7.1   In the preceding section (subsection 9.4.3) we
noted that Quetelet, while applying the methods of the
Theory of Errors for studying data on human popula-
tions, still clung to the astronomical concept of a 'true value'. That led him to
presume the existence of an imaginary true value for each character observable
on the members of such a population and regard the deviation of an individual
observation from the true value as an 'error'. It was Francis Galton who first
emphasized that in the context of biological populations deviations occur intrinsically and variability of a character deserves to be studied for its own sake. That
led to the development of new concepts and a battery of methods and techniques.

Early work
Francis Galton (1822–1911) was born in a well-to-do English family and was well connected by birth to eminent people of learning. His initial training was in medicine but he did not opt for the profession. Instead he devoted his life to the vigorous pursuit of his varied intellectual interests. In his earlier years his interest in geography led him to make explorations in the African continent and the resulting contributions earned him considerable reputation as a geographer. In the next phase of his career he got interested in meteorology. In this area he made noteworthy contributions through organized collection, methodical representation, and imaginative interpretation of numerical data. Such exposures gave Galton an extraordinary capacity for handling masses of observations and discerning patterns underlying them.

Eugenic movement
9.7.2   Around 1860 Galton's interests shifted to the field of anthropology. In 1859 Charles Darwin, who was a first cousin of Galton, published his *Origin of Species*. The Theory of Evolution by natural selection propounded in that book frontally challenged established beliefs of Christian theology and created a commotion in the intellectual world. The traditionalists denounced the theory and the rationalists (naturalists) welcomed it. Many of those, Galton among them, who sided with the latter, took it upon themselves to find evidence and arguments in support of it. But to Galton, it seems, evolution was not only a matter of intellectual assent, but one of faith. He was beholden to the idea that since changes in a biological population occur through evolution by natural selection, the quality of a human population, as represented by the physical and mental abilities of its members, can be significantly improved by promoting scientifically directed artificial selection, i.e. judicious marriages during several consecutive generations. He started and took a leading role in the eugenic movement whose aim was to carry this idea into practice. It should be noted that the driving force behind the movement was the same kind of social activism which was prompted by Quetelet's thesis that conditions in society were produced by underlying causes according to definite laws and can be changed for the better by manipulating the causes suitably (subsection 9.5.2). But whereas for this the social activists banked on manipulation of the environment, the eugenists thought that heredity was the principal factor behind the creation of human beings of exceptional abilities and accordingly advocated artificial selection.[4,5]

---

[4]It is of some interest to note that apart from Galton, Karl Pearson and R. A. Fisher were also more or less motivated by their eugenic persuasions in the pursuit of statistical research. But unlike Galton, who initially developed his methods solely for the measurement of heredity, the latter two from the beginning were conscious about the wider applicability of the tools they developed.

[5]The eugenic movement from its inception was bitterly attacked by believers because of its materialistic view point and from social workers for its neglect of the role of environment. It gained momentum towards the end of the 19th and at the beginning

Galton sought to help the eugenic cause both materially by organizing a Society for its propagation and theoretically by laying down a proper scientific basis for it. For the latter he had to study two interrelated questions: (A) How do abilities of individuals vary in a population? (B) How and to what extent are abilities passed on from parents to offsprings? This brought Galton right into the arena of statistics.

| Distribution of abilities | In the early 1860s Galton became acquainted with Quetelet's use of the 'Law of Error' for describing the variation of anthropometric measurements like those on height and chest circumference—through the writings of Spottiswoode and Herschel (subsections 9.5.1 and 9.5.3). He became persuaded that if the physical traits followed the law of error, so would the natural abilities of men in a homogeneous group. To bolster this idea further, he studied the distribution of marks scored by the candidates in the admission test for the military college at Sandhurst in a particular year and found that the law gave a satisfactory fit. Thus he conceived that the law of error would provide the appropriate tool for tackling question (A).

From the beginning, however, Galton was clear in his mind about the inappropriateness of Quetelet's idea that error on the part of nature in hitting a targeted true value of a trait, was the cause of its variation over the members of a biological population. Variation in such a context is an important feature of the population, which requires to be studied and understood. In fact such a stance was necessitated by Galton's eugenic conviction, since if an exceptionally high or low level of a covetable trait was merely the result of error, there would be no point in trying to perpetuate or eliminate it in future generations. Therefore, Galton used the word 'deviation' for error, called 'the law of error' variously as 'the law of deviations from an average', 'the exponential law' and later coined the term 'Normal Curve' which caught on. (Galton, like Quetelet, generally subscribed to the doctrine that if a group is homogeneous, the variation of a continuous trait over the group would closely conform to the normal distribution. That explains his use of the adjective 'Normal', in the sense of something which is 'standard' or 'commonly found', to describe the distribution.)

To fit a normal curve to an observed frequency distribution, Galton used Quetelet's method of approximating the normal by the binomial with a large n and took the observed median as the estimate of μ and the observed semi-interquartile range $Q = \frac{1}{2}(Q_3 - Q_1)$ as an estimate of the probable error ($= 0.6745\sigma$). In this context he proposed the representation of the observed cumulative frequency distribution by plotting the values of the variate as ordinates against different choices of the percentage cumulative relative frequency $100\,P$ as abscissae and introduced the architectural term 'ogive' for the resulting curve and 'percentile' for the variate values corresponding to $100\,P = 1, 2, \ldots, 99$.

---

of the 20th century, but later got derailed when self-seeking demagogues sought to use it for carrying out racist policies based on questionable assumptions.

His postulate that in a homogeneous group any continuous trait would follow the normal distribution allowed him to devise a method for quantifying a trait which is not directly measurable, but according to which the individuals can be ranked. The method consisted in taking the standard normal deviate corresponding to an individual's relative rank or cumulative relative frequency as his trait-value and obviously amounted to taking Q as the unit (Galton used Q for standardization). The idea caught on with psychologists who made extensive use of it in the next half century and beyond.

| Genealogical studies |

Question (B), however, was more tricky and would require special kind of data relating to successive generations. In his initial attempt Galton sought to tackle it by identifying within different families the most eminent persons in various avocations—judges, statesmen, literary and scientific persons, musicians, military commanders, wrestlers, and so on. He then traced the genealogical trees of such persons to find out the proportions of eminent men among their fathers, uncles, grandfathers, great-uncles, brothers, sons, nephews, grandsons, and so on. He found that the proportion of men of eminence continually decreased as one moved away from the most eminent person roughly according to the law that 'the percentages are quartered at each successive remove, whether by descent or collaterally'. This rough formula, as we will see, adumbrated his later finding that exceptional persons tend to have parents and offsprings who are on an average less exceptional, but it was otherwise too vague to be of any use for the study of heredity.[6]

| Quincunx |

The above developments are mostly contained in Galton's 1869 book *Hereditary Genius* and an 1875 paper entitled 'Statistics by intercomparison'. To explain the wide-spread occurrence of the normal curve as the distribution of a trait over a biological population, Galton drew upon Darwin's theory of gemmules and the De Moivre–Laplace approximation to the symmetric binomial. The measure of a

---

[6]It was at this time that Galton became interested in the problem of extinction of surnames: Given that the probabilities of an adult male having $0, 1, 2, \ldots$ adult sons remain fixed from generation to generation, what is the probability that a particular surname will become extinct after a given number of generations; also under what conditions will there be a positive probability that a surname will survive eventually? Galton gave the problem a mathematical formulation and induced the mathematician H. W. Watson to attempt a solution. That gave rise to the theory of the Galton–Watson branching process. The extinction problem, incidentally, had to wait for more than a half a century before it could be successfully solved by Haldane and Steffenson independently; see Kendall (1966). Galton's main concern in studying the problem seems to have been the question whether the surnames of eminent men became extinct purely due to chance or eminence caused a diminution of fertility and accelerated extinction; see Puri (1976).

trait on an individual was regarded as the resultant of the random contributions
of a large number of gemmules or hereditary particles, inherited from his parents.
Galton supposed that the different gemmules acted independently and made
an equal contribution, positively or negatively, each with probability $\frac{1}{2}$. To
demonstrate the mechanism of generation of the normal law, Galton devised
an apparatus which he called 'the quincunx'. It was so contrived that lead shot
dropped through a funnel at the top of it encountered, in course of their descent,
successive rows of pins arranged on a vertical board in a qunicunx[7] fashion with
their points projecting horizontally forward, the whole thing being covered in
the front with glass. As a shot hit a pin it was deflected to the right or left with
equal probability. The bottom of the apparatus was vertically partitioned into a
number of compartments which would collect the shot. The ultimate displace-
ment of a shot would be the resultant of all the random hits made by it in course
of descent. When a large number of shot is dropped in this way, their accumula-
tion at the bottom AB would tend to generate the contour of the normal curve.
(Figure 9.1(A).)

Questions to be addressed

Galton, however, was alive to the fact that the model
of elementary errors represented by independent, equal,
binary components assumed above was too simplistic
to explain the variation of a trait in a biological population. Firstly, in this
case heredity was an important factor affecting variation and it was not clear
how heredity could be incorporated in the above model. Secondly, one could
always think of a population variously as the aggregate of different homogeneous
subpopulations (say, by region, social group, profession, etc.). If the normal law
holds for each such subpopulation, how could it hold for the aggregate too?
Yet this was the evidence of our empirical experience. Thirdly, how could the
distribution of a trait remain stable from generation to generation of a biological
population as was empirically found? All these questions had bearing on the
broad question (B) about the modality of transmission of a trait from parents
to offsprings, mentioned earlier.

To answer the above questions, Galton initially followed a line of reasoning
(see Stigler (1986) pp. 275–276) which amounted to representing the measures
X and Y of a trait on parent and offspring as

$$X = \mu + \sum_{1}^{m} \epsilon_\ell, Y = \mu + \sum_{1}^{m} \epsilon_\ell + \sum_{m+1}^{n} \epsilon_\ell = X + Z \text{ (say)}, \qquad (9.7.1)$$

where μ is the population mean and ϵ_ℓs are independent elementary errors assuming the values ± 1 with equal probabilities. If $n - m$ is large, given $X = x$,

---

[7]A quincunx is formed by five objects when four of these are placed at the corners
and one at the centre of a square (as in the 'five' of cards). By a quincunx arrangement
is meant one in diagonal crosslines which can be looked upon as successive chains of
adjacent quincunxes.

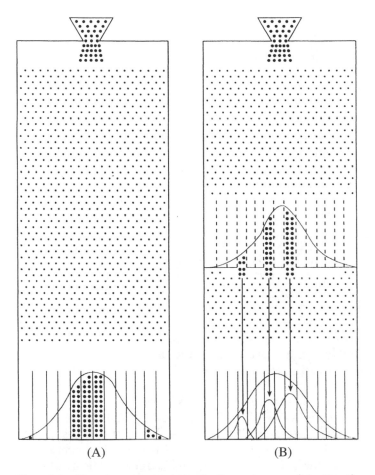

FIG. 9.1. Showing cross sections of a quincunx (idealized).

conditionally Y would be approximately normal with mean x; unconditionally too, Y would be approximately normal with mean μ. The distribution of X would be generally discrete, but close to normal with mean μ if m is large. Such a model may be explained (actually, Galton looked at it in this way sometime later) in terms of a thought-experiment with reference to a conceptual 'two-stage quincunx' formed as follows: Suppose the shot dropped into the funnel of a quincunx are intercepted by placing a replica A′B′ of the partitioned bottom AB at an intermediate position after the m uppermost rows of pins (Figure 9.1 (B)). If m is large, the distribution of the shot on A′B′ would be close to normal. If the interception A′B′ is withdrawn, the released shot from each compartment on it after hitting the lower $n - m$ rows of pins would generate on AB approximately a corresponding (conditional) normal distribution. The aggregation of all these conditional normal distributions with weights equal to the corresponding frequencies on A'B', would plainly be the normal distribution on AB obtained when there is no interception.

The model (9.7.1) can account for, in a sort of way, how for a trait the measure X on a parent can affect the measure Y on an offspring and how the conditional normal distributions of the latter when suitably aggregated can give rise to an unconditional normal distribution. But it cannot explain how the variability of a trait can remain stable from generation to generation (obviously Var(Y) is bound to be greater than Var(X) under (9.7.1)). To solve this puzzle Galton undertook a sustained programme of empirical studies spanning a period of almost fifteen years, starting around 1875 and in course of these developed the concepts of bivariate correlation and regression. The development took place in three distinct phases.

| First phase—heredity | **9.7.3**   Galton was in need of data representing meas-
urements for two or more successive generations. As human data of this type were difficult to procure, initially he planned to collect such data by performing experiments on sweet peas. (Sweet pea seeds after sowing give offspring seeds within a few months; also, sweet peas being predominantly self-fertilizing an offspring can be supposed to have originated from a single parent.) Classifying sweet pea seeds by the character weight (in later studies diameter also was taken) Galton with the assistance of several friends grew 70 seeds of each of 7 different parental grades (x) and measured the individual weights (y) of the offspring seeds produced. Thus he could obtain an array distribution of y for each chosen value of x. The aggregate of all the arrays here could be regarded as a stratified sample from the population of y-values, the strata corresponding to the different parental grades x. As an equal number of parental seeds was deliberately taken for generating the different within-stratum samples, one could not derive the marginal distribution of y in the offspring generation from such data.[8] However, on the strength of past experience, Galton took it for granted (see Hald (1998) pp. 607–608) that the distribution of weights of sweet pea seeds remains stable from generation to generation and is normal with a known mean μ and variance σ^2. (Galton actually worked in terms of the median and the quartile deviation, but in conformity with modern usage, we throughout discuss

---

[8]As each parent seed gave rise to a large number of offspring seeds the underlying population distribution here was more complex than a simple bivariate distribution of (x, y). However, we can think of a unique conditional distribution corresponding to each value of x and for every x the progeny of all seeds with grade x as independent random samples and hence their aggregate also a grand random sample from that distribution. Compounding the conditional distributions of y corresponding to different x-values by the marginal distribution of x, would generate a conceptual bivariate distribution of (x, y).

in terms of mean and variance.) Thus the marginal distributions of x and y both could be taken to be $\mathcal{N}(\mu, \sigma^2)$.

Studying the array distributions of y empirically and generalizing, Galton reached the following conclusions which, he supposed would hold for all such parent-offspring data.

(i) Each array distribution is close to normal.

(ii) The array means when plotted against the parental grade x lie roughly on a straight line.

(iii) The array variance is constant irrespective of the value of x which labels the array.

Further, Galton made the two specific observations: (a) the straightline in (ii) has the form

$$y - \mu = r(x - \mu), \tag{9.7.2}$$

where r is a positive fraction (r was $\frac{1}{3}$ for Galton's sweet pea data) and (b) the constant array variance in (iii) has the value $\sigma^2(1 - r^2)$ where r is same as in (9.7.2).

Galton thought that the observations (a) and (b) have great significance for hereditary transmission. Since for any two jointly distributed random variables X, Y, we have $E(Y) - E\{E(Y|X)\}$ and (a) means

$$E(Y|x) = \mu + r(x - \mu), \tag{9.7.3}$$

clearly $E(Y) = E(X) = \mu$. Also $0 < r < 1$ implies that the average weight of the offspring seeds produced by a parent seed whose weight x is above (below) the population mean μ, is also above (below) μ, but to a lesser extent. Galton called this phenomenon 'reversion to mediocrity': the filial type tends to depart from the parental type (x) and to revert to mediocrity as represented by the ancestral type (μ). Accordingly he interpreted r as a measure of *reversion in heredity*.[9] He sought to give a biological explanation of this phenomenon in terms of differential mortality of the progeny of parents of different grades. Similarly from the general relation

$$\mathrm{Var}(Y) = \mathrm{Var}\{E(Y|X)\} + E\{\mathrm{Var}(Y|X)\}$$

and (9.7.3), if $V(Y|x) = $ a constant σ'^2, we would have

$$\mathrm{Var}(Y) = r^2\sigma^2 + \sigma'^2. \tag{9.7.4}$$

Hence Galton thought that the value $\sigma^2(1 - r^2)$ of σ'^2 is crucial, since it is the only value which would render $\mathrm{Var}(Y) = \mathrm{Var}(X) = \sigma^2$. Thus (a) and (b) seemed to represent a good deal of fine tuning on the part of Nature to keep the distribution stable over successive generations. However, as we know now

---

[9]The symbol r which has since become standard for the correlation coefficient was originally chosen obviously because it was the first letter of the word 'reversion'.

and as Galton himself empirically discovered later on, if the marginal means and variances are same for X and Y, (ii) would imply (a) (generally with $0 \leq |r| \leq 1$) and (ii) and (iii) together would imply (b) automatically.

The above empirical findings were reported in a paper entitled 'Typical laws of heredity', which Galton presented to the Royal Institution in 1977.

Clearly the findings are conformable to a modified version of the elementary errors model (9.7.1) in which $n - m = (1 - r^2)m$ for some large m and

$$X = \mu + \sum_{\ell=1}^{m} \epsilon_\ell, Y = \mu + r \sum_{1}^{m} \epsilon_\ell + \sum_{m+1}^{n} \epsilon_\ell = \mu + r(X - \mu) + \sum_{m+1}^{n} \epsilon_\ell, \quad (9.7.5)$$

where the ϵ_ℓs are as in(9.7.1) and $0 < r < 1$. In fact, Galton implicitly meant something like this when he sought to explain the findings in terms of a more sophisticated version of the two-stage quincunx (of Figure 9.1(B)). In it the distribution of the shot on the intercepting plane A′B′ is subjected to a contraction of scale around the central point before the shot are released for collection on AB. This may be effected by inserting in between A′B′ and AB another partitioned plane A″B″ such that the successive compartments on the latter are proportionately closer to the central point than the corresponding compartments on A′B′. Each compartment on A′B′ is to be linked to the corresponding compartment on A″B″ by a channel sloping suitably inwards so that any shot collected in the former would slide down to the latter. By adjusting the slope of the channels and the number of rows of pins above A′B′ and between A″B″ and AB suitably, one can ensure that the distribution on A′B′ after being contracted on A″B″ regains its original form on AB when the contents of all the compartments on A″B″ are released (see figure on p. 198 in Pearson, K (1920)). The displacements of the same shot on A′B′ and AB represent the realized values of X and Y.

Galton in his empirical study encountered a positive value of r. The above sophisticated version of quincunx also allows for only a positive r determined by the slope of the channels between A′B′ and A″B″. A negative r may be realized by reflecting A″B″ around its central axis, but this is hard to conceptualize and harder still to implement by a mechanical device. Galton, without visualizing this possibility, thought that r in such contexts can only be positive. While this was all right for studying heredity, in the general context this made the model unnecessarily restrictive.

The model (9.7.5) of course means X, Y jointly follow an appropriate bivariate normal distribution with equal means and variances. The bivariate normal distribution, as we have seen, had made its appearance in the work of Laplace and others in early 19th century (subsections 8.3.3, 8.3.4, and 8.4.2) and more explicitly in a paper by Bravais a little later (subsection 8.6.5). But Galton was ignorant about those developments. Further, as he approached the model empirically and through mechanical devices, he had to rack his brain for eight years before he could see the implications of (9.7.5). Before describing that, let us compare (9.7.5) with the representation (8.6.1) which was the starting point of

Bravais's development of the bivariate normal distribution. In (9.7.5), in accordance with the gemmule theory of inheritance and more directly the mechanism of the quincunx, Galton took the ϵ_ℓs to be independent binary variables. If in conformity with Bravais's development we assume that the ϵ_ℓs are i.i.d. $\mathcal{N}(0, \sigma_1^2)$ (which would make the joint distribution of X, Y exactly bivariate normal), clearly (9.7.5) would be a special case of (8.6.1). The former has the advantage that the way fixing X conditionally affects the distribution of Y is immediately seen from the model. It is another matter that the validity of the assumptions of the gemmule theory and, in particular, the independence of the ϵ_ℓs was suspect in practice.

Another point to note is that Galton reached the measure r via the array means by considering the slope of the straight line (9.7.2). Galton's empirical findings suggest an alternative route based on the relation

$$(1 - r^2) = \frac{1}{\sigma^2}(\text{array variance of } y). \qquad (9.7.6)$$

This route would be the natural one to follow if instead of asking how changes in the value of x affect the array mean of y, one were interested in predicting y and wanted to know the gain in precision of prediction from taking the knowledge of x into account. We mention this because in the same year 1877 in which Galton published his analysis of sweet peas data, Richard Strachey, a friend and associate of Galton wrote a paper on meteorology. In it he followed roughly the second approach to study whether the measure of sunspot activity has at all any use for predicting the annual rainfall (see Porter (1986) pp. 277–278). He compared the variability of rainfall around its predicted value based on sunspot activity against the same around its general mean and came out with a negative answer. Of course the variation of sunspot activity was periodic, whereas in Galton's case the parental trait varied randomly. But the parallelism shows that about the same time similar ideas about how variability of one variable affects that of another were sprouting up in different contexts.

| Second phase—family likeness |

9.7.4   In his 1877 sweet peas study Galton was handicapped in that he had but a stratified sample from the population of offspring seeds and not a straightforward sample from the bivariate population of parents and offsprings. That obscured his view to some extent. Within a few years after this, however, through correspondence, public appeal and declaration of incentives, he was able to procure data on the heights of a large number of adult persons, their parents, and other family members. Galton first reported the results of analysis of these data in his Presidential Address to the Anthropology section of the British Association for the Advancement of Science in 1885. A fuller version was published as a paper entitled 'Family likeness in stature' in 1886. The most thorough presentation was contained in his 1889 book *Natural Inheritance*.

In these studies Galton scaled up the heights of females into their male equivalents by multiplying by the factor 1.08 (which he had found to be the ratio the average height of a man bears to that of a woman). Also he replaced the heights of the two parents of a person by a single 'mid-parental height' represented by the mean of father's height and the male equivalent of mother's height. On the basis of some preliminary studies Galton convinced himself that one could assume that the heights of the two parents are independently distributed and the height of an adult child is affected by those of the parents only through the mid-parental height. To study how the heights of parents got transmitted to their children, he then prepared a two-way table representing the bivariate frequency distribution of 928 pairs of values of (x, y) where x was the mid-parental height and y the height of an adult child.

Past experience told Galton that the distribution of height of adult males in a human population is stable and approximately $\mathcal{N}(\mu, \sigma^2)$ with μ, σ^2 known. Applying the procedure followed for sweet peas data to the bivariate frequency table, as earlier he found that

(i) marginally x was distributed approximately as $\mathcal{N}(\mu, \sigma_x^2)$ where by virtue of his assumptions $\sigma_x^2 = \frac{1}{2}\sigma^2$,
(ii) for different x, the array distributions of y were approximately normal with means lying on a straightline of the form.

$$y - \mu = b_{y \cdot x}(x - \mu) \qquad (9.7.7)$$

and variance independent of x,
(iii) the constant array variance of y was given by

$$\sigma_y^2 - b_{y \cdot x}^2 \sigma_x^2, \qquad (9.7.8)$$

where σ_y^2, the marginal variance of y, was of course same as σ^2.

Galton had $b_{y \cdot x} = \frac{2}{3}$ so that, as earlier, he could interpret (9.7.7) as an instance of reversion. But he was quick to see that the two-way frequency table being based on a bivariate random sample, the roles of the two variables in it were logically symmetric and one could make a parallel study by interchanging x and y. Such an interchange, Galton found, gave the same conclusions as above with σ_x^2, σ_y^2 interchanged and a new coefficient $b_{x \cdot y}$, which in the particular case had value $\frac{1}{3}$, replacing $b_{y \cdot x}$. Thus the phenomenon of reversion seemed to work not only from parent to offspring but from offspring to parent and naturally lost its significance as regards heredity!

In the context of these studies on height, Galton replaced his earlier term 'reversion' by the somewhat weaker term 'regression'. Thus (9.7.7) came to be called the regression equation and $b_{y \cdot x}$ the regression coefficient of y on x. Galton now interpreted the regression coefficients $b_{y \cdot x}$ and $b_{x \cdot y}$ as measuring the extent of family likeness in respect of height.

The bivariate normal model

Galton was mildly puzzled by the mismatch between the values of $b_{y \cdot x}$ and $b_{x \cdot y}$. Although he must have seen that the fact that $\sigma_y^2 = 2\sigma_x^2 (= \sigma^2)$ had something to do with it, he wanted a fuller explanation. Examining the bivariate frequency table after suitably smoothing the cell frequencies (by a sort of moving average taken over adjacent cells), he found that the iso-frequency cells (Galton's meteorological experience must have been of use here) tended to be located along similar concentric ellipses centred at the point (μ, μ). He could see that the line (9.7.7) represented the locus of points of contact of the tangents parallel to the y-axis to this family of ellipses and a similar interpretation applied when the roles of x and y are reversed. It struck Galton that all these features might be consequences of an underlying unimodal bivariate frequency surface with mode at (μ, μ), and elliptical equidensity contours. He approached a Cambridge mathematician named J. D. Hamilton Dickson with a question formulated as follows (with X, Y now denoting random counterparts of the translated variables $x - \mu, y - \mu$):

For what joint distribution of the random variables X, Y will

(a) X be marginally distributed as $\mathcal{N}(0, \sigma_x^2)$,
(b) given $X = x$, Y be conditionally distributed as $\mathcal{N}(b_{y \cdot x} x, \sigma_{y \cdot x}^2)$,
(c) the equidensity contours be similar ellipses centred at $(0, 0)$?

As a matter of fact, as Hamilton Dickson immediately showed, from (a)–(b) it followed that the joint density of X, Y is given by

$$f(x, y) = \text{const.}\, e^{-x^2/(2\sigma_x^2)} \times e^{-(y - b_{y \cdot x} x)^2/(2\sigma_{y \cdot x}^2)} = e^{-Q(x,y)/2}, \qquad (9.7.9)$$

where

$$Q(x, y) = x^2 \left(\frac{1}{\sigma_x^2} + \frac{b_{y \cdot x}^2}{\sigma_{y \cdot x}^2} \right) - 2 b_{y \cdot x} \frac{xy}{\sigma_{y \cdot x}^2} + \frac{y^2}{\sigma_{y \cdot x}^2}. \qquad (9.7.10)$$

As $Q(x, y)$ is a positive definite quadratic form, the equi-density contours corresponding to different values of $Q(x, y)$ clearly represent similar ellipses centred at $(0, 0)$. Thus (c) necessarily follows from (a)–(b). Also writing

$$\sigma_y^2 = \sigma_{y \cdot x}^2 + b_{y \cdot x}^2 \sigma_x^2, \quad b_{x \cdot y} = b_{y \cdot x} \cdot \frac{\sigma_x^2}{\sigma_y^2}, \quad \sigma_{x \cdot y}^2 = \sigma_x^2 - b_{x \cdot y}^2 \sigma_y^2, \qquad (9.7.11)$$

$Q(x, y)$ can be recast as

$$Q(x, y) = \frac{y^2}{\sigma_y^2} + \frac{(x - b_{x \cdot y} y)^2}{\sigma_{x \cdot y}^2}.$$

Hence it follows that marginally Y follows $\mathcal{N}(0, \sigma_y^2)$ and given $Y = y$, X conditionally follows $\mathcal{N}(b_{x \cdot y} y, \sigma_{x \cdot y}^2)$. In the case of Galton's parent-child data, (9.7.11) was borne out by the directly determined values of $b_{x \cdot y}$ and $\sigma_{x \cdot y}^2$.

Galton also made a study similar to the above on the bivariate frequency distribution of the heights of brothers in 295 families. In this case as the frequency table was prepared by counting every permutation of 2 brothers in each family as a pair of observations on the two variables (x, y), the distribution was strictly symmetric with regard to the variables. This meant he had a common value for $b_{y \cdot x}$ and $b_{x \cdot y}$ which could be taken as a symmetric measure of family likeness.

It is obvious that for $b_{y \cdot x}$ to be equal to $b_{x \cdot y}$ it is sufficient if $\sigma_x^2 = \sigma_y^2$—a condition which could be taken to hold for two adult members from the same family having any designated kinship. Galton gave a measure of family likeness between father and son, uncle (father's brother) and nephew etc. conceived in this way. But for computation of these he often made the simplistic assumption that whenever we have three variables x, y, z the relation $b_{z \cdot x} = b_{y \cdot x} \times b_{z \cdot y}$ holds. (Apparently this was obtained by substituting the regression equation of y on x in that of z on y.) This permitted him to get a dubious figure for the coefficient for uncle on nephew by multiplying that for father on son by one between two brothers. (As later studies showed, Galton's assumption amounts to the vanishing of the partial correlation coefficient $r_{xz \cdot y}$)

The joint distribution (9.7.9) is of course the bivariate normal derived via the marginal distribution of X and the conditional distribution of Y given X. In this context it seems to us strange that Galton did not immediately take the further small step of reparametrizing from $(\sigma_x^2, b_{y \cdot x}, \sigma_{y \cdot x}^2)$ to $(\sigma_x^2, \sigma_y^2, r = b_{y \cdot x} \sigma_x / \sigma_y)$ and interpret r as a symmetric measure of interdependence or association between X and Y. One reason for this may be that in all these studies X and Y represented measurements on the *same* variable height on different individuals so that the concept of 'likeness' was more direct than that of 'interdependence'. As it happened, this next step dawned upon Galton towards the end of 1888, when he was dealing with a different kind of bivariate data. But this enlightenment came to him not through reparametrization in (9.7.9) but from a somewhat different consideration.

Third phase—correlation

9.7.5 Galton's eugenic mission induced him to maintain a lifelong interest in anthropometry. In 1884, when an International Health Exhibition was held in South Kensington, he set up an anthropometric laboratory there. By subjecting volunteering visitors to measurement, he collected data on nine physical characteristics like height, head length, head breadth, cubit length, knee height, etc. for each of several thousand males and females. In this way he obtained 18 empirical distributions. He computed their standardized percentiles of order $5, 10, 20, \ldots, 90, 95$ and found that for each order the average of the 18 figures was almost identical with the same percentile of the standard normal distribution. This supplied Galton further evidence of the wide applicability of the normal law in practice. But at that time he did not make any use of the data to study the interdependence of the variables.

Towards the end of 1888, however, after sending the final proofs of his book *Natural Inheritance* to the press, he was provoked by the work of a French criminologist of the name of Bertillon (see Porter(1986) pp. 291–292), to consider the problem of identifying a criminal on the basis of a small number of judiciously selected physical measurements. This induced him to investigate to what extent different physical characteristics of individuals are interdependent. Drawing upon his earlier anthropometric data, he prepared bivariate frequency tables for pairs of variables like height and cubit length, height and head length, etc.

To judge interdependence of a pair of variables it had to be judged whether large (small) values of one tend to occur together with large (small) values of the other. As the variables were of diverse nature, to judge the degree of largeness (or smallness) of a variable value, the obvious way, as Galton saw it, would be to standardize it by measuring it from its own mean in units of its own standard deviation (Galton used median and Q.D.). After such standardization Galton proceeded to apply the methods of his family likeness study to the transmuted pair of variables (x, y) and immediately saw what was obscured from his view earlier: in standardized scales the regression slopes $b_{y \cdot x}$ and $b_{x \cdot y}$ have a common value. Denoting this common value by r, he interpreted it as a symmetric measure of interdependence or *correlation*[10] and called it the 'index of correlation'. (The term correlation coefficient was introduced by Edgeworth a few years later. Incidentally Galton thought that r would always lie between 0 and 1.) He now recognized that all through his earlier studies, the general concept that he was groping to catch through terms like 'reversion in heredity' and 'family likeness' was that of correlation. In an address delivered early in 1889 (see MacKenzie (1981) p. 66) he remarked, 'A very little reflection made it clear that family likeness was nothing more than a particular case of the wide subject of correlation, and that the whole of the reasoning already bestowed upon the special case of family likeness was equally applicable to correlation in its most general aspect.'

Galton's work on correlation appeared in two papers published in 1889 and 1890 (see Hald (1998), p. 754). After this, he did not make any contribution of note to statistical theory. (Interestingly, however, he published a book on personal identification by finger prints in 1892.)

| Nature and impact of Galton's contributions | **9.7.6**   Galton's most notable contribution to statistics in our view was breaking the 'true-value syndrome' and pointing out the importance of studying variability. In |

Natural Inheritance he forcefully argued, 'It is difficult to understand why statisticians commonly limit their enquiries to Averages and do not revel in more comprehensive views. Their souls seem as dull to the charm of variety as that of the native of one of our flat English counties, whose retrospect of Switzerland was that, if its mountains could be thrown into its lakes, two nuisances could be

---

[10]The term 'correlation' was already in use in the biological literature to signify the qualitative concept of the tendency of different traits to appear together in an organism. Galton initially used 'co-relation' to distinguish the quantitative concept from the biological, but later fell back on the common spelling.

got rid of at once. An Average is but a solitary fact, whereas if a single other fact be added to it, an entire Normal Scheme, which nearly corresponds to the observed one, starts potentially into existence' (see Pearson (1970)). Once the importance of variability of a character was recognized, the question of studying the joint variability of two or more characters assumed significance. The theory of regression and correlation and its later extensions followed in due course. This changed the orientation of development of statistical thought significantly. Because of this, it is immaterial whether we accord the credit of pioneering the mathematical theory of bivariate normal model and bivariate regression and correlation to Galton or Bravais. Given the concept, the mathematics would have followed in due course. (As we saw in subsection 9.7.3, even someone as peripherally involved with statistics as Strachey approached close to the measure of correlation, motivated by the prediction problem.)

As we mentioned earlier and as the above excerpt from *Natural Inheritance* also shows, Galton, while admitting the importance of variability, remained firmly wedded to the Queteletian doctrine that for a homogeneous group the distribution of a continuous variate would be generally close to normal. In *Natural Inheritance* he made the eloquent remark, 'I know of scarcely anything so apt to impress the imagination as the wonderful form of cosmic order expressed by the "Law of Frequency of Error". The law would have been personified by the Greeks and deified, if they had known of it.' To account for normality Galton (in spite of certain reservations) accepted the postulate of elementary errors. In his 1890 paper on correlation, Galton further sought to explain bivariate normality as occurring due to independent influences, some of which were general and affected both the variables, although not necessarily, to the same degree and some specific to each variable (see Hald (1998) p. 616). This was a generalization of the quincunx model (9.7.5) and had the same structure as Bravais's 'exact' model (8.6.1).

In the second half of the 19th century, positivistic thinking was very much influential, especially in England. The positivists swore exclusively by things which are observable. To them laws were summaries of observed regularities and causation was a metaphysical concept. The concept of correlation provided positivists an escape route for avoiding the issue of causation. As Karl Pearson, a leading positivist of that time put it in his biography of Galton, 'This measure of partial causation was the germ of the broad category—that of correlation, which was to replace not only in the minds of many of us the old category of causation, but deeply to influence our outlook on the universe.... Henceforward the philosophical view of the universe was to be that of a correlated system of variates, approaching but by no means reaching perfect correlation, i.e. absolute causality' (quoted by Hacking (1990) p. 188). The concept of correlation later proved invaluable in many fields, most notably in the field of psychometry. In this context, it may not be out of place to cite the assessment of R. A. Fisher (it may be mentioned that it was Fisher who developed the sampling theory of correlation and was instrumental in popularizing its use) about the utility of correlation. '...no method,' he wrote in his *Statistical Methods for Research*

Workers (Fisher (1944) p. 170), 'has been applied to such various data as the method of correlation. Observational data in particular, in cases where we can observe the occurrence of various possible contributory causes of a phenomenon, but cannot control them, has been given by its means an altogether new importance. In experimental work proper its position is much less central; it will be found useful in the exploratory stages of an enquiry... but it is seldom, with controlled experimental conditions, that it is desired to express our conclusion in the form of a correlation coefficient.'

Was Galton's work concerned with statistical induction? Not directly, for he never bothered about the probabilistic assessment of inductive conclusions. But of course he presumed that the regression or correlation determined from limited (though large) samples of pea seeds or adult human beings would hold for the entire population. In fact, he did not subscribe the Queteletian prejudice against sampling and generally the ideas of statistical regularity and random sampling (at least in a vague sense) remained at the back of his mind. This becomes clear, for instance, from the allusion to the 'axiom of statistics that large samples taken out of the same population *at random* [italics ours] are statistically similar' made in his 1886 paper (see Porter (1986) p. 136). What is more, Galton's studies and their extensions threw up a wide variety of problems of induction whose solution engaged the attention of many later workers for at least the next fifty years.

Apart from his research work, Galton continued to promote the cause of statistics in other ways as long as he lived. In particular, the founding of the journal *Biometrika* in 1901 and creation of a Chair for the study of eugenics and statistics at University College, London were possible because of the support received from him.

9.8 Extension of Galton's work

| Galton's successors |

9.8.1 In the introduction to his book *Natural Inheritance* Galton prophetically wrote, 'This part of the inquiry may be said to run along a road on a high level, that affords wide views in unexpected directions, and from which easy descents may be made to totally different goals to those we have now to reach. I have a great subject to write upon.' Furthermore, in his 1890 paper he remarked, '... there is a vast field of topics that fall under the laws of correlation, which lies quite open to the research of any competent person who cares to investigate it.' As it happened, from 1890 onwards a group of competent persons came forward to fill up the gaps in Galton's work and to extend it in various directions.

The most prominent member of this group was Karl Pearson (1857–1936), who at that time was a Professor of Applied Mathematics at University College, London and had already made a name for himself by writing on various topics including probability and philosophy of science. (In his book *Grammar of Science*, which was first published in 1892, Pearson gave a clear exposition of the aims and methods of science from the viewpoint of a positivist.) He was drawn to Galton's work because of his positivistic leanings and his eugenic convictions. For the next three decades at least, he played a leading role in the development and propagation of statistical thought. Two other members of the group, who actually started working on Galton's problems even earlier than Pearson, were F. Y. Edgeworth and W. F. R. Weldon. Of these Edgeworth's main area of interest was Economics (he became Professor of Economics at Oxford), although he had been writing intermittently on the application of probability to various types of inductive problems for quite sometime before he had exposure to Galton's work on correlation. He got interested in and contributed to the theory of correlation for a brief period after this, but strictly speaking, he cannot be considered as a successor to Galton. Weldon was a Professor of Zoology and a colleague of Pearson at University College. He came to statistics with the definite purpose of finding empirical support to Darwin's Theory of Evolution. He used Galtonian methods for the analysis of varieties of data relating to lower animals, particularly shrimps and crabs (Weldon's area of specialization in zoology was invertebrate morphology) and in the process threw new light on those methods. He was partly instrumental in making Pearson interested in statistics and opening the latter's eyes to the potential of Galton's theory of correlation. Although no mathematician himself, as we will see in this and the next chapter, Weldon drew attention of others to some important problems of statistics, whose solution by others led to considerable advance of statistical theory. Finally, the fourth member of the group was G. U. Yule, who started working on the theory of correlation as an associate of Karl Pearson but later moved away to chart his own course.

| Reification of correlation | **9.8.2** Galton thought that the correlation between two variates would always be positive ('Two variable organs are said to be correlated,' he wrote in his 1888 |

paper, 'when the variation of the one is accompanied on the average by more or less variation of the other *and in the same direction* [italics ours].' Was negative correlation a physical impossibility? If not, this misconception had to be removed. Also his measure of correlation was a sort of intangible one. To compute it, one had to construct a bivariate frequency table from the pairs of standardized observations, compute and plot the array means of one variable against the distinct values of the other on the same graph taking by turn the values of either variable as abscissae, fit a straightline to the plotted points by inspection, and then determine the slope of the fitted line. Was it possible to give a compact algebraic formula for the correlation coefficient?

The first question was empirically settled by Weldon, when he found a particular pair of measurements on shrimps for which the correlation was negative and gave the correct interpretation to the phenomenon. As regards the second question, Weldon's suggestion was to prepare the bivariate frequency table based on the standardized variables, to find for each array of one variable the ratio of its mean to the array label and then take the simple average of all such ratios. But this was still cumbersome and inapplicable to ungrouped data.

Edgeworth gave two methods for the computation of the correlation coefficient. He viewed the problem as one of estimating ρ for the bivariate normal $N_2(0, 0, 1, 1, \rho)$ on the basis of an ungrouped sample (x_ℓ, y_ℓ), $\ell = 1, \ldots, n$. (We henceforth use ρ to denote the population correlation coefficient and r for its sample counterpart.) Using capital letters to denote random variables we have here the observational equations

$$E(Y_\ell | x_\ell) = \rho x_\ell, \text{Var}(Y_\ell | x_\ell) = \text{const.}(= 1 - \rho^2), \quad \ell = 1, \ldots, n. \qquad (9.8.1)$$

Essentially Edgeworth derived from these observational equations two estimates, one a sort of estimate based on the method of averages (subsection 8.3.2) and the other, the least squares estimate. When the sample is from $N_2(\mu_x, \mu_y, \sigma_x^2, \sigma_y^2, \rho)$, one has to replace x_ℓ, y_ℓ by $(x_\ell - \mu_x)/\sigma_x, (y_\ell - \mu_y)/\sigma_y$, in these estimates. Further, if $\mu_x, \mu_y, \sigma_x^2, \sigma_y^2$ are unknown one has to substitute the sample means and variances for these. This would reduce the L. S. estimate of ρ to the product moment correlation coefficient, but Edgeworth did not make this explicit.

The product moment formula for r was explicitly given by Karl Pearson in a paper in 1896 (see Pearson, K. (1948)). Considering the sample (x_ℓ, y_ℓ), $\ell = 1, \ldots, n$ as drawn from $N_2(0, 0, \sigma_x^2, \sigma_y^2, \rho)$, the sample joint probability density is

$$(2\pi)^{-n} \sigma_x^{-n} \sigma_y^{-n} (1 - \rho^2)^{-(n/2)} \exp\left\{ -\frac{n}{2(1 - \rho^2)} \left(\frac{S_{xx}}{\sigma_x^2} - 2\rho \frac{S_{xy}}{\sigma_x \sigma_y} + \frac{S_{yy}}{\sigma_y^2} \right) \right\},$$
$$(9.8.2)$$

where

$$S_{xx} = \sum x_\ell^2, S_{xy} = \sum x_\ell y_\ell, S_{yy} = \sum y_\ell^2. \qquad (9.8.3)$$

Substituting $\sigma_x^2 = S_{xx}/n, \sigma_y^2 = S_{yy}/n$ and writing $\lambda = S_{xy}/(S_{xx}S_{yy})^{1/2}$, (9.8.2) becomes

$$\text{const.}(1 - \rho^2)^{-n/2} \exp\left\{ -\frac{n}{(1 - \rho^2)}(1 - \lambda\rho) \right\}, \qquad (9.8.4)$$

where the const. does not involve ρ. Pearson considered the value of ρ that maximizes (9.8.4), and considering the first derivative with respect to ρ of the logarithm, came out with the solution $\rho = \lambda$. His conclusion was that the observed series of values (x_ℓ, y_ℓ), $\ell = 1, \ldots, n$, is most probable when $\rho = \lambda$ and therefore λ should be taken as the estimate of ρ (see Pearson (1896)). When μ_x, μ_y are unknown, we have of course to replace x_ℓ, y_ℓ by $x_\ell - \bar{x}, y_\ell - \bar{y}$ and that would reduce λ to the sample product moment correlation coefficient r.

On the face of it, Pearson was here applying what we now call, 'the maximum likelihood principle' for the derivation of an estimator of ρ (although the appropriate method here would have been to write the sample probability density as a function of all the unknown parameters $\mu_x, \mu_y, \sigma_x^2, \sigma_y^2, \rho$ and maximize it simultaneously with respect to these). However, that principle (although, as we saw in Chapter 6, it had been introduced by Daniel Bernoulli more than a century ago) was not in general use at that time and Pearson applied it without any preamble. Because of this, doubts have been expressed about this interpretation. We consider this point more closely in subsection 9.8.5 below in the context of discussing a general method for deriving S. E.s of estimates, which Pearson proposed and applied for the determination of the S. E. of r in a later paper.

| Multivariate normal | **9.8.3** Galton encountered linearity of regression along with certain other features in a bivariate population in course of his empirical studies. Trying to find a bivariate distribution which fits in with his empirical findings, he arrived at the bivariate normal model. There was need for extending these concepts to the multivariate case. Galton himself attempted a multivariate generalization of linear regression for predicting the height of an individual on the basis of the heights of his forbears through what he called 'the law of ancestral heredity', but his approach could be justified only under restrictive assumptions. Multivariate generalization of Galton's ideas was successfully made first by Edgeworth and later more thoroughly by Karl Pearson and his associates.

Edgeworth proceeded directly to the problem of developing the multivariate normal model. He considered a set of random variables, say, $X_1, \ldots, X_p$ each of which follows marginally a normal distribution with means $\mu_1, \ldots, \mu_p$ and variances $\sigma_1^2, \ldots, \sigma_p^2$ (we are all through using modern notations and terminology for convenience). Presumably, at the back of his mind he had that each pair of variables follows marginally a bivariate normal distribution with pairwise correlations $\rho_{ij}, i \neq j = 1, \ldots, p$. In analogy with the Galton–Hamilton Dickson form (9.7.9)–(9.7.10) of the bivariate normal density, Edgeworth took the multivariate normal density directly in the form

$$f(x_1, \ldots, x_p) = \text{const. } e^{-Q(x_1, \ldots, x_p)/2}, \tag{9.8.5}$$

where

$$Q(x_1, \ldots, x_p) = \sum_{i,j=1}^{p} c_{ij}(x_i - \mu_i)(x_j - \mu_j) \tag{9.8.6}$$

is a positive definite quadratic form in $x_i - \mu_i, i = 1, \ldots, p$. The problem was to express the c_{ij}s in terms of the σ_i^2s and the ρ_{ij}s. Through a series of papers published during 1892–93 (see Stigler (1986), Chapter 9), Edgeworth reached the expression

$$c_{ij} = \frac{1}{\sigma_i \sigma_j \det R} R_{ij}, \quad i, j = 1, \ldots, p, \tag{9.8.7}$$

where, writing $\rho_{ii} = 1$,

$$R = (\rho_{ij}), \quad R_{ij} = \text{co-factor of } \rho_{ij} \text{ in } R. \tag{9.8.8}$$

He also showed that the conditional distribution of X_1 given that $X_j = x_j$, $j = 2, \ldots, p$, is univariate normal with

$$E(X_1 | x_2, \ldots, x_p) = \mu_1 - \sum_{j=2}^{p} \frac{\sigma_1}{\sigma_j} \frac{R_{1j}}{R_{11}} (x_j - \mu_j), \qquad (9.8.9)$$

$$\mathrm{Var}(X_1 | x_2, \ldots, x_p) = \sigma_1^2 \cdot \frac{\det R}{R_{11}}. \qquad (9.8.10)$$

This was all in conformity with the bivariate case. (9.8.9) provided the linear regression formula for X_1 given $x_2, \ldots, x_p$.

Karl Pearson in his 1896 paper started with an extension of Bravais's model (8.6.1) representing

$$X_i - \mu_i = \sum_{\ell=1}^{n} a_{i\ell} \epsilon_\ell, \quad i = 1, \ldots, p,$$

where $\epsilon_\ell, \ell = 1, \ldots, n(>p)$, are independently normally distributed with means 0 and the set of p n-vectors $(a_{i1}, \ldots, a_{in}), i = 1, \ldots, p$, are linearly independent. By suitably transforming the joint density of the ϵ_ℓs and integrating out the extra variables, he showed that the joint density of $X_1, \ldots, X_p$ reduces to the form given by (9.8.5)–(9.8.8). He then rederived the formulae (9.8.9) and (9.8.10) for the array mean and variance of X_1 given the other variables. He applied (9.8.9) extensively for the prediction of a variable on the basis of readings on a number of variables correlated with it in various studies relating to heredity and anthropometry. By (9.8.10), obviously $(\det R)/R_{11}$ in such contexts can be interpreted as a unit-free measure of the inaccuracy of prediction: it lies between 0 and 1 and closer its value is to 0, the more accurate the prediction. Pearson later showed that the maximum value the bivariate correlation coefficient between X_1 and any linear function of $X_2, \ldots, X_p$ can attain is realized when the linear function on the right of (9.8.9) is taken for the latter and the maximum value is equal to

$$\rho_{1 \cdot 23 \ldots p} = \left(1 - \frac{\det R}{R_{11}}\right)^{1/2}. \qquad (9.8.11)$$

It was called the *multiple* correlation coefficient of X_1 on $X_2, \ldots, X_p$. Obviously it can be interpreted either as a direct measure of accuracy of prediction of X_1, or as a measure of interdependence between X_1 and $X_2, \ldots, X_p$ taken together.

Sheppard, Yule, and Filon were associated with Pearson in his investigation of the multivariate normal distribution. The bivariate conditional distribution of X_1 and X_2 given $x_3, \ldots, x_p$ was shown to be bivariate normal with means linear in $x_3, \ldots, x_p$ and variances and covariance independent of $x_3, \ldots, x_p$. The conditional correlation coefficient was shown to be

$$\rho_{12 \cdot 34 \ldots p} = -\frac{R_{12}}{(R_{11} R_{22})^{1/2}}. \qquad (9.8.12)$$

This came to be called the coefficient of *net* or *partial* correlation between X_1 and X_2 measuring their degree of interdependence when the influence of $X_3, \ldots, X_p$ on them is eliminated.

The formulae (9.8.9)–(9.8.12) above were thus all originally derived assuming that $X_1, \ldots, X_p$ follow the multivariate normal model (9.8.5). Yule pointed out to Pearson a relaxation of this requirement which would still keep the formulae valid with the same interpretations (this was acknowledged by Pearson in a footnote in his 1896 paper). Thus whenever $E(X_1|x_2, \ldots, x_p)$ is a linear function of $x_2, \ldots, x_p$, (9.8.9) would hold. If, in addition $\mathrm{Var}(X_1|x_2, \ldots, x_p)$ is a constant independent of $x_2, \ldots, x_p$, (9.8.10) would remain true. Similarly for the validity of (9.8.12) it is sufficient if $E(X_i|x_3, \ldots, x_p), i = 1, 2$ are linear in $x_3, \ldots, x_p$ and $\mathrm{Var}(X_i|x_3, \ldots, x_p), i = 1, 2$ and $\mathrm{Cov}(X_1, X_2|x_3, \ldots, x_p)$ are independent of $x_3, \ldots, x_p$.[11] Of course it is hard to visualize a multivariate set-up in which normality does not hold and yet all these conditions are realized.

| The least squares |
| connection |

9.8.4   Edgeworth and Karl Pearson, following up as they were the lead given by Galton, envisaged a programme for studying interdependence of variables in the multivariate set-up, which essentially rested on the multinormal model. Their idea was to first determine the marginal means and variances and the pairwise correlation coefficients and substitute these in the formulae (9.8.9)–(9.8.12). The programme could not be justified in situations where the data exhibited pronounced departure from normality. Yule came across a set of bivariate data relating to the extent of pauperism and administration of poor law relief in 580 Poor Law Unions in England and Wales. One variable (x) was the out-pauper ratio. (Among paupers or recipients of relief, some were given free out-of-doors relief and others in-doors relief as payment for labour done in workhouses; the former was supposed to be less stigmatizing.) The other variable (y) was the degree of pauperism as measured by the proportion of paupers in the Union. The bivariate frequency table showed marked skewness so that the bivariate normal model was inapplicable. Yet it was a burning political question whether the extent of pauperism rose as x increased (i.e. as out-relief was doled more liberally).

---

[11]Interestingly, even these conditions are only sufficient but not necessary. Thus if $X_1, X_2, \ldots, X_p$ follow the $(p+1)$-cell multinomial distribution with mass function

$$\frac{N!}{(n - \sum_1^p x_i)! \Pi_1^p x_i!} \left(1 - \sum_1^p \pi_i\right)^{N - \sum_1^p x_i} \prod_1^p \pi_i^{x_i}, \quad \text{where } x_i \geq 0, \quad \sum_1^p x_i \leq N,$$

it can be seen that $\mathrm{Var}(X_1|x_3, \ldots, x_p), \mathrm{Var}(X_2|x_3, \ldots, x_p), \mathrm{Cov}(X_1, X_2|x_3, \ldots, x_p)$ depend on $x_3, \ldots, x_p$, yet the conditional correlation coefficient between X_1, X_2 is given by (9.8.12).

To study the question, Yule went back to Galton's original formulation of the problem of studying dependence in the bivariate case. He considered the array means $\bar{y}_g$ of y corresponding to the distinct values $x_{(g)}$ of x, $g = 1, \ldots, s$ (say). (Galton's discovery of elliptical equi-frequency contours in the case of parent-child stature data and subsequent study of bivariate normal, it may be recalled, came later.) Seeking to obtain the best-fitting straight line $y = a + bx$ to the pairs $(x_{(g)}, \bar{y}_g)$, $g = 1, \ldots, s$ according to the least squares principle, he found that if f_g is the marginal frequency of $x_{(g)}$ and the values in the corresponding y-array are $y_{gh}, h = 1, \ldots, f_g (\bar{y}_g = \sum_{h=1}^{f_g} y_{gh}/f_g)$, then

$$\sum_{g=1}^{s} f_g(\bar{y}_g - a - bx_{(g)})^2 + \sum_{g=1}^{s}\sum_{h=1}^{f_g}(y_{gh} - \bar{y}_g)^2 = \sum_{g=1}^{s}\sum_{h=1}^{f_g}(y_{gh} - a - bx_{(g)})^2. \quad (9.8.13)$$

Since the second term on the left of (9.8.13) does not involve a, b, the minimization of the first term is equivalent to that of the right hand side, i.e. to fitting a straight line to the original pairs $(x_{(g)}, y_{gh}), h = 1, \ldots f_g, g = 1, \ldots, s$. Thinking in terms of the bivariate probability distribution of two random variables X, Y, (9.8.13) becomes

$$E\{E(Y|X) - a - bX\}^2 + E\{Y - E(Y|X)\}^2 = E(Y - a - bX)^2 \quad (9.8.14)$$

and the problem of finding the best-fitting straight line $y = a + bx$ to the conditional means $E(Y|x)$ is equivalent to that of determining a, b so that the right-hand side of (9.8.14) is minimum. The reasoning directly carries over to the case of a p-variate distribution with random variables $X_1, X_2, \ldots, X_p$: the problem of fitting a linear function

$$x_1 = a + b_2 x_2 + \cdots + b_p x_p \quad (9.8.15)$$

to the conditional means $E(X_1|x_2, \ldots, x_p)$ by minimizing

$$E\{E(X_1|X_2, \ldots, X_p) - a - b_2 X_2 - \cdots b_p X_p\}^2 \quad (9.8.16)$$

reduces to that of minimizing

$$E(X_1 - a - b_2 X_2 - \cdots - b_p X_p)^2. \quad (9.8.17)$$

This is achieved when

$$a = \mu_1 - b_2 \mu_2 - \cdots - b_p \mu_p \quad (9.8.18)$$

and $b_2, \ldots, b_p$ are determined from the normal equations.

$$b_2 \rho_{j2} \sigma_j \sigma_2 + b_3 \rho_{j3} \sigma_j \sigma_3 + \cdots + b_p \rho_{jp} \sigma_j \sigma_p = \rho_{j1} \sigma_j \sigma_1, \quad j = 2, \ldots, p. \quad (9.8.19)$$

It is readily seen that thereby we are led back to the expression on the right
of (9.8.9). Thus we may heuristically use the latter which we now call the *least
squares regression*, as the predicted value of X_1 given $x_2, \ldots, x_p$. The minimum
value of (9.8.17) here has the same expression as (9.8.10) and the product moment
correlation coefficient between X_1 and its predicted value is same as the multiple
correlation coefficient (9.8.11). Similarly the partial correlation coefficient given
by (9.8.12) turns out to be the correlation coefficient between the residuals left
over after the predicted values of X_1 and X_2 on the basis of their least squares
regressions on $X_3, \ldots, X_p$ are subtracted from X_1 and X_2. (Obviously, when
$E(X_1 | x_2, \ldots, x_p)$ is linear in $x_2, \ldots, x_p$ and in particular in the multinormal case,
(9.8.16) can be made 0 by choosing $a, b_2, \ldots, b_p$ appropriately and in this case
the least square regression coincides with the array mean or the true regression.)

The above results hold whatever the form of the joint distribution of
$X_1, X_2, \ldots, X_p$. In particular, given a set of observations on $x_1, x_2, \ldots, x_p$ (subject to a mild nonsingularity condition on the coefficient matrix of the normal
equations) we can use it to predict the value of x_1 for a new set of values of
$x_2, \ldots, x_p$. Of course without assuming any underlying probability model, it is
not possible to assess the uncertainty involved in such prediction.

Yule followed the above approach on his pauperism data to study the dependence of the proportion y of paupers in a Union on the out-pauper/in-pauper
ratio x. Later he considered sets of similar data for a number of consecutive
years and related percent changes in y to simultaneous percent changes in x and
two other variables (proportion of the old and size of the population). All these
studies appeared in a number of papers published from 1897 onwards (see Stigler
(1986) pp. 345–358).

Yule's approach widened the scope of regression analysis considerably—one
could follow it even when the underlying distribution is markedly non-normal as
is often the case with social science data. Also as is seen from Yule's pauperism
study, one can use it to analyse in a sort of way various causal factors affecting the variation in the variable of interest. (One could answer such questions
as: Does proportion of paupers increase as out-relief is given more liberally?
Are changes in the former caused by changes in the latter even when the
changes in proportion of the old and size of the population are held fixed?) In
social sciences controlled experiments are generally infeasible and such analysis
applied to observational studies often provides a sort of convenient surrogate for
such experiments. Furthermore, Yule-type regression analysis with simultaneous
observation of multiple causal variables later led to the practice of measurement
of concomitant variables in sample surveys and controlled experiments (when
stratification and blocking according to these cannot be done).

Yule's work showed the Galton–Edgeworth–Pearson regression theory in a
new light and exposed its connection with the least squares theory of Legendre
and Gauss (Chapter 8). It also made available quicker computational methods for
the regression coefficients in (9.8.9). Yule did not consider any kind of inductive
inference about the regression coefficients or the predicted value. However, it is

obvious that the entire apparatus of inference for the Gauss linear model assuming normality of errors, becomes applicable if we adopt a *via media* model for the joint distribution of $X_1, X_2, \ldots, X_p$ under which the conditional distribution of X_1 given $x_2, \ldots, x_p$ is homoscedastic normal with mean linear in $x_2, \ldots, x_p$, although the marginal distribution $X_2, \ldots, X_p$ is left unspecified. Such a model came to be used extensively later on by Fisher and others.

Inductive inference about ρ

9.8.5   Problems of making inductive inference about correlation arose even before the formula for the correlation coefficient became standardized. Galton conjectured that the correlation between measurements on the same two particular organs remains constant for different sub-species of a species of living beings. Weldon found several such correlations for different local races of shrimps and crabs on the basis of large samples. As the differences in such a correlation over various local races seemed to be negligibly small, he tentatively upheld Galton's conjecture. Karl Pearson realized that two correlation coefficients ρ_1 and ρ_2 cannot be judged as equal without taking into account the S. E.s (Pearson considered probable errors) of their sample estimates r_1 and r_2 and comparing the observed difference $r_1 - r_2$ against its S. E. In his 1896 paper he not only derived the product moment formula for the estimate r of a bivariate normal correlation by a maximum-likelihood type argument (subsection 9.8.2), but also gave an expression for the S. E. of r. Two years later in a paper written jointly with Filon, Pearson sketched a general method for the derivation of S. E.s of estimates in such cases and gave a modified expression for the S. E. of r.

The derivation of Pearson and Filon (1898)(see Karl Pearson (1948)) is extremely baffling to present-day readers. Not only does it fail to come up to the standards of modern rigour, but as was often the practice at that time, it uses the same terminology ('frequency constant') and notation for a parameter and its estimate. Besides, apparently Pearson and Filon did not see clearly that their method would apply only in the case of what later came to be called maximum likelihood estimates.[12] Using modern notations and reading between the lines of

---

[12]Possibly because of all this, some authors (Stigler (1986) p. 345, Hald (1998) p. 627) suggest that Pearson was implicitly following the Bayesian posteriorist path to point estimation based on the uniform prior and deriving not the sampling theory S. E. but the posterior S. D. in large samples (cf. subsections 4.4.16 and 7.3.6). But this does not seem correct because nowhere in Pearson (1896) and Pearson and Filon (1898) do we find any hint of an assumed prior and expressions like (9.8.24) are always interpreted as giving the chance of getting the sample under θ. Also Pearson and Filon (1898) (pp. 183, 185) clearly state that they are considering variation over different randomly selected samples. Besides, they obtain the variances and covariances of the estimates by putting $\theta = \hat{\theta}$ in the *expected* second derivatives of $\log f(X|\theta)$ under θ, and not directly as the second derivatives of $\log f(x|\theta)$ at $\hat{\theta}$ as would have been appropriate had the Bayesian approach been followed.

Pearson (1896) and Pearson and Filon (1898) we present below what we think the intended heuristic reasoning behind the derivation was.

Let X be a (possibly vector-valued) random variable with density function $f(x|\theta), \theta = (\theta_1, \ldots, \theta_k)$ and let $x_1, \ldots, x_n$ be a random sample of observations on X. The sample density is

$$p_\theta = p_\theta(x_1, \ldots, x_n) = \prod_1^n f(x_\ell|\theta). \tag{9.8.20}$$

We take the value $\hat{\theta} = (\hat{\theta}_1, \ldots, \hat{\theta}_k)$ for which (9.8.20) is maximum as the estimate of θ. Under appropriate conditions $\hat{\theta}$ is the solution of the θ-equations

$$\frac{\partial \log p_\theta}{\partial \theta_i} = 0, \quad i = 1, \ldots, k. \tag{9.8.21}$$

Consider the ratio $p_\theta/p_{\hat{\theta}}$ where θ is the true value of the parameter. (Pearson and Filon (1898), pp. 181–184, consider here a ratio of the form $p_{\theta+\Delta}/p_\theta$ and state that θ represents 'the observed values' of the 'frequency constants' and $\Delta = (\Delta_1, \ldots, \Delta_k)$, 'the errors made in the determination of frequency constants'.) By Taylor expansion *around* $\hat{\theta}$, neglecting third and higher derivative terms,

$$\log \frac{p_\theta}{p_{\hat{\theta}}} = \sum_{i=1}^k (\theta_i - \hat{\theta}_i)\left(\frac{\partial \log p_\theta}{\partial \theta_i}\right)_{\hat{\theta}} + \frac{1}{2}\sum_{i,j=1}^k (\theta_i - \hat{\theta}_i)(\theta_j - \hat{\theta}_j)\left(\frac{\partial^2 \log p_\theta}{\partial \theta_i \partial \theta_j}\right)_{\hat{\theta}}. \tag{9.8.22}$$

The first term on the right vanishes because $\hat{\theta}$ satisfies (9.8.21). (Pearson (1896) was clear on this point. However, Pearson and Filon (1898) pp. 182–183 wrongly thought that the term could be neglected for large n for any [consistent] estimate because $1/n(\partial \log p_\theta/\partial \theta_i)_\theta \simeq E_\theta(\partial \log f(X|\theta)/\partial \theta_i) = 0$. This is untenable as $n(\theta_i - \hat{\theta}_i)$ would tend to be large in probability.) For large n we have by (9.8.20)

$$\frac{1}{n}\left(\frac{\partial^2 \log p_\theta}{\partial \theta_i \partial \theta_j}\right)_{\hat{\theta}} \simeq E_\theta\left(\frac{\partial^2 \log f(X|\theta)}{\partial \theta_i \partial \theta_j}\right) = -\gamma_{ij}(\theta) \text{ (say)}. \tag{9.8.23}$$

Hence, replacing $\gamma_{ij}(\theta)$ by $\gamma_{ij}(\hat{\theta})$, for large n we can heuristically write

$$p_\theta \simeq p_{\hat{\theta}} \cdot e^{-(n/2)\Sigma_{ij=1}^k(\hat{\theta}_i-\theta_i)(\hat{\theta}_j-\theta_j)\gamma_{ij}(\theta)}. \tag{9.8.24}$$

Pearson and Filon (1898) pp. 184, 186, imply that since in large samples (9.8.24) represents the probability of the sample occurring under θ, the second factor on the right gives 'the frequency distribution for errors in the values of the frequency constants'. In other words, in large samples, $\hat{\theta}_1, \ldots, \hat{\theta}_k$ follow approximately a k-variate normal distribution with means $\theta_1, \ldots, \theta_k$ and dispersion $(n\gamma_{ij}(\hat{\theta}))^{-1}$. (Presumably, they took it for granted that if in (9.8.24) we effect a transformation from $x_1, \ldots, x_n$ to a new set of variables which includes $\hat{\theta}_1, \ldots, \hat{\theta}_k$ with Jacobian

$|J|$, then the integral of $p_{\hat\theta} \cdot |J|$ over the extra variables would give the right constant $\sqrt{\det(n\gamma_{ij}(\hat\theta))}/(2\pi)^{k/2}$ upto the order of approximation considered.)

Pearson and Filon (1898) utilized the above result to find the asymptotic dispersion matrix of the (maximum likelihood) estimates of σ_1^2, σ_2^2, and r. (As the sample means are distributed independently of these estimates, one could confine attention to the latter.) Taking the square root of the relevant diagonal element of $(n\gamma_{ij}(\hat\theta))^{-1}$ here, they found the expression $(1-r^2)/\sqrt{n}$ for the S. E. of r. Earlier, Pearson (1896) had taken in effect just the reciprocal of the relevant diagonal element of $(n\gamma_{ij}(\hat\theta))$ and had come out with the expression $(1-r^2)/\sqrt{\{n(1+r^2)\}}$, which actually represents the conditional S. E. of r given the estimates of σ_1^2, σ_2^2.

The S. E. of r came handy for settling the question of homogeneity of the correlation coefficient between the same two measurements for different local races of a living organism. Thus Weldon had found on the basis of fairly large samples the values of the correlation coefficient between the lengths of total carapace and the post-spinous portion of shrimps for five local races as follows:

Locality	Plymouth	Southport	Roscoff	Sheerness	Helder
Sample size (n)	1000	800	500	380	300
r	0.81	0.85	0.80	0.85	0.83

At first sight it seemed that the differences in r-values were negligible enough to make the hypothesis of homogeneity tenable. But when the differences $r_i - r_j$ were tested against the corresponding estimated S. E.s $\sqrt{\{(1-r_i^2)^2/n_i + (1-r_j^2)^2/n_j\}}$, it was found that not all the differences were insignificant.

In this context it may be mentioned that it was later found that the approximation of the distribution of r by $\mathcal{N}(\rho, (1-r^2)^2/n)$ is poor unless n is very large, particularly if $|\rho|$ is close to 1. The problem of testing homogeneity of several bivariate normal correlations was satisfactorily solved when in 1921, R. A. Fisher showed that if we transform to $Z = \tanh^{-1} r$, then even for moderate sample sizes, Z is distributed as $\mathcal{N}(\tanh^{-1}\rho, 1/(n-3))$ to a close approximation. On the basis of this Fisher suggested that to test the homogeneity of bivariate normal correlations r_i based on independent sample sizes $n_i, i = 1, \ldots, c$, we may find the corresponding transforms Z_i and compute

$$T = \sum(n_i - 3)Z_i^2 - \frac{(\sum(n_i - 3)Z_i)^2}{\sum(n_i - 3)}.$$

Under homogeneity, T would tend to follow the χ^2 distribution with $c - 1$ d.f. In the case of Weldon's problem described above $c = 5$ and the sample sizes are quite large. We find that the observed value of T is 12.71 with d.f. 4. The corresponding right-tail P-value is 1.35%. Hence this test also makes the homogeneity hypothesis suspect and confirms Pearson's conclusion.

9.9   Statistical induction—19th century scenario

Post-Laplacian era   **9.9.1**   In this chapter we have discussed the developments in the field of statistics that took place in the
post-Laplacian era, i.e. roughly in the period from the fourth decade onwards
of the 19th century. (However, we have not included here certain developments
relating to models and fitting of models which began towards the end of this
period; these will be taken up in the next chapter.) It is clear that the main
thrust of the developments in this period was towards expanding the domain of
statistics. So far as principles of statistical induction are concerned, very little
emerged that can be called really new. The two approaches developed by Laplace
continued to rule the day. The first was the posteriorist approach which involved
a complete specification of the model and the assumption of a uniform prior. As
mentioned earlier (section 7.4) there was a lot of outcry against this assumption
particularly from the members of the newly emerging frequentist school of probability. On the other hand, there were those like Edgeworth who held the view
that prior probabilities should be based on knowledge of relative frequency of
occurrence as experienced in the past. When experience does not in any way tell
us that one value of the parameter is more probable than another, Edgeworth
thought that the two must be judged equiprobable. The other approach was that
based on large-sample sampling theory which did not require any hard specification of the model. This was generally used for comparing two proportions or
means. Edgeworth himself though inclined towards posteriorist thinking, was not
averse to such comparison and employed it on demographic and economic data.
In contrast with the error-theorists and in conformity with Galton's emphasis on
the importance of variation, Edgeworth asserted that such comparisons remain
meaningful even when a sample arises not from repeated observation of a single
'true value' but from measurements made on a multitude of 'true values' as in
the case of returns of prices, exports, imports, marriages, births, etc. But as we
noted earlier (subsection 8.6.3) even when a sampling theory test was employed,
generally the interpretation was instantial in spirit—no behavioural significance
was attributed to the measure of inductive uncertainty.

Behavioural approach   **9.9.2**   It seems that there was an inhibition deep-
seated in the 19th century psyche which militated
against adopting the frequentist view of probability straightaway for the assessment of inductive uncertainty. A typical expression of this inhibition is found
in the report of the committee headed by Poisson, which was appointed by the
French Academy of Sciences to deliberate on the efficacy of the statistical comparison of two types of operation for gallstone. The Committee deprecated such
comparison. The burden of their objection was that in statistical affairs we must
strip a man of his individuality and consider him as a member of some population, whereas in medicine we are concerned with the particular patient who
is to be treated; obtaining more data about different individuals is irrelevant

to the particular case at hand. This kind of attitude persisted for a long time (see Hacking (1990) Chapter 10).

The first statistician who protested against this inhibition and adopted openly the behavioural approach to induction was the same C. S. Peirce whom we met as a spokesman for the frequency theory of probability earlier (subsection 9.6.1). To counter the above kind of unwillingness about deciding on an individual case on the basis of statistical studies, Peirce in a paper written in 1878 asserted, 'probability essentially belongs to a kind of inference which is repeated indefinitely ... logicality inexorably requires that our interests shall *not* stop at our own fate, but must embrace the whole community ... To be logical men should not be selfish.' (See Buchler (1955) Chapter 12.) This, it may be noted, is akin to the attitude underlying the interpretation given to measures of assurance in statistical induction in the Neyman Pearson theory (see subsection 4.2.2).

Peirce clearly spelt out the philosophy of the behavioural (he called it 'synthetic') approach by saying that in it we are not concerned with the probability of our conclusion but with 'the degree of trustworthiness of our proceeding'. 'We may ... define the probability of a mode of argument', he wrote, 'as the proportion of cases in which it carries truth with it' (ibid). Also 'we only know that premises obtained under circumstances similar to the given ones (though perhaps themselves very different) will yield true conclusions at least once in a calculable number of times ...' (see Buchler (1955) Chapter 13). At another place (Buchler (1955) Chapter 14) he wrote, '... it is by no means certain that the conclusion actually drawn in any given case would turn out true in the majority of cases, ... but what is certain that in the majority of cases the method would lead to *some* conclusion that was true ...'. Unfortunately Peirce could give only a few illustrations of the application of this principle. One was that of interval estimation of the proportion p of individuals in a large population belonging to a particular class, on the basis of a random sample of size n. If S is the random number of members of the class included in the sample we may choose $d > 0$ to make the probability

$$P \left(\left| \frac{S}{n} - p \right| < d \middle| p \right)$$

high. (We may use the large-sample normal approximation to the distribution of S/n for this with $p(1 - p)$ replaced by its upper bound $\frac{1}{4}$.) Peirce implied that from this knowing the value of p we may get bounds for S/n which would hold with high probability and knowing the value of s of S we may find bounds for p which would hold with high 'probability'. But he was very clear that this latter 'probability', although it is also of the nature of a relative frequency, is of a different kind from the former. (As we continue the drawing of the sample the value of s/n and hence the bounds themselves will become modified.) Another application considered by Peirce was that of testing the equality of two proportions p_1 and p_2 on the basis of random samples (specifically, testing that the proportion of female births is same for white and non-white people on the basis of a population census). He computed the probability of a difference in proportions as large as that observed under $p_1 = p_2 =$ some unspecified p (again using the upper bound $\frac{1}{4}$ of $p(1 - p)$). When this probability was very small he would prescribe the rejection of H_0: $p_1 = p_2$. (In the case of female births he noted that discrepancy in proportions as large as that observed in the census considered would occur under H_0 'only once out of 10^{10} censuses'.) All this shows that Peirce was quite clear in his mind about the interpretation of the measure of uncertainty of an inductive inference in the behavioural sense (see Buchler (1955) Chapters 13–14). However these behaviouralistic ideas of Peirce were not immediately followed up by others—statisticians of the next generation were engaged in the exciting task of developing the exact sampling theory of various statistics and using these for inference and the business of interpretation took a back seat. The behavioural approach was revived by Neyman and Pearson around 1930 without reference to Peirce.

Abstraction of the method

9.9.3 Before concluding this chapter, we note one feature of the application of probabilistic reasoning for the solution of practical problems that gradually developed through the 19th century—abstraction of the method of solution of a problem and the underlying reasoning from the problem itself. This feature was noticeable in the writings of Gauss, but it assumed prominence in the works of Edgeworth and Karl Pearson. These authors, while solving a problem statistically, always remained alive to the potential of the method developed in the general context and stressed its possible use in other problems in diverse fields. (This is also true with regard to the model-fitting methods developed by Karl Pearson towards the end of the 19th century, which we will describe in the next chapter.) This led to the emergence of statistics as a new area of knowledge comprising an increasingly large battery of inductive techniques, which are themselves domain-free but may be applied in widely different contexts in diverse domains. As it happened, in the 20th century the techniques themselves and their rationale, as distinct from particular problems to which they might be employed, became the subject matter of investigation of quite a few mathematically inclined researchers who had no interest in any special area of application.

10

MODERN ERA: THE SUPERSTRUCTURE BUILDS UP

Into the 20th century | **10.1**   The last chapter brought us to the threshold of the 20th century. We had seen in the earlier chapters that many of the basic concepts of probability and statistics had developed through the two centuries which ended roughly with the period of Laplace and Gauss, but their application to statistical induction had remained confined to a narrow area. The last chapter described how towards the middle of the 19th century, these confines were broken and with the extension of application to the field of biometry during the second half of the same century new concepts and methods originated and developed. In the 20th century this process of expansion of domain continued with accelerating momentum. As scope for application of statistical induction was found in newer and newer fields (agriculture, industry, anthropometry, psychometry, medicine, health, population, social science, development and planning, genetics, remote sensing, etc.), further concepts and methods burgeoned and were added to the repertoire of statistics. What is significant is that, following the trend which started towards the end of the 19th century, all these concepts and methods were abstracted from the particular problems which gave rise to them and looked upon as belonging to a new discipline. Also the logical bases of the methods of statistical induction started being thoroughly examined, apart from the particular context of their application. In other words, the identity of statistics as a separate branch of knowledge came to be established.

In Part I of this book we tried to give an integrated broad view of the superstructure of statistics that built up during the better part of the 20th century. A closer look at the developments taking place through this period would show a bewildering variety of embellishments to the general structure. We will not venture to give a brick-by-brick description of the process of building which led to the erection of this structure. (Apart from various practical limitations to undertaking such a stupendous project, we feel the inevitable attention to too much detail and too many foci which such an effort would entail would make us lose sight of the integral whole.) Rather, we will try to recount how each major part of the structure came to be initially conceived and its construction began. Occasionally we will follow up our description of some substructure begun earlier and connect it to later developments nearer our time. But as this will be done only selectively, there are bound to be numerous gaps in our account.

One distinctive feature of developments in statistics during the later part of this period has been the increasingly dominant role of electronic computers in statistical practice. Inevitably this has also affected the growth of the theory during this period. We will, however, touch this aspect only peripherally.

The major participants in the work at the site in this period were initially Karl Pearson and 'Student' (W. S. Gosset). Later, for some time, the towering figure of R. A. Fisher dominated the scene, almost overshadowing others. J. Neyman and E. S. Pearson, and a little later, A. Wald entered the arena to influence the growth of statistics in significant ways. At another corner, Jeffreys, de Finetti, and others went on giving shape to a part of the structure in their own fashion. In Chapter 4 (and also through some anticipatory remarks in Chapters 8 and 9) we referred to and described to some extent the contributions of these major players. For reasons similar to those mentioned above, we will generally not take upon ourselves the chronological narration of the stages through which the final shapes of their contributions were reached.

| Models and parameters | The leitmotif of statistics, as it emerged distinctly through the work of Galton described earlier, is to know the pattern of variation of unpredictable observable characters on the basis of available observations. The variation pattern of characters is represented by their probability distribution. When the characters are quantitative, the probability distribution is given in terms of its pmf or pdf (or more generally cdf). In the objective approach to statistical induction (sections 2.1 and 4.1), it has been customary to classify these distributions into distinct families or *models*. The different distributions under the same model have pmfs or pdfs of the same functional form and are distinguishable only by different values of certain numerical constants.[1] This kind of two-level description of a distribution can be traced back at least to Laplace (subsection 8.2.1). Edgeworth and Karl Pearson both used it explicitly in their work. But it was Fisher who in his seminal paper of 1922 accentuated it most prominently, introducing *inter alia* the term *parameter* to denote the numerical constants (see Stigler's comments in Savage (1976)).

| Types of induction | *Using the term 'model' in the above sense*, we consider the developments in statistical induction that originated in late 19th century and proceeded through the 20th century, under the

---

[1]Which family of distributions would be said to represent a model is, however, largely a matter of convention. Thus the normal model is represented by the family of pdfs const. $\exp\{-(a + bx + cx^2)\}$, $-\infty < x < \infty$, with $c > 0$. But it can be considered as a sub-family of the wider family const. $\exp\{-(a + bx + cx^2 + dx^4)\}$, $-\infty < x < \infty$, where $c > 0$ and $d \geq 0$. The scope for further broadening is also obvious. Generally a family is defined so that the corresponding functional form gives distributions which have similar descriptive features (e.g. frequency curves with similar shape, range etc.) and analytic properties (like sampling distributions of statistics).

following three broad headings:

(A) Model-selecting induction,
(B) Model-specific induction,
(C) Model-free induction.

In the objective approach a problem of induction involving variable characters is comprehensively solved if an appropriate probability model can be specified and suitable values assigned to the parameters in it. Generally to specify the model we have to look into the nature of the phenomenon to which the data relate. Sometimes on the basis of such enquiry we can formulate a set of fairly strong postulates about the phenomenon and this leads to the form of the model. A model arrived at in this way is called a substantively derived or *substantive* model (cf. Cox and Wermuth (1996) p. 14). In such a case there is no problem of model selection as such. More frequently in practice, however, the set of postulates that we can confidently assume (things like continuity of the variate, independence, identity of distribution of different observables, and shape and range of the frequency curve) is too weak to pinpoint a definitive model. It only suggests a class of models, which may be broad or narrow depending on the information carried by the postulates. In this case the model itself must be induced from the data or be part of our inductive concern (section 2.9). The model here is called an empirically selected or *empirical* model.[2] *Model-selecting* induction with all its phases occurs only when our aim is to find such empirical models. However, as we will see in the following, usually whenever a model is set up, the problems of fitting and judging the goodness of fit of the model are there, whether the model has been derived substantively or selected empirically.[3]

Sometimes the problem of selection of a model is solved a priori. It may be that we have enough past experience about the kind of data in a field to

---

[2]Substantive models try to explain the stochastic mechanism generating the data and are supposed to be scientific, whereas empirical models are merely descriptive and utilitarian and are regarded as technological in spirit. However, sometimes a set of postulates is proposed tentatively to generate an apparently substantive model. It may happen that two different sets of postulates, give rise to the same model in this way (we will encounter an instance illustrating this later). Thus the fact that such an apparently substantive model gives a good fit in a particular case is no guarantee that the stochastic mechanism represented by a particular set of postulates is true. A model is truly substantive if its underlying postulates have firm theoretical and empirical bases. As noted above, statistical induction has a lesser role in the choice of a truly substantive model.

[3]There is another process, called 'secondary induction' by Kneale (1949) (pp. 246–250), through which we seek to explain a wide sweep of diverse phenomena. This is exemplified by Newton's 'Law of Gravitation' and Mendel's 'Theory of Inheritance', and generally constitutes what Kuhn described as 'revolutionary science' (see Lehmann (1990)). Statistics has no role to play in it (although statistical induction may be involved in the individual studies encompassed by such secondary induction).

warrant that some particular model holds for it. Or it may be that it is a case of virtue of necessity—the size of the sample is so small that we have no other recourse but to start by assuming some standard model at least as a working hypothesis (model-selecting induction, generally speaking, can be attempted only when the sample is not too small). In all such cases induction is *model-specific*, i.e. concerned with the parameters given a particular model.

The third type of induction, which we call *model-free* arises when the model is not known (or as is the case when the observable characters are attributes, the question of a model does not arise) and we are not concerned with comprehensive induction about the underlying probability distribution. Rather we are interested in inducing about certain special aspects of this distribution and this we want to do without specifying the model in full.

The above division of problems of induction into types is done partly for historical reasons and partly for the sake of convenience of describing the development of their solutions. It is by no means water-tight. In fact, we will see that as it is the usual practice to select an empirical model within a class of models, model-selecting induction can be looked upon as finding the appropriate value of some identifying parameter which varies over the class and thus in a sense is 'model'-specific (with the class playing the role of the 'model'). Similarly model-specific induction can be interpreted as model-selecting in a broad sense. Also, concepts and methods developed for one type of induction readily apply to and are freely drawn upon for others.

10.2   Model-selecting induction

| Beginning |

10.2.1   Problems involving model selection arose towards the end of the 19th century mainly in the context of graduating empirical frequency distributions of variates that can be taken as continuous. ('Graduation' here means smoothing with a view to finding an idealized form which is supposed to represent the population distribution.) In the simplest cases (historically it was such cases which arose first) the obliteration of labels mentioned in section 9.1 is total; the observations relate to members of a group that is ostensibly homogeneous, there being no explanatory variables which vary across the group. In other words, the observations can be regarded as on i.i.d. random variables.

| Frequent encounter with non-normality |

We noted in the last chapter (subsection 9.7.6) that Galton, although he re-attuned the goal of statistical induction to the study of variation, could not shake himself free from the Queteletian doctrine that if a group is homogeneous the distribution of a continuous variate over it would conform to normal. With the expansion of statistical studies, as the empirical distributions of varieties of variates from diverse fields became available, it was discovered that departures from normality were quite common. Thus Edgeworth found that the distribution of prices of a commodity was markedly skewed. Venn cited examples of meteorological variates like pressure and temperature whose distributions were palpably non-normal (see Stigler (1986) p. 331). In the field of biometry, Weldon considered the distribution of relative frontal breadth of a race of crabs found at Naples and the form of the histogram was decidedly tilted to the right. In such cases departure from normality looked like an inherent property of the phenomena observed—it could not be attributed to sampling fluctuation, the samples generally being of substantial size. It gradually dawned upon statisticians that in many situations non-normality may be a fact of life. In such cases selection of a suitable non-normal model became an unavoidable part of statistical induction.

| 'Reservoir of models' |

The strategy followed to tackle the problem has been already mentioned while defining an empirical model. Taking into account the various features of the situation at hand, one starts with a broad reference class of models ('reservoir of models'—Lehmann (1990)) which are distinguished by the values of some identifying parameter. The class is so chosen that it covers a wide variety of shapes. The appropriate value of the identifying parameter is then ascertained on the basis of the data. Once a suitable model is thus selected, there remains the problem of 'fitting' the model, i.e. estimating the specific parameters involved in it. Substantive questions relating to the phenomenon under study (e.g. questions about location, spread, etc.) are usually (but not always) formulated and answered in terms of the specific parameters. (We will see, however, that the distinction between identifying and specific parameters and that between model selection and fitting are not clear-cut in certain approaches.) Historically, the classes of models which were first considered all used the normal distribution as a sort of anchor and the latter was either a member of the class or could be approached as a limit from within the class. We discuss these first. We consider four broad methods which were used for generating the reference class, namely, the methods of (a) transformation, (b) asymptotic expansion, (c) mixture, and (d) formal embedding.

| Method of transformation |

10.2.2 Somewhat paradoxically, one of the earliest applications of the method of transformation to generate a non-normal curve was due to Galton himself. When a variate X with a markedly skew observed distribution cropped up, Galton sought to explain that by assuming that 'the postulate of elementary errors' acts in a multiplicative way so that $\log X$, being the resultant of a large number of additive random components, tends to be normally distributed. At the instance of Galton, around 1879, McAlister made a study of the resulting lognormal distribution. It was found, however, that the logarithmic transformation does not by itself produce a class of models sufficiently rich to cover all types of deviation from normality.

Generally, if we have a variate X, we can standardize it to $U = (X - \gamma)/\delta$ using suitable base and scale parameters γ and δ and then consider a transform

$Z = g_\lambda(U)$. Here for each fixed λ (which may be real- or vector-valued), $g_\lambda(u)$ is some monotonic nonlinear function of u. If, for each λ, Z follows the standard normal distribution, generally X would have a non-normal distribution with parameters λ and γ, δ. In this set-up λ can be looked upon as an identifying parameter and γ, δ as specific parameters, although the distinction here is somewhat blurred. Using a model generated in this way, one can utilize standard normal tables to get probabilities of various interval ranges of values of X. Also, from the theoretical point of view one can suggest that the model occurs because the postulate of elementary errors works on a suitably transformed scale.

An early exploitation of this idea was due to Edgeworth, who in an 1898 paper (see E. S. Pearson (1967)) proposed that $g_\lambda(\cdot)$ be chosen so that $g_\lambda^{-1}(z)$ is a polynomial in z. Of course $g_\lambda^{-1}(z)$ has to be monotonic, at least over the effective range of z, and the number of unknown coefficients in it should be small. All this rendered Edgeworth's proposal to be of limited utility.

After this there were some isolated instances of exploration of this method for graduating empirical distributions. Also there were some applications of the method for approximating certain complicated sampling distributions. Notable among these was Fisher's proposal, made in the early 1920s, that the transform $z = \tanh^{-1} r$ of the correlation coefficient r of a sample from the bivariate normal population be taken as normally distributed.

| Johnson system |

The method of transformation for model-selecting induction was more comprehensively developed much later by Johnson (1949). Johnson's idea was to take $\lambda = (\lambda_1, \lambda_2, \lambda_3), -\infty < \lambda_1 < \infty, 0 < \lambda_2 < \infty, \lambda_3 = 1, 2, 3$ and

$$g_\lambda(u) = \lambda_1 + \lambda_2 h_{\lambda_3}(u), \qquad (10.2.1)$$

where $h_{\lambda_3}(u)$ is $\log u, 0 < u < \infty, \log[u/(1-u)], 0 < u < 1$, and $\log(u + \sqrt{(1 + u^2)}), -\infty < u < \infty$, respectively for $\lambda_3 = 1, 2, 3$.

Varying λ and γ, δ, he got a class of models with a wide variety of shapes and ranges.

| Box–Cox |

Among later developments we mention only the use of a kind of shifted power transformation by Box and Cox (1964). Specializing to the present context of no explanatory variables, their proposal amounts to taking $\lambda = (\lambda_1, \lambda_2, \lambda_3), -\infty < \lambda_1 < \infty, 0 < \lambda_2 < \infty, -\infty < \lambda_3 < \infty$ and $z = g_\lambda(x - \gamma)$ (here there is no loss of generality in taking $\delta = 1$), where

$$g_\lambda(u) = \lambda_1 + \lambda_2 t_{\lambda_3}(u), \quad 0 < u < \infty, \qquad (10.2.2)$$

with $t_{\lambda_3}(u) = (u^{\lambda_3} - 1)/\lambda_3$ or $\log u$ according as $\lambda_3 \neq 0$ or 0.

| Method of asymptotic expansion |

10.2.3 This method has its origin in the postulate of elementary errors (subsection 8.2.8), but here it is supposed that the number of elementary errors is not large enough to make the normal approximation close.

We saw in subsection 8.2.7 that for i.i.d. random variables $X_1, \ldots, X_n$ with mean μ_1, variance σ_1^2, and rth cumulant $k_{1r}(r \geq 3)$, Poisson, Bessel, and later Tshebyshev derived the formal representation of the density function of $Z = \left(\sum_1^n X_\ell - n\mu_1 \right) / (\sqrt{n}\sigma_1)$ as

$$\frac{1}{\sqrt{(2\pi)}} \exp \left\{ \sum_{r=3}^{\infty} \frac{k_{1r}}{n^{(r/2)-1}\sigma_1^r} \frac{(-D)^r}{r!} \right\} \left(e^{-z^2/2} \right), \quad -\infty < z < \infty, \qquad (10.2.3)$$

D standing for d/dz. Now if an arbitrary random variable X with mean μ, variance σ^2, and the rth cumulant $K_r(r \geq 3)$ is *postulated* to have arisen as the sum $\sum_1^n X_\ell$ of a finite number n of i.i.d. random variables, then using notations as above,

$$\mu = n\mu_1, \quad \sigma = \sqrt{n}\sigma_1, \quad K_r = nk_{1r}, \quad r \geq 3. \qquad (10.2.4)$$

Hence denoting the density of $Z = (X - \mu)/\sigma$ by $f(z)$ from (10.2.3) we have the formal representation

$$f(z) = \frac{1}{\sqrt{(2\pi)}} \exp \left\{ \sum_{r=3}^{\infty} \frac{K_r}{\sigma^r} \frac{(-D)^r}{r!} \right\} \left(e^{-z^2/2} \right). \qquad (10.2.5)$$

Taking the partial sum consisting of a few initial terms in the expansion of (10.2.5) we get a model approximating $f(z)$.

| Gram–Charlier expansion |

One way of expanding the operator in (10.2.5) would be a straightforward development in powers of D. This was proposed among others by Tshebycshev, Gram, and Thiele (see Hald (1998) p. 344) and its utility for modelling was studied in detail by Charlier. This expansion is generally known as Gram–Charlier Series. But such an expansion rides rough-shod over the postulate about the genesis of X made at the outset—the orders of the successive terms of the series in powers of n (as determined by the substitution of (10.2.4)) are not monotonically decreasing. This makes a partial sum of the series of doubtful utility as an approximation to $f(z)$ up to a certain order.

| Edgeworth expansion |

For this reason it was proposed that the operator in the original representation (10.2.3) be expanded in powers of $n^{-(1/2)}$ and then K_r/σ^r be substituted for $n^{1-(1/2)r}k_{1r}/\sigma_1^r$. If we retain terms upto $O(n^{-1})$ we thereby get

$$f(z) \simeq \left[1 + \frac{1}{3!} \frac{K_3}{\sigma^3} + \left\{ \frac{1}{4!} \frac{K_4}{\sigma^4}(-D)^4 + \frac{1}{2!(3!)^2} \left(\frac{K_3}{\sigma^3} \right)^2 (-D)^6 \right\} \right] e^{-z^2/2}. \qquad (10.2.6)$$

Such an expansion was originally suggested by Tshebyshev (see Gnedenko and Kolmogorov (1954) p. 195). Its utility for modelling was studied in detail by Edgeworth (after whom the series is usually named) and asymptotic properties were investigated by Cramér (see Cramér (1946) Chapter 17). Generally for the purpose of graduation a model consisting of the first few terms of the Edgeworth series is considered preferable to one consisting of the same number of terms of the Gram-Charlier series. However, it has been found that models of the type (10.2.6) are satisfactory only when the departure from normality is moderate. Also such models are awkward in practice as not infrequently they produce negative ordinates at the tails.

Apart from graduating empirical distributions, the Edgeworth Series representation has been used for approximating complex sampling distributions for which the cumulants can be determined. Cornish and Fisher in 1937 (see Fisher (1950)) inverted such series to derive expansions for the quantiles of various useful statistics in powers of corresponding standard normal quantiles; such expansions facilitate the tabulation of the former quantiles. At another level, the behaviour of various normal-theory statistics, under an assumed Edgeworth-type model has been studied to judge the robustness of standard procedures against various types of departure from normality (see e.g. Gayen (1949)). Among more recent developments, we mention a result of Bhattacharya and Ghosh (1978) which gives an Edgeworth-type expansion for the distribution of a smooth function of sample moments for a multivariate population.

| Method of mixture |

10.2.4 In both the above methods for model generation it is supposed that the group considered is homogeneous and departure from normality arises because all the requirements of the postulate of elementary errors are not directly met for the variate. An obvious alternative would be to postulate that the group is a mixture of more than one homogeneous group for each of which normality holds. Although this possibility had been seen earlier also (see Stigler (1986) pp. 215–218, 268), it was Karl Pearson who first made a comprehensive study of the utility of such mixture models for graduation in a paper (this was Pearson's first statistical paper) in 1894 (see Pearson, K. (1948)). Pearson started with the class of models

$$\lambda \frac{1}{\sigma_1 \sqrt{(2\pi)}} e^{-(x-\mu_1)^2/(2\sigma_1^2)} + (1-\lambda)\frac{1}{\sigma_2 \sqrt{(2\pi)}} e^{(x-\mu_2)^2/(2\sigma_2^2)}, \quad -\infty < x < \infty.$$

$$(10.2.7)$$

Here $\lambda (0 \leq \lambda \leq 1)$ denotes the putative proportion of the first component population in the mixture. Further components may be introduced but that would make the models increasingly complex. Of course in some situations a priori information may allow us to simplify the model by taking some μs or σs equal.

| Two attitudes | There can be two attitudes to a mixture model of the |

There can be two attitudes to a mixture model of the type (10.2.7). Firstly, we can regard that (10.2.7) for varying values of the parameters represents just a class of models out of which one is to be selected for graduation. For this purpose, such models hardly provide a satisfactory answer—with 5 parameters they give us only a limited choice of shapes. Also their likelihood functions are not well-behaved (they are unbounded if σ_1^2 or σ_2^2 can be made arbitrarily small) and create problems of the type mentioned in subsection 4.3.10.[4] The second attitude becomes appropriate when we have reasons to believe that the given population is actually the mixture of more than one physically existent component subpopulation. This is true of many practical situations, e.g. in the case of aggregates of items produced in mass, aggregates of grains in sedimentological studies, heterogeneous biological populations of various types, and especially unclassified fish populations. Here a major aim of induction may be to 'dissect' the composite population, i.e. to empirically identify the components and determine the rates at which they are mixed. In some situations here the process of model-selecting induction may be carried further to include the task of sorting out of the observations into subsets according to subpopulations of origin. In recent years this idea has been pursued, especially in the multivariate context, to develop an approach to *Cluster Analysis* (see Seber (1984) p. 386, Titterington *et al* (1985) pp. 25–27, 104–106, 113–114).

| Compounding | The mixture distribution (10.2.7) is a special type of |

The mixture distribution (10.2.7) is a special type of what is called a *compound* distribution. Generally, if we have an initial family of distributions $f(x \mid \psi)$, where $x \in \mathcal{X}$ for each value of the parameter ψ, a compound distribution may be generated by postulating that ψ itself is random subject to a compounding distribution. By taking expectation of $f(x \mid \psi)$ over ψ for each x, we get a compound distribution. For the compound distribution also $x \in \mathcal{X}$, and when the compounding distribution involves some parameters, these become the parameters of the former. In this way we can generate a new class of models. In the case of (10.2.7), $\psi = (\mu, \sigma^2), -\infty < \mu < \infty, 0 < \sigma^2 < \infty$ and the compounding distribution is the discrete distribution which assigns masses $\lambda, 1 - \lambda$ to the two points $(\mu_1, \sigma_1^2), (\mu_2, \sigma_2^2)$. We will see that the method of compounding is very useful for generating new classes of models starting from various initial families of distributions.

| Method of formal embedding | **10.2.5** The method for the generation of non-normal |

10.2.5 The method for the generation of non-normal models developed by Karl Pearson in his second statistical paper published in 1895 (see Pearson, K. (1948)), was essentially a formal mathematical exercise. In it the normal model is looked

---

[4]Efron (1975) has shown that even in the case $\sigma_1^2 = \sigma_2^2$, the 'curvature' (a measure of departure from linear exponentiality, which he defines) of the family becomes enormous as $|\mu_1 - \mu_2|$ increases, so that MLEs can retrieve only a small part of the information contained in the sample.

upon as the solution of a particular differential equation and then that equation is interpreted as a particular case of a more general one. In this way the normal model is formally embedded in a larger class of models.

| Pearsonian system | The density function of the normal distribution $\mathcal{N}(\mu, \sigma^2)$ satisfies the differential equation |

$$\frac{dy}{dx} = -\frac{y(x-a)}{b_0} \tag{10.2.8}$$

where $a = \mu, b_0 = \sigma^2$. This reflects the fact that the distribution has a single mode at $x = a$ and high degree contact at both ends where $y \to 0$. Generalizing, we may set up the equation

$$\frac{dy}{dx} = -\frac{y(x-a)}{b_0 + b_1 x + b_2 x^2}, \tag{10.2.9}$$

which allows for a single mode and high degree contact at both ends as before, but also admits of other possibilities because of freedom of choice with respect to b_1, b_2. Solving (10.2.9) for different values of b_0, b_1, b_2, we get a system of models with a variety of shapes and ranges.

Pearson sought to motivate the setting up of (10.2.9) by observing that for a hypergeometric distribution the ratio of the difference between two successive ordinates to the mean ordinate has an expression similar to that on the right of (10.2.9). (Recall that long ago Daniel Bernoulli had considered a similar ratio for the binomial distribution to arrive at an equation of the form (10.2.8); cf. (6.5.6).) But the argument was not very convincing (see Stigler (1986) pp. 336–340). However, Pearson demonstrated the utility of his system by showing how it provided satisfactory graduation for distributions of diverse variates ranging from barometric height to carapace measurement on crabs and duration of dissolved marriages to proportion of paupers in poor law unions. Also, according to Pearson's positivistic credo, the function of science was 'not to explain but to describe by conceptual shorthand our perceptual experience'.

Pearson showed that the parameters appearing in equation (10.2.9) can be expressed in terms of the first four moments, or equivalently, in terms of the mean μ_1', the variance μ_2 and the coefficients $\beta_1 = \mu_3^2/\mu_2^3$ (or rather $\gamma_1 = \mu_3/\mu_2^{3/2}$) and $\beta_2 = \mu_4/\mu_2^2$. The type of the solution $y = f(x)$ depends on the nature of the roots of the quadratic in the denominator on the right and could equivalently be characterized by (β_1, β_2). In all, he distinguished twelve different types of models (see Elderton and Johnson (1969) p. 45) and these covered a wide variety of shapes and ranges. (However, if we allow transformation and the taking of limit, the number of distinct types would be much smaller.)

| Main drawback | In the first quarter of the 20th century, Pearson relentlessly propagated the use of his system of curves by |

applying these for the graduation of various kinds of live data. He also organized the preparation of tables of the incomplete gamma and beta functions to facilitate such graduation. However, after an initial period of ascendancy, the system went somewhat out of fashion. Its main drawback is overdependence on the method of moments (to be considered in the next subsection) for selection and fitting. The unified form of density represented by (10.2.9) does not readily lend itself to other methods, nor does it easily allow the incorporation of other information like values of explanatory variables (to be considered in subsection 10.2.8), for which one has to take the individual types separately. However, it has been found that the sampling distributions of many standard statistics (e.g. χ^2, t, F, r when $\rho = 0$) conform to one or other of the Pearsonian types. Also often when the distribution of a statistic is too complicated, it has been the practice to approximate it by a Pearsonian type on the basis of the first few moments.

| Selection and fitting | **10.2.6** In model-selecting induction after selecting a suitable member of the reference class of models we

have to fit the model (i.e. estimate the parameters specific to the selected model). But as noted earlier, the distinction between selection and fitting is not always clear-cut. Often the problem becomes one of simultaneous estimation of all parameters defined for the class of models chosen. In some cases (e.g. for the Pearsonian system and the Johnson system), however, the different types of models within the class correspond to different ranges of values of certain identifying criteria which may be separately computed without going into the estimation of all the parameters. There, we first determine the type of the model on the basis of such criteria and then go for the fitting.

Mainly three different methods have been used to solve the problem of selection-cum-fitting of models. These are (i) the method of moments, (ii) the method of maximum likelihood, and (iii) posterior-based Bayesian method. (There are other methods applicable in particular cases and also methods which are in a sense equivalent to some of those mentioned.)

| Method of moments | When the reference class of models consists of the partial sums of an asymptotic expansion as in sub-

section 10.2.2, the population cumulants themselves appear as parameters in the model. On the basis of standard estimates of these in terms of the sample moments we decide how many terms of the expansion are to be retained in the partial sum. A less trivial application of the method of moments was made by Karl Pearson, for the first time in his 1894 paper, for fitting the mixture model (10.2.7). To estimate the five parameters $(\lambda, \mu_1, \sigma_1^2, \mu_2, \sigma_2^2)$ in the model he set up five equations in these by equating the first five theoretical moments to their sample counterparts. Of course the solution of the equations was quite involved (frighteningly so, in those early times when electronic computers were not even a dream) and as Pearson showed, required finding a root (not always unique)

of a 9th degree polynomial equation. (However, as Rao (1948) has pointed out, if one can make the simplifying assumption $\sigma_1^2 = \sigma_2^2$, the problem reduces to that of solving a cubic equation only.)

The method of moments was extensively employed by Karl Pearson in his 1895 paper for model-selecting induction based on the system of curves (10.2.9). Within the system, the type of the curve is characterized in terms of (β_1, β_2). Entering a diagram marking out regions on the (β_1, β_2)-plane corresponding to the different types (see Elderton and Johnson (1969) pp. 41, 46) with the sample estimates of (β_1, β_2) one can select the appropriate type. Once the type of the curve is determined, one can estimate the parameters specific to it by equating the requisite number out of the first four moments to their sample values. Fisher in his 1922 paper (see Fisher (1950)) showed that this may lead to poor efficiency relative to MLE when the type of the curve deviates markedly from normal. In Pearson's formulation, there is little scope for using any method other than that based on moments, at least upto the stage of model selection. This is one of the reasons why the Pearsonian system later lost its attraction with statisticians.

However, later Johnson (1949) also proposed selection of the value of the identifying parameter λ_3 for the transformation class (10.2.1) on the basis of sample (β_1, β_2). For fitting also, the method suggested by him was generally dependent on moments; see Elderton and Johnson (1969), p. 131.

Generally, even when other estimators are used and their expressions are not available in closed form, the estimates are often reached through iteration, starting with the moment estimates as initial values.

| Maximum likelihood | In situations where for the entire reference class the density is the same smooth function of all the parameters (identifying and specific), the problem of model-selecting induction can be solved in a single stroke by finding their MLEs (subsection 4.3.9). However, for most classes of models having adequate coverage, MLEs cannot be found explicitly and numerical computation involving various forms of iteration have to be resorted to. Rao (1948) followed such a course with regard to the mixture class (10.2.7) when $\sigma_1^2 = \sigma_2^2$. Later others have used the same technique in more complicated mixture situations (see Day (1969)).[5] Box and Cox (1964) also applied the ML technique to their transformation model (10.2.2) (in the general case involving explanatory variables).

We note in passing that in many situations there exist alternative techniques giving estimates which are equivalent to the MLE at least upto the first order of efficiency. One such technique applicable to grouped data is related to the

[5]As noted earlier, in the case of normal mixture models the likelihood function is unbounded; also, often it has multiple stationary points. Generally, the stationary point which maximizes the likelihood among all such points is found to give a satisfactory estimate. Another course proposed here is application of the ML method after an initial grouping of the data.

chi-square test of goodness of fit (to be considered in the next subsection) and
is called the method of minimum chi-square.

| Bayesian methods | In situations where the problems of fitting and selection
are solved by the simultaneous estimation of a number
of parameters, another course which has become increasingly popular in recent
years is to adopt the pro-subjective Bayesian approach. For this the prior is
usually taken so that, as far as possible, it has an impersonal non-informative
structure and the posterior is tractable. The mode, or the mean of the posterior
distribution, or the point where the posterior expectation of some loss function
is minimized, is taken as the estimate. Box and Cox (1964) considered this type
of estimate also for their transformation class of models. Among more recent
applications, we mention Binder (1978) who has handled mixture models in
this way.

The availability of high speed computers in recent years has facilitated and
popularized the use of both maximum likelihood and Bayesian methods for
model-selecting induction in various complicated situations.

| Goodness of fit | **10.2.7**   An important part of model-selecting induc-
tion is validation of the model. Full-scale validation is
a long-drawn process—an induced model can be validated in course of time only
by judging it against fresh samples taken from the same population. But before
that, it is essential, as a first step, to test how well the model fits the very sample
from which it has been induced.

| Nibbling at the problem | As we saw earlier (subsection 9.5.1) the problem of
goodness of fit in the context of fitting the normal distribution was immanent even in Quetelet's studies. Edgeworth made a tentative
attempt to solve the problem, by breaking the range of the variate into two parts,
and testing the binomial frequency for one of these against the corresponding
proportion obtained from the model.[6] In his 1894 paper, while fitting the mixture model (10.2.7) to data by equating the first five theoretical moments to
their sample counterparts, Pearson contemplated the possibility of judging the
goodness of fit by comparing the sixth (i.e., the lowest order unused) moment
of the fitted model with its directly computed value from the sample. But as
higher order moments are notoriously subject to sampling fluctuation, he did
not pursue the idea. In his 1895 paper he vaguely suggested using the percentage

---

[6]In fact, as Stigler (1978b) notes, Edgeworth's treatment was remarkable in that he
considered the problem of testing goodness of fit as part of model-selection. Specifically,
he attached equal prior probabilities to a number of symmetric models (triangular,
parabolic, and double exponential, apart from the normal) and judged the superiority
of one of these on the basis of posterior probabilities.

difference in area between the empirical histogram and that based on the ordinates of the fitted curve at the class-midpoints (see Stigler (1986) pp. 334–335)
as a criterion, but here also he did not proceed further.

Around this time something happened which focussed on the problem of
goodness of fit from another angle. The context was that of judging the authen-
ticity of some dice-throwing data which had been collected by an assistant of
Weldon and supplied by the latter to Pearson. The data represented the outcomes of 7000 throws of 12 dice and were in the form of the frequency distribution
of the number (X) of successes (appearance of 5 or 6 on a die was taken as a success). If the dice were all fair and the throws honestly performed, the frequencies
f_i of i successes would follow a multinomial distribution with total frequency
$n = 7000$ and class probabilities $p_i = \binom{12}{i}(\frac{1}{3})^i(\frac{2}{3})^{12-i}, i = 0, 1, \ldots, 12$.

Pearson thought that the data were fudged. His reasons were as follows.
Denoting the values of i for which $|f_i - np_i|$ is largest and second largest by
ℓ and s, he argued that marginally f_ℓ followed a Binomial (n, p_ℓ) distribution, and
given f_ℓ, conditionally f_s followed a Binomial $(n - f_\ell, p_s/(1 - p_\ell))$ distribution.
As n was large, he used normal approximation for the binomials and computed
that if the data were true the probability of getting frequencies for which f_ℓ and
f_s differed from np_ℓ and $(n - f_\ell)p_s/(1 - p_\ell)$ by as much as the extent observed
was as low as 0.00049. This cast doubt on the genuineness of the data.

Pearson's condemnation disturbed Weldon greatly and he wrote to both
Galton and Edgeworth about the matter; for details see E. S. Pearson (1965).
Edgeworth thought that Pearson's suspicion was unjustified and pointed out that
because the indices ℓ and s were determined by the sample his reasoning was
flawed. The question could not be settled—there was no way one could settle it
objectively with the tools of statistical induction available at that time.

> Pearsonian χ^2—simple
> H_0

Pearson continued to tug away at the problem. In
general terms, the problem was to test how well the
frequencies $f_1, f_2, \ldots, f_k (\sum f_i = n)$ following a k-class
multinomial distribution with class probabilities $p_1, p_2, \ldots, p_k (\sum p_i = 1)$ were in
agreement with a simple null hypothesis H_0: $p_i = p_{i0}, i = 1, \ldots, k$. Edgeworth's
criticism must have made him wise to the fact that a reasonable test criterion
must simultaneously take into account all the differences $f_i - np_{i0}$ and not a
few extreme ones. The solution of the problem which he obtained came out
in an epoch-making paper which was published in the *The London, Edinburgh
and Dublin Philosophical Magazine and Journal of Science* in the year 1900 (see
Pearson, K. (1948)).

As we saw in subsection 9.8.3, Pearson had already studied intensively
the properties of the multivariate normal distribution. In the 1900 paper he
first derived that if X has an r-variate normal distribution $\mathcal{N}_r(\mu, \Sigma)$, then the
quadratic form

$$Q = (X - \mu)' \sum^{-1} (X - \mu) \tag{10.2.10}$$

would follow a χ^2-distribution with d.f. r (the term d.f. was introduced by Fisher much later).[7] Pearson then proceeded to apply the above result to the large-sample distribution of the multinomial frequencies to get a solution 'to the problem of fit of an observed to a theoretical frequency distribution'. We describe his derivation using modern notations.

If

$$U_i = (f_i - np_i)/\sqrt{(np_i)}, \quad U = (U_1, \ldots, U_{k-1})', \qquad (10.2.11)$$

then it readily follows that

$$E(U) = o, \quad \mathrm{disp}(U) = I_{k-1} - \eta\eta', \quad \eta' = (\sqrt{p_1}, \ldots, \sqrt{p_{k-1}}). \qquad (10.2.12)$$

Pearson took it for granted that in large samples U would be approximately normally distributed with mean vector and dispersion matrix given by (10.2.12). In (10.2.10), taking U in place of X, making use of (10.2.12) and replacing p_i by p_{i0}, after some manipulation one gets that when H_0 is true, the symmetric statistic

$$Q_0 = \sum_1^k \frac{(f_i - np_{i0})^2}{np_{i0}} \qquad (10.2.13)$$

is in large samples approximately distributed as χ^2 with $k-1$ d.f. The statistic (10.2.13) has become famous in the statistical literature as frequency χ^2 or Pearsonian χ^2.

It is clear from the above that Pearson's real achievement was to plug together the results on the asymptotic joint normality of multinomial frequencies and the distribution of the quadratic form in the exponent of the multinormal density. Pearson regarded Q_0 as a measure of discrepancy and proposed a right-tailed test based on it. (Incidentally, E. S. Pearson (1965) found that the test applied to the controversial dice-throwing data collected by Weldon's assistant, gave a P-value of 0.088 and hence did not warrant an outright condemnation.) A right-tailed test based on Q_0 would be consistent (cf. subsection 4.2.5) against any alternative under which $f_i/n \xrightarrow{P} p_i$ as n increases, where $p_i \neq p_{i0}$ for at least one i (i.e. at least two is). Broadly it may be used to judge the hypothesis that the frequencies have been generated by independent generalized Bernoulli trials with class probabilities $p_i = p_{i0}, i = 1, \ldots, k$, against any alternative mode of data generation (the possibility of deliberate fudging included) which would tend to make Q_0 large.[8]

---

[7]The distribution of χ^2 as the sum of squares of i.i.d. standard normal variables had been obtained earlier by Helmert and others (see Hald (1998) pp. 634–638), but apparently Pearson while deriving the distribution of (10.2.10) was unaware of that.

[8]However, Fisher thought that too low a value of Q_0 also suggested that the data had been manipulated by someone who knew the theory; see Fisher (1944) p. 81 and also Rao (1989) p. 48 where Fisher's comments on the genuineness of Mendel's genetic data have been discussed.

| Pearsonian |
| χ^2—composite H_0 |

The simple hypothesis $H_0: p_i = p_{i0}$, $i = 1,\ldots,k$ is, however, not directly relevant to the problem of goodness of fit of a model. In the latter case, we want to test a parametric composite hypothesis of the form $p_i = p_i(\theta)$, $i = 1,\ldots,k$, where the functions $p_i(\cdot)$ are specified by the model and θ (of dimension $< k-1$) is an unspecified nuisance parameter specific to the model. For this Pearson proposed that we find any suitable estimate $\hat{\theta}$ of θ and substitute $p_i(\hat{\theta})$ for p_{i0} in Q_0. He thought that when the composite hypothesis is true the resulting statistic $\hat{Q}_0$ would still follow a χ^2 distribution with $k-1$ d.f. and therefore the right-hand tail area beyond $\hat{Q}_0$ of this distribution can be taken as the P-value. Pearson in his 1900 paper and also later applied the test to a variety of model-fitting problems. His use of the test of goodness of fit was in the spirit of what we have called 'extended *modus tollens*' in subsection 4.3.6.

However, there were two serious misconceptions in Pearson's theory of the χ^2 test in the case of a composite hypothesis. He thought that the χ^2 distribution would hold, whatever reasonable estimate $\hat{\theta}$ (including the moment estimate) be substituted for θ and that such substitution would not affect the d.f. Fisher, through a series of papers in the 1920s, showed that both these assumptions are wrong: generally one can take only the MLE (or an estimate which differs from it by a term of order less than $n^{-1/2}$ in probability), and even in that case the d.f. has to be reduced by the number of independent parameters estimated (see Cramér (1946) p. 506). This means for most non-normal models one can not apply the χ^2 test if moment estimates are used and thus detracts from the utility of the Pearsonian system for model-selecting induction. We will see in section 10.4 how this correction has also important implications for certain types of model-free induction to which the χ^2 test is applicable.

| A 20th century wonder |

Pearsonian frequency χ^2 has some attractive features which led to its widespread use in diverse fields including psychometry, biometry, genetics, epidemeology, ecology, medicine, and industry. It provides an omnibus test which is usable whatever the form of the model. The criterion does not involve the class values or the order of the classes, which makes it usable even when the characters under study are attributes (we will see that this accounts for its use in model-free induction). Although the criterion was derived assuming independent and identical trials, it may be employed conservatively in certain other situations also; see e.g. Chatterjee (1970).

Before the advent of Pearsonian χ^2, whether or not a given model gives an adequate fit to an empirical distribution was a matter of subjective judgment. This allowed the personal preference of the experimenter to rule the day. As instances, we can cite Spottiswoode's 1861 study (subsection 9.5.1) on the orientation of mountain ranges (see Stigler (1986) p. 219), C. S. Peirce's 1873 study on the response time to signals (see Stigler (1978)), and the 1875 and 1891 studies respectively of Airy and Merriman on errors of astronomical observations and deviations of shot fired at a line target (see Pearson's 1900 paper in Pearson, K. (1948)). In all these cases empirical distributions which were far from normal were declared to be closely so. The same must have been the case with regard to judgments about homogeneity of groups and independence of attributes (we will consider application of χ^2 to these problems in section 10.4). The χ^2 test gave an objective procedure for reaching conclusion in all such cases. For this reason Pearsonian χ^2 has been hailed as one of the twenty most important discoveries in science, technology, and medicine that have been made in the 20th century and that 'have had a significant impact on the way we live or the way we think about ourselves and world' (see Neter (1986) pp. 1–2).

| Other goodness-of-fit tests |

A test for goodness of fit in model-selecting induction should naturally be based on a measure of discrepancy between the sample and the hypothesized model. Preferably the test should have an omnibus character, i.e. the form of the measure and its null distribution should be same irrespective of the model specified by the hypothesis. This is because traditionally the goodness of fit of different models to the same data has been compared in terms of P-values (cf. subsection 4.3.6) and if different measures and null distributions have to be used for different models such a comparison would be hardly convincing. (Special tests for particular models like normal or exponential are available in the literature; but these lead to conclusions of the type 'this or other' only, and are of limited utility for model selection.) In the case of Pearsonian χ^2, the omnibus character of the test is realized through grouping and the limiting multinormality of the multinomial frequencies.

About thirty years after Pearson proposed his χ^2 test, a number of other tests of goodness of fit which can be applied to an ungrouped sample of small size from a distribution with a continuous cdf $F(x)$, were proposed. All these are based on the 'probability integral transformation' according to which if $X_1, \ldots, X_n$ is a random sample from $F(x)$, then $F(X_1), \ldots, F(X_n)$ can be regarded as a random sample from the uniform distribution $\mathcal{U}[0, 1]$ over $[0, 1]$. For testing a simple hypothesis $H_0: F(x) = F_0(x)$ we can take any measure of discrepancy between the sample empirical cdf $\hat{F}_{(n)}(x)$ and $F_0(x)$, which remains unchanged by monotonic increasing transformations of x as our test-criterion. This would give an omnibus test for H_0, for the transformation $x \rightarrow F_0(x)$ would give the same statistic based on a sample from $\mathcal{U}[0, 1]$, whatever F_0.

We mention here only two such tests, namely, the Cramér–von Mises and Kolmogorov–Smirnov tests (see e.g. Cox and Hinkley(1974), p. 69), based respectively on the criteria

$$W^2 = \int_{-\infty}^{\infty} \left\{ \hat{F}_{(n)}(x) - F_0(x) \right\}^2 dF_0(x), \qquad (10.2.14)$$

and

$$D = \sup_{-\infty < x < \infty} |\hat{F}_{(n)}(x) - F_0(x)|. \qquad (10.2.15)$$

Both the tests are consistent against all alternatives $F \neq F_0$. The null distributions of the criteria have been derived and their limiting forms, which are non-normal, are also known. Tables have been given by Owen (1962).

When the null hypothesis is a parametric composite hypothesis specifying that F belongs to a class $\{F(x \,|\, \theta), \theta \in \Theta\}$, one has to find an estimate $\hat{\theta}$ of θ from the sample and substitute $F(x \,|\, \hat{\theta})$ for $F_0(x)$ in (10.2.20)–(10.2.21). The distributions of the resulting statistics no longer remain tractable for finite n. Some large sample results and results of simulation studies covering special cases are available (see Pearson and Hartley (1972)). Obviously the scope of tests based on the above statistics, compared to the χ^2 test, is considerably limited—they are not applicable (in a strict sense) when the model considered is discrete and the statistics cannot be defined when the character observed is an attribute.

Case of a parametric family

Neither the χ^2 test nor the tests based on criteria like (10.2.14)–(10.2.15) require the specification of the entire reference class of models from which the model under consideration is selected. If the reference class is specified and is a parametric family generated by the variation of certain identifying parameters λ, we can treat the problem of goodness of fit of the model λ_0 as one of testing the composite hypothesis $\lambda = \lambda_0$ (the unspecified specific parameters of the model represent the nuisance parameters). For this, under appropriate conditions we can apply the standard large-sample likelihood ratio test: if R is the ratio of the maximum under $\lambda = \lambda_0$ to the unrestricted maximum of the likelihood, when λ_0 is true, in large samples $-2\log_e R$ would be distributed as a χ^2. Such an approach has been followed by Box and Cox (1964) with reference to their transformation class (10.2.2) (cf. also Cox and Hinkley (1974) p. 327). Obviously, it may be used also for the mixture class considered in subsection 10.2.4 (provided we use suitable dodges to get round the difficulties created by the vagaries of the likelihood function; cf. footnote 7).

Under the above set-up one can also follow a pro-subjective Bayesian approach. For that, as discussed in the context of fitting (subsection 10.2.6), one has to start with a prior distribution and derive the posterior distribution for all the parameters. Hence one can marginalize to get the posterior distribution of λ. Box and Cox (1964) proceeded in this way with reference to the class of models (10.2.2) to get a Bayesian credible set (cf. subsection 4.4.6) for λ. A model $\lambda = \lambda_0$ is acceptable if λ_0 is included in such a set.

Models for other occasions

10.2.8   In the above we confined ourselves generally to model-selecting induction when the observables can be regarded as a random sample from a univariate continuous population. We now briefly consider models that have been developed to meet other situations that may arise in practice. As we will see, the methods for generating classes of models in these other situations are broadly the same as those used earlier. However, the normal distribution can no longer serve as an anchor in such cases.

| Discrete models |

For random samples from discrete populations (when the discreteness must be taken into account), the two original models were the binomial and the Poisson. To meet the need of situations where neither of these is appropriate, new models have been developed, mostly taking the Poisson distribution as an anchor.

As we saw (subsection 8.6.4) the Poisson distribution came to be used towards the end of the 19th century for modelling the distribution of number of rare events. Greenwood and Yule (1920) found that the number of accidents incurred by bus drivers in London in a year was more dispersed than one would expect under the Poisson model. They postulated that this happened because different drivers had different degrees of accident proneness which could be represented by the Poisson mean m. Compounding the Poisson pmf $e^{-m}m^x/x!$ by a gamma distribution of m with density, say, $a^p e^{-am} m^{p-1}/\overline{|p}, 0 < m < \infty$, they arrived at the *negative binomial* model which has pmf

$$ \frac{\overline{|x+p}}{x!\overline{|p}} \left(\frac{a}{1+a} \right)^p \left(\frac{1}{1+a} \right)^x, \quad x = 0, 1, 2, \ldots, \; p > 0, \; a > 0. \qquad (10.2.16) $$

(This distribution with p an integer was implicit in Fermat's solution of the division problem; cf (5.3.1). It has been wrongly attributed to Pascal and called Pascal's distribution; see Hald (1990) p. 61.)

Another form of compounding, which produces a variety of discrete models, starts with a convolution family of discrete distributions $f(x \mid r)$ indexed by the non-negative integral parameter r. Here, for every $r, f(x \mid r)$ represents the pmf of the sum of r i.i.d. random variables each distributed according to a discrete distribution with pmf, say $g(v), v = 0, 1, 2, \ldots$ (with $f(0 \mid 0) = 1$). The model is generated by compounding the distributions $f(x \mid r), r = 0, 1, 2, \ldots$ by the Poisson (m) distribution. (This really amounts taking a composition of two probability generating functions the inner one being of $g(v)$ and the outer one of the compounding Poisson distribution.) We can get different models for different $g(v)$ and m. Thus it can be shown that the Poisson, the negative binomial, and the Neyman Type A contagious distribution are produced in this way if we take the degenerate (with total mass at 1), the logarithimic, and the Poisson distribution for $g(v)$. In fact, taking the negative binomial (10.2.16) itself as $g(v)$, we can generate a broad three-parameter class of distributions whose closure includes all the above and other forms of models as well; see Gurland (1963).

For selection and fitting of such models the methods employed are broadly those described earlier in subsection 10.2.6.

| Life testing models |

Certain special models have been developed for induction on the basis of life testing data. Observations on life (survival time of an individual, time to failure of a component in a manufacturing process, etc.) are non-negative-valued and in life testing often some of the items are censored (i.e. for these observation is prematurely discontinued so that only certain lower bounds to their true lives are known). Further, if $f(x)$ denotes the density function of a life distribution and $F(x)$ its cdf, the age-specific failure rate

$$h(x) = \frac{f(x)}{1 - F(x)} = -\frac{d}{dx}\log\{1 - F(x)\}, \quad 0 < x < \infty, \qquad (10.2.17)$$

which is called the *hazard function*, provides an alternative mode of representing the distribution. It has a readily understandable physical interpretation (as the conditional probability that an individual, who has survived upto age x, would die 'at' that age) and in such contexts often one prefers to define the model so that the hazard function behaves realistically.

Models representing life distributions have been the object of study of actuaries for a long time. Their use for fitting life testing data is a comparatively recent development. The basic model is the exponential whose pdf, cdf, and hazard function when the mean life is $1/\rho$ are given by

$$f(x \mid \rho) = \rho e^{-\rho x}, \ 0 < x < \infty, \rho > 0; \quad F(x) = 1 - e^{-\rho x}; \quad h(x) = \rho, \quad (10.2.18)$$

the hazard being constant for all x.

Various other models may be derived by different techniques starting with the exponential as anchor. For instance, if we assume that $(\rho X)^k, \rho > 0, k > 0$, has an exponential distribution with mean 1, then we get a two-parameter family of distributions known as Weibull distribution which has the hazard function $k\rho(\rho t)^{k-1}$. We may also compound the exponential distribution by various compounding distributions. Thus if we assume that ρ in (10.2.21) is subject to the gamma distribution with pdf proportional to $a^p e^{-a\rho} \rho^{p-1}, a > 0, p > 0$, we get the Pareto distribution whose hazard function $p/(a + x)$ is decreasing.

Many other models for representing life distributions have been formulated. Also a three-parameter general class which includes a number of such life testing models has been proposed. Then there are discrete models which are appropriate when the discreteness of the process of life measurement has to be taken into account.

For selection, fitting, and testing of models belonging to a parametric class, generally methods based on likelihood are those most widely used. In certain cases use of moments and graphical plotting of estimated hazard and related functions are helpful. Special tests like those for exponentiality against particular alternatives have also been developed. For all this we refer to Cox and Oakes (1984).

| Postulational approach | It is always intellectually satisfying to derive a model starting from plausible postulates about the stochastic |

mechanism generating the data and not merely as a mathematical construct. As we saw in subsections 10.2.2–10.2.3, even some of the models proposed by the pioneers were motivated thus, albeit in a feeble way. Several of the models described above as also certain other standard ones can be obtained in this way.

For a Poisson process with rate λ, independently of the history of the process upto time T_1, the probability of the event not occurring in a small time interval $(T_1, T_1 + \tau]$ is $1 - \lambda\tau + o(\tau)$ and the probability of more than one event in such an interval is $o(\tau)$. It is known that the distribution of the number of events in an interval $(T_1, T_2]$ for such a process follows the Poisson distribution. In 1923, Eggenberger and Polya suggested that we relax these conditions by postulating that given the number n of events occurring upto time T_1, the conditional probability of the event not occurring in $(T_1, T_1 + \tau]$ is

$$1 - \lambda\frac{1 + \mu n}{1 + \nu T_1}\tau + o(\tau). \tag{10.2.19}$$

Here μ (the measure of contagion) > 0 and ν (the time effect) > 0, implying that the probability of no events in $(T_1, T_1 + \tau]$ decreases and increases respectively as n and T_1 increase. This indicates that there is a proneness (aversion) for the event (say, accident or epidemic attack) depending on whether many (few) events have occurred in a short (long) time earlier; see Neyman (1963). It has been shown that, under these conditions, the distribution of the number of events in an interval $(T_1, T_2]$ has the negative binomial form (10.2.16).

Another type of model is generated by postulating that the variate under study arises as the largest or smallest order statistic of a sample from some basal continuous distribution whose tail tapers off to zero. Suitably normalized versions of such extreme order statistics are known to have distributions of certain standard forms in large samples (see e.g. Cox and Hinkley (1974) p. 472). These can be used to set up models. Thus Fisher and Tippett in 1928 showed that the smallest sample value when the basal distribution is bounded on the left generates the Weibull model. (The model was later used by Weibull for representing the distribution of breaking strengths of materials.) The largest value when the right tail of the basal distribution tapers off to zero at an exponential rate has been found to generate a model for the yearly high-water mark of flood in a river and the maximum precipitation at a place (see Gumbel (1958) Chapter 6).

An interesting set of postulates interpreting a positive-valued random variable as the time to first passage through a fixed barrier at $a > 0$ of a Brownian motion with drift $\nu > 0$, gives rise to a two-parameter model with a and ν as parameters. This has been called the inverse Gaussian model and has been found useful for graduating various life and other distributions (see Cox and Oakes (1984) p. 22).

Such models generated through the postulational approach may be looked upon as apparently substantive models (footnote 2). Generally, they are not truly substantive. The main purpose of the postulates in their case is to help build a model and these are often proposed in a tentative sort of way. If a model generated by the postulational approach gets empirically confirmed, it does not mean that the stochastic mechanism represented by the postulates necessarily obtains—different stochastic mechanisms may lead to the same model. The negative binomial model, which, as we have seen, can be generated either through the compounding of the Poisson distribution or from the contagion hypothesis corresponding to (10.2.19), illustrates this point (footnote 3).[9]

Non-identical
variables—GLIM

In practice, the problem of modelling in the case of non-identically distributed independent variables arises in particular when we have information on the values of a number of non-stochastic (intrinsically non-random or conditionally fixed) explanatory characters or covariates. The covariates $x = (x_1, \ldots, x_m)$ are observed on each unit on which the study variable Y is observed. Here it is often realistic to assume that the covariates affect the distribution of Y through a linear compound $\eta_x = \sum_1^m \beta_i x_i$ where the coefficients β_i are unknown. If the distribution of $Y - \eta_x$ were same whatever x, we could possibly start with one of the standard models described above and proceed as earlier (of course with added complications because of the presence of the β_is). But often the distribution of Y depends on η_x in a more complex way: Y is supposed to have a standard distribution (say, binomial or Poisson in the discrete case and gamma in the continuous case) with a parameter $\theta_x = d(\eta_x)$, where $d(\cdot)$ is a known invertible function. Of course $d(\cdot)$, which is called the *link function*, has to be chosen here so that $d(\eta_x)$ lies in the appropriate domain and varies with x as expected.

Early studies of such models started piecemeal in the 1920s and 1930s with Fisher's treatment of the dilution direct assay problem (in which the variate Y was Poisson with a mean of the form e^{η_x}) and the probit analysis problem (in which Y was binomial with the value $\Phi(\eta_x)$ of the standard normal cdf at η_x as its probability). Unified analysis of these models which have been called Generalized Linear Models (GLIMs) has come up only in recent years. Since the object of model selection here is to study the effects of the covariates on the distribution of Y, one has to decide not only on the form of the distribution but also on the form of the link function. The most potent methods for all this are those based on the likelihood and their adaptations (note that the choice of $d(\cdot)$ affects the likelihood of the β_is materially). The theory of GLIM deals with these topics; see McCullagh and Nelder (1989) and Sen and Singer (1993).

Dependent variables

In our discussion so far, we have confined ourselves to independent random variables. When we have a sample from a multivariate population a standard model in the discrete case is the

---

[9]In fact how far even a truly substantive model can be confirmed empirically is a moot question of philosophy. As we noted in section 1.5, Popper has argued that a postulated hypothesis can be refuted but never established on the basis of empirical evidence.

multinomial distribution; in the continuous case the model most extensively studied and applied in practice is the multivariate normal. A few special models appropriate for graduating certain particular types of multivariate data (like multivariate life testing data) have also been developed. As mentioned earlier (subsection 10.2.4), in the continuous case mixtures of multivariate normals have been proposed to tackle certain types of departure from normality. Some attempts have been made to develop a general system of nonnormal multivariate models along the lines of the Pearsonian system and to derive series expansions in terms of partial derivatives of the bivariate normal density (see Elderton and Johnson (1969) Chapter 8). But possibly because of inadequate coverage and lack of simplicity, these have not become very popular for graduation. However, the series expansion has been occasionally used in theoretical studies; see e.g. Gayen (1951).

A more radical form of dependence among observables arises when we have unreplicated observations at a subset of time points on a discrete or continuous time stochastic process. Modelling in various special types of situations in such cases has been considered. For instance, for a homogeneous Markov chain a standard problem is that of choosing the order. For time series, the problem of choosing a suitable autoregressive model or some extended version of such a model has been given a lot of attention. But we do not intend to cover such special topics here (see Lehmann (1990) for some references).

| Information criteria | **10.2.9** Instead of viewing model-selecting induction as a three-step process involving selection, fitting, and the testing of goodness of fit, we can look at it in a unified way. The problem then becomes that of choosing from a set of competing models one that is most useful in some plausible sense. Here account has to be taken of the fact that the competing models may have different forms (i.e. may not belong to the same parametric family) and may involve different numbers of unspecified parameters.

When the number of observations is large and the data are given in a grouped form, the classical way of solving the problem is to compare the P-values of the frequency χ^2s corresponding to the various fitted models (subsection 10.2.7). (Note that in this approach, a model involving a larger number of parameters is judged more strictly as the d.f. of the corresponding χ^2 is lesser so that its distribution is more tilted towards the left. Also, as the number of observations increases the number of classes, and hence, the d.f. of χ^2 can be correspondingly increased. This offsets the difficulty that would otherwise arise because even a slight departure from the presumed model would be reflected in a large value of χ^2, and hence, an inordinately small value of P.)

However, the number of observations may not be too large. Also, the data may not always be classifiable into a grouped form, e.g. when covariates have to be taken into account. In such a case the likelihoods of different fitted models (maximal likelihoods, if MLEs are used for model fitting) given the data, have to be suitably adjusted to make them comparable. This is important because none

of the fitted models may be actually true and a model involving a larger number
of parameters, while giving a higher maximal likelihood for the particular data
at hand, has to be weighed against one involving a smaller number, in terms
of other criteria. Simplicity, stability (estimating too many parameters from
limited data would make the model subject to too much sampling fluctuation),
and capability for extrapolation to new ranges of the covariates are some of the
considerations important here and these have to be judged by taking the number
of observations into consideration.

| AIC and BIC | In recent years two criteria, namely the Akaike Inform-
ation Criterion (AIC) and Bayes Information Criterion
(BIC) have been advanced as solutions to this problem. Their basic idea is to
take for each model the maximal likelihood suitably penalized for the number of
estimated parameters. Thus given n independent observations $y_1, \ldots, y_n$, if the
log-likelihood based on a model M involving $p = p(M)$ independent paramet-
ers $\theta_{1M}, \ldots, \theta_{pM}$ is $\Lambda(\theta_{1M}, \ldots, \theta_{pM} \mid y_1, \ldots, y_n; M)$ and if $\hat{\theta}_{1M}, \ldots, \hat{\theta}_{pM}$ are the
MLEs under M, the AIC and BIC are taken respectively as

$$-2\Lambda(\hat{\theta}_{1M}, \ldots, \hat{\theta}_{pM} \mid y_1, \ldots, y_n; M) + 2p(M) \qquad (10.2.20)$$

and

$$-2\Lambda(\hat{\theta}_{1M}, \ldots, \hat{\theta}_{pM} \mid y_1, \ldots, y_n; M) + p(M) \log_e n. \qquad (10.2.21)$$

In (10.2.20) the penalty per each parameter introduced is just 2, where as in
(10.2.21) the same depends on n and is $\log_e n$. In practice, given two competing
models, one has to take the one having a lower value of the preferred information
criterion. AIC has been proposed mainly on intuitive grounds, but it has been
shown that the value 2 of the per parameter penalty has certain advantages in
typical cases. BIC has been justified by showing that of two models M_I and M_{II}
the choice of the one having a lesser BIC can be given a Bayesian interpretation.
For this we have to assume an impersonal prior which assigns equal probabilities
to M_I and M_{II} and under each of these distributes the weight uniformly over
the space of the corresponding θs. We choose M_I if and only if given $y_1, \ldots, y_n$,
the posterior probability of M_I is greater than that of M_{II}, i.e. the Bayes factor
of M_I versus M_{II} (cf. footnote 35 in Chapter 4) is greater than 1 (see Ghosh
and Samanta (2001) for a discussion of these issues and further references).

| Predictive criteria | The spirit of the method based on AIC is that of
likelihood inference, as discussed under the instantial
approach in Chapter 4 (subsection 4.3.5). The BIC-based method is of course
pro-subjective Bayesian. Box (1980) has suggested an approach to the modelling problem which we may call 'predictive' (it is essentially pro-subjective but
behavioural in form). His idea is to consider a comprehensive model comprising the likelihood and a *proper* prior. The parameters are to be estimated as
usual from the posterior. The comprehensive model may be tested by taking any

suitable statistic and judging its observed value against its predictive distribution (i.e. unconditional distribution over the prior). Thus this makes use of the information about both the sampling experiment and the prior. Of course unlike in the above, here, while judging a model, we do not envisage any alternative to it.

Other approaches to model selection have been proposed for special cases. In particular, in the context of regression, expected loss of prediction has been suggested as a criterion (see Rao (1989) p. 60, Lehmann (1990)). But these do not have much relevance for the general problem being considered in this section.

10.3   Model-specific induction

| Beginning |

10.3.1   In this section we consider how the sampling-theory approach to what is generally called 'parametric inference' got under way at the beginning of the 20th century and then rapidly gathered momentum mainly driven by the epoch-making contributions of R. A. Fisher. Stigler in his comments in Savage (1976) asserts that 'it is to Fisher that we owe the introduction of parametric statistical inference (and thus nonparametric inference)' in a literal sense because it was he who first introduced the term 'parameter' in statistics. While it may be contended that the same idea of considering induction about the parameters of a given model separately from the problem of model selection had been peeking earlier through the writings of Edgeworth, Karl Pearson, and as we will presently see, 'Student', there is no denying Fisher's dominant role in the development.

We will see Fisher's development of sampling-theory parametric inference led to the enrichment of statistical thought in at least three ways. Firstly, it gave impetus to the thorough examination of the logic underlying statistical procedures, under the instantial mode by Fisher himself, and under the behavioural mode by Neyman, E. S. Pearson, and later, Wald. Secondly, it opened up the field of multivariate analysis with emphasis on the concept of 'dimensionality'. Thirdly, Fisher's technique of analysis of variance, which was an offshoot of his sampling theory for linear models was instrumental for the development of the entire field of design of experiments, at least in the initial stages.

In model-specific induction we take the form of the model for granted, and given that, induce about the specific parameters in the model. The model here is either deduced from substantive considerations, or plainly assumed because the volume of data available is too small for trying any guess about the model. But even in the latter case, it is essential for the statistician to familiarize himself with the subject matter area to avoid picking a wholly inappropriate model. In this respect both Fisher and Neyman seem to have held more or less the opinion that model selection is more of an 'art' than a 'science'; see Lehmann (1990). In any case, when the data are small it is imperative that we start with a simple model, for if the exact sampling theory is intractable model-specific induction may be a non-starter. (However, in recent times, computer simulation has given statisticians more elbow room in this regard.)

<table>
<tr><td>'Student's' sampling theory</td></tr>
</table>

10.3.2 If we leave out isolated problems tackled by early pioneers like Arbuthnott and Daniel Bernoulli (Chapter 6), the sampling theory approach to model-specific induction may be said to have started in the 20th century. At the turn of the century, the biometric school led by Karl Pearson was dealing with huge volumes of data that permitted model-selecting induction and large sample methods. The need for model-specific induction arose when the methods of statistics were sought to be applied in fields where samples were bound to be limited in size. The path-breaker here was an unassuming man named W. S. Gosset (1876–1937), who wrote under the pen name of 'Student'.

<table>
<tr><td>Encounter with the problem</td></tr>
</table>

After graduating from Oxford around 1904–1905, Gosset had been working as a research chemist in a large brewery establishment in Dublin, Ireland. The brewery had collected a lot of data giving various measurements relating to its product and operations and Gosset was asked to see whether these could be analysed to reach useful conclusions. A typical problem that Gosset was required to solve on the basis of various forms of brewery data was that of inferring about the true mean μ of a variate X on the basis of n independent determinations of its value. Error-theorists at that time used to recommend a sampling theory approach to the problem. Denoting the sample mean by $\bar{X}$ and the sample variance by $\hat{\sigma}^2$, the standard prescription was to use Laplace's result that for large n, $\sqrt{n}(\bar{X} - \mu)/\hat{\sigma}$ is distributed approximately as $\mathcal{N}(0,1)$. When n was small, one assumed the model $\mathcal{N}(\mu, \sigma^2)$ for X, and knowing σ, invoked Gauss's result that under the model $\sqrt{n}(\bar{X} - \mu)/\sigma$ is exactly distributed as $\mathcal{N}(0,1)$. The error-theorists had no solution to offer when n was small and σ was unknown. (In fact, even Galton faced the problem when he was asked by Charles Darwin to analyse the data of a pot experiment in which the heights of corn plants from cross and self-fertilized seeds grown in pairs were sought to be compared. As the number (15) of pairs was too small, Galton was bogged down, failing to see how a reliable estimate of the S. E. of the difference in means could be obtained; see Cochran (1976).) When there was no way out, in such cases one simply used $\hat{\sigma}$ for σ and assumed hopefully that $\sqrt{n}(\bar{X} - \mu)/\hat{\sigma}$ still followed the $\mathcal{N}(0,1)$ distribution.

In the case of brewery data Gosset often encountered small n. While he thought normality of X could be taken for granted, he was acutely conscious of the fact that since $\hat{\sigma}$ itself had a S. E. of $\sigma/\sqrt{(2n)}$, in substituting $\hat{\sigma}$ for σ one must enlarge the interval for μ. 'But enlarge by how much?'—this question continued to occupy Gosset's mind. He posed the query to Karl Pearson, but the latter failed to realize the importance of the issue. In 1906 Gosset had the opportunity of visiting Pearson's biometric laboratory for a year and getting exposure to the methods then being developed by the biometric school. After this, he applied himself to the problem and came out with a solution which appeared in a paper entitled 'The probable error of a mean' in the *Biometrika* in 1908.

| 'Student's' t |

In the paper, Gosset, or as he called himself, 'Student' started with the statement that his object was to infer about the mean μ of the 'population' of 'experiments which may be performed under the same conditions', thus adumbrating the idea of a conceptual population later elaborated by Fisher (see subsection 3.2.3). Also he clearly spelt out the assumption of the model $\mathcal{N}(\mu, \sigma^2)$, made partly on the strength of experience and partly for convenience. Furthermore, contrary to the usual practice at the time, he used a notation different from σ^2 for its sample estimate. If we denote $s^2 = \sum(X_\ell - \bar{X})^2/(n-1)$, the problem addressed by 'Student' in the paper can be described as the derivation of the sampling distribution of

$$t = \frac{\sqrt{n}(\bar{X} - \mu)}{s}. \tag{10.3.1}$$

(Actually, 'Student' used the divisor n in the estimate of σ^2 and worked with $z = t/\sqrt{(n-1)}$; we describe his result in modern notations.)

For this Gosset first worked out the first four moments of s^2, and noting that the (β_1, β_2)-coefficients indicated a Perasonian Type III distribution, derived that $(n-1)s^2/\sigma^2$ follows the distribution of a χ^2 with $n-1$ d.f. He confirmed this for $n = 4, 8$ by model sampling from a large population in which the distribution was close to normal. (Actually, the result had been derived more than 30 years earlier by Helmert and others (cf. footnote 8), but it was at that time generally not known in England.) Gosset then showed that correlation between s^2 and $(\bar{X} - \mu)^2$ for a normal population is zero (he was alive to the fact that because of symmetry s^2 is bound to be uncorrelated with $\bar{X} - \mu$). Hence he reasoned that $(\bar{X} - \mu)$ and s^2 are independently distributed. Of course the reasoning was faulty being based on the false presumption that uncorrelation is sufficient for stochastic independence, but the conclusion was correct. Then starting from the joint distribution of $\bar{X}$ and s^2, by the standard method of transformation and marginalization he arrived at the distribution of t in the familiar form. This distribution has since become immortalized as 'Student's' t distribution with $n-1$ d.f.[10]

Gosset got the theoretical formula confirmed empirically by model sampling experiments with $n = 4, 8$. He found that the distribution was more heavy-tailed than the standard normal distribution and provided tables giving the values of

---

[10]Incidentally, the mathematical form of the t-distribution as the marginal posterior distribution of μ when μ and $1/\sigma$ are subject to independent uniform priors had been derived earlier, see Hald (1998) pp. 396, 423 (cf. Example 4.4.1 in Chapter 4). Of course this was conceptually far from what Gosset derived and surely was unknown to him.

the incomplete integrals of $z = t/\sqrt{(n-1)}$ for $n = 4, \ldots, 10$ (this was extended to $n = 30$ in a later paper published in 1917).

In the last part of the paper he gave four applications of the result to data from small scale experiments one of which was from physiology and the rest from agriculture. Typically, in each case given a sample $X_1, \ldots, X_n$ the problem was to judge whether the population mean μ is significantly positive. (Actually in most applications, pairs of observations $(X_{1\ell}, X_{2\ell}), \ell = 1, \ldots, n$ were given, and the problem was to see whether the means μ_1, μ_2 of the two variates in the population were equal, or μ_2 had a higher value. For this, Gosset took $X_\ell = X_{2\ell} - X_{1\ell}$ and $\mu = \mu_2 - \mu_1$.) In each case Gosset in effect computed t given by (10.3.1) with $\mu = 0$ and found the left-hand probability integral $1 - P$ upto $z = t/\sqrt{(n-1)}$ from the tables.

Although Gosset was ostensibly following the sampling theory approach, the language he used is a bit confusing to us. Thus he stated that his object was to find 'what is the chance that the mean of the population of which these experiments are a sample is positive'. In one case when $1 - P$ was 0.8873 he stated that the odds were about 8 to 1 that the mean is positive and this corresponded in the normal case to a deviation of about 1.2 times the S. E.; in another, when $1 - P$ was 0.9974 the odds were given as about 400 to 1 corresponding to a normal-case deviation of about 2.8 times the S. E. Was Gosset anticipating the concept of a fiducial distribution of μ? Fisher (1959) and even E. S. Pearson (1939) seem to suggest that. But there is nothing in 'Student's' writing which warrants this. To us it seems he was simply employing 'extended *modus tollens*' (subsection 4.3.11) and putting the conclusion in the form of a statement of odds in a loose sense, following a practice that apparently had been customary with the error-theorists of the time.

We note here one point which would have much significance in the context of Gosset's later work on agricultural experiments. In one of his illustrations involving the paired comparison of two sleep-inducing drugs successively applied to each of a group of subjects, he found that the variance of the difference within each pair was smaller than either individual variance. He remarked that this was 'probably due to the different drugs reacting similarly on the same patient'. In other words he visualized that, if two variates have variances σ_1^2, σ_2^2 and a positive correlation ρ, we may have

$$\sigma_1^2 + \sigma_2^2 - 2\rho\sigma_1\sigma_2 < \min(\sigma_1^2, \sigma_2^2). \tag{10.3.2}$$

Work on r

We discussed in subsection 9.8.5 how in 1898 Pearson and Filon arrived at the formula $(1 - r^2)/\sqrt{n}$ for the estimated S. E. of the correlation coefficient r of a sample from the bivariate normal population. It was thought at that time that one could use this and large sample normality of r for inferring about the population correlation ρ. Gosset came to know about the Pearson–Filon formula during his visit of Pearson's laboratory. But he suspected that for the large sample theory of r

to be applicable, n would have to be considerably large and thus realized the importance of studying the distribution of r for small n. On the basis of model sampling experiments for $n = 4, 8$, he was able to empirically guess the correct form const. $(1 - r^2)^{(n-4)/2}$ of the distribution of r for a bivariate normal population when $\rho = 0$. He also studied in the same way the distribution for $n = 4, 8, 30$ in the particular case $\rho = 0.66$, but admitted that his own mathematics was not upto the task of tackling the general problem theoretically. These results appeared in a second paper entitled 'Probable error of a correlation coefficient' in the *Biometrika*, also in the year 1908. In it he expressed the hope that his own results on the distribution of r 'may serve as illustrations for the successful solver of the problem'. As we will see, the successful solver appeared 7 years later in the person of R. A. Fisher.

| Small-scale |
| experiments— |
| background |

10.3.3   Before taking up Gosset's contributions to the theory of agricultural experiments, we briefly discuss the background of small-scale experimentation, although this is not directly linked to model-specific induction. As we saw, the Quetelet–Galton–Pearson tradition in statistics, concerned as it was with the analysis of social and biometric data, occupied itself with the discovery of macro-patterns. Study of cause–effect relationship was not so much the object of their studies. In medicine, physiology, and agriculture, however, our main concern is to find the relation between a cause and its effect, since we want to produce an effect by manipulating its cause or sometimes to identify the cause by observing an effect. Data for such studies have to be collected through small scale experiments in which we intervene before the data are generated. (We are excluding from our consideration experiments in hard sciences like classical physics or chemistry where cause–effect relations are generally of the deterministic type.) In contrast, studies in social sciences, where there is less scope for intervention, are generally of the observational type—the data consist of observations on more or less freely realized values of certain characters. The theory of sample surveys, which we will take up in the context of model-free induction, is an off-shoot of this latter kind of study.

Small scale experiments had been practised in the fields of medicine and agriculture since the days of yore. In the 17th century, Francis Bacon, whom we met in Chapter 1, conducted an experiment to study the effect of steeping wheat seeds before sowing in nine different concoctions (cowdung solution, urine, wine, etc.) on the speed of germination and later growth of the plant and came out with the conclusion that treatment with urine is most effective; see Cochran (1976). As we noted in the same chapter, John Stuart Mill around the middle of the 19th century codified his canons of induction for getting foolproof cause-effect relations under certain ideal conditions. Important among these canons are 'the method of agreement and difference' (co-presence and co-absence of a supposed cause and a supposed effect imply a causal relation) and its graded

version 'the method of concomitant variation' (covariation in intensity of a supposed cause and supposed effect imply a causal relation). These methods can be
seen to underlie many standard practices of experimentation (like comparing a
control and a treatment or different doses of a manurial treatment in the same
experiment).

During 1855–1860, Fechner in Germany conducted his famous psycho-
physical experiments (see Stigler (1986) pp. 242–254) to investigate how different
factors affected an individual's sensitivity to differential stimuli, such as his power
to judge an unknown difference in lifted weights. The response in Fechner's
experiments was of the binary type (right or wrong judgment). He related the
proportion of right judgments to the differential stimuli through a sort of quantal
response model based on the assumption of normality of the implicitly judged
difference. Fechner could study how different factors affected sensitivity, but
there was no scope for assessing the uncertainty of inference in his approach.

Interestingly in 1884, C. S. Peirce, whom we met in Chapter 9 as a cham-
pion of the frequency view of probability, performed (jointly with Jastrow) a
randomized version of Fechner's weight-lifting experiment (see Stigler (1978a)).
Two slightly different weights were presented sequentially to a subject in either
of the two possible orders (heavier first or lighter first). The experiment was
carried out in pairs of blocks of 25 trials, for which the orders of presentation
were determined by drawing cards (red or black) from two well-shuffled packs,
one containing 12 red and 13 black and the other 13 red and 12 black cards
respectively. The proportion of right guesses was related to the relative difference in weight as in Fechner's case. Through such an experiment, Peirce and
Jastrow sought to disprove Fechner's hypothesis that there is a threshold or
least perceptible difference in sensation that can be detected by man. Arguably,
the experiment by Peirce and Jastrow was the earliest instance of a reported
randomized experiment. (However, here the purpose of randomization was to
keep the subject ignorant about the actual order of presentation of weights. In
the case of randomized experiments proposed by Fisher (to be discussed in the
next subsection) the purpose of randomization is to realize the assumptions of
the model more closely.)

Agricultural experiments— accumulated lore

Agricultural experiments have the speciality that in
their case the size and shape of the experimental units
(plots), as also their arrangement on the field, are
chosen by the experimenter. How to conduct an experiment for finding the
best variety or the best farming practice had been the concern of agricultur-
alists for quite a long time; and a number of books had come out dealing with
the topic. Cochran (1976) refers to a three-volume 1771 publication of Arthur
Young, a book by Johnston (published 1849) and translation of a treatise by
Wagner (published 1898). Some principles formulated by these authors were
generally accepted and more or less incorporated in experimental practice by
the end of the 19th century. Thus it was recognized that experiments should be

comparative, i.e. if one or more varieties are to be compared with the control then the varieties as well as the control must be included in the experiment. Since plots near one another tend to give similar results, different treatments should be placed in neighbouring plots and plots receiving the same treatment should be scattered over the field. This led to the proposal of various systematic arrangements or designs like the chess-board, Knight's move, and sandwich arrangements; see Cochran (1976). Then it was agreed that experiments should be replicated over different years and conditions, before making any specific recommendation. But there was one problem, about which the more perceptive among the experimenters were conscious although they could not resolve it. How could one justifiably generalize the results of particular experiments to broader conditions in which we are interested, or in other words, how to induce from the sample to the population? At the beginning of the 20th century some agricultural experts sought the help of error-theorists to resolve the issue, but the problem of getting a reliable estimate of the S. E. remained. A belief that corresponding to each plot size there was a definite value of the S. D. of the yield of a crop grew and comparisons were attempted using such imported values based on uniformity trials to determine the S. E.

| 'Student's' contributions | Around 1910–11 Gosset became interested in the conduct of agricultural experiments. While refereeing a paper for the *Journal of Agricultural Science* at this |

time, in a thoughtful appendix he recommended that for the comparison of two varieties, instead of taking an imported value of the S. E., *we should estimate the S. E. in each case from the set of data itself*. Also he re-emphasized that replications of varieties should be scattered in small plots over the field and 'the changing fertility level or "patchiness" of the experimental field' should be taken advantage of by comparing two varieties in neighbouring plots. A theoretical justification was provided for this—because of inequality (10.3.2), when ρ is highly positive, one is likely to get thereby a lower S. E. for the estimated difference. The brewery with which Gosset was employed had considerable interests in barley farming and in time Gosset was asked to look into the trials with barley. In the mean time he had started collaboration with the noted agricultural scientist E. S. Beaven. Together they exploited the concept of *blocking* for comparing different varieties in neighbouring plots. Generally the varieties were laid out in balanced systematic arrangements in 'chess-board' or sandwich ('half-drill strip') patterns (see E. S. Pearson (1939)).

| Comparison of treatments | When $v(>2)$ varieties were to be compared in b v-plot blocks, Gosset advocated the consideration of the difference in yield $d_{ij.\ell}$ between the ith and jth varieties |

in the ℓth block, $i, j = 1, 2, \ldots, v(i < j)$; $\ell = 1, \ldots, b$. The comparison had to be done separately for each pair on the basis of the ratio

$$t_{ij} = \frac{\sqrt{b}\bar{d}_{ij}}{s}, \tag{10.3.3}$$

where $\bar{d}_{ij} = \frac{1}{b}\sum_{\ell=1}^{b}d_{ij.\ell}$ and the common denominator for all pairs is the square root of the combined estimate of σ^2 given by

$$s^2 = \binom{v}{2}^{-1} \frac{1}{(b-1)} \sum_{i<j}\sum_{\ell=1}^{b}\left(d_{ij.\ell} - \bar{d}_{ij}\right)^2. \tag{10.3.4}$$

As the number of differences on which (10.3.4) is based is large $(=\binom{v}{2}(b-1))$, Gosset suggested using the normal tables for finding the tail probability of each t_{ij}. Apparently this method had been used by Gosset and Beaven during 1912–13, but the procedure was formally published by 'Student' only in 1923 (see E. S. Pearson (1939)). By that time R. A. Fisher had already conducted experiments laid out in blocks at Rothamsted Experimental Station and had proposed the technique of Analysis of Variance (ANOVA) for the simultaneous comparison of several treatments. The s^2 given by (10.3.4) turned out to be same as twice the 'Residual Variance' for a blockwise arrangement of treatments.

In the context of blocked experiments, in contrast with Fisher, 'Student' always preferred a balanced systematic pattern to a randomized arrangement within each block. There was some controversy between the two because of this. We will see later that the issue touches a fundamental question relating to statistical practice.

| Enter R. A. Fisher | **10.3.4** In his 1908 paper on the sample mean, 'Student' pointed out that to extend the application

of statistics to fields like chemistry, biology, and agriculture, where experiments cannot very often be easily repeated, development of exact sampling theory of statistics is required. But the biometric school of Karl Pearson, which was preoccupied with large-sample work, did not pay much attention to this. Actually to carry out such a development, one had to possess both the perspicacity to visualize its scope and the mathematical ability to handle its intricacies. Such a person burst into the scene of statistics around 1912. His name was Ronald Aylmer Fisher (1890–1962).

Fisher's work with regard to model-specific induction until 1930 can be said to have proceeded broadly along three parallel but interlinked lines: (i) sampling distribution, (ii) theory of estimation, and (iii) design of experiments.

| Sampling distribution | **10.3.5** This line of research had 'Student's' 1908 work as its starting point. Even as an undergraduate student

of mathematics and mathematical physics at Cambridge, Fisher was greatly impressed by the idea in 'Student's' derivation of the t-distribution. He conceived a fullscale proof of the joint distribution of $\bar{X}, s^2$ by a geometrical argument and wrote to Gosset about it. Then, in 1915 *Biometrika*, came out his stupendous contribution on the sampling distribution of r for a bivariate normal population with an arbitrary ρ, also derived by a similar geometric argument.[11] (The essence of such an argument is to express the sample joint density solely in terms of a set of relevant statistics. Then the problem of derivation of the joint distribution of such statistics, i.e. the evaluation of the integral of the sample joint density over any infinitesimal region around fixed values of the statistics, reduces to that of the volume element of the region.)

This single work gave Fisher, who during the years of the First World War had been working in an investment company and later in a public school, enough recognition. After this, he was offered the post of a statistician at the Rothamsted Experimental Station, which he joined in 1919. This gave him a unique opportunity to develop and promote the application of statistical techniques appropriate for experiments in agriculture and related fields in particular, and small-scale experiments in general. From 1920 onwards there was a spate of papers by Fisher, in which the normal-model sampling distributions of a large number of statistics and criteria were derived. Specifically, distributions of the sample mean deviation, regression coefficient, two-sample t, correlation ratio, intraclass correlation, variance ratio, analysis of variance F, partial correlation, multiple correlation in the null case, non-central χ^2 and F, and multiple correlation in the non-null case were covered. He pointed out how, when normality holds, these exact sampling distributions can be utilized to formulate tests for various hypotheses of real-life interest, which have validity even in small samples.

| Theory of estimation | **10.3.6** When Fisher started working on problems of statistics, there was no general theory of estimation from the sampling theory point of view, although ideas of unbiased estimation, minimum variance, and asymptotic relative efficiency of rival estimators had appeared in the work of Gauss and others in special contexts. As we saw in subsection 9.8.5, Karl Pearson in his 1898 joint paper with Filon considered something like the MLE of a multidimensional parameter. However, he confused

---

[11]A myth became current that Fisher read 'Student's' papers only after he had solved the problem of the sampling distribution of r to which he had been led by a publication of Soper in the *Biometrika* of 1913 (see Mahalanobis (1938), Kendall (1963)). This impression must have been created by the way the papers of Soper and 'Student' were referred to by Fisher in his 1915 paper. A letter of Gosset to Pearson dated 12 September 1912 reproduced in E. S. Pearson (1968), as also Fisher's obituary article on Gosset (Fisher (1939)), establishes that Fisher had corresponded with Gosset much earlier. Soper actually derived the mean and variance of r to higher order terms. From a letter of Karl Pearson to Fisher, also reproduced in E. S. Pearson (1968), it is clear that the former, after receiving the manuscript of Fisher's paper on r for *Biometrika*, wrote to him about checking with Soper's results.

the issue by mistakenly thinking that the associated formulae for large sample variances and covariances would remain valid for any reasonable estimate.

Later on Edgeworth in the last part of a paper which appeared in four instalments during 1908–09, considered the asymptotic distribution of what we would now call, M-estimators of a location parameter (see subsection 10.3.11) and showed that within this class the MLE (not so called at that time) has the least asymptotic variance; see Pratt (1976). But Edgeworth started his paper considering the pro-subjective (posteriorist) approach to estimation under the assumption of a uniform prior for the parameter, and the MLE was derived as the posterior mode. He studied its sampling variance not so much to supplant the prosubjective point of view but to supplement and vindicate it (cf. subsection 4.4.4).

Fisher, however, took up the problem of estimation with the avowed purpose of finding an alternative to the pro-subjective approach which he considered unsatisfactory and from which he wanted to distance himself. In his first statistical paper, which was published in 1912 when he was still an undergraduate student, he introduced the idea of likelihood (which at that time he called 'relative probability') and pointed out that it may be used for point-to-point comparison of different parameter values but cannot be integrated like an ordinary probability density. He exemplified the maximization of likelihood for estimation by considering the case of normal μ, σ^2, but did not study the sampling behaviour of the estimates.

As noted in subsection 8.4.3, Fisher in 1920 stumbled on the concept of sufficiency while comparing the sampling variances of two alternative estimators of normal σ. Then, in two epoch-making papers in 1922 and 1925 (see Fisher (1950)), he developed a general approach to the problem of estimation from the sampling theory point of view (see Fisher (1950)). In these Fisher formulated the general concepts of sufficiency and efficiency of estimators, amount of information, and the asymptotic sampling theory of the MLE as discussed in Chapter 4 (subsection 4.3.5). He also pointed out the mistakes in Pearson's theory of estimation (subsection 9.8.5) (incidentally, Fisher's heuristic proof of the asymptotic normality of MLE was similar to the development we chalked out in subsection 9.8.5) and showed that the loss of information resulting from the use of the method of moments (subsection 10.2.6) for non-normal populations may be considerable.

In developing sampling theory procedures for point estimation and statistical inference in the 1920s, did Fisher interpret these initially in the behavioural mode, or did he have in mind the instantial mode (sections 4.2 and 4.3) right from the beginning? Fisher's statements at the time incline us to the former view. Thus on p. 327 of the 1922 paper, while considering MLE x/n of the binomial p, he says, 'The reliance to be placed on such a result must depend upon the frequency distribution of x in different samples from the same population'. Even as late as 1933 in the context of finding a 99% fiducial lower bound for a normal S. D. σ, he observed that if s^2 is an unbiased estimate of σ^2 based on 10 d.f., from the χ^2-distribution with 10 d.f. we get that the probability of $s > 2.3209\sigma$ is 0.01. Hence $\sigma < s/2.3209$ with fiducial probability 0.01 in the sense that 'the first inequality will be satisfied in *just 1 percent of random trials, whence we may infer that the second inequality will also be satisfied with the same frequency* [italics ours]' (see Hald (1998) p. 640). From these statements, it appears that Fisher's rigidly instantial stance was a subsequent development. This is also supported by his introduction of a randomization test (to be considered below) for the ANOVA set-up in his 1935 book *The Design of Experiments*. Further, as Savage (1976) p. 449 notes, Fisher's investigation of the noncentral distributions of χ^2, t, and F in the late 1920s can have some *raison d'être* only if one concedes that at that time he attached some importance to the study of the power functions of the corresponding tests.

Fisher's work on designs ┃ **10.3.7** As L. J. Savage (1976) rightly said, 'Fisher is the undisputed creator of the modern field that statisticians call the design of experiments....' One can only wonder whether the field of design of experiments, and for that matter, statistical theory in general would have developed as they did, if in 1919 Fisher had not joined Rothamsted Experimental Station (in fact, at that time he had an alternative offer from Karl Pearson to work in the latter's laboratory, which he declined). As it happened, at Rothamsted Fisher found ready scope for the application of the theoretical results on model-specific induction that he had been deriving at the time. This, on the one hand gave him adequate incentive for deriving further results of the kind, and on the other led him to incorporate new ideas in the planning of agricultural experiments.

Fisher's first substantial paper (jointly with Mackenzie) on field experiments appeared in 1923 (see Fisher (1950)). In it he considered an experiment comparing 12 varieties of potatoes over six 12-plot replications each following a chess-board layout. This meant the idea of replication as also that of blocking, which resulted in the placement of different varieties in neighbouring plots were incorporated in the experiment. (As two types of manures, one having 2 and the other 3 alternative levels giving in all 6 manurial treatments were also accommodated, the actual layout of the experiment was more complex than a simple varietal one. In fact, it was what he later called a split-plot experiment; see Cochran (1980).) Of course Fisher, just like 'Student', required the variance of estimated comparisons to be estimable from the data itself. From his earlier study of the intraclass correlation, he was familiar with the technique of splitting the total sum of squares into components corresponding to different modes of classification and the error, and this he employed on the experimental data. (For some earlier attempts at such splitting see Stigler (1978b) and Cochran (1980).) The technique was initially called 'Analysis of Variation', but soon the name 'Analysis of Variance' (ANOVA) became standard. Although the distribution of the ANOVA F in the null case for the one-way classification readily follows from that of intraclass correlation which Fisher had found, at that time Fisher did not attempt any exact variance ratio test. Instead he employed the large sample test based on the difference in logarithms of the two variances compared. No mention of randomization for allocation of treatments to the field plots was made in the 1923 paper.[12]

In 1924, Fisher derived the distribution of the F statistic (actually Fisher considered $z = \frac{1}{2} \log_e F$; the symbol 'F', as the first letter of 'Fisher', was proposed later by Snedecor, contrary to Fisher's own preference). In 1925 appeared the first edition of Fisher's epoch-making book *Statistical Methods for Research Workers*. Shortly thereafter in 1926 he published a remarkable paper (see Fisher (1950)) entitled 'The arrangement of field experiments' in the *Journal of the Ministry of Agriculture*. In these he gave a detailed exposition of the basic principles for laying out a controlled experiment including replication, randomization, and blocking and introduced the randomized block and Latin square designs (for eliminating heterogeneity in one and two directions respectively) and the split-plot design (for nesting treatments of one type within another). Also he explained the technique of ANOVA as applicable to different layouts and showed how the ratio of the treatment mean square to the appropriate error mean square provides a valid F-statistic for comparing all the treatments. In the 1926 paper he introduced the idea of a factorial experiment and pointed to the possibility of confounding in such experiments for breaking up each replicate into a number of smaller blocks. The idea of analysis of covariance to take account of measurements on concomitant characters with a view to reducing the error variance was also proposed around this time. These and other related topics were included in subsequent editions of *Statistical Methods* and discussed more fully in Fisher's 1935 book *The Design of Experiments* (Fisher (1947)).

| Gauss and Fisher | As we briefly anticipated in subsection 8.3.5, the fixed-effects models underlying Fisher's analysis of various

designs are surely particular forms of the Gauss linear model (8.3.11). (In certain cases like that of the split-plot design observations could be subjected to a prior transformation to generate independent sets of variables, each subject to a Gauss-type model; cf. Nandi (1947).) The estimate appropriate for any contrast between effects is also the least squares estimate proposed by Gauss. Similarly the estimate of the variance of error derived from the ANOVA table is same as the 'residual variance' proposed by Gauss. But the main emphasis in Fisher's ANOVA is on the model-specific F test for the equality of effects and not so much on estimation. (Possibly this is because in agricultural experiments the interest lies in detecting whether a treatment has an effect at all, or

---

[12]Fisher and Mackenzie (1923) considered, apart from the usual additive model, a multiplicative one involving the products of effects; see Cochran (1980). In fact such multiplicative models had been proposed long back in 1835 by Cauchy in the context of analysis of data from an optical experiment; see Hald (1998) pp. 516–518. In his later work, Fisher never again considered the multiplicative model, apparently because such models are not amenable to simple tests like the additive one.

when there are several treatments whether there is one decidedly superior to the others. Incidentally, in the context of ANOVA, Fisher stated his preference for the 5% level of significance, obviously because that makes the test adequately sensitive.) Fisher proposed the different layouts so that the components required for performing such F tests can be worked out neatly in a schematic way. The two most revolutionary ideas of Fisher with regard to the planning of field experiments were those of 'factorial designs' and 'randomization'. We discuss these separately in the following.

| Factorial designs | In our view Fisher's most significant contribution to statistical theory and practice—at least in terms of its impact on scientific thought in general—is the development of factorial designs in which several factors are simultaneously varied in the same experiment. (His other major contribution—the concept of sufficiency—was, as we saw in subsection 8.4.3, anticipated by others in particular contexts, and one can conceive, would have perhaps come even otherwise.) From at least the time of Galileo, the established tradition in physical sciences had been to conduct experiments varying one factor at a time, holding the levels of the others fixed. Even, as late as 1948, Bertrand Russell in his book *Human Knowledge* (pp. 35–36) remarked, 'Speaking generally, scientific progress has been made by analysis and *artificial isolation* [italics ours]. It may be... there are limits to the legitimacy of this process, but if it were not usually or approximately valid scientific knowledge would be impossible.' The process may be illustrated by citing the instance of Boyle's law and Charles's law relating the volume of a fixed mass of gas separately to the pressure and the temperature; the two laws were subsequently combined to deduce a general law. Even when different factors could not be separated in practice (e.g. in studying the motion of a projectile, horizontal, and vertical motion could not be separated) the practice was to develop the theory for the components separately. This tradition had become almost universal, possibly because in physical sciences the factors can be easily controlled and rarely interact with each other; also, generally speaking, elementary physical experiments are relatively less time-consuming and more easily repeatable.

The same tradition held its sway on experimental practice, quite unjustifiably, in fields like agriculture also. Sir John Russell, the Director of Rothamsted Experimental station where Fisher was employed, in a paper published in the *Journal of the Ministry of Agriculture* in 1926, pleaded for observing the tradition faithfully while conducting agricultural experiments and varying one factor at a time, holding the others fixed. Fisher disagreed and reacted by ventilating his views in the paper referred to earlier, which appeared in the same journal later in the same year (see Fisher (1950)). In it he introduced the idea of factorial experiments and stated very pointedly:

No aphorism is more frequently repeated in connection with field trials, than that we must ask Nature few questions, or, ideally, one question at a time. The writer is convinced that this view is wholly mistaken. Nature, he suggests, will best respond to a logical and carefully thought out questionnaire; indeed if we ask her a single question, she will often refuse to answer until some other topic has been discussed.

The reasons for Fisher's advocacy of factorial experiments, elaborated further in his book *Design of Experiments*, were threefold. Firstly, since replication for comparing the levels of each factor are provided by the level-combinations of the others we get greater *efficiency* for all the main effect comparisons at much less cost. Secondly, a factorial experiment gives *information* on factor interactions about which single-factor experiments remain silent. Thirdly, conclusions about a factor from a factorial experiment have a *wider inductive basis* as these are drawn over various levels of the other factors. (Incidentally, this third advantage is related to the requirement of 'instantial variety' in induction, stressed by Francis Bacon three centuries earlier; see section 1.4.) While advocating factorial designs, Fisher also introduced the ideas of confounding and partial confounding, which allowed for the reduction of block size without sacrificing the property of orthogonality. Without these techniques factorial designs would have remained unattractive except in the case of very small experiments. As we will see these ideas proved most fruitful for further research.

| Randomization |

Randomization means performing a subsidiary random experiment, the distribution of whose outcome is fully known. In the present context we are concerned with pre-experimental randomization in controlled experiments for the allocation of treatments to field plots and usually we choose for this purpose a random experiment whose outcome has a uniform distribution over a finite domain. (We will see in the next section that some of the points we make here have relevance also for randomization for sampling from a finite population.) The outcome of randomization here influences the process of generation of the evidence directly and that of inducing from the evidence indirectly. It was Fisher who advocated and popularized such randomization in controlled experiments in a big way.[13,14]

One advantage of randomization is conceded by all: when we allocate the treatments to be compared randomly to the experimental plots, we preclude the experimenter's own bias from affecting the allocation, and, what is more, this impartiality is given the semblance of impartiality. It was with this

---

[13] At about the same time when Fisher was introducing randomized designs, it seems, Neyman was independently considering completely randomized designs in agricultural field trials in Poland; see Scheffé (1959) p. 291, footnote. We will see in the next section that a few years later Neyman made a strong case for random selection in finite population sampling.

[14] Randomization through devices like drawing of lots had been in vogue since ancient times; cf. Chapter 5. To facilitate randomization, Tippett published his *Tracts for Computers* No. XV (Cambridge University Press) giving 41,600 random digits in quartets around 1927.

end that C. S. Peirce introduced randomization in his weight-lifting experiment (subsection 10.3.8). The double-blind experiments for clinical testing, so common nowadays, also have the same purpose.

But Fisher's objective in advocating randomized designs in field trials went much further. When several treatments are to be compared on a number of plots, common sense demands that for the comparison to be meaningful we should balance the other conditions of the plots as far as possible (cf. Mill's method of agreement and difference; subsection 10.3.3). Fisher essentially considered three forms of balancing:

(i) Configurational balancing: this means grouping the plots into blocks of similar plots and then allocating the different treatments to be compared to the plots within each block. Two-way blocking led to the Latin square configuration.

(ii) Probabilistic balancing: this means randomly permuting the treatments allocated to a block over the plots of the block. If all allocations in a block are equiprobable, the conditions of the plots even if variable are being balanced over conceptual repetitions of the allocative process in a long-term sense.

(iii) Model-based balancing: in situations where there is a concomitant variable (like position of the plot along the fertility gradient or the number of plants in the plot) bearing on the variability of the plot yields, we can introduce it explicitly in the model and make suitable adjustments to balance its influence out.

Fisher introduced 'probabilistic balancing' or randomization in agricultural experiments around 1923–25. But the idea was at first resisted by a sizeable group of experimentalists including 'Student', who did not consider it very palatable. They thought that balance could be achieved in a surer way by arranging the treatments within blocks systematically. Thus when the treatments A, B are to be compared over a a number of 2-plot blocks, 'Student' pleaded for the 'sandwich' or 'half-drip strip design' which arranged the treatments like ABBAABBA..., instead of randomizing separately within each block.

| Randomization analysis | Fisher countered this with arguments which essentially meant that under randomization the features of the standard normal-theory model (to which the usual analysis is specific) are closely approximated by a more realistic down-to-earth model, but this advantage does not hold in the case of a systematic arrangement. We describe Fisher's arguments in terms of the randomized block design (RBD) with v treatments in b blocks.

Let Y_{ij} denote the (random) yield of the jth treatment in the ith block. The normal-theory model to which the usual analysis of RBD is specific is

$$Y_{ij} = \alpha + \beta_i + \tau_j + \epsilon_{ij}, \quad i = 1, \ldots, b, \ j = 1, \ldots, v, \quad (10.3.5)$$

where α, β_i, τ_j are respectively fixed, general, block, and treatment effects and the errors ϵ_{ij}s are independently and identically distributed as $N(0, \sigma_\epsilon^2)$.

Now let $\tilde{y}_{ij\ell}$ denote the *conceptual* yield that would be generated if the jth treatment were applied to the ℓth plot of the ith block, $i = 1, \ldots, b, j, \ell = 1, \ldots, v$. We assume

$$\tilde{y}_{ij\ell} = \tau_j + u_{i\ell}, \tag{10.3.6}$$

where τ_j and $u_{i\ell}$ are fixed numbers representing respectively an additive treatment effect and a 'plot error' which does not depend on the treatment. Using standard notations, we write

$$u_{00} = \alpha, \quad u_{i0} - u_{00} = \beta_i, \quad u_{i\ell} - u_{i0} = d_{i\ell}. \tag{10.3.7}$$

We then have the purely algebraic representation

$$\tilde{y}_{ij\ell} = \alpha + \beta_i + \tau_j + d_{i\ell}, \tag{10.3.8}$$

where $\sum \beta_i = 0, \sum \tau_j = 0, \sum_\ell d_{i\ell} = 0, i = 1, \ldots, b$.

Here all of α, β_is, τ_js and $d_{i\ell}$s, and hence, $\tilde{y}_{ij\ell}$s are fixed numbers. Now let the (random) serial numbers of the plots in the ith block to which the treatments $1, \ldots, v$ are assigned through randomization be $L(i, 1), \ldots, L(i, v)$ respectively, so that in our original notation $Y_{ij} = \tilde{y}_{ijL(i,j)}$. Then writing

$$e_{ij} = d_{iL(ij)}, \tag{10.3.9}$$

(10.3.8) gives the model

$$Y_{ij} = \alpha + \beta_i + \tau_j + e_{ij}. \tag{10.3.10}$$

This means $(e_{i1}, \ldots, e_{iv}), i = 1, \ldots, b$ are independently distributed, each having a permutation-symmetric distribution, with $E(e_{ij}) = 0$. Thus the joint distribution of $(e_{i1}, \ldots, e_{iv}), i = 1, \ldots, b$ resembles that of $(\epsilon_{i1}, \ldots, \epsilon_{iv})$ in (10.3.5) to a large extent. Further, Fisher noted that under (10.3.9)–(10.3.10) we have the following:

(a) For any $j \neq j', Y_{0j} - Y_{0j'}$ estimates $\tau_j - \tau_{j'}$ unbiasedly. (Proponents of systematic balancing thought that, if judiciously done, it also would ensure the same thing with a smaller $\text{Var}(Y_{0j} - Y_{0j'})$. Fisher observed that whether systematic balancing actually realizes unbiasedness can never be known—there is no guarantee that balancing a visual pattern would balance the fertility pattern for the design.)

(b) If $s^2 = \Sigma(Y_{ij} - Y_{i0} - Y_{0j} + Y_{00})^2/(b-1)(v-1)$, then $2s^2/b$ gives a valid (i.e. unbiased) estimate of $\text{Var}(Y_{0j} - Y_{0j'})$. Fisher never stopped emphasizing that we cannot generally get such a valid estimate of error from a systematic arrangement. The valid estimate of error allows us to judge whether observed differences in the treatment means are really significant or not.

(c) Furthermore, when there are no treatment effects, the discrete distribution of the ANOVA F-statistic under the model (10.3.9)–(10.3.10), as confirmed by theoretical as well as empirical studies, is closely approximated by the normal-theory F-distribution. The discrete distribution can be used to test what Fisher (1947) p. 43 called a 'wider hypothesis'. This *randomization test* exploits the physical act of randomization and the normal-theory test can be regarded merely as a convenient approximation to it. In other words, randomization ensures the robustness of the standard F-test.

These findings, Fisher thought, made an impregnable case for randomization.

| A theoretical objection | In recent times, the theoretical justification for randomization has been questioned by some statisticians, mostly Bayesian by persuasion (see Savage (1976), p. 464, Good (1983) p. 62, Basu (1988) pp. 279–289). The main burden of their criticism is that our available evidence comprises observations on (i) $Y_{ij}, i = 1, \ldots, b,\ j = 1, \ldots, v$ as well as (ii) $L(i, j), i = 1, \ldots, b,\ j = 1, \ldots, v$, and while invoking randomization, we deliberately ignore the second part of the evidence. Basu (1988), especially, comes down heavily on the randomization test, pointing out that here the $L(i, j)$s represent an ancillary statistic and according to Fisher's own recommendation in the sampling-theory approach we should proceed by conditionally fixing such a statistic (see subsections 4.2.3 and 4.3.5). Under the set up (10.3.9)–(10.3.10), if the $L(i, j)$s are fixed, everything becomes determinate and we have no random variable to induce about! But we should remember that on this ground not only the advantage (c), but also advantages (a) and (b) of randomization will go (Savage (1976) p. 464 is clear on this point).

| An answer | In our view the above reasoning has led us to a dead end because one vital element has been neglected. Actually, when we randomize the treatments over the plots of a block, we decide as a matter of policy to treat the plots in each block at par with each other and to regard the residual intrablock differences among the plots as irrelevant for induction. (This is the same principle of 'obliteration of labels' which in Chapter 9 (section 9.1) we emphasized as essential for fruitful induction). This fact is not taken into account if we build up our model as in (10.3.6)–(10.3.10) on unrealistic assumptions implying a set of fixed $\tilde{y}_{ij\ell}$s. To remedy this we suppose we have for each treatment-plot combination a *random* conceptual yield

$$\tilde{Y}_{ij\ell} = \tau_j + U_{i\ell}, \qquad (10.3.11)$$

where τ_j is as before an additive treatment effect and $U_{i\ell}$ is a *random* 'plot error' not dependent on the treatment. We are hopeful that under the system of blocking adopted, for each $i, U_{i1}, \ldots, U_{iv}$ are interchangeable but are not absolutely sure about that. Writing

$$EU_{i\ell} = \eta_{i\ell}, \quad U_{i\ell} = \eta_{i\ell} + D_{i\ell}$$

so that $ED_{i\ell} = 0$, we denote now

$$\eta_{00} = \alpha, \quad \eta_{i0} - \eta_{00} = \beta_i, \quad \eta_{i\ell} - \eta_{i0} = \delta_{i\ell}.$$

Hence for each i, $\sum_\ell \delta_{i\ell} = 0$ and

$$\tilde{Y}_{ij\ell} = \alpha + \beta_i + \tau_j + \delta_{i\ell} + D_{i\ell}. \tag{10.3.12}$$

(This is a special case of a general model considered in Scheffé (1959), Chapter 9.)

We suppose that $(D_{i1}, \ldots, D_{iv}), i = 1, \ldots, b$ are independently distributed. If our presumption about the $U_{i\ell}$s is true, $\delta_{i\ell} = 0$ for all i, ℓ and the distribution of $(D_{i1}, \ldots, D_{iv})$ would be permutation symmetric for each i. For the randomized experiment generally the representation (10.3.10) holds with the e_{ij}s now given by

$$e_{ij} = \delta_{iL(i,j)} + D_{iL(i,j)}. \tag{10.3.13}$$

Under (10.3.7) and (10.3.13) the conclusions (a)–(c) hold. (This can be seen directly with the distribution of F now generally continuous. Alternatively, this holds for conditionally fixed $D_{i\ell}$s and hence unconditionally; cf. Scheffé (1959), p. 322.) If for each i, the $U_{i\ell}$s are truly interchangeable, randomization would have been superfluous; in that case the ancillary $L(i,j)$s could have been conditionally fixed. Since we are not sure about the interchangeability of the $U_{i\ell}$s and as a policy we have decided not to take account of any asymmetries that may be present in their distribution, we resort to randomization. The ancillarity principle can be applied only when we are sure about our model (cf. Barnard *et al* (1962)). Our random experiment here consists of the random allocation of treatments to plots together with the agricultural part of the experiment. When we visualize repetition we think of repetition of this entire process.

We will see later, a similar question with regard to randomization may arise in the case of finite population sampling also. There too, in some cases, an argument as above would be relevant. Incidentally, not all Bayesians deny the utility of randomization. Thus D. B. Rubin (1978) shows that in the pro-subjective approach randomization can be dispensed with only if we have strong prior information; in the absence of such information randomization leads to a lot of simplification in the analysis. Lad (1996) pp. 212–218, who follows the purely subjective approach, introduces randomization under the new-fangled name 'scrambling', as a tool for ensuring exchangeability of the observables. Thus, considering the case of a completely randomized experiment, scrambling would make each and everybody regard the yields under each treatment as partially exchangeable (Lad, pp. 215, 235–238).[15]

---

[15]Although not directly relevant to our topic, we mention that in recent years the technique of deliberate incorporation of a random noise with a known stochastic

Before concluding our discussion of randomization in controlled experiments we mention two things. First, it may occasionally happen that after carrying out the randomization we end up with an arrangement which is palpably systematic and biased. What should we do then? Discarding the layout and repeating the process will be injurious to the theory. The only rational solution seems to be to exclude such layouts from the beginning and choose randomly from the set of the remaining admissible layouts (see Holschuh (1980) pp. 44–45). That would of course complicate the theory. Fisher saw the problem but did not address it. Second, although Fisher proposed the idea of randomization tests in designs in his 1935 book *Design of Experiments*, in the same year Neyman (1935) considered the randomization analysis of designs from a more general point of view. Without assuming additivity of treatment and plot effects, he showed that under the hypothesis of equality of average treatment effects the mean squares for treatment and error over all possible randomizations have expectations that are same for the randomized block and different for the Latin square design. Fisher failed to appreciate Neyman's point of view and thus began the long series of controversies between the two; see Holschuh (1980) p. 43.

Subsections 10.3.5–10.3.7 above describe broadly Fisher's contributions to model-specific induction in the areas of sampling distributions, theory of estimation, and design of experiments more or less during the period 1915–1930. These were instrumental in generating prolific research activities by a large number of workers in several different directions. In the remaining subsections of this section we give a brief résumé of the initial stages of these further developments under three broad heads: (i) theory of inference (ii) multivariate analysis and (iii) design of experiments.

Further developments—theory of inference

10.3.8   We saw in the preceding chapter (subsections 9.6.1 and 9.9.2) that in the 19th century the frequency theory of probability gradually ensconced itself in the thought world of statisticians and there was a growing dissatisfaction with the Laplacian pro-subjective approach towards statistical induction both for its dependence on subjective probability and for the arbitrariness involved in the choice of the prior. In the 1920s Fisher gave what at that time looked like a decent burial of Laplace by pointedly spelling out the criticisms against the pro-subjective approach (see subsection 7.2.6 and section 7.4), but even more by showing the feasibility of an alternative approach to statistical induction based on sampling theory. But in the sampling theory approach, especially in its behavioural form, the emphasis is on the procedure—the measure of assurance relates to the procedure and not to the conclusion. Therefore it is of the greatest

---

behaviour has been used successfully in certain computer-based control systems for optimization of an objective function or for linearization of the system. Such fostering of a regulated chaos in the system has proved fruitful where direct methods failed, see Wang (1993) pp. 193–198, where further references are given.

importance that the procedure be chosen in the best possible way on the basis of a thorough comparison of the performance of alternative procedures. For the problem of point estimation the issue was largely settled by Fisher by bringing in the concepts of efficiency, sufficiency, and information and developing the method of maximum likelihood. But for the testing (and interval estimation) problem, although Fisher derived a large battery of small sample procedures, these were proposed mainly on intuitive grounds and the fact that they were all based on optimum estimators. When there were several alternative tests available for the same problem, Karl Pearson prescribed that one should choose 'the most stringent of these tests', i.e. the test giving smallest tail area P-value (see E. S. Pearson (1966) p. 459). However, this was little more than sweeping the question under the carpet.[16]

Thus, around 1930 the problem of examination and formulation of 'the philosophy of choice among statistical techniques' with regard to testing and interval estimation cried for solution.

Initially, it were Jerzy Neyman (1894–1981) and Egon S. Pearson (1895–1980) who gropingly took up that task on themselves. At first they thought that, just like in the case of point estimation, Fisher's 'principle of maximum likelihood' would resolve the problem in the case of testing also. In a 1928 joint paper (see Neyman and Pearson (1967)) they proposed the likelihood ratio (LR) λ-criterion (subsection 4.3.5)

$$\lambda = \frac{\max_{\theta \in \Theta_0} L(\theta \mid x)}{\max_{\theta \in \Theta} L(\theta \mid x)}, \quad 0 \le \lambda \le 1, \tag{10.3.14}$$

and a left-tailed test based on it for $H_0: \theta \in \Theta_0$ under the set-up $\theta \in \Theta$. It was found that the criterion produced reasonable tests for many standard problems. But unlike in the case of point estimation where Fisher heuristically reached the optimality property of the MLE, at least in large samples, no general optimal properties of the λ-test were readily discernible. (Large sample optimality of the LR test was established under broad conditions much later by Wald, Lecam, Bahadur, and others; see Bahadur (1967) and LeCam (1970), where further references are given. Of course these later developments drew upon concepts which were formulated in the intervening period.)

| The behavioural approach |

The crucial step that Neyman and Pearson took here was to adopt the behavioural approach (section 4.2).

As we saw in subsection 9.9.2, C. S. Peirce long ago had pleaded strongly in favour of the behavioural point of view. It was implicitly adopted by 'Student', when he studied the sampling behaviour of test criteria through model sampling experiments. In his early writings, Fisher too seemed to have behavioural leanings (although, no doubt, he would have vehemently denied

---

[16]However, much later, this same prescription has been converted into a basis for asymptotic comparison of tests by Bahadur (1971).

any such ascription later). As we saw in section 8.5, Neyman and Pearson got their inspiration from different sources, but both came to the realization that alternative hypotheses must be taken into consideration while choosing among rival tests: one must judge alternative tests meeting the same size requirement in terms of their power or long-term frequency of detecting various alternative hypotheses. Model sampling experiments performed under alternative hypotheses showed how far tests could be compared in terms of their power functions. Soon the Neyman–Pearson Lemma was derived by applying the Calculus of Variations. The concepts of UMP tests, unbiased tests, UMPU tests, locally UMPU tests, similar tests, etc. (subsection 4.2.2) came in due course. Various standard tests, based on the sampling distributions derived by Fisher, were now judged in the light of these new criteria. These results appeared in a series of five joint papers by Neyman and Pearson, the first of which came out in 1933 and the last in 1938 (see Neyman and Pearson (1967)).

| Behavioural–instantial divide |

The problem of interpretation and justification of interval estimates based on the sampling distribution of standard pivotal functions (subsection 4.2.2) remained. Fisher in a 1930 paper had introduced the idea of fiducial probability distribution of a parameter and the fiducial interval with reference to the particular problem of estimation of a bivariate normal correlation coefficient. But no general theory was given and to most people the idea looked rather opaque. The possibility of a behavioural interpretation of interval (set) estimates based on repeated sampling was perceived and vaguely expressed in some writings at this time (Neyman (1976) refers to a 1928 monograph in Polish by Pytkowski in this connection. The word 'confidence' was used with reference to the ellipsoidal set estimate of the multinormal mean vector by Hotelling (1931). However, as we saw in subsection 9.9.2, the idea, although not the mathematics, was spelt out by C. S. Peirce about half a century earlier.) A definitive behaviouralistic interpretation of a confidence interval was given by Neyman (1934) in the context of sampling from a finite population (subsection 10.4.3); a general theory of confidence intervals was first formulated in Neyman (1937). The theory utilizes the ideas of the Neyman–Pearson theory of hypothesis testing on the basis of the simple relation that an interval (or set) estimate of a parameter contains precisely those parameter points which would be accepted by a family of corresponding tests of the same size. At first, it was thought that a fiducial interval and a confidence interval were just two ways of looking at the same thing. But the distinction between the two approaches got sharply focussed through the Behrens–Fisher problem (subsection 4.3.7) in which the fiducial interval of the difference between two normal means, when the variances may be different, was found not to have any assignable confidence level (probability of inclusion) on repeated sampling. We have examined the principal points of distinction between the behavioural approach and the fiducial form of the instantial approach in Chapter 4.

When the behavioural route to statistical induction was chalked out in the 1930s, most statisticians embraced it with alacrity. One reason for this surely was that in it everything was worked out in terms of the frequency theory of probability—no new interpretation of probability was required. But an indirect reason was the growing technological ambience in society with its emphasis on mass production of items.[17] In fact, even before the Neyman-Pearson theory of hypothesis testing took shape, Dodge and Romig in a paper which appeared in the *Bell System Technical Journal* in 1929 considered the choice of plans for sampling inspection of items produced in mass. For sampling inspection of lots by attribute, a lot is definitely acceptable or unacceptable according as the proportion of defective items in it is upto a specified level or exceeds a (higher) tolerance value. For the selection of a suitable plan, Dodge and Romig considered values of the *Producer's Risk* and *Consumer's Risk*. These are respectively the suprema of the probabilities of an acceptable lot being rejected and an unacceptable lot being accepted and correspond to the probabilities of first and second kinds of error in hypothesis testing. What is noteworthy is that in the case of mass production, these risks have tangible interpretations as long term relative frequencies. Neyman and Pearson in their 1933 paper which appeared in the *Proc. Cambridge Phil. Soc.* specifically referred to the relevance of their approach to hypothesis testing in the context of mass production industry. Fisher continued to protest as long as he lived that the context of scientific inference is different from that of mass production. But in the 1930s and 1940s the protagonists of the behavioural approach, who heavily outnumbered others, did not bother to consider such distinctions.

Wald

The two major contributions to statistics, Decision Theory and the Theory of Sequential Procedures of Abraham Wald (1902–1950) are both rooted in the behavioural approach. As discussed in subsection 4.2.4, decision theory is a direct extension of the Neyman–Pearson theory of hypothesis testing. Trying to minimize the risk or long term average loss by choosing the rule can be meaningful only if one is envisaging the possibility of repeated application of the rule in successive experiments. Wald's development of decision theory in his 1939 paper and later in his 1950 book *Statistical Decision Functions* allows looking at the behavioural approach to statistical induction from a broad perspective—it shows that standard problems

[17]There is one industry in which mass production had been practised in many countries at least from the medieval period—the production of coins in mints. Stigler (1977) has unearthed evidence to show that at the Royal Mint in Great Britain sampling inspection in the form of a ceremony called 'Trial of Pyx' used to be practised at least since the 13th century. The object was to check that coins issued met the specifications of the Crown. But tolerances and inspection standards in such trials were not based on any probabilistic considerations, although Newton, who was the Master of the Mint in the first quarter of the 18th century, is known to have emphasized the importance of reducing the variability among the coins while controlling their standard.

like estimation and hypothesis testing and also the hitherto unexplored multiple decision problems all fit into a general framework. What is more, it shows that the real issue in the behavioural mode of solving all such problems is that of identifying the class of all admissible rules. Beyond that, it is a matter of choice within this class and the experimenter has to exercise his subjective preferences at this point. Wald employs prior distributions and Bayes rules only as formal techniques which help one to get to the admissible rules. But no doubt this admission of Bayesian technology indirectly led to the later revival of the Bayesian principle—there was a renewed interest in the pro-subjective Bayesian approach from around 1950 onwards.

The idea of sequential experimentation arose directly in the context of sampling inspection of mass production processes during the Second World War. Wald reached a workable solution to the problem of deciding upon the acceptance or rejection of a lot through sequential sampling, by adapting the likelihood ratio test of a simple hypothesis against a simple alternative. The ideas of Operating Characteristic ($= 1 - $Power) and Average Sampling Number of a procedure extensively studied in such contexts have relevance only in the behavioural setting. However, sequential inference rules, almost since their inception, have been a matter of sharp controversy among theoreticians. On one side are the behaviouralists who believe that the sample space and the way the sample is obtained are relevant for studying the performance characteristics and hence for the choice of the rule and on the other are the protagonists of the likelihood principle (subsection 4.3.4) who think that the way a sample is arrived at is irrelevant for decision taking. In fact the latter group (who of course include the pro-subjective Bayesians) has coined the term Stopping Rule Principle to describe their motto. (The Principle was articulated by Barnard as early as 1947 in his review of Wald's book *Sequential Analysis* in the *Journal of American Statistical Association*; a modern exposition is to be found in Berger and Wolpert (1984) pp. 74–90.) The issue brings into sharp relief the distinctive positions of the behaviouralists and all the other schools who gear their inference to the particular sample at hand.[18,19] The controversy has remained unresolved to this day: the position one takes seems ultimately a matter of personal attitude and preference.[20]

| Subsequent developments | From the 1940s onwards research of the behavioural school has been concerned primarily with the reassessment and consolidation of the standard procedures. |

---

[18]But for the problem of designing the sampling experiment the latter also have to take the stopping rule into account.

[19]Savage (1976) wonders why R. A. Fisher was a 'non participant in sequential analysis'. It seems to us the reason is not far to seek. Sequential analysis with all its aspects fits in well only with the behavioural approach. But by the time sequential theory emerged, Fisher had shed all his earlier behavioural leanings.

[20]However, the problem of choice of an optimum stopping rule remains even under the Bayesian approach; see Ghosh *et al* (1997) pp. 112–125.

For the problem of point estimation, properties like unbiasedness, minimum variance, and admissibility have been investigated to the hilt. In this connection the roles of various lower bounds to the variance estimates and of sufficiency for locating optimal estimators have been brought out. For testing problem also the sufficiency route to optimality, particularly in the case of the exponential family of distributions (subsection 4.4.10), has been thoroughly explored. (Standard references for these results are Rao (1965) and Lehmann (1983, 1986).)

While extending these results to the case of sequential experimentation, initially there was some difficulty. Classical sequential theory was developed for the case of an i.i.d. sequence of random variables $X_1, X_2, \ldots$ and at first it was not very clear how one can pass on to stopping and decision rules based on a sequence of statistics $T_1, T_2, \ldots$, where for every n, T_n is based on $X_1, \ldots, X_n$ only. Nandi (1948) heuristically proposed the use of well-known statistics for this purpose. But the question that begged an answer was essentially how to link the T_ns corresponding to successive n. The tangle was cleared by Bahadur (1954) by bringing in the concept of a transitive sequence of sufficient statistics. $(T_1, T_2, \ldots)$ is such a sequence if for each n, T_n is sufficient, and given T_n, conditionally $(X_1, X_2, \ldots, X_n)$ and T_{n+1} are independently distributed. In most cases this holds because T_{n+1} is a function of T_n and X_{n+1}.) When such a sequence exists, one can confine oneself to rules based on it for deriving optimum procedures; see e.g. Ferguson (1967), pp. 333–338.

As regards decision theory, Wald derived the main results of his theory (which in essence imply that in solving a decision problem we can proceed through the class of Bayes rules) under the assumption of a bounded loss function. This restriction was removed by M. N. Ghosh (1952), who however, had to assume the boundedness of the risk function. The results were later proved under more general conditions by L. LeCam (1955); see Ferguson (1967).

A principle which has proved very fruitful for the development of procedures for estimation and testing as also for the general decision problem, is that of *invariance*. It was implicit in Fisher's 1934 (*Proc. Roy. Soc.*) study on the estimation of a location parameter under the measurement error model and was fully developed by others later. When a problem exhibits certain symmetries (i.e. is invariant under a group of transformations) we want our procedure also to reflect the same symmetries or invariance properties. Apart from the practical desirability of this requirement, it serves two other purposes. Firstly, as such a requirement generally allows us to base our procedures on statistics having certain structures, it often leads to reduction of data (just like the sufficiency principle). Secondly, by restricting the class of available procedures (just like unbiasedness in estimation and testing) it helps us to eliminate procedures with lop-sided performance and thus to identify one which is 'best' in some sense; see Ferguson (1967), Chapter 4.

As regards large sample evaluation of inference rules, Fisher's intuition about the asymptotic properties of MLEs has been largely vindicated under appropriate conditions. We have already mentioned large sample studies relating to the optimality of likelihood ratio tests. Tests based on scores (derivatives of the log likelihood) which are asymptotically equivalent to the likelihood ratio tests and are sometimes easier to work with in practice have also been proposed (see Rao (1965) pp. 349–351). Another kind of investigation which seeks to bridge the gap between the findings of fixed-sample-size and asymptotic studies is concerned with what are called the second order properties of procedures. Broadly speaking, it has been shown that although the MLE is only one of many efficient estimators when variance is considered upto the first order of approximation, in a sense it is *primus inter pares*—a suitably adjusted version of the MLE has higher efficiency than many other efficient estimators when the next higher order term is taken into account (see Efron (1975), Ghosh *et al* (1980), Lehmann (1983) Chapter 5). Similarly for the problem of testing, results obtained under broad conditions suggest that likelihood ratio type tests, when compared in terms of probabilities of large deviations under fixed alternatives, perform better than other tests to which they seem equivalent under traditional modes of comparison (see Hoeffding (1965), Brown (1971)).

As observed above, there has been a revival of interest in the pro-subjective approach to statistical induction from the 1950s onwards. (Before that only isolated workers like Harold Jeffreys had been ploughing their lonely furrows in this area.) The main reason for this revival was the re-examination of the bases of subjective probability that took place in the intervening period. Also, as the behavioural approach expanded and was followed to solve varieties of problems, its shortcomings (subsection 4.2.6) looked more glaring and disillusioned many an enthusiast. The advent of computers and the development of techniques which allow the construction of posterior distributions which are analytically intractable, through iterative simulation (Markov Chain Monte Carlo; see e.g. Gelfand and Smith (1990)) have also helped the process. Strangely, however, among the upcoming pro-subjective Bayesians also there has been a division of ranks. One group consists of Bayesians (the Jeffreysites) who prefer noninformative (possibly improper) public policy priors (see subsection 4.4.5)—they are often not averse to subjecting their methods to frequentist assessment. The other group comprises Bayesians who swear by and admit only proper personal priors (see subsections 4.4.12–4.4.13). Occasionally, a few wayfarers from the latter camp stray far enough to get into the sphere of attraction of the parameter-free purely subjective approach (section 4.5). The general scene in the arena of statistical inference is now quite confusing. To the uninitiated, the assembly would seem like that inside the tower of Babel. One must recognize a speaker's point of view—especially his conception of probability—before one can make any sense of what he is speaking about.

| Further developments— multivariate analysis |

10.3.9   Another field where rapid development took place from 1930s onwards was that of multivariate analysis. We saw earlier that multivariate data started

confronting statisticians when towards the end of the 19th century the methods
of statistics were sought to be applied in the field of biometry. The multivariate normal model was developed to meet the need of such applications.
Various measures quantifying different aspects of the joint variability of several variates came to be formulated in course of studying the properties of this
model. In the 1920's, Fisher derived the sampling distributions of the sample
estimates of many of these measures and thus demonstrated the feasibility of
statistical induction with regard to those aspects in the case of a multivariate
population.

The first step towards developing a full-scale theory of multivariate methodology was of course the extension of the by then standardized univariate inference
procedures to the multivariate case. Towards this end, Wishart in 1928 derived
the joint distribution of the sample variances and covariances by employing the
geometrical technique of Fisher. In 1931, Hotelling considered the null distribution of T^2, the multivariate counterpart of 'Student's' t^2, and showed that it was
distributed like a multiple of the central F. Given a sample from a multivariate
normal population $N_p(\mu, \sum)$ with all parameters unknown, T^2 provided a test
for the hypothesis H_0: $\mu = \mu_0$, as also an ellipsoidal confidence region for μ. By
that time the likelihood-ratio technique for generating test criterion had already
been proposed by Neyman and Pearson (subsection 10.3.8). They, as well as
S. S. Wilks, exploited the technique extensively to derive tests for various hypotheses about mean vectors and dispersion matrices of multinormal populations;
see Anderson (1984).

There is one problem, however, which is special to multivariate situations and
it called for a closer examination—the problem of dimensionality or of choice of
variables. The problem presented itself in two forms in two types of situations.
The first form arises for example in anthropometric studies where we have some
physical populations, on the units of which any number of different characters
can be measured (sometimes only a part of a population is physically at our
disposal, but in that case a random sample from that part is postulated as one
from the whole). There the question is how many and which of the characters
should be measured and whether the choice of characters has any bearing on
our conclusion about the physical populations. The second form of the problem,
which has a somewhat opposite implication, arises when some characters have
already been observed on the sample units. The question is whether we can
select a small number of these characters and work in terms of these *without
losing information substantially.* In the context of the multivariate normal model,
since the original set of characters can be equivalently replaced by a nonsingular
transform, the question becomes whether we can replace the original characters
by a few suitably chosen linear compounds. (Actually, in certain problems at
least, it has been found that involving too many extra variables which do not
carry additional information makes the inference procedure less efficient, see e.g.
Rao (1966). Thus reduction of the number of dimensions, apart from its practical
advantage, has also theoretical justification.)

The first question was implicit in Karl Pearson's 1926 coefficient of Racial Likeness (CRL), but was explicitly represented in Mahalanobis's measure of distance D^2 between two physical populations. CRL was defined as the mean of the standardized squared differences between the sample means of each variable and did not take account of the correlations between the variables (see e.g. Dasgupta (1980)). The final form of D^2, which Mahalanobis reached through a number of papers during 1927–1936, took these correlations into account. Studentized form of D^2 is[21]

$$D^2 = (\bar{x}_{(1)} - \bar{x}_{(2)})'S^{-1}(\bar{x}_{(1)} - \bar{x}_{(2)}), \qquad (10.3.14)$$

where $\bar{x}_{(k)}^{p \times 1}$ is the mean vector of a sample of size n_k from the population $\mathcal{N}_p(\mu_{(k)}, \Sigma)$, $k = 1, 2$ and S is the unbiased estimate of $\sum$ based on the $n_1 + n_2 - 2$ d.f. available from the two samples. D^2 was proposed as an estimate of the corresponding population value.[22]

$$\Delta^2 = (\mu_{(1)} - \mu_{(2)})' \sum^{-1} (\mu_{(1)} - \mu_{(2)}). \qquad (10.3.15)$$

Mahalanobis in a 1936 paper proposed to use the D^2 between pairs chosen out of a collection of populations (races in an anthropometric survey, e.g. Anglo-Indians and the different caste-groups of Hindu society) for clustering these according to their mutual affinities. As the number of characters that could be measured in such a context was potentially very large, the question arose whether the clustering would depend on the particular set of characters chosen. He postulated that, as the number of characters is enlarged, Δ^2 (which would necessarily increase) would remain bounded and therefore converge. This he termed the 'Dimensional Convergence of D^2' (see Rao (1973)).

The statistic D^2 is the two-sample counterpart of Hotelling's T^2. But whereas the two-sample Hotelling statistic is meant to test H_0: $\mu_{(1)} = \mu_{(2)}$, Mahalanobis considered situations where it is known that $\mu_{(1)} \neq \mu_{(2)}$ and one wants an estimate of Δ^2. It was important to study the sampling distribution of D^2 as an estimate of Δ^2. This problem was solved by Bose and Roy (1938), who showed that, apart from a scalar multiplier, the distribution of D^2 is same as that of non-central F.

The measure D^2 has been extensively used for clustering of populations in several large-scale anthropometric surveys; see e.g. Mahalanobis et al (1949).

The second question, we may recall, arose as long ago as 1888, when Galton critically examined Bertillon's prescription for the identification of criminals on

---

[21]We are omitting a divisor p from D^2 and Δ^2 as originally defined, for convenience of presentation.

[22]It is noteworthy that Δ^2 can be arrived at by specializing more general measures of divergence to the case of two populations $N_p(\mu_{(k)}, \sum)$, $k = 1, 2$. Thus the Hellinger–Bhattacharyya distance between two populations (see the review paper Bhattacharyya (1990–91)) here can be shown to be equal to $e^{-\Delta^2/8}$. Also, the Kullback–Leibler information measure of divergence (see (4.4.17)) here reduces to $\frac{1}{2}\Delta^2$.

the basis of a number of anthropometric measurements (subsection 9.7.5). Later, Karl Pearson in his 1898 paper (see Karl Pearson (1948)) showed that the multiple correlation coefficient $\rho_{1.23...p}$, given by (9.8.11), can be interpreted as the correlation between X_1 and the linear compound of $X_2, \ldots, X_p$, which is most highly correlated with X_1. The implication was that for predicting X_1 one could as well replace the set $X_2, \ldots, X_p$ by that single linear compound. The idea of replacing a whole set of variables by a few linear compounds was further pursued by Hotelling in the 1930s in the context of psychometric studies; see Madow (1960). In a 1933 paper he introduced *principal components* whose purpose is to replace the original set of variables by a small number of linear compounds which account for the major part of their joint variability. Later in 1936 Hotelling extended the same idea to the problem of studying the interdependence between two sets of variables in terms of their *canonical correlations*. First, a pair of linear compounds, one of each set, is chosen so that their correlation coefficient is maximized; subsequently further pairs of linear compounds are determined in the same way, subject to these being uncorrelated with the earlier pairs. The sampling theory of principal components and canonical correlations in various situations has been studied by a host of workers later; see Anderson (1984), Chapter 13.

Reduction of dimension in multivariate analysis got a new orientation during 1936–1940, when Fisher through a series of four papers adapted the technique of linear compounding in a new way for solving various multivariate inference problems. Fisher's idea was to reduce a multivariate problem to its univariate counterpart by taking a linear compound of the variables. The compounding coefficients have to be chosen so that the resulting univariate statistic is 'most discriminating', i.e. has the most extreme value possible with reference to the particular data set at hand. Since the compounding coefficients then become data-dependent, the original univariate sampling theory no longer applies—the sampling theory has to be worked out afresh.

In the case of two populations $\mathcal{N}_p(\mu_{(k)}, \sum)$, $k = 1, 2$, if we use the coefficient vector $h^{p \times 1}$ for compounding, using notations as earlier the univariate two-sample t^2-statistic becomes proportional to

$$\{h'(\bar{x}_{(1)} - \bar{x}_{(2)})\}^2 / h'Sh.$$

This is maximized for

$$h = S^{-1}(\bar{x}_{(1)} - \bar{x}_{(2)}) = \hat{h} \text{ (say)}.$$

Substitution of $h = \hat{h}$ in t^2 takes us back to D^2 given by (10.3.14). The problem of sampling distribution is not simplified even a little bit; but the multivariate procedure is 'formally' given a new interpretation as a univariate one (Fisher, in his 1936 paper, gave the correct analysis purely on intuitive grounds; see Dasgupta (1980)). Fisher also suggested that the 'estimated discriminant function' $\hat{h}'x$ may be used for classifying a fresh observation x as one from either the first or the second population by judging whether its value is closer to $\hat{h}'\bar{x}_{(1)}$ or $\hat{h}'\bar{x}_{(2)}$. (Later on Wald (1944) and others treated the problem as a two-decision one and studied the probabilities of mis-classification.) Fisher further considered testing for the hypothesis that an assigned discriminant function $h'_0 x$ is fully adequate for discrimination, or equivalently, that there is no additional information in x over and above that in $h'_0 x$ and arrived at the solution intuitively; see Rao (1965) pp. 467–470.

In the case of more than two populations Fisher's technique of linear compounding followed by maximization of the ANOVA F-statistic led to the roots of a determinantal equation involving the between and within sample matrices of sums of squares and products. For the problem of testing $\sum_1 = \sum_2$ on the basis of two independent samples from $\mathcal{N}_p(\mu_{(k)}, \sum_{(k)}), k = 1, 2$, a similar equation involving the estimates $S_{(k)}$ of $\sum_{(k)}$ was arrived at by Roy (1939) by employing the same technique. Generally, the sampling theory under null hypothesis in such cases requires studying the distribution of the roots of an equation of the form

$$|W_1 - \theta W_2| = 0 \qquad\qquad (10.3.16)$$

where W_1 and W_2 are p-variate Wishart with the same parent dispersion. Fisher in a 1939 paper (see Fisher (1950)) solved the problem in the case $p = 2$ and anticipated the general result. The general solution was independently derived at about the same time by Hsu, Roy, and Girshick; see Anderson (1984) Chapter 13.

The logic underlying Fisher's idea of linear compounding followed by extremization of the corresponding univariate statistic was subsequently abstracted by Roy (1953), who called it the *Union-Intersection Principle*. To apply this principle for the derivation of a test for $H_0: \theta \in \Theta_0$ against $\theta \notin \Theta_0$, we identify a family of hypotheses $H_{0u}: \theta \in \Theta_{0u}, u \in \mathcal{U}$ such that $\bigcap_{u \in \mathcal{U}} \Theta_{0u} = \Theta_0$. The H_{0u}s are chosen such that there are standard tests of size, say, α' with critical regions, say, $\mathcal{X}_{1u}$ for testing the hypotheses $H_{0u}: \theta \in \Theta_{0u}$ against $\theta \notin \Theta_{0u}$. (When testing H_0 is a multivariate problem, often H_{0u} are chosen so that testing H_{0u} can be viewed as a univarate problem for which a standard critical region $\mathcal{X}_{1u}$ is known. Thus the hypothesis $\mu = o$ for $\mathcal{N}(\mu, \Sigma)$ holds if and only if the hypotheses $h'\mu = 0$ hold simultaneously for all $h^{p \times 1} \neq o$. Then $\mathcal{X}_1 = \bigcup \mathcal{X}_{1u}$ is taken as the critical region of a test for $H_0: \alpha'$ is so adjusted that $\mathcal{X}_1$ has the desired size α. This heuristic principle proved very fruitful in generating tests for various multivariate problems—tests which are alternatives to the standard LR tests. These tests are particularly useful for constructing simultaneous confidence intervals for multiple parametric functions; see Roy (1957) Chapter 14. (However as noted in the preceding subsection, their performance as judged by the theory of large deviations falls short of that of the comparable LR tests; see Brown (1971).)

Multivariate analysis is the application of theories of inference for statistical induction from multivariate data. In the above we have recounted the early developments in sampling-theory multivariate analysis. Emergence of various concepts and principles and shift of emphasis in the field of inference have inevitably left their mark in later developments in the area. In the second half of the 20th century a lot of work has been done on the admissibility and other optimality properties of multivariate procedures; the principle of invariance has been invoked in various ways for such studies. Another area where new methods have come up is that of mixed models in which the underlying dispersion matrices have special structures. The revival of interest in the Bayesian approach has led to the reconsideration of various multivariate problems in terms of the posterior distribution of the relevant parameters. At another level, purely subjective, parameter-free induction about unobserved random variables has been considered for various types of multivariate setups; see Lad (1996) Chapters 7 and 8.

Extension of the above ideas and techniques of multivariate analysis leads us naturally to multivariate time series and stochastic processes. But these topics we are keeping outside the scope of this book.

Further developments—design of experiments

10.3.10   Fisher, in his theory of designs, stressed the importance of *balance* (which means equal sampling variance for the estimates of comparable treatment contrasts) and *orthogonality* (which means the absence of correlation between estimates of linear parametric functions which are mutually orthogonal). Ostensibly, under the fixed-effects normal-theory model these properties simplified the analysis of the design and allowed the partitioning of the sum of squares for the hypothesis that all estimable parametric functions are zero into separate components for different sub-hypotheses (e.g. sub-hypotheses about the absence of different types of treatment effects, block effects, etc). As it happened, it was later found that these properties of balance and orthogonality usually ensure maximum efficiency for estimation and sensitivity for testing under plausible conditions. Of course balance and orthogonality have to be realized to the extent possible keeping the block size small. For the case of factorial experiments Fisher proposed the techniques of *confounding* and *partial confounding* for this.

For varietal experiments, Yates, taking the cue from Fisher, proposed the *balanced incomplete block design* (BIBD) and gave the construction of such designs in some particular cases. R. C. Bose attacked the technical problem of construction of BIBDs with the tools of Galois fields and related finite geometries and obtained comprehensive solutions for various situations. Further, as BIBDs are available only under restrictive conditions, he along with K. R. Nair proposed the much wider class of *partially balanced incomplete block designs*. Under the leadership of Bose, a host of workers investigated the problems of existence and construction of these designs over a wide spectrum of situations; see Raghavarao (1971). But admittedly these combinatorial developments were of a technical nature and somewhat removed from the context of model-specific induction from which they originated.

As regards factorial designs, Fisher's ideas were further developed in the case of simple symmetric factorials by Yates (1937). The idea of confounding in such cases was extended to that of *fractional replication* by Finney. This opened up the possibility of application of such designs to situations where there are too many factors—an occurrence especially common in the case of industrial experiments. Again the general theory of symmetric factorials was built up by Bose and his associates, who brought the ideas of finite geometries to bear upon the problem. The key to this was the identification of each treatment (i.e. level combination of factors) for an s^m experiment (i.e. an m-factor experiment with s levels for each factor) where s is a prime power with a point in the m-dimensional finite Euclidean geometry $EG(m, s)$. The theory of *orthogonal arrays* in the symmetric case was proposed by C. R. Rao and this allowed an alternative view of the problem of fractional replication. (For all these topics, we refer to Raghavarao (1971).) This gave rise to a rich literature on combinatorics with ramification in other branches like Coding Theory. But again most of it is of a technical character and far removed from the context of model-specific induction.

Further advance in the theory of design of experiments came when in the second half of the 20th century, multifactor designs started being employed in a big way in industrial experiments. For such experiments often one has quantitative factors whose levels $x_1, \ldots, x_k$ vary continuously and the mean of the response variable $Y, E(Y) = \eta(x_1, \ldots, x_k)$ is a function ('response function') of the continuous factor levels. Unlike what is usually the case with agricultural experiments, one is not interested in picking the best treatment or level combination from amongst those experimented with. Rather, one wants to explore the entire response surface (usually represented by a polynomial in $x_1, \ldots, x_k$) to locate the point of optimum response.

This change of attitude shifted the emphasis from the properties of balance and orthogonality. This is because the kind of efficiency these imply is not so relevant for the exploration of the entire response surface, particularly in regions where the nature of its curvature requires the introduction of higher degree terms in the polynomial formula. Desirable properties like *rotatability* (invariance under orthogonal transformation of $x_1, \ldots, x_k$) and other optimality criteria like those of *D-optimality, A-optimality* and *ϕ-optimality* have been exploited to meet this change of emphasis. (Some of these criteria were introduced earlier in the context of weighing designs. Their basic principle is to order information matrices—which reduce to moment matrices in the case of a normal-theory model—in terms of certain meaningful scalar-valued functions; see Silvey (1980) and Liski *et al* (2001) for reviews of these concepts.)

The condition of orthogonality, however, re-emerged in the context of industrial experiments. Fractional factorial plans are often easier to work on and more cost-effective than arbitrary multifactor combinations. Also according to the motto set up in recent years by Taguchi and his school for designing a production process, the factors affecting a process are classifiable into two groups—'control factors' and 'noise factors'. The latter comprises all factors which may be controlled in the laboratory, but not during the actual use of the product. The object of an industrial experiment is to determine the 'optimum' setting of the control factors—the 'optimum' being that particular level combination for which the product maintains a high stable performance level, irrespective of variation in the noise factors. The design proposed for such a study is usually the Kronecker product of an orthogonal array involving the control factors by another involving the noise factors (see Myers and Montgomery (1995) Chapter 10). Such applications necessitate extension of the concept of orthogonal arrays to the case of asymmetric factors. Luckily such an extension due to C. R. Rao was already available and this has been exploited for the construction of fractional factorial plans of the desired type for industrial experiments (see Dey and Mukerjee (1999) for a unified treatment of orthogonality in the case of asymmetric factorial designs and for further references).

As controlled experiments are being performed in new fields in which new conditions and constraints hold, new types of designs are being evolved. One such field is that of clinical trials. These embrace concepts of life testing, censoring, and repeated significance tests, some of which we have discussed already. Of course basic inductive principles remain essentially same as those considered earlier.

In the above, we have discussed the problem of design from the sampling theory point of view. In recent years there has been some development in this area according to the pro-subjective Bayesian approach also. The main idea of this in the context of design is that when we have to infer about a parameter θ on the basis of an observation on some random variable Y, we should choose the experiment so that the posterior distribution of θ given an observation on Y is most informative. Since Y is yet to be observed, for this we have to adopt the *preposterior* approach. This means we have to consider for some loss function the mean value of the posterior expected loss given Y with respect to the predictive distribution of Y, or what comes to the same thing, the Bayes risk with respect to the prior for θ chosen. We refer to the review article of Chaloner and Verdinelli (1995) and to Toman (1996) for a discussion of this.

Outliers and robustness | **10.3.11**   Model-specific induction hinges on the choice of an appropriate model. But in practice we can never be hundred percent sure that the observations at hand represent a random sample conforming to the assumed model. Doubts may arise because after the observations are taken some of these may seem to be way apart from the main body of the data. But even at the pre-experimental stage we have to be wary about other models which may be true. After all an objective probability model is supposed to represent the pattern of long-term relative frequency distribution and past experience is always limited. There are two schools of thought here and accordingly two different strategies have been proposed to cope with such uncertainties.

Firstly, we may be pretty sure that the model chosen faithfully represents the main data-generating process of interest. The set of observations may not be a random sample from that model because a few of them have been vitiated by extraneous causes. There is no difficulty if these extraneous causes are known. We have simply to discard the rogue observations. But often no such tangible causes are known and yet the internal evidence of the data suggest that there are some aberrant observations or *outliers*. The first school of thought proposes that we apply some tests on the data to identify the outliers, and after weeding these out, work with the rest of the data as if it were a random sample conforming to the assumed model. Such a course was suggested way back in 1852 by Benjamin Peirce (father of C. S. Peirce), who was a professor of mathematics and astronomy at Harvard and the Superintendent of the U.S. Coast Survey. His proposed test for outlier detection in the case of the normal model prescribed rejection of some extreme observations if the joint density (i.e. likelihood) of the entire sample under the assumed model was less than that of the reduced sample multiplied by the probability of getting exactly that many outliers. But since the probability of getting an outlier itself had to be estimated from the data, the argument was somewhat circular and of dubious validity. Nevertheless, Peirce applied the technique extensively on Coast Survey data. A few years later Chauvenet proposed a simpler test in which the cut-off point for an outlier was taken so that the normal tail area beyond it equals $1/(2n)$, n being the sample size. These procedures generated a lot of controversy among error-theorists of the time. (Maistrov (1974) p. 154 attributes this to the then ubiquitous faith in the universality of the normal model under which any observation, however extreme, has a positive probability of occurrence.) After this, from time to time, other tests for outliers have been proposed from various points of view right up to the 1960s.

Some tables to facilitate the application of these tests have also been prepared. In this connection we refer to Tables 24–26b in *Biometrika Tables* Vol. I, where further references are given. Some tests for outlier detection in the case of multivariate data have also been developed; see Seber (1984), pp. 169–173.

But when we reject some observations and retain the others on the basis of internal evidence, there is always a positive probability of some cases of wrong rejection and wrong retention. The above approach fails to take account of this element of uncertainty. For this the second school of thought recommends that instead of outlier hunting, we should try to make our procedure *robust*. (The term 'robust', introduced by Box (see Box and Andersen (1955)), with reference to a procedure means that the procedure is insensitive to departures from the assumptions of the model; see section 2.5. However, the idea is of long standing. As we saw, it was implicit in Edgeworth's recommendation in favour of Boskovich's method for fitting a straight line (subsection 8.3.2) and in Eddington's insistence on estimating a normal σ from the mean deviation about mean (subsection 8.4.3).) The course initially suggested for ensuring this was to embed the given model in a wider class of models and to study the performance of standard inference procedures over it. If the standard procedures are found wanting, we should develop modified procedures that can hold their own in the wider context. The wider classes of models extensively studied for this purpose were again the Edgeworth series class and the mixture class which we introduced while considering model-selecting induction (subsections 10.2.3 and 10.2.4).

Surprisingly, an early piece of work along the above lines, which was done by the American mathematician Simon Newcomb (whom we briefly met in subsection 8.4.3 as the person who anticipated Fisher in conceptualizing sufficiency) dates back to 1886. For estimating a population mean μ, Newcomb assumed that the observations represent a random sample from a mixture of several heteroscedastic normal populations with a common μ. He first took the simple mean $\bar{x}$ and analysed the residuals $x_\ell - \bar{x}, \ell = 1, \ldots, n$ to identify the components of the mixture. Then for the corresponding location family, he assumed a squared error loss function and minimized the posterior expected loss under a uniform prior to estimate μ by the posterior mean; see Stigler (1973). This was really pro-subjective Bayesian decision theory—Simon Newcomb was more than half a century ahead of his time!

But the above line of attack was not pursued by others for a long time. Instead, during the 1930s and 1940s, the standard tests and estimates were subjected to close scrutiny with regard to their performance under various kinds of departure from normality. For this, apart from performing model sampling experiments, the distributions of normal theory test criteria under Edgeworth–type non-normal models were theoretically investigated by Gayen and others. Also their permutation distributions (under the assumption that given H_0 the pooled set of variables is i.i.d.) were derived (see Box and Andersen (1955) where other references are given). Generally the findings were that, so far as validity or distribution under H_0 is concerned, tests for location parameters are broadly robust against certain kinds of departures from the model, but tests for scale parameters are very sensitive to such departures. The permutation theory also suggested modifications in the test criteria which would ensure their validity even when the assumptions of the normal model are violated.

In the second half of the 20th century, however, it was found that, even for the simple problem of estimation of a location parameter, when the normal model is 'contaminated' by mixing a small fraction of a different population with it, efficiency of estimation falls rapidly from its normal theory ideal value; see Tukey (1960). (As can be presumed in the case of testing, even if the size of the test is not affected appreciably by such contamination, its power would be correspondingly affected.) Gradually, interest of researchers shifted from model-specific induction to the adoption of what, according to Anscombe, is the 'philosophy of fire insurance'—the development of procedures in which one pays a small premium in the form of loss of efficiency under the ideal model to have enough efficiency when departures from it occur.

In recent years increasing attention has been paid to the development of robust procedures which have the above insurance-like feature. An approach towards this that has been pursued by a number of workers is to concentrate on a suitably defined neighbourhood of the ideal model and to get at procedures which have adequate performance over such a neighbourhood. One way of generating the neighbourhood has been to follow the idea of Tukey's contamination model as described above, but the contaminating distribution is now left unspecified. In certain simple cases it has been possible to identify procedures which in some sense optimize the minimal performance over a chosen neighbourhood. In more complex problems a standard approach has been to optimize at the ideal model subject to suitable bounds on the local degradation in some performance characteristic for infinitesimal departures from the ideal model (see Huber (1981), Hampel et al (1986)).

In particular, for the point estimation problem, three groups of estimators called M-estimators, L-estimators and R-estimators (see Huber (1981)), which though set up with reference to some particular model, have adequate performance under broad conditions, have been studied from the above angle. In the case of M-estimation of a real parameter θ on the basis of a random sample $X_1, \ldots, X_n$, we start with a real valued kernel function $\rho(x, t)$ such that $E_\theta \rho(X_1, t)$ is minimum for $t = \theta$. Then the M-estimate ('M' for 'maximum likelihood like') is found by minimizing $\sum_1^n \rho(X_i, t)$ with respect to t. When the parent density is $f(x \mid \theta)$, clearly this reduces to the MLE by taking $\rho(x, t) = -\log f(x \mid t)$. Generally, we can choose $\rho(x, t)$ suitably with reference to a putative model so that there is adequate insurance against departures from it and in particular when some of the observations are contaminated by gross errors. L-estimators are linear functions of order statistics; they are generalization of particular forms like the Winsorized mean (mean after replacing the first and the last order statistics by their nearest neighbours) and trimmed means (mean after dropping a pre-fixed proportion of the order statistics from the two ends). The R-estimator of a real parameter θ is appropriate when the joint distribution of $X_1, \ldots, X_n$ involves θ in such a way that for testing the null hypotheses $H_{0t}: \theta = t$ we can set up a rank order statistic $T(X_1, \ldots, X_n; t)$ whose null distribution under H_{0t} is known and same irrespective of t. The value of t for which $T(X_1, \ldots, X_n; t)$ is closest to the central value of that null distribution is taken as the R-estimate of θ. For detailed discussion of the robustness properties of these estimators we refer to Jurecková and Sen (1996).

The problem of developing robust procedures for model-specific induction can also be approached along Bayesian lines. To consider a simple case, for the estimation of the mean μ in a normal population $N(\mu, \sigma^2)$, it may be that we suspect the sample to be really one from the mixture population obtained by mixing $N(\mu, \sigma^2)$ with $N(\mu, k^2\sigma^2)$ at mixing rates $1 - \alpha, \alpha$, where k, α are known. Then assuming a suitable prior for μ, σ^2 we can find the marginal posterior distribution of μ and hence infer about μ. This is similar to model-selecting induction under the mixture set up, but here we are assuming that α is known and giving attention to the estimation of μ; of course after the solution is obtained we can study its sensitivity to changes in α and k.

In the above we have been considering mostly the questions of outliers and robustness when the observations are on i.i.d. random variables. When there are explanatory variables, the observables are no longer identically distributed and interest shifts to regression like parameters. But generally the same considerations apply and the same methods extend themselves. A more complex form of departure from standard assumptions occurs when the random variables observed happen to be serially correlated. This sort of situation is common in the case of time series, but has also been found to obtain in certain long term experiments. Their treatment requires the study of special types of stochastic processes and is beyond the scope of our discussion (see Hampel *et al* (1986) Chapter 8, where further references are given).

10.4   Model-free induction

| Types of problem |

10.4.1   Carrying the idea of robustness of procedures to the extreme, we are led into the arena of model-free induction. As mentioned in section 10.1, in model-free induction we proceed without keeping in mind the possibility of any particular model or class of models. Also, often we are not interested in the model appropriate for the situation at hand in its entirety. Rather, we are concerned only about some particular aspects of it, without bothering about the remaining 'nuisance aspects'. We will discuss model-free induction in two types of situations separately: (i) when sampling is from one or more infinite population in which the units are not distinguishable (ii) when sampling is from a finite population with distinguishable units. The issues that have to be addressed and the considerations that arise are somewhat different in the two cases. Also historically, development of finite population sampling has been along a somewhat independent course.

| Sampling from infinite populations |

10.4.2   Here the populations that are sampled are either conceptually infinite because they are supposed to represent indefinite repetitions of the observational process, or virtually infinite because, though existent, they are very large. In the first case, there is no question of distinctive labels for the units; in the second case, even if such labels exist we ignore them. The characters under study have a distribution in any such population and we want to induce about certain aspects of that distribution in a way which does not require specification of any model.

For most of the current development, we follow the sampling-theory approach and in the sampling-theory approach to model-free induction, more than in any other case, the line to be followed is determined by the problem at hand. We discuss the procedures as solutions to various problems under the two broad heads (a) approximate procedures and (b) exact procedures, according as the measure of assurance (level of significance, confidence coefficient, S. E., etc.) stated with the procedure is determined approximately or exactly. (But the division is by no means water-tight. For certain set-ups, while exact procedures are available for testing and interval estimation, we have to remain satisfied with an approximate determination of the S. E. in the case of point estimation. Also, sometimes for practical reasons an exact procedure has to be replaced by a more tractable approximate one.) However, as it will transpire in course of the development, although the procedures considered will be model-free (at least approximately), their performance characteristics (e.g. power of a test or the distribution of the span of an interval estimate) in a situation will generally depend on the model that holds.

Since in model-free induction we are usually interested in certain aspects of the underlying parent distribution, in all but a few simple problems, the question of elimination of nuisance parameters assumes importance. (In the present context we are interpreting the terms 'parameter' liberally in the sense of any real-, vector-, or function-valued entity defined for the parent distribution; cf. section 2.4.) Although the phrase 'elimination of nuisance parameters' here will have different implications in different contexts—this will be clear as we consider particular types of problems—generally this means that neither the procedure nor the measure of assurance attached to it depends on the nuisance parameters.

Broadly there are two methods by which nuisance parameters may be kept at bay while deriving the procedure.

(i) *By substitution of estimates*: this is a brute force method in which the nuisance parameters are replaced by suitable point estimates.

(ii) *By data shedding*: this means ignoring part of the evidence which is inextricably mixed up with the nuisance parameters. This again can be done in two ways: through *marginalization*, i.e. leaving out the observations on some of the variables (usually after an initial transformation of all the variables),[23] or through *conditionalization*[24] i.e. conditionally fixing some of the variables, often after an initial transformation.[25] (When we hold a random variable conditionally fixed, we leave out of consideration any light that it could have potentially thrown on the parameter of interest.)

As we will see, both (i) and (ii) may be resorted to in the case of approximate procedures (however, here conditioning is often an intermediate stage in the derivation of the procedure, and after the procedure is formulated, it may be

---

[23]What we call 'marginalization' is somewhat relative. Even when we construct a statistic, clearly we leave out part of the data (and also part of the information when the statistic is not sufficient). Conventionally, we call the process marginalization, when it is resorted to at the beginning before the construction of the test statistic or the pivotal function.

[24]In Chapter 4 (subsections 4.2.3, 4.3.5) we discussed conditioning on an ancillary from the point of view of practical relevance. Here the object of conditioning is primarily to get rid of nuisance parameters and the conditioning variables may not be ancillary.

[25]The techniques of marginalization and conditionalization are used in the context of model-specific induction also. There these involve the use of marginal and conditional likelihoods; see e.g. Kalbfleisch and Sprott (1973).

lost sight of). But generally in the case of exact procedures only method (ii) is used. For historical reasons we will first discuss the general techniques for construction of approximate procedures and then take up exact procedures for model-free induction.

| (a) Approximate procedures |

Approximate procedures for model-free induction draw heavily on methods of approximate probability evaluation (section 3.9) and in particular on the Central Limit Theorem (CLT) (and its close relatives) and the Laws of Large Numbers (LLN). We can say that this form of sampling theory induction started with Laplace's large sample approach and Gauss's second approach to linear models (subsections 8.3.4 and 8.3.5; however, Gauss, it may be recalled, was concerned only with point estimation). To illustrate how the approximate procedures are built up and the roles of conditionalization and estimation of nuisance parameters in this, we consider the following adaptation of Laplace's simple regression set-up (8.4.4) with no intercept, to the case of a stochastic predictor.

| Regression with stochastic predictors |

Let $(X_\ell, Y_\ell), \ell = 1, \ldots, n$ represent a random sample from a bivariate distribution whose density function has the structure

$$g(x)f(y - \beta x), \quad -\infty < x, y < \infty. \tag{10.4.1}$$

Here $g(x)$ and $f(y - \beta x)$ are respectively the marginal density of X and the conditional density of Y given $X = x$. Obviously $Z = Y - \beta X$ given $X = x$ has the same distribution irrespective of x, and hence, is distributed independently of X. We assume that $E(Z) = 0 \text{Var}(Z) = \sigma^2$. The forms of the densities $g(\cdot), f(\cdot)$ remain unspecified.

Considering the sequence of observations x_ℓ on $X_\ell \ell = 1, 2, \ldots$, for each n, we denote

$$X_{(n)} = (X_1, \ldots, X_n), \quad x_{(n)} = (x_1, \ldots, x_n). \tag{10.4.2}$$

Just as in subsection 8.4.2, for conditionally fixed $X_{(n)} = x_{(n)}, Y_\ell$ are independently distributed with mean βx_ℓ and variance $\sigma^2, \ell = 1, \ldots, n$, (note that the nuisance 'parameter' $g(\cdot)$ becomes quietly 'eliminated' from the set-up through conditionalization). Hence, writing

$$\hat{\beta}_n = \frac{\sum_1^n Y_\ell x_\ell}{\sum_1^n x_\ell^2}, \tag{10.4.3}$$

by CLT we get that under (8.4.6) (i.e. provided the pattern of growth of x_ℓs is not too lopsided), for large n, $\hat{\beta}_n$ is approximately distributed as $\mathcal{N}(\beta, \sigma^2/\sum_1^n x_\ell^2)$. We now eliminate the nuisance parameter σ^2 by substituting

$$\hat{\sigma}_n^2 = \frac{1}{n} \sum_1^n (Y_\ell - \hat{\beta}_n x_\ell)^2 \tag{10.4.4}$$

for it. By the LLN and standard results, it follows that $\hat{\sigma}_n \overset{P}{\rightarrow} \sigma$ and hence for large n,

$$(\hat{\beta}_n - \beta)\sqrt{\sum_1^n x_\ell^2}/\hat{\sigma}_n \tag{10.4.5}$$

is approximtely distributed as $\mathcal{N}(0,1)$. We can use the pivotal function (10.4.5) for testing an hypothesis specifying β or setting up an interval estimate for β. The size of the test and the confidence coefficient of the interval estimate will be realized approximately whatever the model.

The pivotal function (10.4.5) is of course appropriate for the conditional set-up corresponding to $X_{(n)} = x_{(n)}$. Since here we are sampling from a bivariate population, this may not be considered satisfactory in practice. However, since (10.4.5) is approximately $\mathcal{N}(0,1)$ for *every* $x_{(n)}$, we readily get that under mild restrictions on the distribution of X,[26] the pivotal function (10.4.5) will be distributed unconditionally as $\mathcal{N}(0,1)$ even when x_ℓs everywhere are replaced by X_ℓs. Thus we can say that the approximate procedure finally used does not involve the conditioning.

Large sample procedures for contingency tables

Two important tools for approximate model-free induction got forged in the early part of the 20th century through the work of Karl Pearson, later emended by Fisher. These relate to multinomial set-ups. (As the multinomial frequency set-up is very general and covers varieties of characters and models, we consider it under model-free induction.)

The first was for testing the independence of the two attributes in an $r \times s$ contingency table. Denoting the probability of the (i,j)-cell by p_{ij} and the marginal probabilities by $p_{i\cdot}$ and $p_{\cdot j}$ the problem was to test H_0: $p_{ij} = p_{i\cdot}p_{\cdot j}$, $i = 1,\ldots,r, j = 1,\ldots,s$. Let $f_{ij}, f_{i\cdot}, f_{\cdot j}(\sum_{i,j} f_{ij} = \sum_i f_{i\cdot} = \sum_j f_{\cdot j} = n)$, denote the frequency in the (i,j)-cell and the marginal frequencies of the ith row and jth column respectively. Karl Pearson in a 1904 paper (see Pearson (1948) pp. 443–475) substituted the estimates $f_{i\cdot}/n$ and $f_{\cdot j}/n$ for the nuisance parameters $p_{i\cdot}$ and $p_{\cdot j}$ and extending his goodness-of-fit test for a composite hypothesis (subsection 10.2.7), proposed the criterion

$$\sum_{ij} \frac{n(f_{ij} - (f_{i\cdot}f_{\cdot j}/n))^2}{f_{i\cdot}f_{\cdot j}} \tag{10.4.6}$$

for testing independence. But he wrongly thought that under H_0, (10.4.6) would follow approximately a χ^2-distribution with $rs - 1$ d.f. Fisher in the 1920s removed this misconception and showed that the correct number of d.f. here

---

[26]We require that as $n \to \infty$ condition (8.4.6) holds in probability, when x_ℓ is replaced by X_ℓ, i.e. $\max_\ell X_\ell^2/\sum_1^n X_\ell^2 \overset{P}{\rightarrow} 0$ (see Hoeffding (1952) Theorem 9.1). This holds in particular when $0 < E(X^2) < \infty$.

is $(r-1)(s-1)$. Incidentally, writing χ^2 for this criterion (10.4.6), Karl Pearson also proposed $\phi^2 = 1/n\chi^2$ or its transform $C = \sqrt{(\phi^2/(1+\phi^2))}$ as a measure of association between the two attributes for this set-up.[27]

The other tool proposed by Karl Pearson (1911–12) was for testing the homogeneity of two independent k-class multinomials with class probabilities $p_{1j}, \ldots, p_{kj}$, $\sum_i p_{ij} = 1$ and frequencies $f_{1j}, \ldots, f_{kj}, \sum_i f_{ij} = n_j, j = 1, 2$. Under H_0: $p_{i1} = p_{i2}, i = 1, \ldots, k$, the unspecified common values $p_{i1} = p_{i2} = p_i, i = 1, \ldots, k$ represent the nuisance parameters. Considering the differences $((f_{i1}/n_1) - (f_{i2}/n_2))$, $i = 1, \ldots, k$, and estimating the nuisance parameters p_i by $(f_{i1} + f_{i2})/(n_1 + n_2)$, he arrived at the standard homogeneity χ^2 whose d.f. he correctly identified as $k-1$ (this was a consequence of his working in terms of the differences). The case where we have more than two multinomials followed from Fisher's general result on the frequency χ^2 for testing a composite hypothesis. As is well-known generally, the homogeneity χ^2 has a structure similar to (10.4.6).

Moments, cumulants, U-statistics

At another level, approximate methods for inference about moments and functions of moments can be developed on the basis of CLT without assuming any model for the population. The δ method for finding the sampling variances and covariances of differentiable functions of moments that comes handy here was originally proposed by Gauss. It was extensively employed to various problems by Sheppard in an 1899 paper; see Hald (1998) pp. 627–629. Symmetric functions of sample values called k-statistics, which are unbiased estimates of population cumulants of different order, were proposed by R. A. Fisher who also gave rules for deriving sampling variances (and higher moments) of the statistics. (It was one of the rare occasions when Fisher emphasized the property of unbiasedness in an estimator!) The k-statistics turn out to be linear combinations of U-statistics whose kernels are products of powers. The general concept of a U-statistic was introduced by Hoeffding in 1948 and later extended by others. Given a random sample $X_1, \ldots, X_n$ a U-statistic corresponding to a permutation symmetric kernel $\phi(x_1, \ldots, x_m)$ is defined as

$$U_n = \binom{n}{m}^{-1} \sum_{1 \le \ell_1 < \ldots < \ell_m \le n} \phi(X_{\ell_1}, \ldots, X_{\ell_m}). \qquad (10.4.7)$$

---

[27] Karl Pearson arrived at C by postulating that the two attributes, if they could be quantified, would follow jointly a bivariate normal distribution with correlation ρ. He showed that if the classifications could be indefinitely refined, in the limit, C would equal $|\rho|$. Another measure appropriate for the 2×2 case derived by Pearson from the same postulate was the tetrachoric r. But these attempts to convert the model-free set-up into one based on a particular parametric model were not liked by all his contemporaries. A notable dissenter was Yule who proposed his own model-free measures of association in the 2×2 case (see Mackenzie (1981) Chapter 7 for an account of the controversy).

It has been shown that if the class of possibilities for the parent distributions is broad enough, U_n has minimum variance among all unbiased estimates of $\xi = E\phi(X_1, \ldots, X_m)$ (see e.g. Fraser (1957) pp. 141–142). Furthermore, it has been shown that CLT holds for U_n in the sense that when $E\phi^2(X_1, \ldots, X_m) < \infty$, as $n \to \infty$, $\sqrt{n}(U_n - \xi)$ is asymptotically distributed as $\mathcal{N}(0, m^2\zeta_1)$, where

$$\zeta_1 = \mathrm{Cov}\{\phi(X_1, X_2, \ldots, X_m),\ \phi(X_1, X_{m+1}, \ldots, X_{2m-1})\} \tag{10.4.8}$$

(see e.g. Puri and Sen (1971) pp. 51–66). The result has been extended to the multisample case and to the case of several U-statistics. This property clearly makes U-statistics very useful for developing approximate procedures for inferring about parameters ξ for which suitable kernels $\phi(\cdot)$ are available. Of course for this one has to get consistent estimates for nuisance parameters of the form (10.4.8). For illustration of the technique we refer to Puri and Sen (1971) pp. 167–172, 201–211.

Cox's partial likelihood methods

An interesting application of conditionalization has been done by Cox (1972) to develop approximate model-free procedures for life testing experiments in which the main object is to study the dependence of the study variate on certain covariates. Writing Y for the study variate life (taken to be continuous for simplicity) and $(x_1, x_2, \ldots, x_k) = x$ for k covariates, it is assumed that the hazard function (cf. (10.2.17)) has the proportional form

$$h(y; x) = e^{\sum_1^k \beta_i x_i} h_0(y), \tag{10.4.9}$$

where $h_0(y) = h(y; 0)$ is an *unspecified* base line hazard rate which represents the nuisance 'parameter' in the model. Let $(x_{1\ell}, \ldots, x_{k\ell})$ denote the covariate values of the ℓth individual. Suppose r among the experimental individuals are actually observed to fail, say, at the successive failure times $y_{(1)} < y_{(2)} < \cdots < y_{(r)}$ (others are censored). Let $\mathcal{R}(y_{(j)})$ denote the set of individuals at risk 'at' time $y_{(j)}$ (i.e. who have not failed or been censored earlier) and let $(x_{1(j)}, \ldots, x_{k(j)})$ denote the covariate values at the time of failure of the individual who actually fails at $y_{(j)}$. Given $\mathcal{R}(y_{(j)})$, under (10.4.9) the conditional probability that the individual in the set with covariate values $x_{1(j)} \ldots x_{k(j)}$ will fail at $y_{(j)}$ is

$$L_{(j)} = e^{\sum_1^k \beta_i x_{i(j)}} \bigg/ \sum_{\ell \in \mathcal{R}(y_{(j)})} e^{\sum_1^k \beta_i x_{i\ell}} \tag{10.4.10}$$

irrespective of the base line hazard function h_0. Note that (10.4.10) holds even when as is often the case in practice (e.g. in clinical trials), the covariates change with time.

Cox (1975) called the product $L^* = \Pi_1^n L_{(j)}$ *partial likelihood* and proposed that this may be handled just like an ordinary likelihood and first and second

derivatives of $\log L^*$ may be used to develop approximate large sample procedures for deriving point estimates as also for setting up tests and internal estimates
for the β_is. Subsequent studies suggest that the loss of information involved in
such procedures in which conditionalization is carried to the extreme, is generally not substantial. Models such as (10.4.9) are known in the literature as
semiparametric.

| (b) Exact procedures |

The exact procedures for model-free induction are
known as *nonparametric* procedures. Even if the model
is unspecified, often when the parameter of interest is given the joint distribution of the observables possesses certain known aspects which are brought out
under suitable transformations. Sometimes these known aspects take the form
of special types of symmetries (invariance under particular transformations).
Exact model-free procedures are developed by utilizing these known aspects. We
illustrate this by considering the solutions to some typical problems.

| Use of probability integral transforms |

When $X_1, \ldots, X_n$ is a random sample from some population with continuous $F(x)$, $F(X_1), \ldots, F(X_n)$ is distributed like a random sample from $\mathcal{U}[0,1]$. Earlier
(subsection 10.2.7) we saw that this fact can be used to formulate goodness-
of-fit tests when $F(\cdot)$ is fully specified by the hypothesis. We now show how
the same probability integral transformation leads to exact solutions to certain
simple problems of inference when $F(\cdot)$ is unspecified.

First consider the problem of determining a confidence band for the entire
function $F(x), -\infty < x < \infty$. Denoting the sample empirical cdf by $\hat{F}_{(n)}(x)$ we
can take the Kolmogorov-Smirnov statistic (cf. (10.2.15)

$$D = \sup_{-\infty < x < \infty} |\hat{F}_{(n)}(x) - F(x)| \qquad (10.4.11)$$

as the pivotal function here. As D remains invariant under the monotonic transformation $x \to F(x)$ of the scale of x, the distribution of D under F is same
irrespective of F. We can invert any statement specifying an upper probability
bound of D to get a two-sided band for $F(x)$.[28]

Another problem which lends itself to a simple solution is that of determination of a tolerance interval. If $X_{(1)} < \cdots < X_{(n)}$ denote the sample order

---

[28]The naïve estimator $\hat{F}_{(n)}(x)$ of $F(x)$ may not be considered suitable when $F(x)$ is
known to possess special smoothness properties. Thus when $F(x)$ admits of a density
$f(x)$, we may want an estimator of $F(x)$ which also has the same property, so that
density corresponding to the latter can be taken as an estimate of $f(x)$. Among various
such special estimators we mention the 'window' estimators proposed by Rosenblatt
and Parzen (see Parzen (1962)). The corresponding density estimates, however, are
less well-behaved than the cdf estimates—they converge more slowly and allow for only
approximate inference.

statistics (for simplicity we ignore ties which are improbable), for any $1 \le r < s \le n$, the distribution of the probability content or coverage $F(X_{(s)}) - F(X_{(r)})$ of the interval $(X_{(r)}, X_{(s)})$ is same irrespective of F. Hence given a lower bound L to the coverage we can choose r, s suitably to ensure that the measure of confidence

$$P\left(F(X_{(s)}) - F(X_{(r)}) > L\right)$$

has a high level.

Inference about a quantile

We next consider inference about the median ξ (assumed unique) of the unspecified cdf $F(\cdot)$. Given ξ, the transformed observables $X_\ell - \xi, \ell = 1, \dots, n$ are known to be i.i.d. with $P(X_1 - \xi > 0) = \frac{1}{2}$. To utilize this known fact, we set up the pivotal function

$$U(\xi) = \#\{\ell : X_\ell - \xi > 0\}, \qquad (10.4.12)$$

which, irrespective of ξ, follows a Binomial $(n, \frac{1}{2})$ distribution. Hence for any $1 \le r < s \le n$ we can evaluate

$$P(r \le U(\xi) \le s | \xi) = \sum_{j=r}^{s} \binom{n}{j} \frac{1}{2^n}. \qquad (10.4.13)$$

Writing $X_\ell, \ell = 1, \dots, n$ for the order statistics of the sample the event $r \le U(\xi) \le s$ is readily seen to be equivalent to $X_{(n-s)} \le \xi < X_{(n-r+1)}$. We can choose r and s so that the probability in (10.4.13) is at least equal to a preassigned level (it is customary to take $s = n - r$).

If we want to test the hypothesis $H_0 \colon \xi = 0$, we can take $U(0)$ as our test statistic. The resulting test, known as *sign test* takes only the number of positive signs among the observations and ignores the rest of the information in the sample.

The case when we are interested in the quantile ξ_p of known order $0 < p < 1$ can obviously be dealt with similarly with $P(X_1 - \xi_p > 0) = 1 - p$ taking the place of $\frac{1}{2}$.

Association in a 2 × 2 table

Suppose a population is cross-classified in two ways according to two attribues A and B of the yes-no type.

In the classical example of this, A and B stand respect-ively for 'inoculated against' and 'attacked by' a particular disease. The cell probabilities and their marginal totals are as follows.

	B	not-B	
A	p_{11}	p_{12}	$p_{1\cdot}$
not-A	p_{21}	p_{22}	$p_{2\cdot}$
	$p_{\cdot 1}$	$p_{\cdot 2}$	1

The model for the joint distribution of the characters here is unspecified. In fact the set-up applies even if the characters are qualitative so that no question of a model arises. Here the two attributes are independent if $p_{11}p_{22} = p_{12}p_{21}$ (which is equivalent to $p_{ij} = p_{i.}p_{.j}$, $i.j = 1, 2$) and

$$\gamma = \frac{p_{11}p_{22} - p_{12}p_{21}}{p_{11}p_{22} + p_{12}p_{21}}$$

can be taken as a measure of association between the attributes.[29] The cell probabilities can be expressed in terms of $p_{1.}, p_{.1}$ and γ using

$$p_{11} = p_{1.}p_{.1} + \frac{2\gamma}{1 - \gamma}p_{12}p_{21}.[30] \qquad (10.4.14)$$

The probability distribution of a random sample of size n from the population in which the frequencies are given by

	B	not-B	
A	f_{11}	f_{12}	$f_{1.}$
not-A	f_{21}	f_{22}	$f_{2.}$
	$f_{.1}$	$f_{.2}$	n

clearly involves the parameter of interest γ, as also the nuisance parameters $p_{1.}$ and $p_{.1}$. No exact procedures for inference about γ can be set up by proceeding directly. However if we fix $f_{1.}$ and $f_{.1}$ conditionally, the conditional distribution of the sample involves γ and γ alone (see Lehmann (1986) p. 157) and hence we can set up an exact test for any hypothesis specifying γ and also an interval estimate for γ. Further, while estimating γ in terms of the f_{ij}s we can take the conditional S. E. given $f_{1.}, f_{.1}$ (cf. subsection 4.2.7). Note that here $f_{1.}, f_{.1}$ carry some information about γ (though a small amount). Since that part of the evidence is inextricably mixed up with the nuisance parameters $p_{1.}, p_{.1}$, we deliberately shed it through conditioning to get exact model-free procedures. (The hypothesis $\gamma = 0$ corresponds to the independence of the two characters. We have considered earlier an approximate test of the counterpart of this for a general $r \times s$ table. We can formulate an exact test for independence for the general case also along the above lines. However, there is no simple counterpart of γ in the general case.)

| Course of development | It seems that interest of statisticians in exact model-free procedures as above was aroused in the 1930s. Fisher

---

[29]γ corresponds to one of the measures of association proposed by Yule; see Mackenzie (1981) p. 155 (cf. footnote 28).

[30]We can use (10.4.14) to express each of $p_{ij}, i, j = 1, 2$, in terms of $p_{1.}, p_{.1}, \gamma$, and $c = p_{12}p_{21}$. Ultimately the expressions for p_{12}, p_{21} may be used to eliminate c.

gave the exact test for independence in a 2×2 contingency table in the 5th edition (1934) of his *Statistical Methods for Research Workers*. Cochran (1957) made a detailed examination of the sign test and Fisher in his 1939 obituary article on 'Student' gave some hints about model-free interval estimation of a quantile. Other tests like the tests for independence based on a bivariate sample, the two-sample counterpart of the Kolmogorov–Smirnov test, and the Wald–Wolfowitz run test (see Hajek *et al* (1999)) were also proposed around the same time. These tests, being based on ranks only, proceeded through what we have called marginalization. Vigorous work went on during the 1940s–1960s with the development of exact model-free tests for single-sample location under symmetry, two-sample location, two-sample scale, etc. and related estimation procedures (see Hajek *et al* (1999)). These procedures all utilized certain known symmetries in the joint distribution of the observables, which hold when the aspect of interest is known. We illustrate by considering a typical problem.

| Two sample location problem |

Given independent random samples $X_1, \ldots, X_{n_1}$ and $X_{n_1+1}, \ldots, X_n, n = n_1 + n_2$, from two populations with continuous cdfs $F(x)$ and $F(x-\delta)$, where $F(\cdot)$ is unspecified, suppose we want to infer about the location difference δ. Given δ, the joint distribution of $X'_\ell(\delta)$, where

$$
\begin{aligned}
X'_\ell(\delta) &= X_\ell, \quad \ell = 1, \ldots, n_1 \\
&= X_\ell - \delta, \quad \ell = n_1 + 1 \ldots, n,
\end{aligned}
\tag{10.4.15}
$$

is clearly permutation-symmetric. Let $R_\ell(\delta)$ be the rank of $X'_\ell(\delta)$ in the entire set. Given $\delta, (R_1(\delta), \ldots, R_n(\delta))$ would assume all permutations of $1, \ldots, n$ with the same probability $1/n!$, and thus the distribution of any statistic

$$
T_\delta = T(R_1(\delta), \ldots, R_n(\delta))
\tag{10.4.16}
$$

would be fully known and free from δ.

If $T(\delta)$ is nondecreasing in δ, we can invert a statement of the form

$$
T_L \leq T_\delta \leq T_U,
$$

where T_L, T_U are two values suitably chosen with reference to the known distribution of T_δ, to get a confidence interval for δ with a preassigned confidence level. Standard choices of (10.4.16) here correspond to the rank score statistic

$$
T(\ell_1, \ldots, \ell_n) = \sum_{j=n_1+1}^{n} a(\ell_j),
\tag{10.4.17}
$$

where $a(1) \leq a(2) \leq \cdots a(n)$ is a set of nondecreasing scores, not all equal (see e.g. Hajek *et al* (1999) Chapter II). To get a point estimate we can take the value of δ for which T_δ is closest to the central point of the distribution of T_δ (R-estimate; cf. subsection 10.3.11). For testing a hypothesis specifying δ, say H_0: $\delta = 0$, we can use T_0. Particular choices of the rank scores $a(\ell)$ in (10.4.17) here give particular tests. For example the Wilcoxon test is obtained for $a(\ell) = \ell$. (This was proposed by Wilcoxon in 1945. The corresponding estimate was derived by Hodges and Lehmann in 1963 and independently by Sen in the same year for a slightly different problem; see Puri and Sen (1971) p. 228, where further references are given.)

Note that even though (10.4.16) is a function of $R_\ell(\delta)$, $\ell = 1, \ldots, n$ only, since δ here has to be adjusted with reference to the sample values, the interval and point estimates so obtained do not involve any kind of real marginalization. However, for testing H_0: $\delta = 0$, if we use the test statistic T_0 which depends on the observations solely through the ranks $R_\ell = R_\ell(0), \ell = 1, \ldots, n$, this involves sacrificing the information in the sample order statistics $X_{(1)} < \cdots < X_{(n)}$.

Since $X_\ell = X_{(R_\ell)}$, the order statistics $X_{(\ell)}$ together with the ranks $R_\ell, \ell = 1, \ldots, n$ carry the same information as the original data. For the testing problem we can here conditionally fix the order statistics $X_{(\ell)} = x_{(\ell)}, \ell = 1, \ldots, n$ and make the scores $a(\ell)$ depend on these (one particular choice is $a(\ell) = x_{(\ell)}$). Then we can make use of the fact that, under $\delta = 0$, $(R_1, \ldots, R_n)$ assumes all possible permutations of $1, 2, \ldots, n$, *independently* of the order statistics, to construct a conditional test. But such an approach also involves an element of data-shedding through conditionalization.

So far we have considered exact model-free procedures only for univariate problems. In the multivariate case making the procedure model-free becomes much more troublesome. For testing, generally one has to resort to some form of conditionalization. But even then application of the exact procedures in practice becomes too laborious except in the simplest cases. Often the only course open is to replace these by large sample approximations. For a discussion of these topics we refer to Puri and Sen (1971).

Another possible area of extension is the development of model-free sequential procedures. For development of tests, we can formulate large sample procedures or shed some information to work in terms of ranks. The sufficiency route to construction of such procedures which is so handy under parametric set-ups is now no longer available. But a counterpart of the property of transitivity has to be invoked to link the statistics for successive sample sizes. This is usually done on intuitive grounds so that the graph traced by plotting suitable versions of the statistics as the sample size progressively increases asymptotically behaves like a stable process (usually some form of the Brownian movement process). Corresponding estimation problems can be solved by proceeding along standard lines. Generally all such model-free procedures have to be used in their asymptotic forms either for theoretical reasons or because of practical intractability of the exact procedures. For a detailed treatment of all these topics we refer to Sen (1981).

<table>
<tr><td>

Approximate boot-strap procedures

</td><td>

In recent years a new type of computer-friendly approximate procedures for model-free induction called *boot-*

</td></tr>
</table>

strap procedures have been proposed. To illustrate their general idea, suppose given a sample $X_1, \ldots, X_n$ from a population with an unspecified cdf F, our problem is to infer about a parameter $\tau(F)$ which is defined as a functional of F. Let $T(X_1, \ldots, X_n)$ be a standard estimator of $\tau(F)$. Had we known the sampling distribution of $D(X_1, \ldots, X_n; F) = T(X_1, \ldots, X_n) - \tau(F)$, our problem would be solved. But usually the sampling distribution of $D(X_1, \ldots, X_n; F)$ involves certain nuisance parameters, and otherwise also, its form depends on the unknown F. In the classical approximate procedures considered earlier, we 'Studentize', i.e. replace the nuisance parameters by suitable estimates and set up a modified criterion based on D, whose limiting distribution is free from F. In the bootstrap approach we estimate $F(x)$ in its entirety (parameter of interest, nuisance parameters, model and all) by an estimator $\hat{F}(x)$ based on $X_1, \ldots, X_n$ (the naive estimator represented by the empirical cdf $\hat{F}_{(n)}(x)$ is of course a possible choice here). Denoting a random sample from an infinite population with cdf $\hat{F}(x)$ by $X_1^*, \ldots, X_n^*$ (the bootstrap sample), the sampling distribution of $D(X_1, \ldots, X_n; F)$ under F is estimated by that of $D(X_1^*, \ldots, X_n^*; \hat{F})$ *under* $\hat{F}$. As $\hat{F}$ is known the latter is determinate; what is more, if the distribution is theoretically troublesome to derive, we can determine it through simulation. Treating this known estimated distribution, as if it were the true sampling distribution of $D(X_1, \ldots, X_n; F)$ under F, we can solve our problem of inference about $\tau(F)$.

The above bootstrap technique was proposed by Efron (1979). Its consistency as $n \to \infty$ and other properties have since been examined (see Beran's *Introduction* to Efron's paper in Kotz and Johnson (1992), where other references are given) and generally it has been found that it provides an approximate model-free solution to many an otherwise intractable problem.

<table>
<tr><td>

Comparison of performance

</td><td>

We have been considering approximate and exact procedures for model-free induction under the sampling-

</td></tr>
</table>

theory approach, and as we have stressed all along, in this approach it is of paramount importance to compare the relative performance of alternative solutions to a problem, with a view to identifying an optimum procedure. But in the present case we face a stumbling block: although the procedures are model-free, their performances are not. Indeed it would be foolish to expect that a model-free procedure would have the same performance irrespective of the model.

Thus we have to study the performance of a model-free procedure under particular models. Even when the procedure is an exact one, most often we can make headway in such a study only in the large-sample scenario. So far as point and interval estimation are concerned, we can follow standard approaches for this. For a point estimate which is asymptotically unbiased, we can derive the asymptotic S. E. For an interval estimate (suitably defined for different sample sizes) study of the asymptotic distribution of the interval span would provide a sensible solution. However, difficulty arises when we set out to study the performance of a test in large samples. This is because the power of a test of fixed size, for any fixed alternative hypothesis against which it is consistent, tends to one as the number n of observations becomes large. Thus the comparison of the limiting powers of two tests in such a case would throw no light on their sensitivity.

Several artifices have been proposed to get round this difficulty. We mention two of these. The first was originally proposed by Pitman in 1948. In it we consider the limiting power of a fixed-size test against a shifting alternative, e.g. a sequence of alternatives which approaches H_0 as n increases. The closing in of the alternative on H_0 tends to reduce the power, just as increase in n tends to pull it towards one. The rate of approach of the alternative to H_0 is so adjusted in relation to increasing values of n that the limiting power has a meaningful value less than one. Given two tests 1 and 2 of the same size, under certain conditions, for a sequence of alternatives approaching H_0, we can ensure that for a number $e_{12} > 0$ this limiting power has the same value for test 2 as for test 1, whenever the former is calculated for a sequence of n values which in the limit is e_{12} times the same for the latter. If this e_{12} is same for a broad class of alternative sequences (under the model assumed), it is interpreted as the asymptotic relative efficiency (ARE) of test 1 versus test 2 with reference to that class (cf. the interpretation of ARE in the point estimation context given by Gauss and Laplace; section 8.4). In the second approach to the problem, for increasing n, two tests are compared against a fixed alternative but in terms of their P-values or rather a monotonic increasing transform of their P-values. The idea is that lower the value of P a test tends to give under an alternative, more sensitive it would be against it; cf. footnote 19. Bahadur, who introduced this idea in 1960, chose a transform such that the per-observation value of the transform under suitable conditions converges stochastically to a positive limit (which he called 'slope') under the fixed alternative. The ARE of a test versus another against the alternative is computed as the direct ratio of their slopes. Under the conditions assumed, such an ARE also can be interpreted as the inverse limiting ratio of sample sizes as before. (For more detailed discussion of these concepts and techniques see Puri and Sen (1971) and Hajek *et al* (1999).)

All such comparisons of model-free procedures are with reference to particular models. We can make use of these to choose a suitable procedure in inductive set-ups where we have some hunch about the model but do not want our procedure to depend too heavily on that.

| Question of robustness | The model-free inductive procedures described above hinge heavily on the assumptions of internal homogeneity of the samples and independence of the observables. Hence, although these procedures are not specific to any particular model, the question of robustness against heterogeneity and non-independence remains very much relevant. So far

as sample homogeneity is concerned, it has been found that, for certain problems at least, the requirement can be relaxed without affecting the validity of the procedures adversely. For example, for the single sample problem of testing for location (univariate or multivariate, with or without the assumption of symmetry) many standard tests remain valid even when the random variables are not identically distributed (see Puri and Sen (1971), Chapter 4). But the assumption of independence is more crucial. Isolated studies suggest that the validity of standard procedures may be seriously affected (realized value of a size (confidence coefficient) may grossly overshoot (fall short of) the stipulated value) when the observables are serially correlated. In fact here also the kind of study based on special stochastic processes to which we referred at the end of subsection 10.3.11 (Hampel *et al* (1986)) seem to be very much relevant. In the absence of definitive guidelines, we can only sound a general note of caution against recommending these procedures when the assumption of independence is suspect.

Model-free Bayesian approach

In our discussion of model-free procedures so far, we have been adhering to the sampling theory approach.

As the pro-subjective Bayesian approach that we considered in section 4.4 is too dependent on the form of the likelihood, first thought suggests that no model-free development would be possible for it. However, in recent years some progress has been made in this direction. The initial breakthrough came out of the work of Ferguson (1973) and Blackwell and Macqueen (1973). The idea these authors advanced was that the distribution of the observable X over the space $\mathcal{X}$ can be conceived as providing a generalized Bernoulli distribution for every finite (measurable) partition of $\mathcal{X} = \cup_1^k \mathcal{X}_i$ irrespective of the model representing the distribution. The class probabilities of the generalized Bernoulli distribution are $p_i = p_i(\cdot) = p(X \in \mathcal{X}_i|\cdot)$, where we have used $(\cdot)$ to indicate that these probabilities relate to a particular distribution of X. The prior over the space of such distributions on $\mathcal{X}$ has to yield the finite-dimensional prior joint distributions of the class probabilities of all such multinomials in a consistent way. Ferguson proposed a prior (Dirichlet process or Ferguson prior) under which these finite-dimensional prior joint distributions are Dirichlet distributions (cf. (4.5.30))

$$\text{const.} \prod_1^{k-1} p_i^{\alpha_i} \left(1 - \sum_1^{k-1} p_i\right)^{\alpha_k}, \quad p_i > 0, \ i = 1, \ldots, k-1, \ \sum_1^{k-1} p_i \leq 1.$$

$$(10.4.18)$$

(For a particular partition this leads to an extension of the generalized rule of succession (7.3.16) mentioned in the context of Laplace's pro-subjective theory of induction.)

Of course the prior over the space of distributions has to be specified in such a way that for any partition of $\mathcal{X}, \alpha_1, \ldots, \alpha_k$, the hyper-parameters of the corresponding finite-dimensional prior joint distribution (10.4.18) are fully determined and this can be done consistently for all such partitions. In Ferguson's case the problem is solved as the Dirichlet process prior involves a 'meta-hyper-parameter' in the form of a *finite* measure $\alpha(\cdot)$ over $\mathcal{X}$. For the partition $\mathcal{X} = \cup_1^k \mathcal{X}_i$, this gives $\alpha_i = \alpha(\mathcal{X}_i), i = 1, \ldots, k$. Given a set of observations $x_1, \ldots, x_n$ the posterior of the distribution over $\mathcal{X}$ can be determined and this also turns out to be a member of the Dirichlet process family with $\alpha(\cdot)$ replaced by a modified meta-hyperparameter $\alpha(\cdot|x_1, \ldots, x_n)$ which is as earlier a finite measure over $\mathcal{X}$. (Thus the Dirichlet process prior is a conjugate prior; cf. footnote 34 in Chapter 4.) Ferguson (1973) gave some applications of the 'model-free posterior' to simple problems of inference.

Incidentally, the problem of estimation under squared error loss of the probability of an event $X \in A$ (where $A \subset \mathcal{X}$) given a number of independent observations $X_1, \ldots, X_n$ on X under the Dirichlet process prior can be looked upon from the purely subjective point of view. The problem then becomes that of prevising for the event $X_{n+1} \in A$ given $X_1, \ldots, X_n$ for a certain exchangeable sequence $X_1, X_2, \ldots, X_n, X_{n+1} \ldots$. For a discussion of this we refer to Skyrms (1996).

The Dirichlet proess prior is of course not a non-informative prior. What subjective intuition of the experimenter it represents is not very easy to see. This and other issues as also further developments on the topic are given in the recent book by Ghosh and Ramamoorthi (2002).

Sampling from a finite population

10.4.3   We consider sampling from a finite population under model-free induction, because the bulk of its theory has been developed without assuming any model. Also, usually in the case of finite population sampling we are concerned only with the estimation of one or more population measure and not with any kind of comprehensive induction. Special considerations arise in this context only when the units in the population are distinguishable and can be labelled. Because of the distinguishability of units, we can make draws without replacement i.e. avoid duplication of units in the sample. Also knowing the label of a unit, with practically no additional cost, we can know the values of various qualitative and quantitative auxiliary characters, e.g. locality, group, size, previous records, etc. But of course to know the value of the character y under study we have to expend some effort. However, the auxiliary characters and hence indirectly the labels bear some relations to the values of y. These relations may be weak or strong, and in any case, our knowledge about them is only partial.

In finite population sampling we have liberty to select the labels of the units on which y would be observed and in probability sampling we make this selection in accordance with some stochastic mechanism. As the selected labels affect the corresponding values of y via the auxiliary characters, by suitably choosing the stochastic mechanism we can control the distribution of the sample y-values to a certain extent. Thus, although we do not here have any power to influence the value of y on a unit as we can do in the case of a controlled experiments (subsection 10.3.3), we have some power to control the sampling distribution of a statistic. In the sampling theory approach, we exploit this fact as also, whatever knowledge we can manage to gather about the underlying relations, to choose a suitable stochastic mechanism for sample selection and to set up a suitable estimator.

| Early developments |

Initially the theory of finite population sampling or as it is called, *sample surveys* developed outside the ambit of main-stream statistical theory. Yet, this development could be considered as an extension of the kind of study that Quetelet undertook on human populations (although, as we noted in subsection 9.4.5, Quetelet himself harboured little love for sampling and based his studies on complete enumeration). Arguably the person who first put into practice in any significant way the idea of basing a study of a large population on sampling was A. N. Kiaer, Director of the Norwegian Central Bureau of Statistics. In the 1890s he organized some sampling investigations in Norway to gather information on such topics as the number of persons eligible for a proposed retirement benefit and sickness insurance scheme, the income distribution of adult males classified according to occupation, etc. In each case the sample taken was subjectively staggered over the entire population across different localities and groups, according to some objective procedure. The general idea was to have a sort of systematic balance in the choice of the sample (cf. the systematic arrangement of treatments in controlled agricultural experiments—subsection 10.3.3). No probability was involved either in the choice of the sample or in the analysis of the data collected. The method followed was called by Kiaer the *representative method*, the object being to have a miniature replica of the population.

The representative method was critically examined when in 1895 Kiaer presented a paper on it at the Berne meeting of the International Statistical Institute (ISI). Kiaer's ideas were disapproved of by many of the official statisticians present at the the meeting, some of whom thought that it was 'dangerous' to think of substituting a partial enquiry for a complete census. As one of them voiced it, 'A sample provides statistics for the units actually observed, but not true statistics for the entire terrain' (see Kruskal and Mosteller (1980) p. 174). (This was in essence Hume's objection to induction (section 1.4), specialized to the concrete case of sampling from a finite population.) Still Kiaer relentlessly propagated the representative method and pleaded for it at successive meetings of the ISI, and through various writings.

There was, however, as yet no significant attempt to amalgamate statistical theory with finite population sampling. A noteworthy step in this direction was taken by the British economist–cum–statistician A. L. Bowley around 1915. Bowley investigated the economic conditions of working class households in a number of English towns by the sampling method. He considered both random and systematic sampling and studied the margin of error as given by the S. E. in the case of the former. A real advance in this regard took place, a few years later when the ISI appointed a Committee to look into the proper

mode of application of the representative method of enquiry and interpretation of its results, with Bowley as one of its members. In the report of the
Committee, which was presented at the ISI meeting in Rome in 1925 and published in 1926, Bowley contributed by providing a long theoretical discussion
of sampling methods. He considered random sampling—both unrestricted and
stratified (with proportional allocation)—and discussed estimation from both
the sampling-theory and the Bayesian points of view. He also gave some attention to purposive sampling in which a set of large clusters are chosen so that
the sample averages of certain control variables equal the known values of the
corresponding population averages.

Surprisingly, parallel developments in sampling theory took place in (Soviet)
Russia at about the same time. Specifically, these started a few years before
the First World War and continued through several years after the War and the
Revolution. There were a number of contributors to these developments, notable among them being Tchuprow. A treatise entitled *Basic Theory of Sampling
Methods* by Kowalsky came out in 1924! It is remarkable that from the beginning the Russian experts emphasized probability sampling. Different forms of
simple random, stratified, systematic, and cluster sampling were considered.
Also these sampling methods were employed in practice for studying populations of individuals, households, agricultural holdings, etc. However these early
developments could not continue after the death of Lenin, apparently because
the Soviet authorities thereafter did not encourage such studies; see Zarkovich
(1956), Kruskal and Mosteller (1980).

| Neyman |

A real shot in the arm was given to the development of
the theory of sample surveys by Neyman's 1934 paper
on the representative method. By that time, Fisher's sampling theory approach
to statistical induction had become well-established in many areas. Also the
principle of randomization introduced by Fisher in the design of experiments had
been broadly accepted. Neyman, since some years earlier, had been intimately
associated with the design and conduct of large scale sample surveys in Poland.
In the paper he drew upon the ideas of Fisher and his own experience to give
a clear exposition of the theory of sample surveys from the sampling theory
point of view and in the process subjected the report of the ISI Committee and
Bowley's development to a critical appraisal. Also, Neyman took this occasion
to expound the philosophy of the behavioural approach with its emphasis on the
method of sampling and estimation. Furthermore, as noted earlier, Neyman's
first clear enunciation of the concept of a confidence interval appeared in that
paper.

In the paper Neyman unequivocally laid down that the object of a sampling
enquiry is to obtain confidence intervals for the parameters of interest. To achieve
this the sample must be taken according to some probabilistic mode of selection
(Neyman specifically considered unrestricted and stratified random selection of
units as well as clusters, but of course any form of probability selection will do).

Furthermore, he emphasized that we should take some optimum estimate of the parameter of interest and get an estimate of the S. E. of the estimate. (In the case of sampling of units, within each stratum Neyman took the classical best linear unbiased estimate ignoring the labels.) In the case of stratified random sampling of units, Neyman gave the standard formula for optimum allocation of a fixed total sample size to the different strata so as to minimize the S. E. of the overall estimate. (However, unknown to Neyman, this result had been derived a few years earlier by Tchuprow; see Kruskal and Mosteller (1980).) A major part of Neyman's paper was, however, devoted to debunking the method of purposive sampling which until then was regarded as being at par with random sampling. (Around 1927 two Italian statisticians Gini and Galvani had fared badly in trying to use the purposive method on the population of census records in Italy and one of the purposes of Neyman's paper was to explain that failure.) Neyman showed that the purposive method can at best be interpreted as a sort of stratified sampling coupled with a particular method of estimation, neither of which can be optimal except under dubious and restrictive assumptions.[31] This study of Neyman for the time firmly established that in the case of a finite population the only rational way to sample would be to adopt some form of probability selection (at least from the sampling theory point of view).

Expansion of applications

After this there was a rapid development of sample survey theory during the next 15–20 years. Much of it was application-driven. Different forms of probability sampling such as stratified, multistage, multiphase, systematic with random start, unequal probability selection, etc. were developed in quick succession. Different forms of estimators for these, making use of auxiliary information in different ways, were also proposed and studied *in extenso*.

The two major types of application initially were for the estimation of production of crops on the basis of agricultural surveys and the extent of unemployment and labour force participation on the basis of human population surveys. In the former case significant work was done at the beginning in England at Rothamsted under the leadership of Yates; in India, especially at the Indian Statistical Institute, Calcutta under the leadership of Mahalanobis, and in the United States at Ames, Iowa under the leadership of Cochran. The unemployment and labour force survey, initially conducted by the US Work Project Administration, was taken over by the Census Bureau in the early 1940s, where their scope was

---

[31] However, in recent times Royall and Herson (1973) have re-examined purposive sampling under a form of superpopulation set-up (see later). They have shown that for a purposive sample, in which the sample mean of an auxiliary variable x is made equal to the corresponding population mean, under suitable conditions the usual estimator not only provides best linear prediction for the population total but also ensures robustness against certain kinds of departure from the set-up. But of course the superpopulation set-up would be hardly relevant in the situation considered by Gini and Galvani.

enlarged. Ultimately these have taken the form of a continuing multi-subject survey known as the Current Population Survey. In India, to meet the needs of planning and evaluation of development programmes, a wide-ranging continuing survey called the National Sample Surveys was instituted in 1950. The conduct of these surveys has ultimately devolved on a separate wing of the Department of Statistics of the Government of India. For an account of these early developments, we refer to Hansen and Madow (1976). It so happened that the above pioneering surveys have served as prototypes for similar surveys in many other countries. United Nations agencies like the Food and Agricultural Organization and the International Labour Office have been helping in the dissemination and standardization of the methodology.

| Developments in theory | Initially the methods and estimation procedures for probability sampling from a finite population were

developed on an *ad hoc* basis. These utilized the information carried by the unit labels only indirectly for stratification or selection and for the setting up of estimates, using the values of auxiliary variates. The widespread application of sample survey methods in practice led to a deeper examination of their theoretical basis from around 1950 onwards.

In the formulation that gradually became standard for such studies, the population is regarded as a collection of N units numbered $1, 2, \ldots, N$. The values $y_\ell, \ell = 1, \ldots, N$ of the study variable y associated with these units are supposed to be unknown fixed numbers, $\theta = (y_1, y_2, \ldots, y_N)$ being regarded as an unknown parameter vector. Generally we are interested in estimating some function $\xi(\theta)$ such as the population total $G = G(\theta) = \sum_1^N y_\ell$ or mean $\bar{y} = G/N$. The sample is represented by a subset $s = (\ell_1, \ldots, \ell_n), 1 \leq \ell_1 < \cdots \ell_n \leq N$ of $\{1, 2, \ldots, N\}$ and the collection of all possible samples by $\mathcal{S}$. For the sake of simplicity we consider sampling with a fixed sample size n. In this case $\mathcal{S}$ consists of the $\binom{N}{n}$ possible choices for s. The sampling design is given by a *known* probability distribution over $\mathcal{S}$ represented by a set of probabilities $p(s) \geq 0, s \in \mathcal{S}, \sum_\mathcal{S} p(s) = 1$. (For simplicity, we are considering non-sequential designs only.)

The sampling experiment consists in taking a random member $S = (L_1, \ldots, L_n)$ of $\mathcal{S}$ so that $P(S = s) = p(s), s \in \mathcal{S}$. After we have chosen a realization $s = (\ell_1, \ldots, \ell_n)$ of S we observe the values $y_{\ell_1}, \ldots, y_{\ell_n}$ corresponding to the selected units. Thus an estimator T here is generally a function of S and $y_{L_1}, \ldots, y_{L_n}$. When $\xi = \sum_1^N y_\ell$, a homogeneous linear estimator T here is defined by specifying, for each $s \in \mathcal{S}$, coefficients $b_\ell(s)$ for each unit ℓ in the population such that $\ell \in s$, and setting up

$$T = \sum_1^n b_{L_i}(S) y_{L_i}.^{32} \tag{10.4.19}$$

---

[32]This is to be contrasted with the linear estimator defined for the classical Gauss–Markov set-up. In the latter case we have some serially numbered random variables

It has been proved that, excluding trivial sampling designs, within the sub-class of estimators (10.4.19) which are unbiased for $G(\theta)$, there is none which has minimum variance for all θ (see Godambe (1955) where a more general result is proved). Various weaker forms of optimality requirements have been proposed and studied to sidetrack this difficulty (see e.g. Basu (1971), Cassel *et al*, Chaudhuri and Vos (1988), Part A).

The randomization imbroglio

We now come to the question of randomization or artificial injection of probability from outside in finite population sampling. The set of labels S representing the sample in any form of probability sampling has a known distribution over S and hence represents an ancillary statistic (see Basu (1988) Chapter IX). Therefore, if we go by the principle of conditionalization (subsections 4.2.3 and 4.3.5), we should keep S conditionally fixed at its realized value. But if we fix $S = s = (\ell_1, \ldots, \ell_n)$, the sample observation $y_{\ell_1}, \ldots, y_{\ell_n}$ become totally fixed and statistical induction is impossible by the sampling theory approach. Also in that case there is little point in selecting s by a random mechanism.

This is similar to the difficulty we faced with randomization in the context of a randomized experiment (subsection 10.3.7). But whereas in the case of the latter the assumption of fixed plot errors was unrealistic, in the case of sample surveys when there are no observational errors, it may be realistic to assume that $y_\ell, \ell = 1, \ldots, N$ are predetermined. (This holds, for example, when the y_ℓs are the claims made by a population of insurants.)[33] It seems that in such cases, unless the principle of conditioning on an ancillary is abandoned, in the sampling theory approach we would get stalled. Very often, however, each y_ℓ can be regarded as the realization of a random varible Y_ℓ, rendered unpredictable due to measurement errors or otherwise. (This is the case, for example, in surveying for yields of plots or for consumption expenditures of households.) There we can argue as in subsection 10.3.7. Thus in the case of simple random sampling we can say that we take to equiprobability selection because we want to regard the units in the population to be at par with each other. This means we have in mind a representation such as

$$Y_\ell = \mu + D_\ell, ED_\ell = 0, \quad \ell = 1, \ldots, N, \tag{10.4.20}$$

where we presume the D_ℓs have a permutation-symmetric joint distribution, but we are not absolutely sure about this presumption. Our object is to estimate the true per-unit mean μ (e.g. average plot yield or consumption). We take to the equiprobabilistic form of sampling because as a matter of policy we want to

---

$Y_1, \ldots, Y_n$ and a linear estimator has the form $\sum_1^n c_i Y_i$, where $c_1, \ldots, c_n$ are predetermined numbers. The Gauss–Markov Theorem gives the best estimate within the class of unbiased linear estimators of this form.

[33]However, nowadays with the expansion of computers, the need for sampling in such cases is gradually shrinking.

disregard any inter-unit heterogeneity and want to insure against any possible departure from permutation symmetry of the distribution of the D_ℓs.

Measurement error models of the type (10.4.20) are often called *super-population* models (it is as if the random Y_ℓs arise because the finite population is a random sample from an infinite superpopulation). As we saw in subsection 7.3.9 the idea of such a model was first conceived by Laplace. In modern times various forms of such models have been occasionally brought in, often to establish the optimality of standard estimation procedures (an early instance is Godambe (1955)). More recently superpopulation models with specific assumptions about the form of the distribution of the errors D_ℓ, have been examined by Royall and others (see e.g. Royall (1970)). Such an approach takes the theory of sample surveys back to the realm of model-specific induction. Under such a model we can take the liberty of using estimators which do not depend on the selection probabilities of the sample, i.e. on how the sample has been selected. However, although such an approach may lead to gain in efficiency when the underlying model holds, when there is departure from the model, the corresponding estimators may perform poorly and may even involve biases that are non-negligible.

| Non-sampling-theory approaches |

Going back to the case of fixed unit values y_ℓ, the question of randomization has been examined from a non-sampling-theory point of view by Basu (1969). He emphasizes that, whatever the sampling design, given $s = (\ell_1, \ldots, \ell_n)$, the likelihood function is a constant over the subset of values of $\theta = (y_1, \ldots, y_N)$ for which $y_{\ell_1}, \ldots, y_{\ell_n}$ are as observed and is 0 outside this subset. Thus if one accepts the likelihood principle (subsection 4.3.4) there would be no scope for using the selection probabilities of the design for estimation, after the sample has been taken, and hence, no justification for selecting the sample probabilistically. A natural consequence of the likelihood function being independent of the sampling design is that, if we adopt the pro-subjective Bayesian approach, the sampling design becomes irrelevant. It has even been suggested that in such an approach one should use prior information only to identify the subclass of samples s which are most informative according to some reasonable criterion. After that it does not matter if one selects a sample purposively out of that subclass; cf. Basu (1969).

In fact, nowhere in statistical theory the contrast between the sampling theory approach and the non-sampling-theory approaches (likelihood or Bayesian) is as sharply focused as in the area of finite population sampling (this is because the parametric function to be estimated has a physically existent value in this case). In the former, before choosing the sample, we adopt a probabilistic sampling design which gives a narrow interval estimate (confidence or fiducial interval) with a high probability (long-term relative frequency—repetitional or instantiated). In the latter, after choosing a sample from subjective consideration, we subjectively judge what credible limits we can set for the parametric function of interest on the basis of the subjective probability model we believe to be true. Note that in the pro-subjective Bayesian approach objective probability will come into the picture only if we assume a superpopulation model as in (10.4.20). In this connection we note that, as Godambe (1982) has pointed out, even if we follow a pro-subjective approach, it is usually judicious to select the sample in accordance with a probabilistic sampling design. Thereby with a high probability we get a sample which safeguards us against possible departures from the assumptions of the model.

In recent years there has been a good deal of activity towards developing the theory of sample surveys from the Bayesian point of view (see e.g. Ghosh and Meeden (1997)). It is remarkable that in the pro-subjective approach with a superpopulation model, often one can make considerable advance here within the model-free set-up, making assumptions only about certain aspects of the joint distribution of the Y_ℓs and without assuming its full form. However, if the y_ℓs are supposed to be objectively fixed, we are led into the realm of the purely subjective approach (section 4.5): given s and $y_{\ell_1}, \ldots, y_{\ell_n}$ we want to previse, say, about $\sum_1^N Y_\ell$, or equivalently, $\sum_{\ell \notin s} Y_\ell$. The y_ℓs here are only subjectively stochastic. Lad (1996) pp. 211–227 considers some finite population sampling problems from this angle. As comparison with Ghosh and Meeden (1997) p. 9 shows, the distinction between the pro-subjective and purely subjective approaches here is not well-marked. However, in the purely subjective approach, for reasons that we elaborated in section 4.5, one prefers to have exchangeability. It is remarkable that Lad, pp. 212, 215 suggests randomization (randomly numbering the population units) or 'scrambling' for this purpose (cf. subsection 10.3.7). Of course in such contexts one is interested only in prevising some symmetric function of the population values $y_1, \ldots, y_N$.

10.5   Epilogue

This brings us to the conclusion of our chronicle. We are fully conscious that in the account given in this chapter there are some glaring gaps. For reasons stated earlier, we have not touched such important areas as induction for stochastic processes and time series, or paid attention to topical issues like those involving 'observational studies'. Nor have we discussed, except making passing references to, the very important question of the impact of computers on the current state of the art and science of statistical induction.

As we look back at the long track covering about three and half centuries, it seems to us that the history of statistical thought is the history of the ongoing struggle of the human mind to catch hold of the meaning of 'induction'. As the ancient philosophers knew and as Locke (section 1.4) articulated it clearly, the one principle which is inviolable in induction is 'the principle of total evidence'. But in no circumstances are all the details available relevant for induction and in fact some of the details must be obliterated before one can see through the data. Yet as we noted briefly in Chapter 1 (footnote 5) and in more detail in Chapter 4 (section 4.6), there is no unanimity among statisticians about which part of the evidence must be taken into account, and which not. Most pointedly, the issue which has queered the pitch in statistical induction, is how far and in what form should the opinion of experts be regarded as part of the evidence.

We have reached the end of our story. But the story does not have any end— at least, it seems, it is not going to have any end in the forseeable future. Is there a true mode of induction, or is it, as Basu (1988) (Chapter IX) observed, we statisticians are like blind men in a dark room groping for a black cat that is perhaps not there. But we think that the situation, although not fully reassuring, is not as thankless as that. Karl Popper at one place likened the search for truth to mountain climbing amid heavy cloud. As one scales successive peaks one knows that one is moving higher and higher, but one is never sure whether one has reached the summit. Only the history of statistical thought shows that here we have several groups of climbers following different routes, who can occasionally have a glimpse or hear the voices of each other. It is doubtful whether any of these groups will ever pip the others to the summit and know and let others know that they have done it. In fact here we are not absolutely sure even about whether a summit, i.e. a 'correct' mode of codification of the process of statistical induction exists at all. But the quest goes on!

Under the circumstances, we feel that the best attitude, as we said it at the end of Chapter 4, is to shun dogmatism and be an eclectic.

REFERENCES

Anderson, T. W. (1984): *An Introduction to Multivariate Statistical Analysis*, 2nd edn., Wiley, New York.

Anscombe, F. J. (1957): Dependence of the fiducial argument on the sampling rule. *Biometrika*, **44**, 464–9.

——and Aumann, R. J. (1963): A definition of subjective probability. *Ann. Math. Statist.*, **34**, 199–205.

Bahadur, R. R. (1960): Asymptotic efficiency of tests and estimates. *Sankhyā*, A, **22**, 229–52.

——(1967): An optimal property of the likelihood ratio statistic. *Proc. 5th Berkeley Symp.* Vol. I, ed. Lecam, L. and Neyman, J., University of California Press, 13–26.

——(1971): *Some Limit Theorems in Statistics*. NBS–CBMS Monograph No. 4, SIAM, Philadelphia.

Barnard, G. A. (1949): Statistical Inference. *J. Roy. Statist. Soc.*, B, **11**, 115–49.

——(1958): Thomas Bayes's essay towards solving a problem in the doctrine of chances. *Biometrika*, **45**, 293–315.

——(1967): The use of the likelihood function in statistical practice. *Proc. 5th Berkeley. Symp.*, University of California Press, **1**, 27–40.

——, Jenkins, G. M., and Winsten, C. B. (1962): Likelihood inference and time series. *J. Roy. Statist. Soc.*, A, 321–72.

Bartlett, M. S. (1949): Fitting a straight line when both variables are subject to error. *Biometrics*, **5**, 207–12.

Basu, D. (1969): Likelihood principle and survey sampling. *Sankhyā*, **31**, 441–54.

——(1971): On the logical foundations of survey sampling. *Foundations of Statistical Inference*, ed. Godambe, V. P. and Sprott, D. A. Holt, New York. Reprinted as Chapter XII in Basu, D. (1988): *Statistical Information and Likelihood*, ed. Ghosh, J. K., Springer–Verlag, New York.

——(1975): Statistical information and likelihood. *Sankhyā*, A, **37**, 1–71.

——(1978): Partial sufficiency. *J. Statist. Plan. Infer.*, **2**. Reprinted as Chapter VI in Basu, D. (1988): *Statistical Information and Likelihood*, ed. Ghosh, J. K., Springer–Verlag, New York.

——(1988): *Statistical Information and Likelihood*, ed. Ghosh, J. K., Springer–Verlag, New York.

Berger, J. O. (1984): The robust Bayesian viewpoint. *Robustness of Bayesian Analysis*, ed. Kadane, J. B., Elsevier Sc., Amsterdam.

——(1985a): *Statistical Decision Theory and Bayesian Analysis*, Springer–Verlag, New York.

Berger, J. O. (1985*b*): The frequentist viewpoint and conditioning. *Proc. Berkeley Conference in Honor of Jerzy Neyman and Jack Kiefer*, Vol. I, ed. Lecam, L. and Olshen, R., Wadsworth, Belmont.

—— and Wolpert, R. (1988): *The Likelihood Principle*, IMS Lecture Notes Monograph Series, **9**, Hayward, California.

Bernardo, J. M. (1979): Reference posterior distributions for Bayesian inference. *J. Roy. Statist. Soc.*, B, **41**, 113–47.

Bhattacharyya, A. (1990–91): On a geometrical representation of probability distributions and its use in statistical inference. *Calcutta Statist. Assoc. Bull.*, **40**, 23–49.

Bhattacharyya, J. V. (1953): Navya Nyāya. *The Cultural Heritage of India*. Vol. III, 2nd edn., R. K. M. Institute of Culture, Calcutta, 125–50.

Bhattacharya, R. N. and Ghosh, J. K. (1978): On the validity of the formal Edgeworth expansion. *Ann. Statist.*, **6**, 434–51.

Billingsley, P. (1968): *Convergence of Probability Measures*, Wiley, New York.

Binder, D. A. (1978): Bayesian cluster analysis. *Biometrika*, **65**, 31–8.

Birnbaum, A. (1962): On the foundations of statistical inference. *J. Amer. Statist. Assoc.*, **57**, 269–326.

Blackwell, D. and Macqueen, J. B. (1973): Ferguson distributions via Polya urn schemes. *Ann. Statist.*, **1**, 353–5.

Box, G. E. P. (1980): Sampling and Bayes's inference in scientific modelling and robustness. *J. Roy. Statist. Soc.*, A, **143**, 383–430.

—— and Andersen, S. L. (1955): Permutation theory in the derivation of robust criteria and the study of departure from assumption. *J. Roy. Statist. Soc.*, B, **17**, 1–26.

—— and Cox, D. R. (1964): An analysis of transformations. *J. Roy. Statist. Soc.*, B, **26**, 211–52.

—— and Tiao, G. C. (1973): *Bayesian Inference in Statistical Analysis*, Addison–Wesley, Reading, Mass.

Brown, L. D. (1971): Non-local asymptotic optimality of appropriate likelihood ratio tests. *Ann. Math. Statist.*, **42**, 1206–40.

Buchler, J. ed. (1955): *Philosophical Writings of Peirce*, Dover, New York.

Carnap, R. (1947): On the application of inductive logic. *Philosophical and Phenomenological Research*, **8**, 133–48.

—— (1950): *Logical Foundations of Probability*, Chicago University Press, Chicago.

Cassel, C. M., Sarndal, C. E., Wretman, J. H. (1977): *Foundations of Inference in Survey Sampling*, Wiley, New York.

Chaloner, K. and Verdinelli, I. (1995): Bayesian experimental design: a review. *Statist. Sc.*, **10**, 273–304.

Chatterjee, S. K. (1970): Robustness of frequency chi-square. *Calcutta Statist. Assoc. Bull.*, **19**, 145–54.

—— (1972): Rank approach to the multivariate two-population mixture problem. *J. Multivariate Anal.*, **2**, 261–81.

—— (1989): Some thoughts on the foundations of statistical inference. Presidential address delivered at the Statistics Section of the 76th session (Madurai) of Indian Science Congress. Reprinted in *Calcutta Statist. Assoc. Bull.*, **38**, 1–26.

—— and Chattopadhyay, G. (1992): Detailed statistical inference—an alternative non-Bayesian approach: two-decision problem. *Calcutta Statist. Assoc. Bull.*, **42**, 41–74.

—— and —— (1993): Detailed statistical inference—multiple decision problem. *Calcutta Statist. Assoc. Bull.*, **43**, 155–80.

Chaudhuri, A. and Vos, J. W. E. (1988): *Unified Theory and Strategies of Survey Sampling*, North Holland, Amsterdam.

Cochran, W. G. (1937): The efficiencies of the binomial series tests of significance of a mean and of a correlation coefficient. *J. Roy. Statist. Soc.*, **100**, 69–73.

—— (1976): Early development of techniques in comparative experimentation. *On the History of Statistics and Probability*, ed. Owen, D. B., Marcel–Dekker, New York, 1–25.

—— (1977): *Sampling Techniques*, 3rd edn., Wiley Eastern, New Delhi.

—— (1978): Laplace's ratio estimator. In *Contributions to Survey Sampling and Applied Statistics*, ed. David, H. A., Academic, New York.

—— (1980): Fisher and the analysis of variance. *R. A. Fisher: An Appreciation*, ed. Fienberg, S. E. and Hinkley, D. V., Springer–Verlag, New York, pp. 17–34.

Cohen, L. J. (1989): *An Introduction to the Philosophy of Induction and Probability*, Clarendon Press, Oxford.

Copi, I. M. and Cohen, C. (1996): *Introduction to Logic*, 9th edn. Prentice–Hall of India, New Delhi.

Cox, D. R. (1958): Some problems connected with statistical inference. *Ann. Math. Statist.*, **29**, 357–72.

—— (1972): Regression models and life tables. *J. Roy. Statist. Soc.*, B, **34**, 187–220.

—— (1975): Partial likelihood. *Biometrika*, **62**, 269–76.

—— and Hinkley, D. V. (1974): *Theoretical Statistics*, Chapman–Hall, London.

—— and Oakes, D. (1984): *Analysis of Survival Data*, Chapman–Hall, London.

—— and Wermuth, N. (1996): *Multivariate Dependencies*, Chapman–Hall, London.

Cramér, H. (1946): *Mathematical Methods of Statistics*, Princeton University Press, Princeton.

Dasgupta, S. (1980): Discriminant analysis. *R. A. Fisher: An Appreciation*, ed. Fienberg, S. E. and Hinkley, D. V., Springer–Verlag, New York, pp. 161–70.

Datta, D. M. (1953): Indian epistemology. *The Cultural Heritage of India*, Vol. III, 2nd edn., R. K. M. Institute of Culture, Calcutta, 548–61.

David, F. N. (1955): Dicing and gaming (a note on the history of probability). *Biometrika*, **42**, 1–15.

—— (1962): *Games, Gods and Gambling*, Griffin, London.

Dawid, A. P. (1984): Statistical theory: The prequential approach. *J. Roy. Statist. Soc.*, A, **147**, 278–92.

Day, J. P. (1961): *Inductive Probability*, Routledge and Kegan Paul, London.

Day, N. E. (1969): Estimating the components of a mixture of normal distributions *Biometrika*, **68**, 463–74.

de Finetti, B. (1937): Foresight, its logical laws, its subjective sources. *Studies in Subjective Probability*, ed. Kyburg, H. E. (Jr.) and Smokler, H. E. (1963). John Wiley, New York & London, pp. 99–157.

—— (1974): *Theory of Probability*, Vol. 1, Wiley, New York.

—— (1975): *Theory of Probability*, Vol. 2, Wiley, New York.

De groot, M. H. (1970): *Optimal Statistical Decisions*, McGraw–Hill, New York.

Dempster, A. P. (1963): Further examples of inconsistencies in the fiducial argument. *Ann. Math. Statist.*, **34**, 884–91.

—— (1968): A generalization of Bayesian inference. *J. Royal Statist. Soc.*, B, **30**, 205–47.

Dey, A. and Mukerjee, R. (1999): *Fractional Factorial Plans*, Wiley, New York.

Diaconis, P. and Freedman, D. (1986): On the consistency of Bayes estimators. *Ann. Statist.*, **14**, 1–26.

Durbin, J. (1970): On Birnbaum's theorem on the relation between sufficiency conditionality and likelihood. *J. Amer. Statist. Assoc.*, **65**, 395–8.

Edwards, A. W. F. (1972): *Likelihood*, Cambridge University Press, Cambridge.

Eerola, M. (1994): *Probabilistic Causality in Longitudinal Studies*, Springer–Verlag, New York.

Efron, B. (1975): Defining the curvature of a statistical problem (with applications to second order efficiency). *Ann. Statist.*, **3**, 1189–242.

—— (1979): Bootstrap methods: another look at the jackknife. *Ann. Statist.*, **7**, 1–26.

—— and Hinkley, D. V. (1978): Assessing the accuracy of the maximum likelihood estimator: observed versus expected Fisher information. *Biometrika*, **65**, 457–87.

Eisenhart, C. (1961): Boskovich and the combination of observations. *Studies in the History of Statistics and Probability* Vol. II, ed. Kendall, M. and Plackett, R. L., Griffin, London.

Elderton, W. P. and Johnson, N. L. (1969): *Systems of Frequency Curves*, Cambridge University Press, Cambridge.

Epstein, R. A. (1967): *The Theory of Gambling and Statistical Logic*, Academic, New York.

Feller, W. (1950): *An Introduction to Probability Theory and Its Applications*, Vol. 1, Wiley, New York.

Ferguson, T. S. (1967): *Mathematical Statistics*, Academic Press, New York and London.

—— (1973): A Bayesian analysis of some nonparametric problems. *Ann. Statist.*, **1**, 209–30.

—— (1982): An inconsistent maximum likelihood estimate. *J. Amer. Statist. Assoc.*, **77**, 831–4.

Festa, R. (1993): *Optimum Inductive Methods*, Kluwer Academic, Dordrecht and London.

Finney, D. J. (1952): *Probit Analysis*, Cambridge University Press, Cambridge.

Fisher, R. A. (1935): The logic of inductive inference. *J. Roy. Statist. Soc.*, **98**, 39–82.

—— (1944): *Statistical Methods for Research Workers*, 9th. edn., Oliver and Boyd, London.

—— (1947): *Design of Experiments*, 4th edn., Oliver and Boyd, London.

—— (1950): *Contributions to Mathematical Statistics*, Wiley, New York.

—— (1956): *Statistical Methods and Scientific Inference*, Oliver and Boyd, London.

—— (1959): Mathematical probability in the natural sciences. *Technometrics*, **1**, 21–9.

—— and Yates, F. (1963): *Statistical Tables*, 6th. edn., Longman, London.

Fraser, D. A. S. (1957): *Nonparametric Methods in Statistics*, Wiley, New York.

—— (1961): The fiducial method and invariance. *Biometrika*, **48**, 261–80.

—— (1968): *The Structure of Inference*, John Wiley, New York.

Gayen, A. K. (1949): The distribution of 'Student's' t in random samples from non-normal universes. *Biometrika*, **36**, 353–69.

—— (1951): The frequency distribution of the product moment correlation coefficient in random samples of any size drawn from non-normal universes. *Biometrika*, **38**, 219–47.

Geisser, S. (1980): A predictive primer. *Bayesian Analysis in Econometrics and Statistics*, ed. Zellner, A., North Holland, Amsterdam.

Gelfand, A. E. and Smith, A. F. H. (1990): Sampling-based approaches to calculating marginal densities. *J. Amer. Statist. Assoc.*, **85**, 398–409.

Ghosh, J. K. (1971): A new proof of the Bahadur representation of quantiles and an application. *Ann. Math. Statist.*, **42**, 1957–61.

—— and Ramamoorthi, R. V. (2002): *Bayesian Nonparametrics*, Springer, New York.

—— and Samanta, T. (2001): Model selection—an overview. *Current Sc.*, **80**, 1135–44.

——, Sinha, B. K. and Wiend, H. S. (1980): Second order efficiency of the MLE with respect to any bounded bowl-shaped loss function. *Ann. Statist.*, **8**, 506–21.

Ghosh, M. N. (1952): An extension of Wald's decision theory to unbounded weight functions. *Sankhyā*, **12**, 8–26.

Ghosh, M. and Meeden, G. (1997): *Bayesian Methods for Finite Population Sampling*, Chapman–Hall, London.

——, Mukhopadhyay, N. and Sen, P. K. (1997): *Sequential Estimation*. Wiley, New York.

Ghosal, S., Ghosh, J. K. and Samanta, T. (1995): On convergence of posterior distributions. *Ann. Statist.*, **23**, 2145–52.

Gigerenzer, G., Swijtink, Z., Daston, L., Beatty, J., Krüger, L. (1989): *The Empire of Chance*, Cambridge University Press, Cambridge.

Gnedenko, B. V. and Kolmogorov, A. N. (1954): *Limit Distributions for Sums of Independent Random Variables*, Addison–Wesley, Cambridge, Mass.

Godambe, V. P. (1955): A unified theory of sampling from finite populations. *J. Roy. Statist. Soc.*, B, **17**, 269–78.

—— (1976): A historical perspective of the recent developments in the theory of sampling from actual populations. *J. Indian Soc. Agri. Statist.*, **38**, 1–12.

—— (1982): Likelihood principle and randomization. *Statistics and Probability: Essays in Honor of C. R. Rao*, ed. Kallianpur, G., Krishnaiah, P. R., Ghosh, J. K., North Holland, Amsterdam, pp. 281–94.

Good, I. J. (1950): *Probability and the Weighing of Evidence*. Charles Griffin, London.

—— (1983): *Good Thinking*, University of Minnesota Press, Minneapolis.

Greenwood, M. (1941): Medical statistics from Graunt to Farr. *Biometrika*, **32**, 101–27.

Guglielmi, A. and Melilli, E. (1998): Non-informative invariant priors yield peculiar marginals. *Commun. Statist.—Theor. Meth.*, **27**, 2293–306.

Gumbel, E. L. (1958): *Statistics of Extremes*. Columbia University Press, New York.

Gurland, J. (1963): A method of estimation for some generalized Poisson distributions. *Classical and Contagious Discrete Distributions*, ed. Patil, G. P., Pergamon Press, Oxford, 141–58.

Hacking, I. (1965): *Logic of Statistical Inference*, Cambridge University Press, Cambridge.

—— (1975): *The Emergence of Probability*, Cambridge University Press, Cambridge.

—— (1990): *The Taming of Chance*, Cambridge University Press, Cambridge.

Hájek, J., Sidak, Z. and Sen, P. K. (1999): *Theory of Rank Tests*, 2nd. edn., Academic Press, New York.

Hald, A. (1990): *A History of Probability and Statistics and Their Applications before 1750*, John Wiley, New York.

—— (1998): *A History of Mathematical Statistics from 1750 to 1930*, John Wiley, New York.

Hampel, F. H., Ronchetti, E. M., Rousseeuw, P. J., Stahel, W. A. (1986): *Robust Statistics*, Wiley, New York.

Hansen, M. H. and Madow, W. G. (1976): Some important events in the historical development of sample surveys. *On the History of Statistics and Probability*, ed. Owen, D. B., Mercel–Dekker, New York, pp. 73–102.

Hasover, A. M. (1967): Random mechanisms in Talmudic literature. *Biometrika*, **54**, 316–21.

Hawking, S. (1988): *A Brief History of Time*, Bantam, Toronto.

Heath, D. and Sudderth, W. (1978): On finitely additive priors, coherence, and expected admissibility. *Ann. Statist.*, **6**, 333–45.

Hinkley, D. V. (1980): Fisher's development of conditional inference. *R. A. Fisher: An Appreciation* ed. Fienberg, S. E. and Hinkley, D. V., Springer–Verlag, New York.

Hoeffding, W. (1952): The large-sample power of tests based on permutations of observations. *Ann. Math. Statist.*, **23**, 169–92.

—— (1965): Asymptotically optimal tests for multinomial distributions. *Ann. Math. Statist.*, **36**, 369–400.

Holschuh, N. (1980): Randomization and design: I. In *R. A. Fisher: An Appreciation*, ed. Fienberg, S. E. and Hinkley, D. V., Springer–Verlag, New York, pp. 35–45.

Hotelling, H. (1931): The generalization of Student's ratio. *Ann. Math. Statist.*, **2**, 360–78.

Howson, C. (1995): Theories of probability. *Brit. J. Phil. Sc.*, **46**, 1–32.

Huber, P. J. (1981): *Robust Statistics*, Wiley, New York.

Hume, D. (1739): *Treatise of Human Nature*, ed. Selby-Bigge, L. A., (1888), Clarendon Press, Oxford.

Ibragimov, J. A. (1956): On the composition of unimodal distributions. *Theor. Prob. and Its Applicns.* (English), **1**, 255–60.

Jaynes, E. T. (1976): Confidence intervals vs. Bayesian intervals. In *Foundations of Probability Theory, Statistical Inference, and Statistical Theories of Science*, Vol. II, ed. Harper, W. L. and Hooker, C. A., Reidel, Dordrecht.

—— (1983): *Papers on Probability, Statistics and Statistical Physics*, ed. Rosenkrantz, R. D., Reidel, Dordrecht.

Jeffreys, H. (1931, 2nd ed. 1957): *Scientific Inference*, Cambridge University Press, Cambridge.

—— (1939, 3rd ed. 1961): *Theory of Probability*, Clarendon Press, Oxford.

—— (1940): Note on the Behrens–Fisher formula. *Ann. Eug.*, **10**, 48–51.

Johnson, N. L. (1949): Systems of frequency curves generated by methods of translation. *Biometrika*, **36**, 149–76.

Johnson, W. E. (1921): *Logic*, Part I, Cambridge University Press, Cambridge.

—— (1924): *Logic*, Part III, Cambridge University Press, Cambridge.

—— (1932): Appendix to 'Probability: deductive and inductive problem'. *Mind*, **41**, 421–3.

Johnston, J. (1984): *Econometric Methods*, 3rd edn., McGraw–Hill, London.

Jurecková, J. and Sen, P. K. (1996): *Robust Statistical Procedures*, Wiley, New York.

Kadane, J. B. and Schum, D. A. (1996): *A Probabilistic Analysis of the Sacco and Vanzetti Evidence*, John Wiley, New York.

Kalbfleisch, J. D. and Sprott, D. A. (1970): Application of likelihood methods to models involving large numbers of parameters. *J. Royal Statist. Soc.*, B, **32**, 175–208.

—— —— (1973): Marginal and conditional likelihoods. *Sankhyā*, A, **35**, 311–28.

Kass, R. E. and Wasserman, L. (1996): The selection of prior distributions by formal rules. *J. Amer. Statist. Assoc.*, **91**, 1343–70.

Kempthorne, O. and Folks, L. (1971): *Probability, Statistics, and Data Analysis*, Iowa State University Press, Ames.

Kendall, D. G. (1966): Branching processes since 1873. *J. London Math. Soc.*, **41**, 385–406. Reprinted in *Studies in the History of Statistics and Probability* Vol. II, ed. Kendall, M. and Plackett, R. L., Griffin, London.

Kendall, M. G. (1956): The beginnings of a probability calculus. *Biometrika*, **43**, 1–14.

—— (1963): Ronald Aylmer Fisher. *Biometrika*, **50**, 1–15.

Keynes, J. M. (1921): *A Treatise on Probability*, Macmillan, London.

Kiefer, J. (1977): Conditional confidence statements and confidence estimators. *J. Amer. Statist. Assoc.*, **72**, 789–827.

Kneale, W. (1949): *Probability and Induction*, Oxford University Press, London.

Kolmogorov, A. N. (1933): *Foundations of the Theory of Probability*, English edn. (1950), Chelsea, New York.

Koopman, B. O. (1940a): The axioms and algebra of intuitive probability. *Ann. Math.*, **41**, 261–92.

—— (1940b): The bases of probability. *Bull. Amer. Math. Soc.*, **46**, 763–74, reprinted in *Studies in Subjective Probability* ed. Kyburg, H. E. (Jr.) & Smokler, H. E. (1963). John Wiley, New York and London, pp. 161–72.

—— (1941): Intuitive probabilities and sequences. *Ann. Math.*, **42**, 169–87.

Kotz, S. and Johnson, N. L. (1992): *Breakthroughs in Statistics* Vol. II, Springer–Verlag, New York.

Kruskal, W. and Mosteller, F. (1980): Representative sampling, IV: the history of the concept in statistics, 1895–1939. *International Statist. Rev.*, **48**, 169–95.

Lad, F. (1996): *Operational Subjective Statistical Methods*, John Wiley, New York.

Lazarsfeld, P. (1961): Notes on the history of quantification in sociology—trends, sources and problems. *Isis*, **52**, 164–81. Reprinted in *Studies in the History of Statistics and Probability* Vol. II, ed. Kendall, M. and Plackett, R. L., Griffin, London.

Lecam, L. (1955): An extension of Wald's theory of statistical decision functions. *Ann. Math. Statist.*, **26**, 69–81.

—— (1970): On the assumptions used to prove asymptotic normality of maximum likelihood estimates. *Ann. Math. Statist.*, **41**, 802–28.

—— (1977): A note on metastatistics or 'An essay toward stating a problem in the doctrine of chances'. *Synthese*, **36**, 133–60.

—— (1986): *Asymptotic Methods in Statistical Decision Theory*, Springer–Verlag, New York.

—— and Yang, G. L. (1990): *Asymptotics in Statistics*, Springer–Verlag, New York.

Lehmann, E. L. (1983): *Theory of Point Estimation*, Wiley, New York.

—— (1986): *Testing Statistical Hypotheses*, 2nd edn., Wiley, New York.

—— (1990): Model specification: the views of Fisher and Neyman, and later developments. *Statist. Sc.*, **5**, 160–68.

—— (1993): The Bertrand–Borel debate and the origins of the Neyman–Pearson theory. *Statistics and Probability—a Raghuraj Bahadur Festscrift*, ed. Ghosh, J. K., Mitra, S. K., Parthasarathy, K. R. and Prakasa Rao, B. L. S. P., Wiley Eastern, New Delhi, pp. 371–80.

Lindley, D. V. (1965): *Introduction to Probability and Statistics*, Parts 1 & 2, Cambridge University Press, Cambridge.

Liski, E. P., Mandal, N. K., Shah, K. R., Sinha Bikas K. (2001): *Topics in Optimal Designs*, Springer, New York.

Loeve, M. (1960): *Probability Theory*, 2nd edn., Van Nostrand, New York.

Mackenzie, D. A. (1981): *Statistics in Britain 1865–1930*, Edinburgh University Press.

Madow, W. G. (1960): Harold Hotelling. *Contributions to Probability and Statistics—essays in honor of Harold Hotelling*, ed. Olkin, I., Ghurye, S., Hoeffding, W., Madow, W. G., Mann, H. B., Stanford University Press, Stanford, pp. 3–5.

Mahalanobis, P. C. (1938): *Sankhyā*, **4**, 265–72. Reprinted in *Contributions to Mathematical Statistics*, Fisher, R. A. (1950), Wiley, New York.

—— (1950): Why Statistics? *Sankhyā*, **10**, 195–228. Reprinted in *Why Statistics and Other Essays*, ed. Bose, P. K., Statist. Publishing Soc., Calcutta.

—— (1957): The foundations of statistics. *Sankhyā*, **18**, 183–94.

—— (1986): *Why Statistics and Other Essays*, ed. Bose, P. K., Statist. Publishing Soc., Calcutta.

——, Majumdar, D. N. and Rao, C. R. (1949): Anthropometric survey of the United Provinces, 1941: a statistical study. *Sankhyā*, **9**, 89–324.

Maistrov, L. E. (1974): *Probability Theory. A Historical Sketch*, Academic, New York.

Mauldon, J. G. (1955): Pivotal quantities for Wisharts' and related distributions and a paradox in fiducial theory. *J. Roy. Statist. Soc.*, B, **17**, 79–85.

McCullagh, P. and Nelder, J. A. (1995): *Generalized Linear Models*, 2nd edn., Chapman–Hall, London.

Mill, J. S. (1843): *System of Logic*. Longmans–Green edn., London, 1896.

Murti, T. R. V. (1953): Rise of the philosophical schools. *The Cultural Heritage of India*. Vol. III, 2nd edn. R. K. M. Institute of Culture, Calcutta, pp. 27–40.

Myers, R. H. and Montgomery, D. G. (1995): *Response Surface Methodology*, Wiley, New York.

Nair, K. R. and Srivastava, M. P. (1942): On a simple method of curve fitting. *Sankhyā*, **6**, 121–32.

Nandi, H. K. (1947): A mathematical set-up leading to analysis of a class of designs. *Sankhyā*, **8**, 172–6.

—— (1948): Use of well-known statistics in sequential analysis. *Sankhyā*, **8**, 339–44.

Neter, J. (1986): Boundaries of statistics—sharp or fuzzy? *J. Amer. Statist. Assoc.*, **8**, 1–8.

Neyman, J. (1934): On the two different aspects of the representative method: the method of stratified sampling and the method of purposive selection. *J. Roy. Statist. Soc.*, A, **97**, 558–606.

—— (1935): Statistical problems in agricultural experimentation. *J. Roy. Statist. Soc.*, Suppl. **2**, 107–80.

Neyman, J. (1937): Outline of a theory of statistical estimation based on the classical theory of probability. *Phil. Trans. Roy. Soc. London,* **236**, A, 333–80.

—— (1950): *First Course in Probability and Statistics,* Henry Holt, New York.

—— (1952): *Lectures and Conferences on Mathematical Statistics and Probability,* 2nd edn., Graduate School, U. S. Dept. of Agriculture, Washington.

—— (1963): Certain chance mechanisms involving discrete distributions. *Classical and Contagious Discrete Distributions,* ed. Patil, G. P., Pergamon Press, Oxford, pp. 4–14.

—— (1976): The emergence of mathematical statistics. *On the History of Statistics and Probability,* ed. Owen, D. B., Mercel–Dekker, New York, 146–93.

—— and Pearson, E. S. (1967): *Joint Statistical Papers,* Cambridge University Press, Cambridge.

Parzen, E. (1962): On estimation of a probability density function and mode. *Ann. Math. Statist.,* **33**, 1065–76.

Pearson, E. S. (1939): 'Student' as statistician. *Biometrika,* **30**, 205–50.

—— (1965): Some incidents in the early history of biometry and statistics 1890–94. *Biometrika,* **52**, 3–18.

—— (1966): The Neyman–Pearson story: 1926–34. *Festscrift for J. Neyman,* Wiley, New York. Reprinted in *Studies in the History of Statistics and Probability* (1970), ed. Pearson, E. S. and Kendall, M. G., Griffin, London.

—— (1968): Some early correspondence between W. S. Gosset, R. A. Fisher and Karl Pearson, with notes and comments. *Biometrika,* **55**, 445–57.

—— (1970): Some incidents in the early history of biometry and statistics, 1890–94. *Studies in the History of Statistics and Probability* (1970), ed. Pearson, E. S. and Kendall, M. G., Griffin, London, pp. 323–38.

—— and Hartley, H. O. (1972): *Biometrika Tables for Statisticians* Vol. I, Cambridge University Press, Cambridge.

Pearson, K. (1920): Notes on the history of correlation. *Biometrika,* **13**, 25–45. Reprinted in *Studies in the History of Statistics and Probability,* ed. Pearson, E. S. and Kendall, M. G. (1970), Griffin, London, 185–205.

—— (1948): *Karl Pearson's Early Statistical Papers,* Cambridge University Press, Cambridge.

—— (1968): *Tables of the Incomplete Beta-Function,* 2nd edn., ed. Pearson, E. S. and Johnson, N. L., Cambridge University Press, Cambridge.

—— (1978): *The History of Statistics in the 17th and 18th Centuries,* ed. Pearson, E. S., Griffin, London.

Popper, K. (1959a): *The Logic of Scientific Discovery,* Hutchinson, London.

—— (1959b): The propensity interpretation of probability. *Brit. J. Phil. Sc.,* **10**, 25–42.

—— (1963): *Conjectures and Refutations,* Routledge, London.

Porter, T. M. (1986): *The Rise of Statistical Thinking 1820–1900*, Princeton University Press, Princeton.

Pratt, J. W. (1976): F. Y. Edgeworth and R. A. Fisher on the efficiency of maximum likelihood estimation. *Ann. Statist.*, **4**, 501–14.

Press, J. S. (1989): *Bayesian Statistics*, John Wiley, New York.

Pukelsheim, F. (1993): *Optimal Design of Experiments*, John Wiley, New York.

Puri, M. L. and Sen, P. K. (1971): *Nonparametric Methods in Multivariate Analysis*, Wiley, New York.

—— (1985): *Nonparametric Methods in General Linear Models*, Wiley, New York.

Puri, P. S. (1976): Biomedical applications of stochastic processes. *On the History Statistics and Probability*, ed. Owen, D. B., Marcel–Dekker, New York, pp. 377–99.

Pusalker, A. D. (1962): The Māhabhārata: its history and culture II. *The Cultural Heritage of India*, Vol. II, 2nd edn., R. K. M. Institute of Culture, Calcutta.

Rabinovich, N. L. (1969): Probability in the Talmud. *Biometrika*, **56**, 437–41.

Raghavarao, D. (1971): *Constructions and Combinatorial Problems in Design of Experiments*, Wiley, New York.

Ramsey, F. P. (1926): Truth and probability. Reprinted in *Philosophical Papers* (1990), ed. Mellor, D. H., Cambridge University Press, Cambridge.

Rao, C. R. (1948): The utilization of multiple measurements in problems of biological classification. *J. Roy. Statist. Soc.*, B, **10**, 159–203.

—— (1965): *Linear Statistical Inference and Its Applications*, John Wiley, New York.

—— (1966): Covariance adjustment and related problems in multivariate analysis. In *Multivariate Analysis*, ed. Krishnaiah, P. R., Academic Press, New York.

—— (1973): Mahalanobis era in statistics. *Sankhyā*, **35** suppl., 12–26.

—— (1989): *Statistics and Truth*, C. S. I. R., New Delhi.

Reichenbach, H. (1935): *The Theory of Probability*, English ed. (1949), University of California Press, Berkeley.

Renyi, A. (1970): *Probability Theory*, Elsevier, New York.

Robert, C. P. (1994): *The Bayesian Choice*, Springer–Verlag, New York.

Roy, S. N. (1939): *p*-statistics or some generalizations in analysis of variance appropriate to multivariate problems. *Sankhyā*, **4**, 381–96.

—— (1953): On a heuristic method of test construction. *Ann. Math. Statist.*, **24**, 220–38.

—— (1957): *Some Aspects of Multivariate Analysis*, Asia Publishing, Calcutta.

Royall, R. M. (1970): On finite population sampling theory under certain linear regression models. *Biometrika*, **57**, 379–87.

Royall, R. M. (2000): On the probability of observing misleading statistical evidence. *J. Amer. Statist. Assoc.*, **95**, 760–68.

Royall, R. M. and Herson, J. (1973): Robust estimation in finite populations I. *J. Amer. Statist. Assoc.*, **68**, 880–89.

Rubin, D. B. (1978): Bayesian inference for causal effects: the role of randomization. *Ann. Statist.*, **6**, 34–58.

Russell, B. (1912): *The Problems of Philosophy*, Oxford University Press, London.

——(1948): *Human Knowledge: Its Scope and Limits*, Simon and Schuster, New York.

——(1961): *History of Western Philosophy*, Allen and Unwin, London.

Savage, L. J. (1954): *The Foundations of Statistics*, John Wiley, New York.

——(1961): The foundations of statistics reconsidered. *Proc. Fourth Berkeley Symp.* reprinted in *Studies in Subjective Probability*. ed. Kyburg, H. E. (Jr.) and Smokler, H. E. (1963), John Wiley, New York.

——(1962): *The Foundations of Statistical Inference*, Methuen, London.

——(1976): On re-reading R. A. Fisher. *Ann. Statist.*, **4**, 441–500.

Scheffé, H. (1943): On solutions of the Behrens–Fisher problem, based on the *t*-distribution. *Ann. Math. Statist.*, **14**, 35–44.

——(1956): Alternative models for the analysis of variance. *Ann. Math. Statist.*, **27**, 251–71.

——(1959): *The Analysis of Variance*, John Wiley, New York.

Schwarz, G. (1978): Estimating the dimension of a model. *Ann. Stat.*, **6**, 461–4.

Seal, H. L. (1967): The historical development of the Gauss linear model. *Biometrika*, **54**, 1–24.

Seber, G. A. F. (1984): *Multivariate Observations*, Wiley, New York.

Sen, P. K. (1981): *Sequential Nonparametrics*, Wiley, New York.

——and Singer, J. M. (1993): *Large Sample Methods in Statistics: an Introduction with Applications*, Chapman–Hall, New York.

Shafer, G. (1982): Belief functions and parametric models. *J. Roy. Statist. Soc.*, B, **44**, 322–52.

Shastri, D. R. (1953): Materialists, Sceptics and Agnostics. *The Cultural Heritage of India*, Vol. III, 2nd edn., R. K. M. Institute of Culture, Calcutta, pp. 168–83.

Silvey, S. D. (1980): *Optimal Design*, Chapman–Hall, London.

Skyrms, B. (1996): Carnapian inductive logic and Bayesian statistics. *Statistics, Probability and Game Theory*, IMS Lecture Notes Monograph Series, Vol. 30.

Stein, C. (1959): An example of wide discrepancy between confidence intervals and fiducial intervals. *Ann. Math. Statist.*, **30**, 877–81.

Stigler, S. M. (1973): Simon Newcomb, Percy Daniell, and the history of robust estimation, 1885–1920. *J. Amer. Statist. Assoc.*, **68**, 872–79. Reprinted in *Studies in the History of Statistics and Probability*, Vol. II, (1977), ed. Kendall, M. G. and Plackett, R. L., Griffin, London.

——(1977): Eight centuries of sampling inspection: the trial of the Pyx. *J. Amer. Statist. Assoc.*, **72**, 493–500.

——(1978*a*): Mathematical statistics in the early States. *Ann. Statist.*, **6**, 239–65.

—— (1978b): Francis Ysidro Edgeworth, statistician. *J. Royal Statist. Soc.*, A, **141**, 287–322.

—— (1982): Poisson on the Poisson distribution. *Statist. Prob. Letters.*, **1**, 33–5.

—— (1983): Who discovered Bayes's theorem? *Amer. Statist.*, **37**, 290–6.

—— (1986): *The History of Statistics: The Measurement of Uncertainty before 1900*, Belknap Press of Harvard University Press, Cambridge, Mass.

Teicher, H. (1961): Maximum likelihood characterization of distributions. *Ann. Math. Statist.*, **32**, 1214–22.

Titterington, D. M., Smith, A. F. M., Makov, U. E. (1985): *Statistical Analysis of Finite Mixture Distributions*, Wiley, New York.

Toman, B. (1996): Bayesian experimental design for multiple hypothesis testing. *J. Amer. Statist. Assoc.*, **91**, 185–90.

Tukey, J. W. (1960): A survey of sampling from contaminated distributions. *Contributions to Probability and Statistics—essays in honor of Harold Hotelling*, ed. Olkin, I., Ghurye, S., Hoeffding, W., Madow, W. G., Mann, H. B., Stanford University Press, Stanford, pp. 448–85.

Uspensky, J. V. (1937): *Introduction to Mathematical Probability*, McGraw–Hill, New York.

Von Mises, R. (1928): *Probability, Statistics and Truth*, Second English edition (1957), Allen and Unwin, London.

von Neumann, J. and Morgenstern, O. (1947): *Theory of Games and Economic Behavior*, 2nd edn., Princeton University Press, Princeton.

Wald, A. (1940): The fitting of straight lines if both variables are subject to error. *Ann. Math. Statist.*, **11**, 284–300.

—— (1944): On a statistical problem arising in the classification of an individual into one of two groups. *Ann. Math. Statist.*, **15**, 145–62.

—— (1950): *Statistical Decision Functions*, John Wiley, New York.

Walley, P. (1991): *Statistical Reasoning with Imprecise Probabilities*, Chapman and Hall, London.

Wang, C. (1993): *Sense and Nonsense of Statistical Inference*, Marcel–Dekker, New York.

Weiss, L. and Wolfowitz, J. (1967): Maximum probability estimation. *Ann. Inst. Statist. Math.*, **19**, 193–206.

Welch, B. L. and Peers, H. W. (1963): On formulae for confidence points based on integrals of weighted likelihoods. *J. Roy. Statist. Soc.*, B, **25**, 318–29.

Weyl, H. (1949): *Philosophy of Mathematics and Natural Science*, Princeton University Press, Princeton.

Whittaker, E. and Robinson, G. (1952): *The Calculus of Observations*, 4th edn., Blackie, London.

Wilkinson, G. N. (1977): On resolving the controversy in statistical inference. *J. Roy. Statist. Soc.*, B, **39**, 119–71.

Yates, F. (1937): *The Design and Analysis of Factorial Experiments*, Imperial Bureau of Soil Science, Harpenden, England.

Zarkovich, S. S. (1956): Note on the history of sampling methods in Russia. *J. Royal Statist. Soc.*, A, **119**, 336–8.

INDEX